Lyell: the Past is the Key to the Present

Lithograph of Charles Lyell made in Philadelphia by J. E. Mayall during Lyell's visit to America in the 1840s.

Geological Society Special Publications
Series Editors
A. J. FLEET
A. C. MORTON
A. M. ROBERTS

Acknowledgements

The Geological Society, Palaeontographical Society and Palaeontological Association would like to thank the following companies for their generous sponsorship of the 1997 Lyell Meeting and Bicentennial Conference:

Amerada Hess Ltd
Amoco Services Ltd
ARCO British Ltd
BP Exploration Operating Company Ltd
Chevron Europe Ltd
Conoco (U.K.) Ltd

IKODA Ltd
Kerr McGee Corporation
Oryx U.K. Energy Company
Ready Mixed Concrete (U.K.) Ltd
Shell U.K. Exploration and Production
Statoil (U.K.) Ltd

They are also grateful for the valuable help and support of INHIGEO, the International Commission on the History of Geology, and especially its President, Dr Hugh Torrens.

The editors are especially grateful to the following who refereed the papers contained in this book, and whose help and advice have improved this work immeasurably:

R. D. Adams, P. Allen, E. Belt, D. W. J. Bosence, G. S. Boulton, C. W. A. Browitt, J. H. Calder, W. G. Chaloner, R. Cocks, P. J. Coney, I. Davison, D. Dean, R. Fortey, J. Francis, P. W. Francis, B. Fritscher, M. Fuller, R. Gayer, M. Gibling, D. A. Gray, R. Hall, M. Hasselboro, A. Hirn, C. H. Holland, G. A. L. Johnson, G. Kelling, M. King Jr, J. Lowe, P. Lyons, W. J. McGuire, E. G. Nesbit, G. J. Nichols, E. P. F. Rose, M. S. Rudwick, W. A. S. Sarjeant, R. Savage, J. Secord, R. H. Silliman, A. Smith, R. S. J. Sparks, J. Thackray, H. G. Torrens, A. B. van Riper, L. G. Wilson and N. Woodcock.

GEOLOGICAL SOCIETY SPECIAL PUBLICATION NO. 143

Lyell: the Past is the Key to the Present

EDITED BY

DEREK J. BLUNDELL

&

ANDREW C. SCOTT

Department of Geology, Royal Holloway, University of London,
Egham, UK

1998
Published by
The Geological Society
London

THE GEOLOGICAL SOCIETY

The Society was founded in 1807 as The Geological Society of London and is the oldest geological society in the world. It received its Royal Charter in 1825 for the purpose of 'investigating the mineral structure of the Earth'. The Society is Britain's national society for geology with a membership of around 8500. It has countrywide coverage and approximately 1500 members reside overseas. The Society is responsible for all aspects of the geological sciences including professional matters. The Society has its own publishing house, which produces the Society's international journals, books and maps, and which acts as the European distributor for publications of the American Association of Petroleum Geologists, SEPM and the Geological Society of America.

Fellowship is open to those holding a recognized honours degree in geology or cognate subject and who have at least two years' relevant postgraduate experience, or who have not less than six years' relevant experience in geology or a cognate subject. A Fellow who has not less than five years' relevant postgraduate experience in the practice of geology may apply for validation and, subject to approval, may be able to use the designatory letters C Geol (Chartered Geologist).

Further information about the Society is available from the Membership Manager, The Geological Society, Burlington House, Piccadilly, London W1V 0JU, UK. The Society is a Registered Charity, No. 210161.

Published by The Geological Society from:
The Geological Society Publishing House
Unit 7, Brassmill Enterprise Centre
Brassmill Lane
Bath BA1 3JN
UK
(*Orders*: Tel. 01225 445046
            Fax 01225 442836)

First published 1998

British Library Cataloguing in Publication Data
A catalogue record for this book is available from the British Library.

ISBN 1-86239-018-5

Distributors

USA
  AAPG Bookstore
  PO Box 979
  Tulsa
  OK 74101-0979
  USA
  (*Orders*: Tel. (918) 584-2555
             Fax (918) 560-2652)

Australia
  Australian Mineral Foundation
  63 Conyngham Street
  Glenside
  South Australia 5065
  Australia
  (*Orders*: Tel. (08) 379-0444
             Fax (08) 379-4634)

India
  Affiliated East-West Press PVT Ltd
  G-1/16 Ansari Road
  New Delhi 110 002
  India
  (*Orders*: Tel. (11) 327-9113
             Fax (11) 326-0538)

Japan
  Kanda Book Trading Co.
  Cityhouse Tama 204
  Tsurumaki 1-3-10
  Tama-shi
  Tokyo 206-0034
  Japan
  (*Orders*: Tel. (0423) 57-7650
             Fax (0423) 57-7651)

Typeset by E&M Graphics, Midsomer Norton, Bath, UK.

Printed by The Alden Press, Osney Mead, Oxford, UK.

Contents

It is recommended that reference to all or part of this book should be made in one of the following ways:

BLUNDELL, D. J. & SCOTT, A. C. (eds) 1998. *Lyell: the Past is the Key to the Present.* Geological Society, London, Special Publications, **143**.

RUDWICK, M. J. S. 1998. Lyell and the *Principles of Geology. In:* BLUNDELL, D. J. & SCOTT, A. C. (eds) *Lyell: the Past is the Key to the Present.* Geological Society, London, Special Publications, **143**, 3–15.

Introduction

Charles Lyell was appointed to the Chair of Geology at King's College London in February 1831, and proceeded to present a series of 12 lectures based on his recently published book *Principles of Geology*. According to his biographer, Leonard G. Wilson (1972), Lyell took a lot of trouble to prepare illustrations and demonstrations in support of his geological arguments. The lectures were open to the general public and attracted large audiences, including a number of his peers in the Geological Society, their wives and friends, who outnumbered the College students. Public interest in this newly emerging science was intense, fired by the wide readership of *Principles of Geology* which, remarkably, sold over 15 000 copies – far more than we can expect from this present volume! Lyell's lecture course began on 1 May 1832 and continued until 12 June, with audiences generally of around 100 but, on several occasions, rising to an estimated 300 (Wilson 1972). His lectures created a great deal of interest and were generally well received. However, the following year the College authorities decided not to allow women to attend any of the lectures other than the introduction because they were regarded as a distraction to the students. Lyell's audience dwindled to around 15. Lyell decided that he could spend his time more profitably in other ways and consequently resigned from his Chair in October 1833. Despite the brevity of his tenure of the post, Lyell established geology as a scientific study at King's College. The Lyell Chair was established and a succession of distinguished geology professors held the Chair at King's until 1985. In a major reorganization of the University of London, the King's College Geology Department joined with those from Bedford College and Chelsea College and moved to the new, purpose-built Queen's Building at Royal Holloway and Bedford New College, with its campus on the western fringe of London, near Windsor. The Lyell Chair thus came to Royal Holloway where it is currently held by Prof. John Mather.

With this direct lineage from Sir Charles Lyell, Royal Holloway was the natural choice to host the Bicentenary Lyell Meeting and for Prof. Derek Blundell (like Lyell, a former President of the Society) and Prof. Andrew Scott to act as conveners. The Lyell Meeting was organized jointly by the Geological Society, the Palaeontographical Society and the Palaeontological Association. The Meeting formed part of an international conference from 30 July to 9 August 1997 to mark the birth of Lyell and death of Hutton, and celebrate the lives and work of these two great geologists. It also marked the sesquicentennial anniversary of the Palaeontographical Society which was founded in 1847. It is one of the less widely sung jewels in the British crown that it is the oldest palaeontological society in the world. Its chief purpose was and is the production of palaeontological monographs which have the aim of illustrating and describing the entire British fauna and flora throughout geological time, hence the 'graphical' part of the title of the Society.

The international conference included the Lyell Meeting in London, followed by the Hutton/Lyell Meeting in Edinburgh hosted by the Royal Society of Edinburgh. The Lyell Meeting was preceded by an excursion to visit Lyell's boyhood home, Bartley Lodge, Cadnam, in the New Forest, where Mr Paul Clasby explained its historical significance for the young Charles Lyell, and to visit coastal exposures of Eocene strata at nearby Barton-on-Sea, led by Dr Margaret Collinson, to collect fossil molluscs of particular interest to Lyell when he set up the classification of the Periods within the Tertiary. The meeting was followed by excursions to localities across the Weald of southeast England, led by Dr Hugh Torrens and Mr John Cooper, to examine the sequences of dinosaur-bearing Cretaceous rocks which were first unravelled by John Farey and Gideon Mantell, whose work proved to be an important early influence on Charles Lyell. Prof. Gerald Friedman led a walking tour to visit Lyell's London home in Wimpole Street. On Saturday, 2 August, the Lady Lyell laid a wreath on the tomb of Sir Charles Lyell in Westminster Abbey during a ceremony to mark the bicentenary of his birth. During the meeting, the Linnean Society kindly invited participants to a Special Lecture given by Dr Jim Secord on 'Placing Lyell in history: the Principles of Geology in an Age of Revolutions' which was followed by a reception jointly hosted by the Linnean and Geological Societies. The Meeting concluded with a formal dinner in the Picture Gallery at Royal Holloway at which the guest speaker was Sir Charles' great grand nephew Mr Malcolm Lyell. This University of London setting was particularly appropriate as diners were surrounded by paintings of Sir Charles' period, many of which he had probably seen when they were first exhibited in London at the Royal Academy.

The Bicentenary Lyell Meeting was made special through our ability to invite key speakers from around the world and bring them to London. This was wholly due to the generosity of our sponsoring companies, acknowledged inside the front cover, whose contributions allowed us to pay towards the speakers travel costs and, for overseas speakers, towards their accommodation at Royal Holloway during the meeting. For this, we are profoundly grateful and we trust that this book provides a worthy consequence of their generosity.

Held at the Geological Society apartments in Burlington House, the Lyell Meeting attempted to examine the influence of Charles Lyell on the science of geology, both in his own day and subsequently to the present. We invited an interesting blend of historians and earth scientists to speak about Lyell's work, his extensive travels across Europe and North America, and his interactions with others of his time (including Charles Darwin) and to bring Lyell's concepts and ideas across a wide range of geological disciplines into a modern context. The title of the meeting was thus 'Lyell: the Past is the Key to the Present'. Using the same title, this book is the outcome of the Bicentenary Lyell Meeting, made up of the papers presented by invited speakers. The book has been organized in three parts. Part 1 begins with an examination of Lyell's work and his role in the society of the day, the impact of his travels and his influence on his contemporaries. Part 2 develops Lyell's ideas on geological time, stratigraphy, climatic change, and volcanic phenomena and contemporary ideas on hydrogeology. Part 3 concludes by illustrating the legacy of Lyell's 'Principles' into the modern contexts of plate tectonics, stratigraphy, sedimentology, salt tectonics, and environmental monitoring and management.

Reference

WILSON, L. G. 1972. *Charles Lyell. The years to 1841: the Revolution in Geology*. Yale University Press, New Haven and London.

Derek J. Blundell
Andrew C. Scott

Part 1.   The life and influence of Lyell

For geology, Lyell proved to be the right man at the right time in the right place. He grew up in comfortable surroundings, encouraged by his parents to develop a keen observation of the natural world around him. He learned the importance of geological observations in the field and the precise and accurate recording of geological information, which led him to travel extensively across Europe and North America to examine a wide range of geological phenomena. The knowledge and ideas that he gained from field observations and the work of fellow geologists were clearly the foundations for his great work *Principles of Geology*. In the first paper of this book, Martin Rudwick analyses *Principles of Geology* through successive editions as a dynamic development of geology 'as one long argument'. Lyell, more than anyone, established geology as a *scientific* discipline, hence the word 'Principles' in the title of his book, and gave it credibility both amongst practitioners of the subject and, equally importantly, amongst the wider, literate public. It is his legacy that nowadays we take for granted that geology can be treated as a science, and that we use modern analogues to interpret and quantify past geological events and processes.

Rudwick demonstrates how important it was to Lyell to be thoroughly conversant with the geological literature of his day – and that he learnt French, German and Italian in order to do so – and to engage in debate with his peers, many of whom were Fellows of the Geological Society. Lyell played an active part in its development, serving in turn as Secretary, Foreign Secretary and President. John Thackray, the Society's Archivist, writes about Lyell's role as a Fellow between 1819 and his death in 1875. John gives us an insight of how the Society operated in those days, with constraints on the presentation of formal papers at evening meetings largely to factual descriptions, although greater latitude was given to theorize and speculate in Presidential Addresses. The real value of the evening meetings, however, came in the unrecorded, informal debates that followed the formal presentations. The encouragement of controversy endured in the Society through the debating chamber layout of the Meeting Room until 1972 when it was modernized into the current lecture room.

In the next paper, Leonard Wilson gives us an insight into Lyell's life, the people he met who influenced his thinking, and the development of his ideas on Uniformitarianism. In particular, he demonstrates Lyell's tenacity and determination in working out from field evidence the structure and formation of volcanoes like Vesuvius, Mount Etna, Tenerife, Grand Canary and Madeira and disproving the theory that they had formed, by catastrophic upheaval, as 'craters of elevation'. Wilson staunchly defends Lyell's strict interpretation of uniformitarianism against those who regarded the Earth as having cooled from an originally hot body, not least Lord Kelvin who calculated the age of the Earth from an estimate of its rate of cooling. Later on, as Wilson explains, Lyell became increasingly interested in the human timespan on Earth, culminating in the *Antiquity of Man* published in 1863.

Ezio Vaccari discusses the reception to Lyell's ideas in Europe as *Principles of Geology* and *Elements of Geology* were reviewed and translations of them began to be published in the 1830s. Vaccari is able to demonstrate the wide acceptance of most of Lyell's views apart from the same reservations as his English colleagues (explained in Rudwick's paper) about the constancy of geological history, most preferring the model of a cooling Earth with a diminishing level of tectonic activity through geological time. It would seem that De Beaumont's synthesis of mountain building processes linked with the cooling of the Earth's crust may have been more influential. None the less, Lyell's personal influence was strong as a consequence of his travels and his meetings with most of the leading geologists of Europe. He was recognized as far as Moscow and was elected a member of the Moscow Naturalist Society in 1855. Russian translations of his works began to appear from 1859. Bringing things up to date, Prof. E. E. Milanovsky, as current Vice President, delivered a formal Jubilee Address to the Geological Society during the Bicentenary Lyell Meeting.

Robert Dott gives us an account of Lyell's travels in North America. Lyell visited North America four times from 1841 to 1853 not only to travel and see the geology, but also to give public lectures. Dott traces the emergence of geology through a study of Lyell's changing lecture topics. Dott demonstrates that there was a true interaction between American geologists and Lyell, to their mutual benefit. Clearly Lyell brought stature to the new science of geology which increasingly gained public acceptance. Lyell gained, however, much from American geologists, especially in relation to field work. As Dott points out, many of his North American field examples are used to illustrate new concepts in successive editions of his books. Lyell's exposure to American life not only led to new geological understanding but also developed his interest in aspects of education, public reforms and religious belief. On his return to London, Lyell soon became the resident expert on North America, fuelled by the publication of two books on his travels in North America and numerous letters and articles.

Lyell's visit to America is further amplified by Gerald Friedman who concentrates on his visit to New York State in 1841. Friedman discusses Lyell's visit to Amos Eaton, an influential American geologist, just before he died. Together with Eaton's former student James Hall, Lyell visited Niagara Falls, details of which he was to use later in his publications. Much of Lyell's activity in New York did not relate to geology, but Friedman also discusses a resentment that some felt about the use by Lyell of their unpublished work in their volumes. It is clear, however, that Lyell did much to aid the public understanding of this new and exciting science.

The relationship between Lyell and Darwin has always been a fascinating one. Clearly Darwin was highly influenced by Lyell and indeed it is well known that he took volume one of the '*Principles*' with him on his voyage of the Beagle. Indeed, on his return from the Beagle, Darwin and Lyell became close friends. It was Lyell who suggested that Darwin and Wallace should publish a joint paper on the origin of species. Clearly Lyell did not readily accept Darwin's idea on the origin of species, but over time was gradually converted to the idea of evolution but not necessarily by natural selection. It was inevitable that Lyell should consider the geological evidences of the antiquity of man. As Claudine Cohen points out in the final paper in this section, the publication of his book on the *Antiquity of Man* in 1863 contributed to the founding of the two scientific disciplines of Pre-historic Archaeology and Palaeoanthropology. In his book Lyell brings to bear aspects of archaeology, palaeontology, anthropology and geology which were used to demonstrate conclusively the contemporaneity of extinct animals and man-made objects. Cohen emphasizes the strong relationship between Lyell and French scientists who mutually benefited from their interactions.

Derek J. Blundell
Andrew C. Scott

Lyell and the *Principles of Geology*

M. J. S. RUDWICK

Department of History and Philosophy of Science, University of Cambridge,
Free School Lane, Cambridge CB2 3RH, UK

Abstract: Lyell's *Principles of Geology* is still treated more often as an icon to be revered than as the embodiment of a complex scientific argument rooted in its own time and place. This paper describes briefly the origins of Lyell's project, in the international geological debates of the 1820s and in his own early research; the structure of argument of the first edition (1830–1833), and its relation to its intended readership; and the modification of the work in subsequent editions, and the transformation of its strategy in response to its critical evaluation by other geologists. The fluidity of the *Principles* (and its offshoot the *Elements of Geology*) reflects the ever-changing interaction between Lyell and his fellow geologists, and between them and a much wider public, during one of the most creative periods in the history of geological science. Lyell's scientific stature is best appreciated if he is placed not on a pedestal but among his peers, in debates at the Geological Society and elsewhere.

Charles Lyell's *Principles of Geology* (first published in 1830–1833) is one of the most significant works in the history of the Earth sciences, but it is still more often cited than read, more often revered than analysed. Above all, it needs and deserves to be understood in the context of its own time, before being recruited to support modern arguments, or repudiated for its failure to support them.

Unlike many comparably important works in the history of art or literature, the *Principles* was not launched into the public realm as a once-and-for-all achievement, thereafter to be admired or criticized but not significantly modified. On the contrary, Lyell reissued it in successive editions over almost half a century, and during that period it changed radically in character. The *Principles* was in effect one side of a dialogue between Lyell and his contemporaries: not only the other leading geologists in Britain and abroad, but also much wider groups in British society. Like some other important scientific works from the same period – Charles Darwin's *Origin of Species* (1859) is a good example – Lyell's *Principles* was a work with multiple goals and a diverse range of intended readers (Secord 1997). It was not only an original scientific treatise directed at other geologists; it was also at the same time a work in what the French recognize as a distinctive genre, that of *haute vulgarisation* or authoritative high-level popularization. The construction of the first edition, and its modification in subsequent editions, should therefore be seen as Lyell's lifelong contribution to continuing debates, not only about the changing profile of geological knowledge and its basic

'principles', but also about the place of such knowledge in a scientific view of the world.

While he was an undergraduate at Oxford, Lyell's adolescent interests in natural history were channelled into geology by the charismatic presence and famously entertaining lectures of William Buckland (Rupke 1983; Wilson 1972). When in 1819 Lyell graduated and moved to London to begin a legal career, Buckland sponsored him for membership of the Geological Society (itself little more than a decade old). This put Lyell at the heart of geological debate in Britain (Morrell 1976), and in 1823 he became one of the Society's secretaries. He then took an early opportunity to visit Paris, making himself known in the world centre of scientific research and improving his fluency in the international language of the sciences. Back in London, as a member of the Athenaeum – a new social club for intellectuals of all kinds – he also became known far beyond scientific circles: some of his most important geological ideas were first published in essay–reviews for the influential Tory periodical *Quarterly Review*, which was widely read by the social, political and cultural elites in Britain.

It was a time of intense public interest in all the sciences; elementary books were selling well and proving profitable to their authors. Lyell therefore planned to use that well tried format to propagate his own emerging conception of geology; like many young barristers he urgently needed other sources of income, in his case to supplement a modest allowance from his father. But as he began to put his ideas on paper, he found them expanding beyond the bounds of an introductory work.

RUDWICK, M. J. S. 1998. Lyell and the *Principles of Geology*. *In*: BLUNDELL, D. J. & SCOTT, A. C. (eds) *Lyell: the Past is the Key to the Present*. Geological Society, London, Special Publications, **143**, 3–15.

Eventually he realized they would need to be expounded in a full-length treatise, and he began to talk about writing a book that would establish the basic 'principles' of the science. In other words, Lyell's proposed work was modified profoundly before it was ever published. The word 'principles', when used at that time in a book title, carried connotations not of an elementary textbook but of Isaac Newton's monumental *Principia*; adopted by Lyell, a man barely into his thirties, it signalled substantial scientific ambitions. Nonetheless, Lyell's book never completely lost its initially intended character as a work of *haute vulgarisation*, and its later offshoot the *Elements of Geology* (1838) returned explicitly to that genre.

Resources for the *Principles*

During the later 1820s, Lyell formulated his own conception of geology, constructed the argument for his work, and assembled empirical material to support it. Occasionally he did so in the field, in direct contact with the natural world, but far more often it was in interaction with his scientific contemporaries, either face to face or through their publications. However, rather than being passively 'influenced' by them, he treated them as his scientific *resources*; like any great scientist, he actively selected from all the past and present research that was available to him, choosing what best served his own intellectual goals and reshaping it for his own purposes.

To help make sense of this rich body of empirical material, Lyell drew on two contrasting theoretical models for the study of the Earth. The importance of one is well known, but the significance of the other has been widely overlooked. The first was the Huttonian theory of the Earth. This was taken by Lyell, as it was by most of his contemporaries, not from James Hutton himself but from John Playfair's bowdlerized *Illustrations* (1802). Like the many other works using the title 'theory of the Earth', Hutton's had been a hypothetical model (1788) followed – in his case, years later – by a mass of empirical 'proofs and illustrations' (1795). But Playfair had shorn it of its pervasive deistic metaphysics and anthropocentric teleology, and had recast it in an effort to turn geology into a branch of physics. As such, geology would be above all a science of *causes* of natural processes the same yesterday, today and for ever. The Earth would be treated as a theatre of ceaseless and essentially repetitive change, played out on a timescale left conveniently indefinite but treated in practice as virtually infinite. The Earth was assumed to have been roughly the same kind of place at all times: as a mathematician, Playfair had found the cyclic

stability of the Newtonian solar system an appealing parallel.

Such theories were commonplace in Enlightenment ideas about the natural world in general, and Playfair's application of them to the then new science of geology was attractive to many in Lyell's generation. Specifically, Lyell followed Playfair in putting the Newtonian notion of a *vera causa* or 'true cause' at the heart of his method: phenomena should be explained only in terms of causal agencies that are observably effective, both in kind and in degree (Laudan 1982, 1987). For a science like geology, faced with understanding the past as well as the present, that meant in practice that explanations should always be in terms of 'modern' or 'actual' causes ('actual' was used in a sense now obsolete in English, meaning current or present-day, as in the modern French *actualités* for the day's news). In geology, *verae causae* were necessarily also actual causes, since only the present (and, by extension, the reliably documented human past) could be directly observed and witnessed.

The other theoretical model, equally important for understanding Lyell's work, was that of the Earth as a product of *history*. This was quite compatible with the physical model, since all the events in the history of the Earth and its life – with the possible exception of the origins of new species, and particularly the human species – were assumed to have had natural causes of some kind. But on the historical model such causes were taken to have produced an Earth that had *not* always been the same kind of place; it had had a complex contingent history that could not be predicted from any physical model, and that could only be discovered and reconstructed by attending to the presently observable traces of specific past events. The historical model had first been formulated, a century and a half before Lyell, within the framework of biblical history and its traditional short timescale. But by the time of Buffon's *Époques de la Nature* (1778) it had been fully secularized; Buffon's experiments with cooling globes gave it a quantified timescale that was modest by modern standards but already literally beyond human imagination (Roger 1989). Just after the French Revolution, the anatomist Georges Cuvier had greatly deepened the historical model, and highlighted the 'otherness' of the Earth's past, by using fossil bones to claim the reality of extinction as a perennial natural phenomenon (Rudwick 1997). Cuvier had become convinced that no physical cause known to him could have wiped out fossil species as well adapted as he believed them to have been. So he had adopted the common idea that some kind of natural but catastrophic agency must have been responsible for extinctions (e.g. Cuvier 1812). He had remained uncertain about the physical character of these putative catastrophes; but others, for example Hutton's old friend James Hall (1814), had suggested they could have been mega-tsunamis, as it were, caused by sudden major bucklings of the oceanic crust. In other words, the historical model for the Earth was here interpreted with resources drawn from the physical model: an observable kind of natural event, albeit of a magnitude unwitnessed in human history, was held responsible for otherwise inexplicable effects in the Earth's history.

The historical model, and Cuvier's version in particular, was mediated to Lyell by his mentor Buckland. Buckland's fieldwork in the Oxford region convinced him that the puzzling deposits he termed 'diluvium' were quite distinct from the more recent alluvium along the present rivers: for example, the 'diluvial' boulder clay and gravels had apparently been swept from the Midlands plain over the watershed of the Cotswolds into the Thames valley. Buckland's inference of a geologically recent mega-tsunami or 'geological deluge' gave concrete form to what Cuvier had inferred mainly from fossil bones. It then seemed to be confirmed by the chance discovery of Kirkdale cave in Yorkshire, which Buckland (1822, 1823) interpreted geohistorically as a den of extinct hyenas that had been annihilated by the geological deluge. His 'diluvial' interpretation became a powerful scientific theory, which most geologists found highly persuasive as an explanation of some extremely puzzling phenomena; it won Buckland the Royal Society's Copley Medal, and Lyell defended it when he visited Paris in 1823.

Buckland's 'diluvialism' was presented, however, within a context that made it far more than a scientific theory; its reception was moulded by societal issues more prominent in England than in France, and more pressing in Oxford than in London. Cuvier had sought to link the new geohistory to the far briefer annals of human history by searching the literature of *all* ancient civilizations for possible early human records of the most recent catastrophe; as a cultural relativist typical of the Enlightenment he had treated the biblical story of Noah's Flood as just one of many obscure folk-memories of some such event (Rudwick 1997). Buckland, on the other hand, needed to prove to his colleagues that geology was a legitimate field of academic study in the intellectual centre of the Church of England, and that it would not undermine Oxford's traditional textual forms of learning (Buckland 1820; Rupke 1983). He therefore invoked the geological deluge as decisive evidence for the historicity of the biblical Flood in particular; he interpreted it as a violent event of global extent and colossal magnitude, although in fact this entailed taking substantial

liberties with the literal meaning of the text (English scholars were still largely ignorant of the textual criticism pioneered in the German universities, which rendered biblical literalism obsolete, not least in religious terms). In other words, in Britain – though never in the same way on the Continent – an effective geological theory was entangled with issues of biblical interpretation, ecclesiastical authority and, above all, the political power of the established Church.

Once Lyell moved from Oxford to London, away from generally conservative dons and into the company of politically liberal lawyers, he became uneasy about this mixing of science and divinity, although (as mentioned already) he continued for some time to champion Buckland's diluvial theory as a good explanation for many otherwise puzzling phenomena. His shift away from Buckland's kind of theorizing was unquestionably powered by his growing hostility, not to religion in general, but to specific aspects of the contemporary Church of England, and particularly its virtual monopoly on higher education. But what eventually convinced him that the geological deluge was a chimera was the cumulative weight of specific empirical cases, in which the phenomena could be still better explained without recourse to any recent catastrophe. These cases were drawn from the published literature of the multinational geological community, backed up by Lyell's own first-hand confirmation of some of the more striking examples.

For example, when he visited Paris in 1823 Lyell was taken into the field by Constant Prévost, a former student of Cuvier's colleague Alexandre Brongniart. Cuvier and Brongniart's joint monograph on the Paris Basin (1811) was the outstanding exemplar of how a detailed geohistory could be inferred from a pile of formations and their fossils: it was a local history of tranquil sedimentation on a vast timescale, punctuated by occasional sudden alternations between marine and freshwater conditions (Rudwick 1996, 1997). But Prévost was already arguing against such sudden changes, and claiming that the ancient environments were paralleled in every respect in the modern world, for example around the estuary of the Seine (Bork 1990). Lyell was duly convinced by Prévost's arguments (Wilson 1972), and the next year he followed the Frenchman's example by studying a recently drained lake near his family home in Scotland. In his first major paper to the Geological Society, Lyell (1826) argued that the modern sediments were identical in every detail to some of those that Prévost had shown him in France. For Lyell, it was a powerful vindication of the claim that actual causes might be adequate to explain *everything* in the geohistorical record.

In the wake of Cuvier's claims to the contrary, many geologists had realized that any such conclusion about actual causes depended on a better knowledge of just what effects such agencies were actually having in the present world. In 1818 the Royal Society of Göttingen had therefore offered a valuable prize for a monograph on the subject. The prize was won by the civil servant (and part-time geologist) Karl von Hoff of Gotha. Hoff's exhaustive search of the geographical and historical literature (1822–1834) showed that ordinary agencies, such as sedimentation and erosion, volcanoes and earthquakes, had indeed effected major physical changes, even within the two or three millenia of reliably recorded human history (Hamm 1993). The implication was that in the far vaster tracts of geological time there might be no limit to what they could produce. Lyell quickly saw the significance of this for his own project. He learnt German specifically to exploit the empirical riches of Hoff's work (and thereby also gained access to a major new scientific literature). He also followed Hoff's example by searching through the accounts of the voyages and expeditions of all the scientific nations, finding copious further evidence of the power of actual causes within recorded history (Rudwick 1990–1991).

A third example of great importance to Lyell came, like Prévost's, from fieldwork studies of the geohistory of a specific region. George Scrope, a geologist who had joined Lyell as a Secretary of the Geological Society, published a speculative 'theory of the Earth' (1825) based like Buffon's on a cooling model of the Earth's history. He backed it up, however, with a detailed study of the extinct volcanoes of Auvergne (1827), illustrated vividly by a set of geologically interpreted panoramas of the landscape there. Scrope's work borrowed extensively – with scant acknowledgement – from earlier French writers, but it presented anglophone geologists with persuasive evidence for a continuous geohistory of this famous region, uninterrupted by any diluvial event. The occasional episodes of volcanic activity, stretching far back into the geological past, served in effect as markers recording successive stages in the continuous slow fluvial erosion of the topography; and Scrope claimed that what was needed for the correct interpretation of the region was simply 'Time! Time! Time!' (Rudwick 1974).

Lyell was deeply impressed by Scrope's case study, and explained its significance to a wider audience in the *Quarterly Review* (Lyell 1827). He suggested that it was a model for how geology ought to be done, using observable actual causes to penetrate from the known present back into the more obscure past. This was in fact just the strategy that Cuvier had famously advocated, but Lyell now

intended to use it to undermine Cuvier's argument for a recent catastrophe. He extended Scrope's argument to cover the organic world as well as the inorganic, claiming that the fossils being found by local geologists in Auvergne proved that there had been no recent mass extinction or marine incursion there. When in 1828 he had an opportunity to do fieldwork on the Continent, his very first destination was Auvergne, where he duly confirmed Scrope's interpretation and his own extension of it.

A fourth and final example of Lyell's use of the rich contemporary geological literature strengthened that conclusion about gradual faunal change; it extended to Tertiary formations and their fossils what Scrope had claimed for the volcanic and erosional history of the same era. Cuvier had suggested that the formations on the flanks of the Apennines, with their superbly preserved fossil molluscs, might help to bridge the taxonomic and geohistorical gap between the Parisian fossil faunas and the living molluscs of present seas. Giovanni-Battista Brocchi of Milan had pursued that idea with a fine monograph on the Italian fossils (1814); but to interpret them he had formulated a model of piecemeal faunal change, which dispensed with the need for any Cuvierian catastrophes, at least for marine faunas (Marini 1987). Extracts had been translated by Leonard Horner (Lyell's future father-in-law) in the *Edinburgh Review* (1816), the Whig counterpart of the *Quarterly*, and Lyell later learnt Italian, partly in order to read Brocchi's work more thoroughly; it became the basis for his own more extensive use of *all* the known Tertiary molluscan faunas in the reconstruction of Tertiary geohistory.

These four examples are sufficient to suggest how, when writing the *Principles*, Lyell appropriated a massive and mature body of contemporary geological literature, published not only as books but also in a growing number of scientific periodicals. Lyell exploited that literature to the full, by being as international in outlook – and as multilingual – as the leading geologists in the rest of Europe (the word 'Europe' is used in this paper in its proper pre-Thatcherite sense, to include the British Isles). It was no coincidence that when in 1826 Lyell resigned as a Secretary of the Geological Society he was promptly appointed its Foreign Secretary.

The strategy of the *Principles*

Immediately after his fieldwork in Sicily, the culmination of his long expedition on the Continent in 1828–1829, Lyell described his proposed book in a letter to Roderick Murchison (his companion on the earlier part of the trip). It was to vindicate his 'principles of reasoning' in geology, which were that '*no causes whatever* have ... ever acted, but those *now acting*, and that they never acted with different degrees of energy from that which they now exert' (Lyell 1881, vol. 1, p. 234). In other words, actual causes were wholly adequate to explain the geological past, not only in kind but also in degree. The first proposition was relatively uncontroversial, but with the second Lyell distanced himself from a virtual consensus among his fellow geologists (Bartholomew 1979). Even those like Buckland who claimed there had been a drastic catastrophe in the geologically recent past thought it might have been analogous to the tsunami that had famously destroyed Lisbon in 1755; but they insisted that the physical evidence of its action proved that it must have been on a far larger scale. Likewise, even those like Prévost and Scrope who most strongly supported Lyell's claims for the power of actual causes believed that volcanic and tectonic events, for example, must have been on a far larger scale in the Earth's hotter and more active infancy. But Lyell was now convinced that if geology were to be truly 'philosophical' (or in modern terms, scientific), the Newtonian *vera causa* principle would need to be applied rigorously, confining causal explanation to agencies observably active in the modern world, acting *at their present intensities*.

Lyell therefore had a major task of persuasion ahead of him, if he was to convince other geologists – let alone the general educated public – that everything in the geological record could be adequately explained in these terms. This is where Lyell's legal training stood him in good stead: as his contemporaries noticed at once, the *Principles* was a barrister's brief from start to finish, rhetorical in character throughout (using the term 'rhetoric' in its proper and non-pejorative sense). Lyell set out to present a good case for the explanatory adequacy of actual causes: as he put it in his carefully phrased subtitle, the work was 'an attempt to explain the former changes at the Earth's surface, by reference to causes now in operation'. Lyell's *Principles*, as Darwin later said of his own book, was 'one long argument'.

The core of Lyell's argument was his exhaustive survey of the whole range of actual causes. Before embarking on that survey, he inserted important preliminary chapters. In a brief introduction, he stressed not only the causal goals of geology but also its close analogy with human history (Rudwick 1977), thus showing that he proposed to combine the physical and the historical models for the science. He then launched into a long account of the history of geology, with much of the factual material taken straight from Brocchi (McCartney 1976). But Lyell's history was – as he admitted privately – a discreet cover for criticisms of his contemporaries (Porter 1976, 1982). He rewrote the history of geology to imply that those who now insisted on the reality of catastrophic events in the Earth's past were following a tradition that had consistently retarded the progress of the science. It was, he claimed, a tradition that invoked catastrophes arbitrarily whenever the explanatory going got difficult, that failed to appreciate the sheer scale of geological time, and that concocted speculative accounts of the very origin of the Earth as well as its subsequent history. In fact this was highly unfair to Buckland and others like him, as they were quick to point out once the *Principles* was published. But for Lyell it had the polemical advantage of associating them with the quite distinct genre of 'scriptural geology', which was currently enjoying great popularity among the general public. It implied that some of the most prominent English geologists were as wrong-headed as the popular writers (none of them a member of the Geological Society) who were trying to reconcile the new science with a traditional and literalistic exegesis of Genesis. In effect, Lyell equated his own conception of 'the uniformity of nature' with being *scientific* in matters of geology, thereby branding his anticipated geological critics as unscientific.

Lyell then attacked those same geologists for what they regarded as an increasingly well founded inference in geology. This was that the immensely long history of the Earth had been broadly *directional,* from a hot and highly active origin to a cool and less turbulent present, with a broadly progressive history of life to match (Rudwick 1971). As foreshadowed in his letter to Murchison, Lyell set against this his own Huttonian vision of a broadly stable, or at most a cyclic, history of both the inorganic and organic realms. He had reluctantly conceded that there was good evidence for major climatic change in the course of geohistory: the tropical-looking Carboniferous flora, for example, was difficult to explain away. At the eleventh hour, as it were, he devised a new theory that accounted for such climatic changes without conceding anything to the directional theory of a cooling Earth (Ospovat 1977). Drawing on the climatology of the great Prussian geographer Alexander von Humboldt (1817), Lyell argued that local climates in the distant past, as at present, were products not only of latitude but also of the configuration of continents and oceans, winds and ocean currents. If – as he argued – those configurations had changed continuously through the ordinary agencies of geological change, any given region might have had highly diverse climates at different periods of geohistory.

From this ingenious explanation Lyell drew consequences that his geological readers found nothing less than startling. Since organisms were closely tied to their environments, he claimed that the history of life was likewise in the long run directionless, so that organisms long extinct might recur in at least approximately the same forms: to illustrate the point he indulged in a speculative thought experiment about how the ecosystem of the Jurassic reptiles might reappear in the far distant future. This undercut the foundations of the historical model of the Earth, as Lyell's contemporaries understood it: it denied the reality of the unrepeatedness of the fossil record. Lyell had to explain away that appearance, by stressing the extreme imperfection of the record, as a consequence of the extreme chanciness of preservation. The recent discovery of rare mammalian fossils in the (Jurassic) Stonesfield 'slate', for example, gave him welcome evidence that mammals were not after all confined to the Tertiary, and enabled him to argue that they might have been in existence as far back as the fossil record could then be traced (in modern terms, well into the Lower Palaeozoic). Such were the lengths to which the consistent application of his extremely rigorous 'principles' required Lyell to go.

After these preliminary discussions, Lyell embarked on his survey of actual causes both inorganic and organic, describing with a wealth of detailed examples how much change they had effected even within the short span of recorded human history. The inorganic causes alone took up the whole of the rest of the first volume (1830). Although Lyell borrowed extensively from Hoff's factual material, he rearranged it to illustrate his own theoretical model of the Earth as a system of *balanced* physical agencies. Erosion was balanced by sedimentation, for example, elevation by subsidence, and so on; it was in fact the Newtonian model of the solar system applied to the Earth, as it had been by Hutton and Playfair before him. This explains Lyell's surprising choice for the frontispiece of his opening volume (at that time the marketing equivalent of a modern dustjacket). Instead of a picture of, say, a violently erupting volcano or an idealized section through the Earth's crust, Lyell chose a picture of a human artefact: a view of the famous Temple of Serapis near Naples, which he had seen on his travels (Fig. 1). It was in fact a highly effective epitome of his argument. It linked the human world with the natural, human history with geohistory, classical scholarship with scientific research; it illustrated the power of actual causes, and the reality of crustal movements even within the short span of time since the Roman period; and it was a vivid demonstration that those movements had been in both directions, both elevation and subsidence, thus illustrating his conception of the Earth as a system in dynamic equilibrium.

Lyell's repertoire of actual causes continued in his second volume (1832) with a survey of processes in the organic world. As with inorganic agencies, his argument was that even the largest features in the geological record could be explained by the summation of smaller effects over vast spans of time. So his discussion centred on the reality of species, and the processes by which their appearances and disappearances, origins and extinctions might be explained. He rejected Lamarck's notion of the imperceptibly gradual 'transmutation' of one species into another, claiming instead that species were real and stable units of life (Coleman 1962). Lamarck's evolutionary views, increasingly in vogue on the Continent, were widely suspect in Britain for their supposed implications of materialism, irreligion and even revolutionary politics (Corsi 1978; Desmond 1989). Lyell's explicit repudiation of such ideas signalled that his conception of geology, by contrast, should be acceptable to his intended readers, however conservative their opinions might be (Secord 1997); privately, he also found deeply repugnant the implication that human beings had evolved from animal ancestors (Bartholomew 1973). But Lamarck's kind of evolution – denying the reality of extinction altogether – was also unacceptable for strictly scientific reasons: Lyell's method of reconstructing geohistory was going to depend on the stability of species as natural units, each with a definable time of origin and of extinction; and above all, Lamarck's ideas embodied no observable *vera causa*.

In the course of a discussion of biogeography and the means by which organisms may be dispersed, Lyell tried to define the circumstances in which new species have originated; but he left the process itself unexplained. He told his friends privately that he thought it was a natural process of some kind, but publicly his reticence was prudent, deflecting any suspicion that he was a closet evolutionist (Secord 1997). With extinction there were no such problems: Lyell was free to argue that species become extinct gradually and one by one, by equally gradual changes in their environments, without recourse to sudden catastrophes or mass extinctions. So he outlined a model of organic change, developed from Brocchi's, in which species originated and became extinct in piecemeal fashion, in both time and space.

Lyell's survey of actual causes concluded with those in which the inorganic and organic worlds interact. Much of his discussion was about taphonomy, analysing the circumstances in which organisms may – or may not – be preserved as potential fossils. This served to underpin his insistence on the extreme imperfection of the fossil record; but he also pointed out that some groups, such as the molluscs, were likely to have a less

Fig. 1. The frontispiece of the first volume of Lyell's *Principles of Geology* (1830). This engraving of a view of the ruined 'Temple of Serapis' (now thought to have been a market) at Pozzuoli near Naples was redrawn from one published by a local archaeologist (Jorio 1820). The zone bored by marine molluscs on the surviving pillars was well understood to be evidence of changes in relative sea-level since the Roman period; for Lyell, the site epitomized in miniature his model of the Earth as being in a dynamic non-directional equilibrium expressed in repeated small-scale changes in the Earth's crust, even within human history.

imperfect record than others, such as mammals. This foreshadowed his crucial use of fossil molluscs in the reconstruction of geohistory.

Such applications followed in the third and culminating volume (1833). Lyell emphasized that the first two volumes were 'absolutely essential' to the third: actual causes were merely the 'alphabet and grammar' of geology, making it possible to decipher nature's 'language', in which the past history of the Earth had been recorded. The construction of geohistory and its causal explanation were the goal of geology; understanding actual causes was simply the means to that end. Lyell therefore began his final volume by discussing how the ordinary succession of formations and their fossils should be interpreted in order to yield reliable geohistory. Underlying this was not only

his profound sense of the sheer magnitude – and therefore the explanatory power – of the timescale itself, but also his conviction that the continuous piecemeal change in faunas and floras could provide the basis for *quantifying* geochronology (Rudwick 1978). On the assumption that the rates of species origins and extinctions had been statistically uniform and constant, the age of any formation should be directly related to the proportion of its fossil species that were still extant. Lyell had therefore recruited (and paid for) the taxonomic expertise of Paul Deshayes in Paris, who arranged all the known Tertiary molluscan faunas according to their percentages of extant species. The faunas fell into three distinct groups, for which the Cambridge philosopher William Whewell supplied Lyell with the appropriately classical names 'Eocene', 'Miocene' and 'Pliocene', to express respectively the 'dawn', minority and majority of modern species in the successive faunas (Lyell split the Pliocene into 'Older' and 'Newer', making four groups). For Lyell, however, these were not four contiguous *periods* of time; they were merely random samples from the deep past, separated perhaps by far longer spans of unrecorded geohistory (Rudwick 1978). So, as he penetrated still further back from the present, the *total* disjunction between the Eocene fauna and that of the youngest part of the Chalk was attributed not to any sudden catastrophe or mass extinction but to an even longer span of gradual but wholly unrecorded faunal change. Such was the kind of conclusion to which Lyell was led, by his insistence on the vast timescale of geohistory, on the extremely gradual character of physical and organic change, and on the extreme imperfection of the natural record of geohistory.

The bulk of Lyell's last volume was devoted to a detailed study of the Tertiary era, presented as an exemplar of how the whole of geohistory should be reconstructed. As Cuvier had recommended, the rocks and fossils were analysed retrospectively, from the known present back into the unknown past. Lyell analysed each of the preserved fragments of Tertiary geohistory in turn, from the Newer Pliocene (which he later renamed the 'Pleistocene') back to the Eocene, in order to demonstrate that the same processes had been operative at each time, processes the same in kind as those in the present world, acting at much the same intensities. Throughout the unimaginable span of Tertiary time, the Earth had been much the same kind of place. Lyell tried to demonstrate just how vast the timescale was, in the longest part of his discussion of the Newer Pliocene. He used his own fieldwork in Sicily to suggest a rough calibration of his mollusc-based geochronology in terms of the far briefer timescale of human history (Rudwick 1969). He presented evidence to suggest that the many eruptions of Etna recorded in the past two millennia had added very little to its bulk, so that the huge cone as a whole must be extremely ancient in human terms. Yet it appeared to be built on formations that he identified as Newer Pliocene in age, with a molluscan fauna almost identical to that still living in the Mediterranean. So a volcano that was unimaginably ancient in human terms was also extremely recent on the timescale of even the Tertiary formations, let alone geohistory as a whole.

Having established his methods with respect to the Tertiary, Lyell could afford to treat the record of the 'Secondary' and 'Transition' eras (roughly, in modern terms, the rest of the Phanerozoic) briefly and programmatically. Finally, he knocked the bottom out of the consensual reconstruction of geohistory, by claiming that the 'Primary' rocks, so called because they appeared to be the most ancient (in modern terms, they were mostly Palaeozoic or Precambrian), were not primary at all, but either igneous or what he called 'metamorphic' (coining that term for the first time). In any case, geohistory was left without any recoverable beginning; so Lyell ended on an unmistakeably Huttonian note, by claiming that the Earth was a complex physical and biological system in dynamic equilibrium, for which there were no traces of a beginning and no foreseeable end (Gould 1987).

Transformations of the *Principles*

Lyell's *Principles* was published by John Murray, the leading British scientific publisher at the top end of the market (and also the publisher of the *Quarterly Review*); Murray's impeccable Tory credentials, and the handsome format and high price of the work, indicated both the readership to which it was directed and also its social and political respectability. The process of dialogue between Lyell and his contemporaries began as soon as the first volume was published, as it generated reviews in a wide range of periodicals, and of course much unpublished comment too. (It would have had a much wider impact beyond Britain if Prévost had not become embroiled in the politics of the July Revolution, abandoning his plan to publish an edition in the international language of the sciences.) British geologists generally admired Lyell's demonstration of the surprising power of actual causes, but continued to doubt whether they were adequate to account for *all* the traces of apparent catastrophes – particularly the most recent – however generous the time allowed. Reviewers representing a broader public gave it a more mixed reception, depending largely on their position on religious and political issues (Secord 1997). Some welcomed its application to geology of the same kind of naturalistic rationality that was proving so successful in other spheres. Others were alarmed at its apparent implications for traditional ways of understanding the relation between the human and the divine, and above all between the natural and the social; it should not be forgotten that the work was published at the height of the political unrest surrounding the passing of the great Reform Act of 1832.

Reviewing the second volume in the *Quarterly*, Whewell (1832) coined the terms 'uniformitarians' and 'catastrophists' for what he called 'two opposing sects' in geology, predicting correctly that Lyell would have a harder task than he imagined to persuade the sceptics on the other side. But in fact there were few if any strict uniformitarians apart

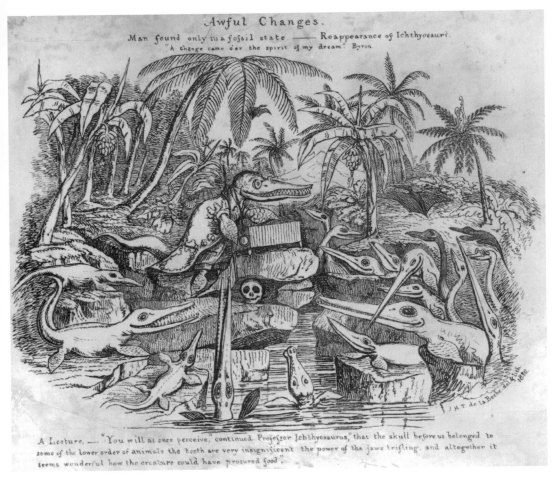

Fig. 2. Henry De la Beche's lithographed cartoon (1830) showing a 'Professor Ichthyosaurus' lecturing an audience of other Liassic reptiles on the subject of a fossil human skull (Lyell had just been appointed part-time Professor of Geology at the new King's College in London). The cartoon was designed to ridicule Lyell's notion of a cyclic pattern of geohistory, as expounded in the *Principles*, in which the Jurassic past – or something much like it – would eventually recur in the distant post-human future.

from Lyell himself; and anyway those famous labels confused the issues. Lyell had subtly amalgamated several distinct meanings of 'uniformity', about some of which there was little argument; and his critics were less concerned at his rejection of catastrophes than at his denial of any direction to geohistory (Rudwick 1971). It was the latter that aroused the most vigorous criticism among geologists. Henry De la Beche epitomized their opposition in his famous and widely circulated cartoon of a 'Professor Ichthyosaurus' lecturing to a reptilian audience on a human skull (Fig. 2). This ridiculed Lyell's ideas by turning his own verbal thought-experiment into visual form: De la Beche portrayed a distant post-human *future*, on the next

round of Lyell's cycle of directionless geohistory (Rudwick 1975, 1992). Lyell's attempt to explain away the increasingly solid evidence for an unrepeated sequence of distinctive geohistorical periods seemed utterly unconvincing to geologists, even to those such as Scrope who were most sympathetic to his project (Scrope 1830). Unlike Lyell's neo-Huttonian model of a steady-state Earth, however, Lyell's reconstruction of the Tertiary portion of geohistory was generally admired, and was prominent in the citation for the Royal Medal that the Royal Society awarded him in 1834.

Lyell began to respond to his critics promptly, even before the *Principles* was complete. In a polemical introduction to the third volume, he again set out the reasons for his rejection of any catastrophes greater in intensity than those authenticated by reliable human history. He accused his opponents of arbitrarily 'cutting the Gordian knot' when they had recourse to catastrophes, instead of engaging in the hard work of 'patiently untying' it; he thereby claimed the high ground morally as well as methodologically. But by this time he was mainly concerned to bring out a new edition of the whole work. The first edition (and the so-called 'second', reprints of the first two volumes) had sold well, by the standards of a large work of non-fiction, so Murray brought out the next (the 'third' edition, 1834) in the same cheap format as his series of popular scientific books, making it accessible to a much wider range of readers (Secord 1997). It was full of small but cumulatively important revisions, many of them made in the light of the criticisms the work had received both publicly and privately. In the subtitle, for example, Lyell's confident 'attempt' to explain everything in terms of actual causes became more modestly an 'inquiry into how far' such explanations could go. In this more popular format the work sold well enough for Murray to call for two more editions in quick succession, incorporating further modifi-

cations in the light of new research, including Lyell's own.

However, by the later 1830s Lyell's project was running into serious problems. He had failed to convince highly competent colleagues such as Scrope (1830) and De la Beche (1834) that the appearance of directionality in the geohistorical record was an artefact of differential preservation, and his neo-Huttonian steady-state theory was therefore threatened with a total collapse in plausibility (Bartholomew 1976). Even his model of uniform faunal change, on which his reconstruction of the Tertiary depended, was being undermined by the divergent opinions of the taxonomists, on whose expertise his quantitative estimates depended.

It was therefore no coincidence that just at this time Lyell decided to recast his work in a radically different form. In 1838 Murray published his *Elements of Geology* as a small and inexpensive volume, summarizing his interpretation of geohistory for the general public. Its diagrammatic frontispiece epitomized his claim that all the main classes of rocks had been produced – and all the agencies responsible had been active – at every period of geohistory (Fig. 3). The volume focused on the rocks and fossils themselves, which Lyell,

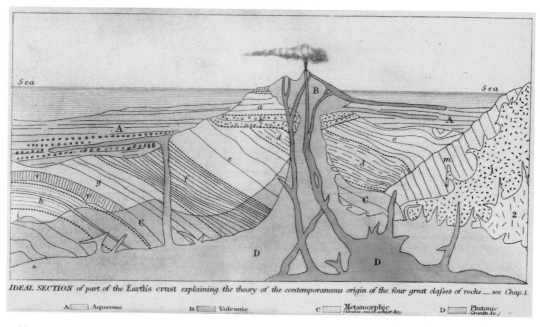

Fig. 3. The frontispiece of Lyell's *Elements of Geology* (1838). This engraved and hand-coloured 'ideal section' through the Earth's crust illustrated Lyell's claim that *all* the main classes of rock (A, aqueous; B, volcanic; C, metamorphic; D, plutonic) are still being formed in the present world; earlier products of the same processes are also shown (a–i, aqueous [in retrospective order]; m, metamorphic; v, volcanic; 1, 2, plutonic).

like his readers, regarded as the subject matter of 'Geology proper'. Conversely, the next (sixth) edition of the *Principles* (1840) was transformed into a treatise on actual causes. In effect it was an enlarged and updated version of the first two volumes of the original work (and some theoretical chapters from the third); in the words of the subtitle – changed once more – it was on 'modern changes', now considered merely as 'illustrative of geology'. Lyell planned a parallel treatise on the Tertiary formations and their faunas, but that work never appeared; instead, the second edition of the *Elements* (1841) was greatly enlarged, partly to incorporate a more adequate summary of his interpretation of the Tertiary. So a decade after he launched his work, Lyell's argument had been split into two distinct parts: a treatise on actual causes and a more popular work on geohistory. The unitary conception of the original *Principles* had been lost.

In the years that followed, Lyell's insistence on the explanatory power of actual causes came increasingly to be taken for granted by other geologists: his repertoire of persuasive examples seeped into the ordinary practice of others, so that on many (though not all) issues most geologists became in effect Lyellians. On the other hand, his 'uniformitarian' geohistory remained deeply implausible. Indeed it became more so, as new discoveries cumulatively reinforced the sense of directionality (Bartholomew 1976). For example, even allowing for the imperfection of the record – which geologists now conceded, though not in Lyell's extreme form – the resolution of the famous Devonian and Cambro–Silurian controversies (Rudwick 1985; Secord 1986) greatly clarified the distinctive character of the earlier part of the fossil record. So Lyell's continued denial of the directionality of geohistory left him isolated in his own community (Bartholomew 1979). Underlying his dogged reluctance to concede directionality – as his private notebooks have revealed (Wilson 1970) – was Lyell's continuing repugnance at the implications of evolution for human dignity; for a directional fossil record was now being used by others as the strongest evidence for evolution, and Lyell knew that Darwin, his closest geological ally, was working privately to establish evolution as a respectable and persuasive scientific theory.

After Darwin published his theory in the *Origin of Species* (1859), he was deeply disappointed that Lyell failed to support him unambiguously. In the *Antiquity of Man* (1863), a work occasioned by the coincidental discovery of clear evidence of prehistoric human life in the Pleistocene, Lyell did review the case for an evolutionary interpretation of geohistory, but hardly announced his own conversion in the ringing tones that Darwin wanted (Bynum 1984). Only in the next (tenth) edition of the *Principles* (1868) did Lyell at last concede the case for evolution and for the directionality of the fossil record. This entailed a major recasting of the argument of the book, at least in its treatment of actual causes in the organic world. But the work continued to sell well, and Lyell saw a twelfth edition through the press shortly before his death in 1875. The *Elements* ended its equally successful career as a textbook explicitly designed for students, having become a set book in the new regime of Victorian examinations.

Conclusions

In retrospect, Lyell's *Principles* may seem paradoxical. On the one hand, its eloquent argument for the adequacy of actual causes in geological explanation was largely successful, by the criterion most appropriate in the sciences, that of collective amnesia. The systematic use of physical and biological processes observable in the present world came to be taken for granted by geologists as the necessary first step towards interpreting the always fragmentary traces of past geohistorical events; and later generations of geologists were often unaware of Lyell's role. Likewise his persuasive explanatory use of a timescale of humanly unimaginable magnitude came to be taken for granted in the everyday work of geologists; Lyell showed them the implications in practice of what they already conceded in principle (and that practice was scarcely affected either by Kelvin's subsequent restriction of the timescale or by its still later expansion after the discovery of radioactivity).

On the other hand, Lyell's claim that his methods entailed a denial of any kind of directionality in geohistory became increasingly implausible, and was abandoned eventually even by Lyell himself. Likewise his principled rejection of any explanations invoking causal processes more intense than those observed in human history became increasingly a straitjacket on legitimate theorizing. Even in his own lifetime, for example, Lyell's stance made it difficult for geologists to recognize the full severity of the Pleistocene glaciations; and a century later Lyell was still being invoked to justify the dogmatic dismissal of any conjectures about episodes of mass extinction, let alone their possibly extraterrestrial causation.

However, such a two-edged legacy is just what we should expect from even the greatest scientists: like lesser mortals, they do not get everything right, nor should we expect them to. In fact the history of Lyell's *Principles* is a good illustration of the way that true progress in the Earth sciences is generally the product of a synthesis between initially opposed positions (Rudwick 1985). In retrospect, we can see that our modern understanding of geohistory and its causal explanation is a synthesis between Lyell's approach and that of his critics. We inherit the product of nineteenth-century debates, not least the famously vigorous arguments that took place at the Geological Society.

References

Note. The first edition of the *Principles* (1830–1833) is now easily accessible and inexpensive, both as a facsimile reprint of the entire three-volume work and in a sensitively abridged one-volume format. The full reprint (Lyell 1990–1991) has an introduction analysing Lyell's argument, and a bibliography of Lyell's own sources, listing in full the items that he himself often cited obscurely in abbreviated form (Rudwick 1990–1991). The abridged reprint (Lyell 1997) has a valuable introductory essay, stressing particularly the context of the work in British intellectual and cultural history (Secord 1997). Copies of editions from late in Lyell's life are still relatively easy to find on the antiquarian market, but it should be remembered that they are radically different works.

BARTHOLOMEW, M. 1973. Lyell and evolution: an account of Lyell's response to the prospect of an evolutionary ancestry for Man. *British Journal for the History of Science*, **6**, 261–303.
—— 1976. The non–progression of non–progression: two responses to Lyell's doctrine. *British Journal for the History of Science*, **9**, 166–174.
—— 1979. The singularity of Lyell. *History of Science*, **17**, 276–293.
BORK, K. B. 1990. Constant Prévost (1787–1856): the life and contributions of a French uniformitarian. *Journal of Geological Education*, **38**, 21–27.
BROCCHI, G.-B. 1814. *Conchiologia fossile subapennina con osservazione geologiche sugli Apennini e sul suolo adiacente*. 2 vols. Stamperia Reale, Milano.
BUCKLAND, W. 1820. *Vindiciae geologicae; or the connexion of geology with religion explained, in an inaugural lecture*. W. Buckland, Oxford.
—— 1822. Account of an assemblage of fossil teeth and bones belonging to extinct species of elephant, rhinceros, hippopotamus, and hyaena, and some animals discovered in a cave at Kirkdale, near Kirkby Moorside, Yorkshire. *Philosophical Transactions of the Royal Society*, **122**, 171–236.
—— 1823. *Reliquiae diluvianae; or, observations on the organic remains contained in caves, fissures, and diluvial gravel, and on other geological phenomena, attesting to the action of an universal deluge*. Murray, London.
BUFFON, G. DE 1778. Des époques de la nature. *In*: BUFFON, G. de. *Histoire naturelle*, Supplément 5. Imprimérie royale, Paris, 1–254.
BYNUM, W. F. 1984. Charles Lyell's *Antiquity of Man* and its critics. *Journal of the History of Biology*, **17**, 153–187.
COLEMAN, W. 1962. Lyell and the "reality" of species, 1830–1833. *Isis*, **53**, 325–338.

CORSI, P. 1978. The importance of French transformist ideas for the second volume of Lyell's *Principles of Geology*. *British Journal for the History of Science*, **11**, 221–244.
CUVIER, G. 1812. *Recherches sur les ossemens fossiles de quadrupèdes, où l'on rétablit les caractères de plusieurs espèces d'animaux que les révolutions du globe paroissent avoir détruites*. 4 vols. Déterville, Paris.
—— & BRONGNIART, A. 1811. Essai sur la géographie miné ralogique des environs de Paris. *Mémoires de la Classe des Sciences Mathématiques et Physiques de l'Institut Impérial de France*, vol. for 1810, 1–278.
DARWIN, C. 1859. *On the Origin of Species by Means of Natural Selection, or the Preservation of Favoured Races in the Struggle for Life*. Murray, London.
DE LA BECHE, H. T. 1834. *Researches in Theoretical Geology*. Knight, London.
DESMOND, A. 1989. *The Politics of Evolution: Morphology, Medicine, and Reform in Radical London*. University of Chicago Press, Chicago.
GOULD, S. J. 1987. *Time's Arrow, Time's Cycle*. Harvard University Press, Cambridge, MA.
HALL, J. 1814. On the revolutions of the Earth's surface. *Transactions of the Royal Society of Edinburgh*, **7**, 139–211.
HAMM, E. P. 1993. Bureaucratic *Statistik* or actualism? K. E. A. von Hoff's *History* and the history of geology. *History of Science*, **31**, 151–176.
HOFF, K. E. A. VON. 1822–1834. *Geschichte der durch Überlieferung nachgewiesenen natürlichen Veränderungen der Erdoberfläche: ein Versuch*. 3 vols. Perthes, Gotha.
HORNER, L. 1816. [Review of] *Conchiologia fossile Subapennina... di G. Brocchi*, Milano, 1814. *Edinburgh Review*, **26**, 156–180.
HUMBOLDT, A. VON. 1817. Des lignes isothermes et de la distribution de la chaleur sur le globe. *Mémoires de la physique et de la chimie de la Société d'Arceuil*, **3**, 462–602.
HUTTON, J. 1788. Theory of the Earth; or an investigation of the laws observable in the composition, dissolution and restoration of the land upon the globe. *Transactions of the Royal Society of Edinburgh*, **1**, 209–304.
—— 1795. *Theory of the Earth, with Proofs and Illustrations*. 2 vols. Creech, Edinburgh.
JORIO, A. DE. 1820. *Ricerche sul Tempio di Serapide, in Puzzuoli*. Società Filomatica, Napoli.
LAUDAN, R. 1982. The role of methodology in Lyell's science. *Studies in the History and Philosophy of Science*, **13**, 215–249.
—— 1987. *From Mineralogy to Geology: The Foundations of a Science, 1650–1830*. University of Chicago Press, Chicago.
LYELL, C. 1826. On a recent formation of freshwater limestone in Forfarshire, and on some recent deposits of freshwater marl. *Transactions of the Geological Society of London*, 2nd ser., **2**, 72–96.
—— 1827. [Review of] *Memoir on the geology of central France, including the volcanic formations of Auvergne, the Velay, and the Vivarrais*, by G.P.

Scrope, London. 1827. *Quarterly Review*, **36**, 437–483.

—— 1830–1833. *Principles of Geology, Being an Attempt to Explain the Former Changes of the Earth's Surface, by Reference to Causes Now in Operation*. 3 vols. Murray, London.

—— 1834. *Principles of Geology, Being an Inquiry How Far the Former Changes of the Earth's Surface are Referable to Causes Now in Operation*. 3rd edn, 4 vols. Murray, London.

—— 1838. *Elements of Geology*. Murray, London.

—— 1840. *Principles of Geology, or, the Modern Changes of the Earth and Its Inhabitants Considered as Illustrative of Geology*. 6th edn, 3 vols. Murray, London.

—— 1841. *Elements of Geology*. 2nd edn, 2 vols. Murray, London.

—— 1863. *The Geological Evidences of the Antiquity of Man with Remarks on Theories of the Origin of Species by Variation*. Murray, London.

—— 1868. *Principles of geology, or, the modern changes of the Earth and its inhabitants considered as illustrative of geology*. 10th edn, 2 vols. Murray, London.

—— 1881. *Life Letters and Journals of Sir Charles Lyell, Bart* (ed. K. Lyell). 2 vols. Murray, London.

—— 1990–1991. *Principles of Geology*. (Facsimile reprint of 1st edn), 3 vols. University of Chicago Press, Chicago.

——1997. *Principles of Geology* (abridged from 1st edn). Penguin, London.

McCARTNEY, P. J. 1976. Charles Lyell and G.B. Brocchi: a study in comparative historiography. *British Journal for the History of Science*, **9**, 177–89.

MARINI, P. (ed.) 1987. *L'opera scientifica di Giambattista Brocchi (1772–1826)*. Città di Bassano del Grappa.

MORRELL, J. B. 1976. London institutions and Lyell's career, 1820–41. *British Journal for the History of Science*, **9**, 132–146.

OSPOVAT, D. 1977. Lyell's theory of climate. *Journal of the History of Biology*, **10**, 317–339.

PLAYFAIR, J. 1802. *Illustrations of the Huttonian Theory of the Earth*. Creech, Edinburgh.

PORTER, R. 1976. Charles Lyell and the principles of the history of geology. *British Journal for the History of Science*, **9**, 91–103.

—— 1982. Charles Lyell: the public and private faces of science. *Janus*, **69**, 29–50.

ROGER, J. 1989. *Buffon: un philosophe au Jardin du Roi*. Fayard, Paris.

RUDWICK, M. J. S. 1969. Lyell on Etna, and the antiquity of the Earth. *In*: SCHNEER, C. J. (ed.) *Toward a History of Geology*, MIT Press, Cambridge, MA, 288–304.

—— 1971. Uniformity and progression: reflections on the structure of geological theory in the age of Lyell. *In*: ROLLER, D. H. D. (ed.) *Perspectives in the History*

of Science and Technology. Oklahoma University Press, Norman, OK.

—— 1974. Poulett Scrope on the volcanoes of Auvergne: Lyellian time and political economy. *British Journal for the History of Science*, **7**, 205-242.

—— 1975. Caricature as a source for the history of science: De la Beche's anti-Lyellian sketches of 1831. *Isis*, **66**, 534–560.

—— 1977. Historical analogies in the early geological work of Charles Lyell. *Janus*, **64**, 89–107.

—— 1978. Charles Lyell's dream of a statistical palaeontology. *Palaeontology*, **21**, 225–244.

—— 1985. *The Great Devonian Controversy: The Shaping of Scientific Knowledge among Gentlemanly Specialists*. University of Chicago Press, Chicago.

—— 1990–1991. Introduction [and] Bibliography of Lyell's sources. *In*: LYELL, C. *Principles of Geology*. (Facsimile reprint). University of Chicago Press, Chicago, vol. 1, vii–lxviii; vol. 3, Appendices, 113–160.

—— 1992. *Scenes from Deep Time: Early Pictorial Representations of the Prehistoric World*. University of Chicago Press, Chicago.

—— 1996. Cuvier and Brongniart, William Smith, and the reconstruction of geohistory. *Earth Sciences History*, **15**, 25–36.

—— 1997. *Georges Cuvier, Fossil Bones, and Geological Catastrophes: New Translations and Interpretations of the Primary Texts*. University of Chicago Press, Chicago.

RUPKE, N. A. 1983. *The Great Chain of History: William Buckland and the English School of Geology (1814–1849)*. Clarendon, Oxford.

SCROPE, G. P. 1825. *Considerations on Volcanos... Leading to the Establishment of a New Theory of the Earth*. Phillips, London.

—— 1827. *Memoir on the Geology of Central France, Including the Volcanic Formations of Auvergne, the Velay and the Vivarrais*. 2 vols. Longman, London.

—— 1830. [Review of] *Principles of Geology* by Charles Lyell, London, 1830. *Quarterly Review*, **43**, 411–469.

SECORD, J. A. 1986. *Controversy in Victorian Geology: the Cambrian–Silurian Dispute*. Princeton University Press, Princeton.

—— 1997. Introduction. *In*: LYELL, C. *Principles of Geology* (abridged). Penguin, London.

WHEWELL, W. 1832. [Review of] *Principles of Geology* by Charles Lyell, vol. 2, London, 1832. *Quarterly Review*, **47**, 103–132.

WILSON, L. G. (ed.) 1970. *Sir Charles Lyell's Scientific Journals on the Species Question*. Yale University Press, New Haven.

—— 1972. *Charles Lyell, the Years to 1841: The Revolution in Geology*. Yale University Press, New Haven.

Charles Lyell and the Geological Society

JOHN C. THACKRAY

Department of Library and Information Services, The Natural History Museum,
Cromwell Road, London SW7 5BD, UK

Abstract: Charles Lyell was a loyal and committed Fellow of the Geological Society from 1819 until 1875, serving as Secretary, Foreign Secretary and President. He was principally a theorist, his uniformitarian outlook strongly influenced by the work of James Hutton. The Society on the other hand was devoted to the collection and dissemination of facts; on the whole, theorizing was unwelcome. In line with this tradition the papers Lyell read to the Society are largely factual, and do not add a great deal to the ideas put forward in the *Principles of Geology*. However, his theories were not ignored by the Society, being noticed in the annual presidential address and, more importantly, in the unscripted and largely spontaneous discussions that took place after many of the papers read at the Society's evening meetings.

Charles Lyell was elected a Member of the Geological Society on 19 March 1819. His admission certificate, which is number 498 in a series which had reached 18 000 by 1990, is signed by the Revd William Buckland, the Revd William D Conybeare and the Revd William Hony. Lyell was then 22 years old and a student at Exeter College, Oxford. He remained a Member (converting to Fellow in 1826 after the granting of the Royal Charter) until his death 56 years later. Lyell served as Secretary, Foreign Secretary, Vice President and President. He attended Council Meetings in one capacity or another for an astonishing 52 years.

The Geological Society was not, of course, Lyell's only society, for he belonged to the Royal, Linnean, Zoological and Geographical societies and probably others (Morrell 1976). But there is little doubt that Lyell loved the Geological Society above all others. This was partly of course because geology was his favourite science, but also because he subscribed to its ethos and way of doing things. One sees his pride when he reports to his friends: 'a splendid meeting', 'an excellent meeting' and 'all the best men present'. Certainly in the 1820s and 1830s most observers agreed that the Geological was easily the brightest and liveliest of London's scientific societies, and even so critical a writer as Charles Babbage sang its praises (Babbage 1830, pp. 45–46).

Lyell was not however prepared to devote his whole life to the Society, and occasionally resented the amount of time that his offices took up. He managed to avoid being President for 1833–1835, though he succumbed for 1835–1837. He wrote to

Charles Darwin on 26 December 1836:

> Don't accept any official scientific place, if you can avoid it, and tell no one I gave you this advice, as they would all cry out against me as the preacher of anti-patriotic principles. I fought against the calamity of being President as long as I could ... (Burkhardt & Smith 1985, p.532)

We have a pen picture by John Ruskin of Lyell at a Society meeting during his first presidency which, although distinctly unflattering, is well worth reproducing:

> One finely made, gentlemanly-looking man was very busy among the fossils which lay on the table, and shook hands with most of the members as they came in. His forehead was low and not very wide, and his eyes were small, sharp, and rather ill-natured. He took the chair, however, and Mr. Charlesworth, coming in after the business of the meeting had commenced, stealing quietly into the room, and seating himself beside me, informed me that it was Mr. Lyell. I expected a finer countenance in the great geologist. (Cook & Wedderburn 1909, vol. 1, p. 9)

Charles Lyell was a great theorist. His *Principles of Geology* (12 edns, 1830–1875) offers his readers a world in equilibrium, where organic and inorganic processes interact and balance; where things move slowly and steadily and naturally within a long timescale. He had a deep and long term commitment to what became known as uniformitarianism, and the presentation of observations

THACKRAY, J. C. 1998. Charles Lyell and the Geological Society. *In*: BLUNDELL, D. J. & SCOTT, A. C. (eds) *Lyell: the Past is the Key to the Present*. Geological Society, London, Special Publications, **143**, 17–20.

17

and field examples are part of Lyell's strategy for promoting this theory (Rudwick 1970). This outlook owed a lot to James Hutton, or at least to Playfair's interpretation of Hutton, and Hutton was one of those geological system builders that the Geological Society deeply disapproved of. The Geological Society had been founded in 1807 to facilitate the collection and communication of new facts, and not for the discussion of systems and theories (Rudwick 1963). G. B. Greenough (1778–1855), the first President, was still active through the 1820s and 1830s and was fiercely sceptical of all attempts at theorizing. The typical paper presented during the first 50 years of the Society's life was a piece of regional geology in which strata were described and mapped, and their age deduced from a study of their fossil remains. The deduction of theories from facts was certainly acceptable, but anything that smacked of the all-inclusive theories of the previous century was deeply suspect.

Were the theoretical tenets to which Lyell was so committed simply ignored by the Society to which he gave his allegiance?

Lyell read 45 papers to the Geological Society between 1824 and 1854. Twenty derive from his visits to North America, 15 relate to the Tertiary and Quaternary strata of Britain and the Continent of Europe, five to other aspects of British geology, and five to overseas affairs. None relates explicitly to Lyell's uniformitarian programme. They are all papers based on geological fieldwork which lie quite easily alongside those of his less philosophical colleagues such as R. I. Murchison (1792–1871), H. T. De la Beche (1796–1855) and the rest. On closer inspection, however, many of the papers do relate to subsidiary themes covered in the *Principles of Geology*, and many contained observations that were added to the ever-developing text of the *Principles* and *Elements of Geology* (six edns, 1838–1865). Three papers will serve as examples of the ways in which this happened.

In April 1834 Lyell read a paper on the loamy deposit called 'loess' in the valley of the Rhine, which was based on fieldwork carried out the previous year. Among the distinctively Lyellian features of the paper are the use made of the modern sediments of the Rhine to deduce the environment of deposition of the loess, and the emphasis laid on the considerable geographic changes that must have taken place since the deposition of this geologically recent deposit, and yet with no sign of violence or catastrophe. The paper was published in the Society's *Proceedings* in 1834 (Lyell 1834), and the observations and ideas it contained were incorporated into Lyell's discussion of the Newer Pliocene in the fourth edition of the *Principles of Geology* (Lyell 1835, vol. 4, p. 29), where it helped to emphasize the similarities between the world of today and the world of the geological past.

The second paper was read in April 1839 and concerned some fossil and recent shells collected in Canada by Captain Bayfield. Lyell used these to deduce that the climate was colder in the recent past and slightly less cold at a more distant period. Once again, the paper was published in the *Proceedings* (Lyell 1839). The shells are cited in the second edition of the *Elements of Geology* (Lyell 1841, vol. 1, p. 236), where they join the mass of evidence that Lyell gathers to show that the climate of particular locations has changed during the Earth's history, but that the overall temperature of the Earth has remained constant.

The third paper was read on Lyell's behalf in January 1842 and deals with Niagara Falls, which he had visited the previous autumn. Lyell's enthusiasm for this great marvel of nature is clear as he describes the strata of the district and then the phenomena of the falls themselves (Lyell 1842). Niagara had featured in the *Principles of Geology* from the very first edition to demonstrate the role of running water in erosion, with a description based on the work of Robert Bakewell (Bakewell 1830) and Captain Basil Hall (Hall 1829). In the seventh edition the account is rewritten on the basis of Lyell's own fieldwork and this paper, but in essence it is unchanged (Lyell 1847). The importance of Lyell's visits to America in the context of his total field experience has recently been demonstrated (Dott 1997).

In sum then, Lyell's Geological Society papers add some depth and supporting detail to the argument set out in the *Principles of Geology*, but their overall effect is quite small.

There was, however, another forum where Lyell's ideas featured more prominently: the annual presidential address. By tradition the President was given more latitude in dealing with speculative or theoretical issues than were speakers at the Ordinary General Meetings. So, in his address of February 1831, the Revd Professor Adam Sedgwick, having reviewed the papers read to the Society through the year, turned his attention to Lyell's *Principles* (Sedgwick 1831, pp. 298–306). He first gave an account of his own views of the origin of the Earth as a molten globe and the effects of its steadily decreasing temperature through geological time, before showing how Lyell had misunderstood and 'sometimes violated' these basic tenets of geological science. He expressed regret that 'in the language of the advocate' Lyell sometimes forgot the character of the historian. Sedgwick made it clear that he accepted the unchanging nature of the fundamental physical laws, but could not believe that 'those secondary combinations' that we call geological processes had remained constant throughout time. This was a thoroughly critical review, given in front of a large and distinguished audience, with no immediate opportunity for any response. Two years later Roderick I. Murchison gave a more generous description of Lyell's second volume (Murchison 1833, pp. 441–443), but in 1834 George B. Greenough was critical when he considered the question of elevation in relation to ideas put forward in the *Principles of Geology* (Greenough 1834, pp. 60–70).

But even more significant than the presidential addresses in bringing Lyell's theoretical ideas before the Society were the discussions that followed the reading of papers at the evening meetings. These unscripted and spontaneous debates set the Geological apart from its fellow societies, where questions were either submitted in advance or not allowed at all. Not only did the Society not publish any record of these discussions, it forbade others from doing so, preferring to maintain the myth that science was uncontroversial. Our sources are therefore contemporary letters and diaries. Although the discussions took many forms, very often they served to incorporate a factual and apparently innocuous paper into some lively contemporary debate, and so to draw out the theoretical significance of the facts.

Lyell himself preserved a record of the extended discussion that followed Dr Turnbull Christie's paper 'On certain younger deposits in Sicily' on 2 November 1831, when attention seems to have been focused firmly on his book:

A paper by Dr. Turnbull Christie, on Sicily, led Murchison to call me up, and I had the field much to myself. I told you that a temporary cloud came over me in Edinburgh at seeing the controversial storm gathering against me in the horizon; but I must say it was dispelled by this meeting, for never on the first meeting, after my first volume appeared, did I hear it so much spoken of. Fitton declared to the meeting his conviction that my theory of earthquakes wd. ultimately prevail. ... [Basil Hall] then eulogised my book, and after the meeting two American gentlemen came up to be introduced, and poured out a most flattering comment on its popularity in the United States; but what was much more to the purpose than this sort of incense, they gave me many facts bearing on my theories respecting part of North America. (Lyell 1881, vol. 1, p. 349; Wilson 1972, p. 326)

Greenough attacked Lyell's theories whenever he had an opportunity, notably on 18 November 1835, after Charles Darwin's paper on 'Geological notes made [in] South America'. Charles Bunbury wrote to his father: 'Mr Greenough made a speech of some length, in his usual spirit of scepticism, attacking the Lyellian theory, with a good deal of humour, but not much argument' (Lyell 1906, vol. 1, p. 78).

But by March 1838, when Darwin read his paper 'On the connexion of certain volcanic phenomena ...', Lyell at least sensed that the tide had turned, for he wrote:

I was much struck with the different tone in which my gradual causes were treated by all, even including de la Bêche, from that which they experienced in the same room 4 years ago when Buckland, de la Bêche, Sedgwick, Whewell and some others treated them with as much ridicule as was consistent with politeness in my presence. (Wilson 1972, p. 456)

This was no final victory of course, as Lyell had geological battles to fight throughout his life. In the 1850s the reality of progress and progression in the fossil record was the hot topic, with Lyell eagerly looking for evidence that might upset the generally accepted progression from fish (Palaeozoic) to reptile (Mesozoic) to mammals (Cenozoic). A note in A. C. Ramsay's diary for 19 March 1856 reveals the significance of a little paper by the Revd Mr Dennis, 'On some organic remains from the bone-bed at the base of the Lias at Lyme Regis', in this debate:

Dined at the Geological Club & sat between Sir Phillip & Galton, who is a gentleman of great good humour & small capacity. Jukes was there as Sir Roderick's guest & Sir Charles Lyell also had another, a country parson who afterwards read a paper on some supposed Mammalian bones from the Lias, & whom Owen regularly 'chewed up' after. While Owen was demolishing the clergyman it was curious to see Lyell winking and blinking & feeling ashamed of his protégé. (Ramsay 1856)

It would be good to be able to report the discussions that followed Lyell's own papers, to see to what extent they drew out the underlying theoretical concerns, but few descriptions have survived, and those are not revealing. Lyell was clearly a disappointing speaker. Ruskin commented that there was 'too much talking and verbiage' in February 1843, while in December 1849 Ramsay found him 'so long winded' that he had to leave at ten, before the paper was over.

The final forum where Lyell's ideas were discussed was the Geological Society Dining Club. The Club was an inner circle, where many of the leading fellows met to dine before the Society's evening meeting. Lyell was a founder member of the Club in 1824, and regularly attended its dinners. Although the Club's primary purpose was social, geological topics were freely discussed, and Lyell lost no opportunity of making his views known (Gray 1996). However, as many meetings only attracted ten or twelve Fellows, any conversations that did take place had a limited impact on the fellowship as a whole.

Whatever Lyell's own failings as a speaker, it is clear that uniformitarianism, gradualism, the efficacy of modern causes and all the other things that Lyell held dear were not ignored at the Geological Society, and that there was a place for the discussion of theoretical issues within its primarily Baconian ethos. These discussions were not published at the time, either by the Society or others, but they are crucial to any clear understanding of the Society's attitude to theory and to the personal and scientific dynamics of the day.

References

BABBAGE, C. 1830. *Reflections on the Decline of Science in England, and on Some of its Causes*. B. Fellowes, London.

BAKEWELL, R. 1830. On the Falls of Niagara and on the physical structure of the adjacent country. *Magazine of Natural History,* **3**, 117–130.

BURKHARDT, F. & SMITH, S. 1985. *The Correspondence of Charles Darwin*. Cambridge University Press, Cambridge, vol. 1.

COOK, E. T. & WEDDERBUN, A (eds) 1909. *The Letters of John Ruskin*. 2 vols. Allen, London.

DOTT, R. H. 1997. Lyell in America – his lectures, field work, and mutual influences, 1841–1853. *Earth Sciences History,* **15**, 101–140.

GRAY, D. A. 1996. *A Review of the History of the Geological Society Dining Club*. The Geological Society Club.

GREENOUGH, G. B. 1834. Address delivered at the anniversary meeting of the Geological Society on the 21st of February 1834. *Proceedings of the Geological Society of London,* **2**, 42–70.

HALL, B. 1829. *Travels in North America in the Years 1827 and 1828*. 3 volumes. Cadell & Co, Edinburgh.

LYELL, C. 1834. Observations on the loamy deposit called 'loess' in the valley of the Rhine. *Proceedings of the Geological Society of London,* **2**, 21–22.

—— 1835. *Principles of Geology*. 4th edn, 4 volumes. Murray, London.

—— 1839. Remarks on some fossil and recent shells collected by Capt. Bayfield in Canada. *Proceedings of the Geological Society of London,* **3**, 119–120.

—— 1841. *Elements of Geology*. 2nd edn. 2 volumes. Murray, London.

—— 1842. Memoir on the recession of the Falls of Niagara. *Proceedings of the Geological Society of London,* **3**, 595–602.

—— 1847. *Principles of Geology*. 7th edn. Murray, London.

LYELL, K. M. (ed.) 1881. *Life, Letters and Journals of Sir Charles Lyell, Bart*. 2 volumes. Murray, London.

—— 1906. *The Life of Sir C. J. F. Bunbury Bart., with an Introductory Note by Sir J. Hooker*. 2 volumes. Murray, London.

MORRELL, J. B. 1976. London institutions and Lyell's career. *British Journal for the History of Science,* **9**, 132–146.

MURCHISON, R. I. 1833. Address to the Geological Society, delivered on the evening of the 15th of February 1833. *Proceedings of the Geological Society of London,* **1**, 438–464.

RAMSAY, A. C. 1856. Unpublished diary held in the Ramsay papers. Imperial College Archives, London.

RUDWICK, M. J. S. 1963. The foundation of the Geological Society of London: its scheme for co-operative research and its struggle for independence. *British Journal for the History of Science,* **1**, 325–355.

—— 1970. The strategy of Lyell's 'Principles of Geology'. *Isis,* **61**, 5–33.

SEDGWICK, A. 1831. Address to the Geological Society, delivered on the evening of the 18th of February 1831. *Proceedings of the Geological Society of London,* **1**, 281–316.

WILSON, L. G. 1972. *Charles Lyell. The Years to 1841: The Revolution in Geology*. Yale University Press, New Haven & London.

Lyell: the man and his times

LEONARD G. WILSON

Department of History of Medicine, Medical School, University of Minnesota,
Box 506 Mayo, 420 Delaware Street SE, Minneapolis, MN 55455, USA

Abstract: Born in Scotland at the end of the eighteenth century, Charles Lyell spent his early life in Hampshire in the midst of the kind of comfortable rural society described in Jane Austen's novels. Influenced by his father, who was a keen botanist, the young Lyell collected butterflies and studied natural hisory. As a student at Oxford, he attended the Revd William Buckland's lectures on geology and continued to pursue geology while studying for the bar at Lincoln's Inn. In 1824 Lyell wrote his first scientific paper on the freshwater limestones and marls of Scottish lakes, demonstrating their detailed similarity to ancient freshwater formations among the Tertiary strata of the Paris Basin. Throughout his life Lyell remained an active field geologist, travelling throughout Europe from Sicily to Scandinavia. He made repeated geological tours through the Swiss and Austrian Alps and studied intensely the Tertiary strata of France, Belgium and England. He examined the geology of North America from Nova Scotia to the Mississippi Delta, working in the field in all weather. In 1853–1854 he spent several months studying the volcanic geology of Madeira and the Canary Islands and in 1857 and 1858 revisited Sicily to spend arduous weeks studying Mount Etna. In his seventy-fifth year and nearly blind, he travelled to the south of France to examine the caves of Aurignac and the Dordogne.

In 1830 in the *Principles of Geology* Lyell challenged the assumption of a cooling Earth and of the greater former intensity of volcanic action and other geological forces. He separated geology from cosmology, pointing out that the crust of the Earth told nothing of its origin. What geology did reveal was an unending cycle of changes such as were still going on. In his metamorphic theory developed in the *Elements of Geology*, Lyell showed how the various classes of igneous and metamorphic rocks were formed in the gradual elevation of mountains. Lyell's theories brought about a revolution in geology, changing the meaning of geological data. In successive editions of the *Principles* and the *Elements*, two books of American travel, and the *Antiquity of Man*, Lyell's stature as an author and as a scientist grew steadily throughout his lifetime, the sale of his books often exceeding the publisher's expectations. Liberal in his sympathies, Lyell worked effectively for the reform of English university education and during the American Civil War defended the Union against its British critics

Charles Lyell was born at Kinnordy House in Forfarshire, Scotland on 14 November 1797, the son of Charles Lyell, laird of Kinnordy. Although he would live most of his life in the nineteenth century, becoming one of the eminent scientists of the Victorian age, Lyell began his childhood in the long eighteenth century which ended only with the defeat of Napoleon in 1815 – a century of wars, naval battles, and blockades, and of the industrial revolution, of the rising power and prosperity of Great Britain, and epidemic disease. At the age of two months the young Charles Lyell was inoculated with smallpox, which for two weeks made him quite ill, just a half year before Edward Jenner would announce his discovery of vaccination.

In June 1798 Lyell's father took his wife and infant son to the south of England and that autumn leased Bartley Lodge, a country house with 80 acres of land at Lyndhurst on the edge of the New Forest in Hampshire. There Lyell grew up amidst the kind of quiet comfortable society described by Jane Austen – comfortable but anxious, menaced by the threat of invasion from France, by the possibility of social unrest, and by disease.

At the age of seven Lyell was sent with his younger brother Tom to a small private school at Ringwood. There in November 1805 Lyell and his school fellows celebrated the battle of Trafalgar with bonfires on the hills and mourned the death of Nelson. In 1807 the boys were transferred to another school at Salisbury. Dreamy and absent-minded, Charles did not do well at school. In December 1808 he fell severely ill with pneumonia and his father had to bring him home. For the 11-year-old boy this seeming calamity proved a formative influence. During the four months that he remained at home convalescing, Charles pored over the coloured plates of Donovan's *Natural History of British Insects* (Donovan 1793–1813), using which he identified various species of butterflies

WILSON, L. G. 1998. Lyell: the man and his times. *In*: BLUNDELL, D. J. & SCOTT, A. C. (eds) *Lyell: the Past is the Key to the Present.* Geological Society, London, Special Publications, **143**, 21–37.

Fig. 1. Sir Charles Lyell in 1853. Chalk drawing by George Richmond. Courtesy of Lord Lyell of Kinnordy.

and moths. As he recovered his strength he began to collect butterflies and to watch the metamorphosis of the caterpillar first into a pupa and then into the adult butterfly. Throughout his life he would delight in butterflies. Aquatic insects also fascinated him and he spent mornings by a pond watching them.

In 1810 Charles and his brother Tom went to school at Midhurst in Sussex, but in 1813 Tom left Midhurst to enter the navy as a midshipman. Charles remained at Midhurst until 1815 and in his final year there distinguished himself as a Latin scholar. In 1816 he entered Exeter College, Oxford, where he pursued the usual classical course of study. Nevertheless, he had already developed another interest. At Bartley Lodge in 1816 Lyell had read Robert Bakewell's *Introduction to Geology* (Bakewell 1815), from which he learned of Dr James Hutton's theory that the Earth was indefinitely old. The succession of stratified rocks showed no sign of a beginning because the oldest rock strata represented sediments formed by the wearing down of pre-existing lands, long vanished. At Oxford in 1817 Lyell attended William Buckland's lectures on mineralogy. During July, while on a visit to Dawson Turner at Yarmouth, Lyell studied the delta of the Yare. He learned how the river channel had shifted and sand dunes

accumulated in the recent past from the interactions of the river with the sea and the wind. In September Lyell visited the island of Staffa to see its famous basalt columns and Fingal's Cave.

At Oxford in the spring of 1818 Lyell attended his second course of Buckland's lectures on mineralogy and geology. During the summer he accompanied his family on a tour on the Continent. At Paris, Lyell visited the Jardin des Plantes to see the bones of fossil animals described by Cuvier and to read Cuvier's books on fossil remains (Cuvier 1812) and on the geology of the Paris Basin (Cuvier & Brongniart 1811). At Chamonix in Switzerland he saw how the glacier of Bosson had recently advanced, treading down pine forests in front of it. While the Lyell family travelled along the roads in Switzerland, Lyell made parallel excursions in the Alps, often walking 35–40 miles a day. In the autumn he returned to Oxford for his final year and graduated there in the spring of 1819. In March 1819 Lyell was also elected a Member of the Geological Society of London.

In 1820 Lyell began to study law at Lincoln's Inn, but his eyes, which had begun to trouble him the year before while he was studying for examinations at Oxford, again became so inflamed that he had to suspend his studies temporarily while continuing to keep his terms. During the summer of 1820 he made a second tour on the Continent with his father, travelling through Belgium, up the Rhine Valley and across the Alps into Italy as far as Rome. While on vacation at Bartley Lodge in October 1821 Lyell rode on horseback to Midhurst in Sussex and from there rode across the South Downs to Lewes to make the acquaintance of the Lewes surgeon Gideon Mantell, who was actively collecting fossils of the Sussex strata. Lyell was excited to learn of Mantell's discovery in the strata of the Weald, beneath the Chalk, of the remains of land plants, the bones of land vertebrates, and freshwater shells, such as might have lived in a river delta. The ancient delta in which the Weald strata had accumulated had later sunk beneath the sea to be buried under the thick Chalk formation before being re-elevated and denuded. The Weald strata offered dramatic evidence for upward and downward movements of the land. They supported the views of the eighteenth-century Scottish geologist, Dr James Hutton of Edinburgh, who had argued that sedimentary strata had been elevated from beneath the sea by the internal heat of the Earth. The German mineralogist Abraham Gottlob Werner had presented the opposite view that, instead of the land's having risen, the sea had receded. According to Werner, a universal ocean had formerly extended over the Earth, covering even the highest mountains, and stratified rocks had been deposited from it. By providing clear evidence of downward and upward movements of the land, the Weald strata supported Huttonian geology rather than the Wernerian view.

In May 1822 Lyell was called to the bar at Lincoln's Inn. He also began seriously to study geology in the field. At various places in southeast England, Lyell studied the succession of beds beneath the Chalk, revealed by the denudation of the anticlinal fold of the Weald. In June 1822, he visited the Isle of Wight, where he confirmed the succession of strata described there earlier by Thomas Webster (Webster 1814, 1821). In a second visit to the Isle of Wight in June 1823 Lyell found in the cliffs of Compton Bay on the western side of the island a more complete series of strata below the Chalk, containing all the same beds as in the Weald formation.

During the summer of 1823 Lyell went to Paris, bearing gifts and introductions to many of the scientists there, including Georges Cuvier, Alexander von Humboldt and Constant Prévost. At Paris he attended Cuvier's weekly soirées, where he met many French scientists, and attended lectures at the Jardin du Roi. Alexandre Brongniart invited him to visit the pottery at Sèvres, and Constant Prévost entertained Lyell at his country house. From the Tertiary upper freshwater formation of the Paris Basin, Lyell collected fossil fruiting bodies of the aquatic green alga *Chara*, called 'gyrogonites' because of their spiral markings. In various strata of the Paris Basin, Prévost had found in strata along the boundary between freshwater and marine formations an intermingling of freshwater and marine shells that suggested that the change from fresh to salt water had been gradual rather than abrupt. Prévost decided that the Paris Basin had formerly been a large inlet of the sea, which at times had been cut off from the sea and converted gradually by the inflow of rivers into a freshwater lake. Prévost thought that the distribution of fossils among strata could be understood only by comparison with conditions in modern lakes, estuaries and seas (Prévost 1825). Prévost thought that geologists who believed that during the geological past the natural order had been different from that of the present had formed such ideas without studying causes presently at work in modern waters.

In Scotland the following year Lyell studied marls formed in two small lakes near Kinnordy, the lochs of Bakie and Kinnordy, which had been drained to obtain marl to fertilize the land. The Bakie loch, covering 200 acres, had been drained completely. The marl of the lake bottom, varying from 9 to 16 feet in thickness, was covered by a thin layer of hard limestone in which Lyell found fossilized stems of the green alga *Chara* and fossil gyrogonites similar to those from the Tertiary freshwater limestones of the Paris Basin. The Bakie limestone also contained valves of the minute aquatic crustacean *Cypris ornata*, Lamarck, which particularly impressed Lyell, because Alexandre Brongniart had described valves of a related species of *Cypris* from an ancient Tertiary freshwater formation in central France. Finally, the Bakie limestone contained various species of freshwater shells and Lyell showed that the thick layer of marl beneath it was an accumulation of broken and crumbled shells. The source of calcium carbonate needed to form so much marl and limestone was the water of springs rising through the underlying strata of Old Red Sandstone, rich in calcium (Wilson 1962, 1972).

Cuvier and Brongniart had assumed that the Tertiary freshwater limestones of the Paris Basin had no counterparts in modern lakes. Lyell had now found in the Bakie and Kinnordy lochs hard limestones that corresponded in detail to ancient Tertiary limestones (Lyell 1829). Thereafter, when Lyell encountered freshwater strata, he would always see in imagination some ancient counterpart to the Scottish lakes. Lyell was now passionately eager to discover new geological facts. When Joseph Hooker visited Kinnordy in 1827 as a boy of ten, he looked out of the window to see Lyell pushing a wheelbarrow load of marl up to the house from the loch to look through for shells (Huxley 1918, vol. 2, p. 475). Throughout his life Lyell spared no effort in the pursuit of geology, whether it was pushing a wheelbarrow load of marl or travelling to remote places to collect fossils and rock samples in all kinds of weather.

The beginning of Lyell's *Principles of Geology*

In March 1827 Lyell started work on a book on geology to demonstrate the correspondence between ancient and modern causes. He had begun to see in his own work and that of others many examples of similarity between conditions during the geological past and those of the modern world, but his ideas were still vague and tentative. He decided that geology was not concerned with the origin of the Earth, but with its history. The history of a nation, he noted, did not attempt to explain the origin of mankind. Later in 1827 Lyell reviewed George Poulett Scrope's recent book on the volcanic district of central France (Scrope 1827; Lyell 1827). Scrope described a succession of volcanic mountains that had been raised up in central France and then worn down. The country was dotted with extinct volcanoes in varying stages of decay. Repeatedly currents of lava had poured down valleys and repeatedly rivers had cut through the sheets of lava in the valleys so that their banks showed sections through multiple layers of lava often separated by thin layers of river gravel. Among the volcanic rocks were also Tertiary freshwater strata, which showed that through a long period of the geological past, during which many species had become extinct, central France had been a district of volcanic mountains interspersed with freshwater lakes. The youngest volcanoes had erupted so recently that their cones and craters remained perfectly preserved, although no eruptions had occurred within the historic period. As he read Scrope's work, Lyell was struck by the profound difference between the freshwater strata of central France and those of Hampshire and the Isle of Wight. The former were part of a district whose geography could be reconstructed in imagination through a long period of the geological past; the latter represented areas of land and sea whose outlines had disappeared completely. Lyell became eager to see central France – the Auvergne particularly – because it offered an unbroken history of geological events from early in the Tertiary period. He thought that if one were able to travel back through time to view the Tertiary lakes of France, they would seem as calm and beautiful as modern lakes, even though surrounded by volcanoes subject to occasional eruption.

From his reading of Scrope and of Lamarck's *Philosophie zoologique* (Lamarck 1809) in 1827, many ideas began to ferment in Lyell's mind. He had not progressed very far with his projected book on geology when in March 1828 he decided to go with Roderick Murchison to visit the volcanic districts of central France, described so vividly by Scrope. Early in May, Lyell joined Murchison and Mrs Murchison in Paris and together they travelled by carriage southward into central France, to spend almost three months studying the volcanic rocks and the Tertiary freshwater formations. At first sight some of the Tertiary freshwater strata looked like much older Secondary rocks in England, but their fossils showed them to be Tertiary. The thinly foliated marls, hundreds of feet thick, had accumulated slowly in the clear waters of a Tertiary lake over hundreds of thousands of years. Each paper-thin layer of marl represented a single year's accumulation of the valves of the minute crustacean *Cypris*. Among the Tertiary freshwater strata, Lyell found repeated analogies to modern formations in Scottish lakes. Lyell and Murchison also found conclusive evidence that the valleys among the volcanic mountains had been excavated slowly over long periods of time by the rivers flowing in them. At Aix-en-Provence they found the great valley of the river Arc cut out of Tertiary freshwater strata containing fossil *Chara* and freshwater shells.

At Nice in August 1828, Lyell decided to go on to study the active volcanoes of Vesuvius and Etna in southern Italy and Sicily. On Mont Dore in Auvergne he had seen freshwater strata high on the sides of an ancient volcano and he wished to learn whether still younger stratified rocks existed around modern volcanoes. Murchison told him that Sicily would answer many questions that their work among the old volcanoes of central France had raised.

From Nice, Lyell and the Murchisons travelled along the coast road to Genoa and thence inland to Turin. At Turin, Professor Franco Bonelli showed them a large collection of Tertiary fossil shells from the Subapennine beds of Italy, about 20 per cent of which belonged to living species. He told them that the fossils from the strata of the hill of Superga at Turin were quite different from those of the Subapennine beds, but similar to those of Tertiary strata at Bordeaux in France. From Turin, Lyell and the Murchisons travelled to Milan and thence to Padua, where they parted, the Murchisons to turn north and Lyell to go to southern Italy. He intended to look for signs of elevation or subsidence in the vicinity of active volcanoes and for the effects of the earthquakes that occurred frequently in southern Italy and Sicily.

When he reached Naples, Lyell visited the island of Ischia, where high on the sides of the extinct volcano he found clay beds containing fossil shells all belonging to living Mediterranean species. He also found on Ischia elevated beaches and former sea cliffs, indicating the recent uplift of the island. At Puzzuoli near Naples, Lyell visited the Temple of Serapis, which had undergone both subsidence and re-elevation since ancient times, as shown by the borings of *Lithodomi* on its pillars.

In Sicily, Lyell visited the Valle del Bove, the great caldera on the east side of Mount Etna. Its walls provided sections through many layers of lava, including sections through former lateral volcanic cones on the sides of Etna. The Valle del Bove revealed how the great mass of Etna had accumulated through a long succession of ordinary volcanic eruptions, occurring over an immense period of time. On the outer slopes of Etna, several hundred feet above the sea, Lyell found clay strata containing fossil sea shells with their original colours. The clay strata underlay the volcanic lavas, indicating they were older than Etna and that their age would be a measure of the age of the great volcano. From near Primosole on a low plateau overlooking the plain of Catania, Lyell made a sketch of Etna, showing the great mass of the volcano, resting on a platform outlined by the clay hills above Catania (Fig. 2). It showed how Etna lay above and was, therefore, younger than the sedimentary strata around its base.

From Primosole Lyell rode over the hard white limestone formation of the Val di Noto, a rock containing casts of shells and corals that looked like an ancient Secondary limestone (Fig. 3). At first Lyell thought this great formation, hundreds of feet thick, was probably Tertiary, but if so among the oldest of Tertiary formations. He was, therefore, startled to find at Syracuse, strata of blue clay, containing beautifully preserved fossils of living Mediterranean species, that appeared to extend beneath the limestone hills of the Val di Noto. If it did, the hard white limestone containing casts of shells must be younger than the blue clay at Syracuse with its well preserved shells of living species. Instead of being old, the Val di Noto limestone must be a very recent formation. The idea impressed Lyell profoundly, shaking assumptions implicit in the geology he had learned – assumptions of which previously he had hardly been aware. He had taken for granted that a hard white limestone containing casts of shells, not unlike the Oolite of the Cotswold hills, must be very old. Now he learned it could be very recent. The hard compact appearance of a limestone containing casts of shells, therefore, bore no relation to its age. Lyell found such conclusions so startling that he wondered whether he was mistaken in the order of the strata. At Agrigento (Girgenti) on the south coast of Sicily, Lyell became convinced that the Val

di Noto Limestone lay above the blue marl, and in the interior of the island in a great escarpment at Enna, he found the whole succession of Sicilian strata exposed in order, from the white limestone down through the blue clay to a gypsum formation. When Lyell had his collection of fossil shells from Sicily identified at Naples, he found that most of them belonged to living species, confirming the recentness of the Sicilian formations. They were younger than the Subapennine formation of Italy, but had been elevated slowly over an immense period of time, with many intervals of rest and subsidence. Recent though they were, the Sicilian strata were older than Mount Etna, which had taken hundreds of thousands of years to accumulate. The vista of past time thus opened transformed Lyell's ideas of the length of geological time. Yet if southern Sicily had been elevated from the sea bottom by the intrusion of lava from below, a volume of molten rock equal to the whole bulk of southern Sicily from the bottom of the Mediterranean to the heights of Enna, 3000 feet above sea level, must have been intruded, in successive events, into the strata.

In southern Italy and Sicily, Lyell found so many striking analogies between the geological past and the present that his outlook was transformed. He learned also that Tertiary formations were not all of the same age but represented a succession of long

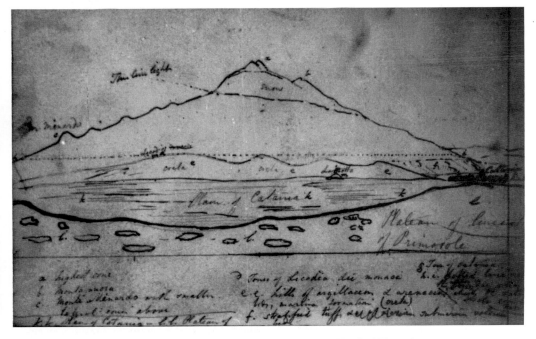

Fig. 2. Lyell's sketch of Mount Etna from Lentini, 1828. Courtesy of Lord Lyell of Kinnordy.

Fig. 3. Val di Noto Limestone, Sicily. Photograph by the author, 1995.

geological periods that he was to name Eocene, Miocene and Older and Newer Pliocene. From 1830 to 1833 he published the three volumes of his *Principles of Geology* (Lyell 1830–1833), in which he developed the idea that geological causes in the past were identical with those acting at present and had always acted with the same degree of energy as at present. That work, published in 11 editions during Lyell's lifetime – at the time of his death he was at work on a twelfth – brought about a revolution in geology. His view of Earth history was opposed fundamentally to that of Cuvier, or Buckland, or even the followers of Hutton. Lyell demonstrated that the history of the Earth was continuous and unbroken; the geological past was uniform with the present. The long continued action of the geological forces currently at work had brought about past geological changes. Lyell viewed the Earth as a stable system, its surface subject to slow constant change brought about by changes in the great laboratory of its interior. No universal floods, violent convulsions or sudden upheavals of mountains punctuated its history.

In 1830 Lyell's theory was new, startling and revolutionary. James Hutton had remained silent on how sedimentary strata were raised up from the ocean bed to form hills and mountains, but his loyal followers John Playfair and Sir James Hall had postulated great convulsions of the Earth's surface to account for the elevation and dislocation of strata. (Playfair 1802, p. 49; Hall 1815; cf. Wilson 1967, pp. 38–40; 1980, p. 181). Lyell transformed Playfair's and Hall's version of Huttonian theory of a succession of stratified formations extending indefinitely into the past, but punctuated by periods of convulsive upheaval, into a theory of slow continuous geological change. Where Cuvier and Buckland saw sharp contrasts between the geological past and the present, Lyell saw fundamental identity. Lyell's theory was reasoned from geological evidence and it drove him relentlessly to study the effects of modern changes on the Earth's surface – the limestones and marls deposited in Scottish lakes and the plants and animals that lived in freshwater. Lyell found that in each epoch of the Tertiary period – the Eocene, Miocene and Older and Newer Pliocene – volcanoes were active. In central France some volcanoes erupted during the Miocene, others during the Pliocene. Southern Italy was littered with extinct volcanoes, but Vesuvius was still active, as was Etna in Sicily. Volcanic activity shifted its location over time, but did not change noticeably in intensity. The supposed decline in volcanic activity since earlier geological periods, assumed by many of Lyell's contemporaries, was an illusion. Each period of the past had been marked by volcanic activity, similar in scale and intensity to that in the modern world.

The evolution of the *Principles*

In 1833 in the third volume of the *Principles of Geology,* Lyell introduced the term 'metamorphic' to describe sedimentary strata that had been transformed into crystalline rocks. Wherever granite was intruded among sedimentary rocks the adjacent strata were converted more or less completely to crystalline rocks. He cited examples from the Highlands of Scotland, Cornwall, the Alps, and Table Mountain at the Cape of Good Hope. Lyell emphasized that such conversions of sedimentary strata could occur only under conditions of great heat and pressure, deep beneath the surface, and suggested that granite and metamorphic rocks should both be called 'hypogene' rocks to indicate their origin within the depths of the Earth.

Up to the fifth edition of the *Principles,* published in 1837, Lyell did not change significantly his discussion of metamorphic rocks, but after 1834 he was also writing the *Elements of Geology* (Lyell 1838). In the latter work he elaborated what he called the 'metamorphic theory' on the basis of his further study of granite intrusions in the Swiss Alps, the isle of Arran, and at Christiania in Norway. He also used Charles Darwin's observations of masses of granite in the central axis of the Andes mountains in South America (Lyell 1838; cf. Wilson 1980, pp. 196–201). Wherever granite was intruded among sedimentary rocks, Lyell found the strata converted more or less completely to crystalline rocks.

In 1833 Lyell had shown that volcanic activity had occurred throughout the Tertiary period; by 1838 he had accumulated evidence to show that it had occurred also through the Transition and Secondary periods. He described the occurrence of trap rocks in the Cambrian, Silurian and Old Red Sandstone formations, as well as through the whole series of Secondary formations. In each geological period volcanic activity was restricted to particular areas, usually mountainous. Over time it shifted its location.

Earlier geologists, particularly Werner and his followers, had thought that crystalline rocks in mountains were Primary and therefore among the oldest rocks. Hutton had recognized that such crystalline strata were ordinary sedimentary strata altered by heat. Lyell demonstrated that such crystalline rocks might be metamorphosed Secondary or even Tertiary strata. Their crystalline structure alone told nothing of their age, which could be inferred only from their relationship to neighboring unaltered sedimentary strata. He now

connected the rocks of volcanic districts with the rocks seen in trap districts, and the trap rocks with the metamorphosed strata and masses of granite in mountains. All were effects of a connected series of geological processes at work within the Earth throughout its history and still continuing.

Lyell's metamorphic theory of 1838 in a sense completed the revolution that he had worked to bring about in geology since 1830. It provided a theoretical framework to link volcanic rocks formed at the surface of the Earth with granite and metamorphic rocks formed under intense heat and pressure beneath the Earth's surface. In place of the assumption of a cooling Earth, a decline in volcanic activity, and a violent geological past, opposed to a tranquil present, Lyell offered a closely reasoned theory, supported by ever more extensive geological evidence, that geological change had proceeded gradually and uniformly (Lyell 1830–1833, vol. 3, pp. 97–102).

Craters of elevation

In 1835 three Continental geologists, Leopold von Buch, Léonce Élie de Beaumont and Armand Dufrénoy, launched a concerted attack on Lyell's theory centred upon the question of the age of Mount Etna. In opposition to Lyell, Buch and Élie de Beaumont insisted that Etna was quite young. It had arisen suddenly, they said, as a crater of elevation, at the same time as the equally sudden appearance of the Alps and the Pyrenees (Élie de Beaumont 1835).

The theory of craters of elevation is little known today. In a 1980 article on Graham Island and Lyell, Dennis Dean discussed it briefly (Dean 1980). Much earlier, in a eulogistic essay on the founder of the theory, Leopold von Buch, Sir Archibald Geikie quite failed to mention Buch's theory of craters of elevation, an omission that would have sadly puzzled Buch (Geikie 1905, pp. 245–252). During an extended visit to the Canary Islands in 1815, Buch decided that the great caldera on the island of Palma had been formed by a sudden upheaval of accumulated layers of lava, poured out originally in a horizontal position on the sea floor. In 1825 Buch published the theory at Berlin in his book on the Canary Islands (Buch 1825) and in 1836, to make the theory of craters of elevation better known, he published in Paris a revised and expanded French translation of his Canary Island book, accompanied by an account of major volcanoes of the world (Buch 1836a). The same year Élie de Beaumont published in four installments a long paper, totalling 188 pages of text, with illustrations and maps, on the structure and origin of Mount Etna. With Buch and Dufrénoy, Élie de Beaumont argued that the highly inclined lavas on the sides of Mount Etna could not have solidified into solid rock where they lay. Instead he insisted that the lavas must have been poured out on a nearly horizontal surface and upheaved later.

In 1830 Lyell discussed Buch's theory of craters of elevation in the *Principles of Geology*, particularly Buch's example of the island of Santorin in the Aegean Sea. Lyell noted that if the volcanic rocks of Santorin and its associated islands had originally been poured out as submarine lavas on the sea floor they should be interstratified with at least thin layers of marine sediments, but they were not. Lyell also mentioned Buch's example of great circular valleys surrounded by cliffs of volcanic rocks on Palma, the Grand Canary and Teneriffe in the Canary Islands (Lyell 1830–1833, vol. 1, pp. 386–395). Lyell continued his analysis and refutation of Buch's theory in successive editions of the *Principles,* enlarging it considerably in the fifth edition in 1837 (Lyell 1837, vol. 2, pp. 152–177), after the publications by Buch and Élie de Beaumont in 1835 and 1836 (Buch 1836a,b; Élie de Beaumont 1835, 1836).

In 1848, in recognition of his scientific work, Lyell was knighted, but it was not until 1854 that, accompanied by Lady Lyell and Charles and Frances Bunbury, he was able to visit Madeira and the Canary Islands to examine the evidence from which Buch in 1815 had derived his theory of craters of elevation. In Madeira he met a young German, Georg Hartung, who joined him in studying the geology of the island. Lyell and Hartung found that Madeira consisted entirely of lavas and basalts produced by a long series of volcanic eruptions from a multitude of centres on the island. Major cones of eruption in the centre of the island had buried smaller cones of eruption around its periphery. The many streams of water flowing down from the central mountains had cut deep narrow valleys, providing sections through the accumulated lavas. Frequently they included sections through old volcanic cones buried under later flows of lava from higher centres of eruption. Such sections revealed multitudes of vertical dikes, marking channels where molten rock had risen through layers of lava and scoriae (Fig. 4). They showed also that lavas could form sheets of compact rock on steep slopes, and these had not been disturbed since they solidified. The valleys showed Lyell how much more powerful rivers were in eroding the land than he had thought previously. The volcanic richness of Madeira fascinated and excited Lyell. On a trip through the central mountains he astonished his guides and companions by standing 'on the edge of a precipice with his glass to his eye, expounding the structure of the rocks in a sort of extempore lecture,

seemingly perfectly forgetful that a slight movement would send him sheer down several hundred feet' (Bunbury 1894, vol. 3, p. 66). From fossil plants in a bed of clay among volcanic rocks at Sao Jorge, hundreds of feet above the sea, Lyell discovered that when Madeira was only half its present size, it was already covered with plants. The additional lavas that had built up Madeira had therefore flowed from centres of eruption on land. They had not been poured out over the sea bottom. Some volcanoes had become extinct before others began to erupt, but the whole island was the product of a long succession of volcanic eruptions.

After about two months on Madeira, where Lyell's unceasing activity had astonished his companions, the Lyells and Bunburys sailed for the Canary Islands, accompanied by Georg Hartung. While the Bunburys remained on Teneriffe, the Lyells and Hartung went to the Grand Canary, where Lyell and Hartung spent ten days studying lava flows and calderas in the centre of the island. After returning briefly to Teneriffe to leave Lady Lyell with the Bunburys at Puerto de la Orotava, Lyell and Hartung sailed for Palma, Lyell taking with him a copy of Buch's book on the Canary Islands (Buch 1836a).

Buch had written that in the craters of elevation on Palma and Teneriffe, the rim of the crater was marked by fissures extending radially like the spokes of a wheel. On Palma, Lyell and Hartung found that no such radial fissures existed. The rim of the great caldera of Palma was unbroken except for a single deep narrow gorge, evidently created by the stream running in it (Fig. 5). The ravines on the outer slopes of Palma widened and deepened as they descended and had been excavated similarly by streams of running water. The precipices encircling the caldera consisted of layers of lava and scoriae dipping outward, with no trace of marine sediments among them (Fig. 6). Buch's evidence that the caldera of Palma was a crater of elevation was non-existent.

After two weeks of intense geological field work on Palma, Lyell and Hartung sailed for Teneriffe on 27 March. Two days later they set out at 3.00 a.m. on muleback to explore the volcanic geology of the great peak of Teneriffe. After a ten-hour ride they stopped for lunch at an altitude of 8000 feet at Degolliado de Cedro on the escarpment overlooking Las Canadas, the caldera surrounding the base of the great peak (Fig. 7). Lyell decided that the precipices encircling Las Canadas were the remains of an older volcanic cone. They were not the rim of a crater of elevation, as Buch had believed them to be. After three more days of riding and geologizing, Lyell and Hartung rejoined the rest of the party at Santa Cruz, bringing to an end more than three months of constant, physically

Fig. 653.

SECTION OF MADEIRA FROM NORTH TO SOUTH, OR FROM POINT S. JORGE TO POINT DA CRUZ, NEAR FUNCHAL.

South.   North.

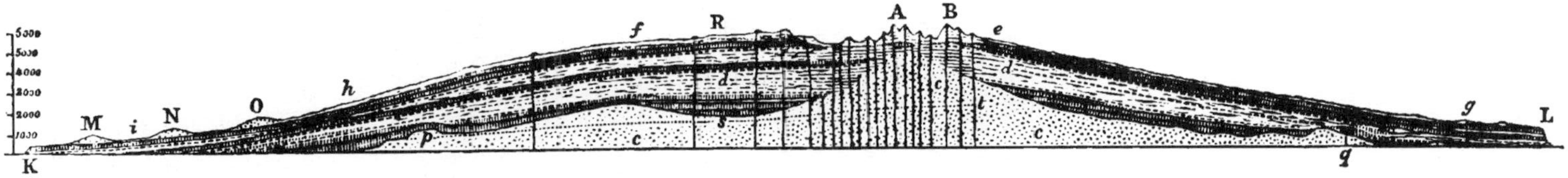

Length of section twelve miles.   Drawn on a true scale of heights and horizontal distances from the observations of C. Lyell and G. Hartung, 1853—4.

A. Pico Torres (or Pico do Gatto), about 6050 feet high.
B. Pico Ruivo; the highest mountain in Madeira; about 6060 feet high.
c. Scoriæ, agglomerate, lapilli, tuff, and ejectamenta, with some highly scoriaceous lava.
d. Alternations of lava with tuff and lapilli, or with parting layers of red clay (laterite).
   Under this same head of " alternations " must be included all the beds between
   R and s.
e. Commencement of more highly inclined lavas on north side of Madeira ; slope
   usually 10 degrees.
f. Commencement of more highly inclined lavas on southern slope, usually at angle
   at 15 degrees.
g. Dike of Jogo da Bola, in Ribeiro S. Jorge.
h. Slope of beds 15 degrees, occasionally but rarely 20 degrees.

i. Slope or dip of lavas 5 degrees.
K. Point da Cruz, near Funchal.
L. Point S. Jorge. on north coast.
M. Pico da Cruz, 813 feet high ; modern cone.
N. Pico S. Martinho. 1100 feet high.
O. Pico S. Antonio, 1440 feet.
p. Buried cone in Ribeiro do Torreão.
q. Lignite and leaf-bed.
R. Pico S. Antonio, 5706 feet high.
p, s, t. Line below which the rocks are not exposed to view. All below this line is given
   conjecturally.

The beds indicated by the sign No. 1. consist of lavas more or less
stony, under which occur red clays or laterites, probably ancient
soils (see p. 475.), represented by the interrupted lines, No. 2.   These
red bands, as well as the lavas, No. 1. are very numerous in nature,
and for want of space a few only are introduced into the diagram.

Fig. 4. Section of Madeira. From Lyell 1865. p. 641.

demanding geological field work in Madeira and the Canary Islands.

In 1857 and again in 1858 Lyell returned to southern Italy and Sicily to re-examine the streams of lava on Vesuvius and Etna that Élie de Beaumont claimed could not have solidified into sheets of solid rock in the steeply inclined positions where they lay (Élie de Beaumont 1835, 1836; Buch 1836b). Lyell was in his sixtieth year. In 1857 his eyesight had become sufficiently bad that Lady Lyell did not want him to go into the field alone. The young Gaetano Gemellaro, then 23, accompanied him on expeditions on muleback around the eastern flank of Etna and into the Val del Bove, covered with lava streams from the great eruption of 1852. Together with their guide, Lyell and Gemellaro climbed from the floor of the Val del

Bove up the 2000 foot precipice of Solfizio over a thick layer of compact lava inclined at an angle of about 35 degrees. At the top they were caught in a sleet storm and had to shelter overnight in a hut. In 1857 Lyell reported to the Royal Society of London that Élie de Beaumont's supposed evidence for 'craters of elevation' was either false or non-existent. Lavas could solidify on steep slopes on compact rock and Etna had been built up by a long succession of ordinary volcanic eruptions (Lyell 1857).

In September 1858 Lyell returned to Sicily to check Baron Sartorius von Waltershausen's opinion, based on nine years' study of Etna, that some elevation and disturbance of the layers of lava had occurred as a result of the later injection of lava from below. Lyell ascended to the summit of Mount

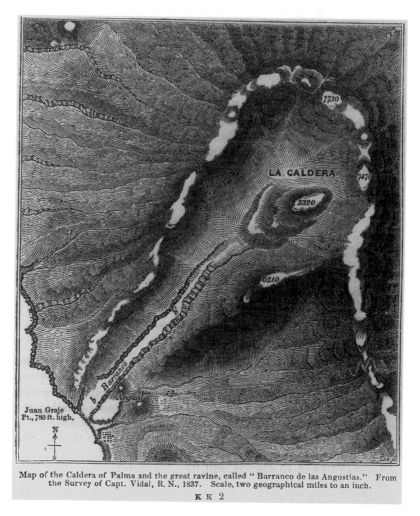

Map of the Caldera of Palma and the great ravine, called " Barranco de las Angustias." From the Survey of Capt. Vidal, R. N., 1837. Scale, two geographical miles to an inch.

K K 2

Fig. 5. Map of the Caldera of Palma. From Lyell 1865, p. 622.

Etna, where he sketched the two craters and spent several nights uncomfortably in the stone hut called the Casa Inglese at an altitude of 9000 feet, just below the summit cone. From the Casa Inglese he

made two descents at different places into the Valle del Bove to determine the inclinations of the lava beds in its walls. In the middle portions of the cliffs, more than 3000 feet high, the lava beds were horizontal, whereas near the bottom they were steeply inclined. In the summit ridge of the south wall of the Valle del Bove, Lyell discovered a deep notch at the head of a ravine, indicating that the ravine must formerly have extended higher on the sides of a great volcano that once rose over the present site of the Valle del Bove. Lyell rode on muleback completely around the base of Etna to study the volcano from all sides. In a month of extended exploration of Etna, Lyell found many sections of lava currents hardened into sheets of compact rock on very steep slopes. He showed that Etna had formerly possessed two major centres of eruption. A centre, now extinct, at a spot called Trifoglietto in the Valle del Bove had been overwhelmed and buried by lavas flowing from the present centre of eruption on Etna. In his long paper on Etna for the *Philosophical Transactions* (1858), Lyell incorporated his new observations and reasoning of 1858. He also dissected every assumption, every misconstruction of evidence and every element of reasoning in the theory of craters of elevation to bring the whole preposterous edifice crashing to the ground (Lyell 1858).

In November 1858 the Royal Society of London awarded Lyell the Copley Medal, its highest honour, mentioning his demonstration that the present geographical distribution of plant and animal species reflected earlier geological changes, his use of the proportion of living to extinct fossil species to correlate Tertiary formations on the two sides of the Atlantic Ocean, and his discussion of 'craters of elevation' in connection with his recent studies of Etna (Hooker 1858). In awarding the Copley Medal, the Royal Society was acknowledging Lyell's decisive victory over his scientific opponents. By the 1850s 'craters of elevation' had remained the only serious scientific challenge to his theory of uniformity. Belatedly the Paris Academy of Sciences elected Lyell a corresponding member in 1863.

Lyell and the *Origin of Species*

In September 1859 Lyell read Charles Darwin's *Origin of Species* in proof sheets before its publication in November (Darwin 1859). He had known of Darwin's doubts about the fixity of species for more than twenty years and in 1856 urged Darwin to publish his theory of natural selection before he was forestalled by some other scientist (Wilson 1970, pp. xlvi–xlix). In 1858, with Joseph Hooker, Lyell had arranged for the presentation to the Linnean Society of the joint

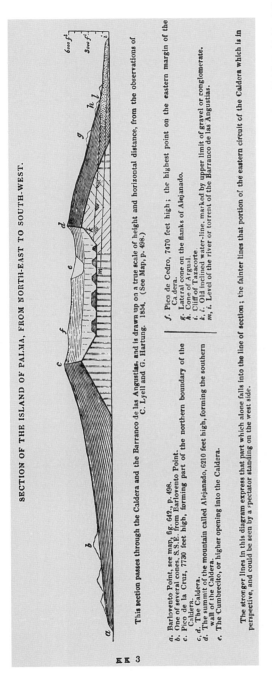

SECTION OF THE ISLAND OF PALMA, FROM NORTH-EAST TO SOUTH-WEST.

This section passes through the Caldera and the Barranco de las Angustias, and is drawn up on a true scale of height and horizontal distance, from the observations of C. Lyell and G. Hartung. 1854. (See Map, p. 498.)

a. Barlovento Point, see map, fig. 642, p. 498.
b. One of several cones, S.S.E. from Barlovento Point.
c. Pico de la Cruz, 7730 feet high, forming part of the northern boundary of the Caldera.
c, d, The Caldera.
d. The summit of the mountain called Alejanado, 6210 feet high, forming the southern wall of the Caldera.
e. The Cumbrecito, or higher opening into the Caldera.

f. Pico de Cedro, 7470 feet high; the highest point on the eastern margin of the Caldera.
g. Lateral cone on the flanks of Alejanado.
h. Cone of Argul.
i. Cliff of Tazacorte.
k. l. Old inclined water-line, marked by upper limit of gravel or conglomerate.
m, i. Level of the river or torrent of the Barranco de las Angustias.

The stronger lines in this diagram express that part which alone falls into the line of section; the fainter lines that portion of the eastern circuit of the Caldera which is in perspective, and could be seen by a spectator standing on the west side.

Fig. 6. Section of the island of Palma. From Lyell 1865, p. 624.

Fig. 7. Lyell and Hartung with their guides at Degolliado di Cedro, Teneriffe, about 2.00 p.m. on 30 March 1854; looking southeast over las Canadas. Watercolour by Georg Hartung. Courtesy of Malcolm Lyell.

papers of Darwin and Alfred Russel Wallace on the modification of species by natural selection. During the autumn of 1859 Lyell decided that he must state his opinions on the species question in a new edition of his *Manual of Geology*, on which he had begun to work (Lyell 1865). He wanted also to discuss the question of the antiquity of man raised by the discovery in 1856 of a fossil human skeleton in a cave in the Neanderthal in Germany and in 1858 of flint implements intermingled with the bones of extinct animals in Brixham Cave near Torquay in Devonshire. In January 1860, Lyell found that his discussion of the two questions was growing much too long to include in the *Manual* and might require a separate work on the geological evidences for the antiquity of man. In April Lyell visited Torquay, where Hugh Falconer and William Pengelly took him to see Brixham Cave and showed him the flint tools and fossil bones found there. During the summer of 1860 Lyell travelled to Liège in Belgium to examine some of the caves in which Philippe-Charles Schmerling had found human bones together with those of extinct animals in the 1830s. He also obtained a cast of the human skull that Schmerling had found in a cave at Engis near Liège. From Liège, Lyell went on into Germany to visit the site on the Neanderthal river where the remarkable fossil human skeleton had been found and obtained a cast of the Neanderthal skull to take back to England. In March 1861 Lyell visited Paris where he conferred with Eduard Lartet on Lartet's discoveries in caves in the south of France. From Paris, Lyell visited various sites near Amiens and Abbeville where flint implements had been found in the gravels of the Somme Valley, implements showing that humans had lived in France as early as the glacial period. Shortly after his return to England, Lyell went with John Evans and Joseph Prestwich to visit a gravel pit near Bedford, where flint implements had been found. The flint tools at Bedford demonstrated the presence of humans in England at a time before such extinct animals as the elephant and hippopotamus, but after the period of the boulder clay, a glacial deposit. By contrast, in the coastal cliffs of Norfolk the bones of elephants and other extinct animals lay in beds beneath and, therefore older than, glacial beds. Lyell concluded that Europe must have experienced more than one glacial period and that various species of extinct animals, including the fossil elephant, and humans had lived in Europe both before and after periods of glaciation. Lyell struggled to correlate the successive faunas with a succession of glacial periods and interglacial periods, some possibly warmer than the present. He saw human prehistory as linked inextricably to the geological and climatic history of Europe.

In November 1861, Lyell, finding that the *Manual* was growing impossibly large, decided finally to publish a separate work on the evidence for early human life in Europe in relation to the glacial periods and post-Pliocene (i.e. Pleistocene) geology. He planned to include a discussion of Darwin's theory. Lyell's *Antiquity of Man* appeared in February 1863 (Lyell 1863). Throughout Lyell's career his reputation as a scientific author had grown steadily. Successive editions of the *Principles of Geology* and the *Elements of Geology* were printed in ever larger numbers and continued to sell steadily. His two works on his American travels were read widely. But the success of *Antiquity of Man* surpassed that of any of Lyell's previous books. Four thousand copies of the first edition disappeared so rapidly that within two months the publisher issued a second edition and, before the end of the year, a third.

In *Antiquity of Man*, Lyell presented a long sustained plea for consideration of the evidence for the early appearance of man in Europe and for his animal ancestry. He withheld his own opinions, leaving the readers to draw their own conclusions from the evidence. Hostile readers could not attack Lyell's conclusions for he offered none. They could only attack the evidence, but in order to do so they had to examine it and, in examining it, to be confronted by its meaning. Lyell thereby succeeded in presenting to the public the question of human origins, a question so peculiarly charged at the time that Darwin had deliberately avoided it in the *Origin. Antiquity of Man* drew together previously isolated kinds of evidence – the succession of stone, bronze and iron ages described by Danish archeologists, the discovery of ancient pile dwellings in Swiss lakes, the association of flint implements with bones of extinct animals in cave and river deposits, and the discovery of fossil human bones – and connected them with recent geological history, especially the glacial period.

In his discussion of Darwin's theory, Lyell summarized the evidence for the change of species by natural selection, presenting Darwin's arguments clearly and effectively and adding original arguments of his own, especially a comparison of the origin and development of languages with the origin of species. Yet, to Darwin's disappointment, Lyell did not declare himself convinced of the truth of Darwin's theory, a failure that evoked protests from both Darwin and Hooker. In the second edition of *Antiquity of Man*, which appeared only two months after the first, Lyell modified his conclusion to say that he fully expected that Darwin's theory would become generally accepted by men of science (Lyell 1863, 2nd edn, p. 469). In January 1865 in the sixth edition of the *Elements of Geology*, Lyell finally acknowledged that, despite the great incompleteness of the fossil record, its broad features supported the theory of a development of animal and plant life through successive geological ages (Lyell 1865, pp. 580–586).

In the summer of 1865, after several weeks of rest at the German spa of Bad Kissingen, Lyell went to Botzen in the Austrian Tyrol to see the unusual Earth pillars there, each capped by a large glacial boulder. By protecting the underlying pillar from the rain that had washed away the surrounding clay, the boulder caps distinguished strikingly the erosive power of rain from that of running water in streams. From Botzen, Lyell went to Eggischorn in Switzerland to study the Marjelen See, a small lake formed by the Aletsch glacier. The terrace around the Marjelen See was an exact counterpart of the parallel roads of Glen Roy in Scotland. It demonstrated to Lyell how the level of a glacial lake might change at different times, its level being determined by the level of its drainage outlet.

Lyell included discussion of both the Earth pillars of Botzen and the Marjelen See in the tenth edition of the *Principles of Geology* (Lyell 1866–1868). For the tenth edition he undertook a major revision of the book to incorporate the great changes that had occurred in geology since the ninth edition had appeared in 1853. Darwin's theory of the origin of species by natural selection was the most fundamental change. Lyell again acknowledged that, despite its fragmentary nature, the fossil record did show a development of plant and animal life through successive geological periods. In his discussion of past climates, Lyell distinguished a general cooling trend since the beginning of the Tertiary period, from tropical climates in the Eocene, and warm temperate climates throughout the northern hemisphere in the Miocene, to cool temperate conditions in the Pliocene, culminating in the Arctic chill of the glacial period.

Lyell and the unchanging nature of Earth history

In the second volume of the tenth edition of the *Principles,* not completed until 1868, Lyell refuted the theory of a cooling Earth, showing that the Earth behaved throughout its history as a body steadily producing heat within its interior. The assumption that the Earth was merely losing residual heat had distracted physicists from attempting to discover the source of heat produced constantly within the Earth. Lyell's discussion of heat was in answer to William Thomson, professor of natural philosophy in the University of Glasgow, who, without knowledge of geology or interest in the subject, had argued that since the Earth was losing heat it must formerly have been much hotter

than it is now (Thomson 1861, 1866). Thomson considered the Sun to be similarly a cooling body that had been losing heat steadily since the original creation.

Although he was a brilliant physicist, justly famous for his work in thermodynamics, in his arguments based on the heat lost from the Sun and the Earth, Thomson was presenting Old Testament theology in the guise of theoretical physics. In his early calculations of the quantity of heat radiated from the Sun in 1854, Thomson assumed that the Sun had been giving out heat at its present rate for 6000 years, a figure in accordance with Archbishop Usher's calculation from the Bible that the world was created in 4004 BC (Thomson 1854; Smith & Wise 1989, p. 506). He treated the Earth and Sun as though they were gigantic steam boilers radiating heat previously supplied to them. But Thomson never asked whence came the heat that was supplied to the Sun and the Earth at the original creation. His assumption that God had created such heat was a theological assumption, not a scientific one. Thomson's attack on uniformitarianism in the 1860s was primarily a response to Darwin's theory of the origin of species by natural selection, which required for the Earth an indefinitely long and stable history. Thomson perceived accurately that Darwin's theory rested upon a foundation of uniformitarian geology (Burchfield 1975, pp. 71–86).

In his discussion of the species question in the tenth edition, Lyell reprinted unchanged his critical analysis of Lamarck's theory of transmutation, which he had first published in the *Principles* in 1832. From 1832 through to the ninth edition of the *Principles* in 1853, Lyell had steadily upheld the fixity of species, although he had also shown 'that the gradual extinction of species was part of the constant and regular course of nature.' (Lyell 1866–1868, vol. 2, p. 267). After discussing briefly the anonymous *Vestiges of the Natural History of Creation* (Chambers 1844) and Alfred Russel Wallace's 1855 paper on the species question (Wallace 1855), Lyell described the events leading to the presentation of the joint papers by Wallace and Darwin to the Linnean Society in 1858. He thought that Wallace's and Darwin's independent development of the theory of natural selection was a strong argument for its truth. In the *Origin of Species,* said Lyell, Darwin had shown not only how new races and species might arise by natural selection, but also how this theory illuminated many different and previously unconnected biological phenomena. In new chapters, Lyell discussed at length Darwin's study of variation among domesticated animals and plants (Darwin 1867) and compared the effects of natural selection with those of artificial selection. Lyell rewrote his discussion of the geographical distribution of species, giving new examples of the means of dispersal of plants and animals. In a new and original treatment of the flora and fauna of oceanic islands, Lyell showed that island species resembled the species of neighboring continents in proportion to the ease with which they might cross over water. Thus the birds of Madeira were almost all of European species, but no mammals were native to the island except for one species of bat. The insects, land shells, and plants of Madeira were mostly distinct species. The species of land shells were restricted to individual islands, often very small ones. The theory of variation and natural selection, wrote Lyell, would explain such relationships, but no other hypothesis for the origin of species would do so (Lyell 1866–1868, vol. 2, p. 432). Darwin, Hooker and Wallace all considered Lyell's discussion of species on oceanic islands to provide powerful new evidence in support of Darwin's theory.

Lyell also introduced a new discussion of the origin and geographical distribution of man, in which he showed that the distribution of human races corresponded to the main zoological provinces of the world. Although human beings had spread over almost the whole of every continent, many oceanic islands were uninhabited when first discovered by Europeans. By contrast, islands scattered over large areas of the Pacific Ocean had become inhabited by people who sailed to them in canoes. Lyell thought that the close similarity among human races meant that the human species had been essentially what it is now when it began to spread over the Earth. Like other species, man must have originated at a single geographical centre, although not necessarily from a single pair. He compared the gradual emergence of a new species to the gradual emergence of a new language, noting how difficult it would be to establish a year when the English language first appeared. Lyell thought that in bodily structure humans resembled most closely the apes of Africa and Asia, but the fossil forms intermediate between apes and humans remained unknown, because the Pliocene and post-Pliocene (Pleistocene) formations of Africa and Asia had not yet been explored.

Lyell accepted Darwin's theory of natural selection with one fundamental reservation. The theory would account for the selection of variations favourable to a species in the struggle for existence, but it could not account, Lyell thought, for the origin of the variations to be selected. For years Lyell wrestled with the implications of natural selection. He saw clearly that Darwin's theory destroyed the foundations of natural theology. By explaining the origin of the adaptations of plants and animals to particular modes of life, adaptations that had been cited so often as evidence of God's design, natural selection removed divine design from the natural world. Nevertheless, Lyell accepted organic evolution, for, he wrote, 'the amount of power, wisdom, design, or forethought required for such a gradual evolution of life, is as great as that which is implied by a multitude of separate, special, and miraculous acts of creation' (Lyell 1866–1868, vol. 2, p. 492). So far as himans were concerned, Lyell thought that their dignity must rest not on their origin, but on their search for the truth.

Lyell, the man (Fig. 1)

The search for truth was a central directing force in Lyell's own life. Throughout his career, even until his final illness, Lyell continued to pursue the development of geology, seeking constantly to place such questions as those of the origin of species or of humanity within a framework of uniformitarian geology, that is, within a history of slow continuous change. Lyell's critics have charged that he was extreme in his uniformitarian views. Such criticisms were made in the late nineteenth century, when, under the influence of William Thomson (later Lord Kelvin), the age of the Earth was thought to be strictly limited. The discovery of radioactivity at the beginning of the twentieth century provided the steady continuing source of heat within the Earth which, from the geological record, Lyell believed must exist. The development of plate tectonics since mid century has confirmed Lyell's confidence in the gradual nature of geological change. Continental plates, moving at a rate of 1–2 cm per year, can fold strata and elevate mountains only very slowly. It is an unacknowledged tribute to uniformitarian geology that today those who would introduce sudden marvellous events into Earth history are usually obliged to summon meteorites from outer space. Although meteorite impacts have left their traces in the form of craters at various places on the Earth's surface, their possible effects on world climate and biological extinction remain obscure. Meanwhile, the Earth itself works slowly but relentlessly to bring about geological change.

Charles Lyell was a man of courage, integrity and kindness. His own family were devoted to him. Among scientists, Charles Darwin, Joseph Hooker and Gideon Mantell were lifelong friends, as were also Edward Forbes, Andrew Ramsay and Alfred Wallace. Lyell became quite close to Prince Albert, collaborating with him in the reform of English universities and in the Great Exhibition of 1851. Prince Albert may have been the primary influence responsible for Lyell's knighthood in 1848. Queen Victoria's respect for her late husband's opinion gave Lyell his baronetcy in 1864. Lyell himself sought neither honour. Lyell also became a friend of Prince Albert's daughter, Victoria, Crown Princess of Prussia. The outbreak of the Civil War in the United States in 1861 tested Lyell's moral courage. Although he had friends on both sides of the conflict, Lyell firmly supported the North in the face of virulent British hostility to the United States. He befriended Charles Francis Adams, the US minister in London, when Adams was shunned by the British upper classes. Through the discouraging defeats suffered by Federal forces early in the war, Lyell remained confident that the North would ultimately be victorious and that the Union would be preserved. Sympathetic to younger geologists, Lyell often worked to promote their careers. He was especially helpful to John William Dawson of Nova Scotia, who became principal of McGill University at Montreal and did pioneering work on Devonian fossil plants (Sheets-Pyenson 1996).

To the end of his life Lyell retained his freshness of mind. He delighted in every new geological fact and was willing to examine every new theory, no matter how strange it might seem to him. Yet, while willing to consider any geological theory, and frequently to balance in his mind the relative weights of evidence for opposing theories, in the end Lyell formed his opinions on the basis of geological evidence and scientific reasoning. When evidence and reasoning appeared to him conclusive, he was unshakable. For decades he upheld the uniformity of geological causes in the face of opposition and ridicule from many of his fellow geologists. For more than a quarter century Lyell opposed the theory of craters of elevation upheld by some of the most famous Continental geologists and accepted even by his friend Charles Darwin. In the end he destroyed the theory of craters of elevation utterly. Lyell was slow to accept Darwin's theory of the change of species by natural selection, but once he accepted it, Lyell not only supported Darwin's theory, but contributed evidence and arguments to support it. For his part, Darwin always acknowledged that his theory rested solidly on a foundation of Lyell's geology. As an original, creative and independent thinker, Charles Lyell stands as the greatest geologist of the nineteenth century.

References

ALBRITTON, C. C., JR (ed.) 1967. *Uniformity and Simplicity, a Symposium on the Principle of the Uniformity of Nature.* Special Paper 89. Geological Society of America, New York.

BAKEWELL, R. 1815. *An Introduction to geology Illustrative of the General Structure of the Earth; Comprising the Elements of the Science and an Outline of the Geology and Mineral Geography of England.* 2nd edn. J. Harding, London.

BUCH, L. VON 1825. *Physikalische Beschreibung der Canarischen Inseln.* Royal Library, Berlin.

—— 1836a. *Description physique des Iles Canaries, suivie d'une indication des principaux volcans du globe* (trans. C. Boulanger, revised and expanded by the author). F. G. Levrault, Paris & Strasbourg.

—— 1836b. On volcanoes and craters of elevation. *Edinburgh New Philosophical Journal,* **21**, 189–206.

BUNBURY, C. J. F. 1894. *Life, Letters and Journals.* 3 vols. Privately printed, London.

BURCHFIELD, J. D. 1975. *Lord Kelvin and the Age of the Earth.* Science History Publications, New York.

CHAMBERS, R. 1844. *Vestiges of the Natural History of Creation.* J. Churchill, London.

CUVIER, G. 1812. *Recherches sur les ossemens fossiles de quadrupèdes.* 4 vols. Déterville, Paris.

—— & BRONGNIART, A. 1811. *Essai sur la géographie minéralogique des environs de Paris avec une carte géognostique et des coupes de terrain.* Baudouin, Paris.

DARWIN, C. 1859. *On the Origin of Species by Means of Natural Selection, or the Preservation of Favoured Races in the Struggle for Life.* Murray, London.

—— 1867. *The Variation of Plants and Animals under Domestication.* 2 vols. Murray, London.

DEAN, D. R. 1980. Graham Island, Charles Lyell, and the craters of elevation controversy. *Isis,* **71**, 571–588.

ÉLIE DE BEAUMONT, L. 1835. Recherches sur la structure et l'origine du Mont Etna. *Comptes rendus des Séances de l'Académie des Sciences, Paris,* **1**, 429–433.

—— 1836. Sur la structure et sur l'origine du Mont Etna. *Annales des Mines,* ser. 3, **9**, 175–216, 575–630; **10**, 351–370, 507–576.

GEIKIE, A. 1905. *The Founders of Geology.* 2nd edn. Macmillan & Co, London.

HALL, J. 1815. On the revolutions of the Earth's surface. *Transactions of the Royal Society of Edinburgh,* **7**, 139–211.

HOOKER, J. 1858. [Award of the Copley Medal]. *Proceedings of the Royal Society of London,* **9**, 511–514.

HUXLEY, L. 1918. *Life and Letters of Sir Joseph Dalton Hooker.* 2 vols. Murray, London.

LAMARCK, J. B. A. P. DE M. DE 1809. *Philosophie zoologique, ou exposition des considerations relatives á l'histoire naturelle des animaux.* 2 vols. Dentu, Paris.

LYELL, C. 1827. Art. IV. Memoir on the Geology of Central France. By G. P. Scrope F.R.S., F.G.S., London, 1827. *Quarterly Review,* **36**, 437–483.

—— 1829. On a recent formation of freshwater limestone in Forfarshire, and on some recent deposits of freshwater marl; with a comparison of recent and ancient freshwater formations; and an appendix on the Gyrogonite or seed vessel of the Chara. *Transactions of the Geological Society of London,* ser. 2, **2**, 73–96.

—— 1830–1833. *Principles of Geology, Being an Attempt to Explain the Former Changes of the Earth's Surface by Reference to Causes Now in Operation.* 3 vols. Murray, London.

—— 1837. *Principles of Geology.* 5th edn, 4 vols. Murray, London.

—— 1838. *Elements of Geology.* Murray, London.

—— 1857. On the formation of continuous tabular

masses of stony lava on steep slopes; with remarks on the origin of Mount Etna, and on the theory of "Craters of Elevation". *Proceedings of the Royal Society of London*, **9**, 248–254.

—— 1858. On the structure of lavas which have consolidated on steep slopes with remarks on the mode of origin of Mount Etna, and on the theory of "Craters of Elevation". *Philosophical Transactions of the Royal Society of London*, **148**, 703–786.

—— 1863. *The Geological Evidences of the Antiquity of Man, with Remarks on Theories of the Origin of Species by Variation.* Murray, London.

—— 1865. *Elements of Geology; or the Ancient Changes of the Earth and Its Inhabitants as Illustrated by Geological Monuments.* 6th edn. Murray, London.

—— 1866–1868. *Principles of Geology.* 10th edn, 2 vols. Murray, London..

PLAYFAIR, J. 1802. *Illustrations of the Huttonian Theory of the Earth.* W. Creech, Edinburgh.

PRÉVOST, C. 1825. De la formation des terrains des environs de Paris. *Bulletin de la Société Philomathique, Paris*, 74–77, 88–90.

SCROPE, G. P. 1827. *Memoir on the Geology of Central France; Including the Volcanic Formations of Auvergne, the Velay and the Vivarais, with a Volume of Maps and Plates.* Longman, London.

SHEETS-PYENSON, S. 1996. *John William Dawson: Faith, Hope, and Science.* McGill–Queen's University Press, Montreal & Kingston.

SMITH, C. & WISE, M. N. 1989. *Energy and Empire: A Biographical Study of Lord Kelvin and the Age of the Earth.* Science History Publications, New York.

THOMSON, W. 1854. On the mechanical energy of the solar system. *Philosophical Magazine*, **8**, 409–430.

—— 1861. On the secular cooling of the earth.

Transactions of the Royal Society of Edinburgh, **23**, 157–170.

—— 1866. The "doctrine of uniformity" briefly refuted. *Proceedings of the Royal Society of Edinburgh*, **5**, 512–513.

WALLACE, A. R. 1855. On the law which has regulated the introduction of new species. *Annals and Magazine of Natural History*, ser. 2, **16**, 184–196.

WEBSTER, T. 1814. On the freshwater formations in the Isle of Wight, with some observations on the strata over the Chalk in the south-east part of England. *Transactions of the Geological Society of London*, **2**, 161–254.

—— 1821. On the geognostical situation of the Reigate Stone and of the Fuller's Earth at Nutfield. *Transactions of the Geological Society of London*, **5**, 353–357.

WILSON, L. G. 1962. The development of the concept of uniformitarianism in the mind of Charles Lyell. *In*: *Proceedings of the 10th International Congress on the History of Science.* Hermann, Paris, 993–996.

—— 1967. The origins of Charles Lyell's uniformitarianism. *In*: ALBRITTON, C. C., JR (ed.) *Uniformity and Simplicity, a Symposium on the Principle of the Uniformity of Nature.* Special Paper 89, Geological Society of America, New York, pp. 35–62.

—— (ed.) 1970. *Sir Charles Lyell's Scientific Journals on the Species Question.* Yale University Press, New Haven & London.

—— 1972. *Charles Lyell, The Years to 1841: The Revolution in Geology.* Yale University Press, New Haven & London.

—— 1980. Geology on the eve of Charles Lyell's first visit to America, 1841. *Proceedings of the American Philosophical Society*, **124**, 168–202.

Lyell's reception on the continent of Europe: a contribution to an open historiographical problem

EZIO VACCARI

*Centro di Studio sulla Storia della Tecnica, CNR Via Balbi 6,
16126 Genova, Italy*

Abstract: After the publication of the *Principles of Geology* the name of Charles Lyell became
well known among European geologists. However, the diffusion of his works and the intensity of
the debates on his ideas varied throughout the continent. This paper attempts to analyse some
aspects of Lyell's reception in Continental Europe, emphasizing the role of the translations of
fundamental works such as the *Principles of Geology* and the *Elements of Geology*, as well as
their influence on the first textbooks of geology published in Europe, especially from the 1840s
onwards. Some significant examples of nineteenth-century European reactions to Lyell's geology,
and in particular to his uniformitarian statements, will be pointed out, in order to show the extent
of Lyell's influence.

The question of Charles Lyell's reception in Europe is undoubtedly a major topic in the history of geology, which encompasses several stimulating historiographical and scientific issues. In 1990, at the end of his new introduction to the facsimile reprint of the first edition of Lyell's *Principles of Geology*, Martin Rudwick (1990, p. lv) remarked that 'the nature and timing of Lyell's influence outside the English-speaking world is an important topic on which little historical research has yet been done'. Indeed, in most works published since the time of the Lyell Centenary Symposium held in 1975 – but also in earlier literature – the diffusion and reception of Lyell's geological ideas in the various European national and scientific contexts have been studied only patchily, for example in Germany (Guntau 1975), Poland (Maslankiewicz & Wojcik 1975), Sweden (Frängsmyr 1976) and Spain (Ordaz 1976). In general, as Rachel Laudan (1987, p. 221) remarked on the question of Lyell's impact, there is still need for 'further historical scholarship to reveal the developments in geology after 1830'.

In this paper I do not claim to give a final answer to the various gaps in this open historiographical problem. Besides my preliminary conclusions, I would like to propose an agenda for further research, in order to indicate the remarkable quantity and the variety of the primary sources that should be explored in detail. These are, in my opinion, the necessary tools for overcoming the historiographical impasse and for achieving the most complete picture possible of such a complex cultural phenomenon as the 'reception' of a scientific author in a continental context.

The different kinds of sources may be briefly described as follows :

(1) Translations of Lyell's major works, such as the *Principles of Geology* (1830–1833), the *Elements of Geology* (1838) and *Antiquity of Man* (1863a), but also other writings. Research may allow us to understand the editorial story of some translations, i.e. to evaluate the role of the possible promoters, the scientific standing of the translators and the extent of diffusion of these publications within their linguistic areas.

(2) Reviews of Lyell's writings in scientific periodicals that were more or less specialized in the geological sciences. Research here should address the problem of surveying systematically all the European periodicals that may have treated subjects related to the Earth sciences during the second half of the nineteenth century – not only the first specialized journals of geology which started to appear in this century, but also the periodicals of general natural history and the bulletins, proceedings or transactions of societies, academies and other institutions. This is an impressive body of literature (Bolton 1885; Gascoigne 1985), which has rarely been investigated in relation to the geological sciences.

(3) The early textbooks of geology which followed the publication of Lyell's *Principles* and were published, especially in France and Germany, from the late 1830s onwards.

(4) The numerous copies of *Quarterly Journal of the Geological Society of London* which were sent to overseas Fellows and Correspondents.

VACCARI, E. 1998. Lyell's reception on the continent of Europe: a contribution to an open historiographical problem. *In*: BLUNDELL, D. J. & SCOTT, A. C. (eds) *Lyell: the Past is the Key to the Present.* Geological Society, London, Special Publications, **143**, 39–52.

(5) The university courses (published or unpublished) and other public lectures given by European geologists, especially in the second half of the nineteenth century.

(6) Debates during the meetings of geological and scientific societies during the early and middle nineteenth century, as already pointed out by Rudwick (1986) in the cases of the Société Géologique de France and the Gesellschaft Deutscher Naturforscher und Arzte.

(7) Correspondence between geologists, including Lyell himself – although a preliminary analysis of his published letters (Lyell 1881) has revealed only a few comments about the reception of his ideas on the continent. Nevertheless, during his travels in Italy, France, Germany, Switzerland and Scandinavia, Lyell met several scientists and gave lectures in some institutions: consequently the heritage of contacts and scientific exchanges he experienced in Continental Europe cannot be ignored.

It is evident that any historical research arranged according to this agenda is destined to deal for a long time with a large amount of material, hence the risk of studying in depth some aspects more than others. Therefore, in this paper I present a selection of sources, considered as significant samples of a larger potential field of research and related to Lyell's reception in France, Germany, Italy, Spain, Belgium and Switzerland. I do not treat here the British context, where the response to Lyell's scientific ideas has been discussed in detail in several historical works, often with particular attention to the role of Lyell's influence on Charles Darwin (Gillispie [1951] 1996; Rudwick 1967, 1975, 1990; Cannon 1960, 1976; Wilson 1972, 1996; Bartholomew 1976, 1979; Page 1976; Morrell 1976; Ruse 1976, 1979; Lawrence 1978).

The first reviews of Lyell's *Principles*

Although this paper is a preliminary report of work in progress, I hope that the historical data used to show Lyell's influence, or lack of it, will provide a useful contribution to the analysis of a topic that has been treated in the past mainly in terms of the 'uniformitarian/catastrophist' controversy (Hooykaas 1959; Cannon 1960; Hallam 1983) or in the light of the historiographical debate on Lyell's revolutionary or non-revolutionary impact (Elena 1988).

Without underestimating the role of various scientific travels throughout the continent undertaken by Lyell during the late 1820s, especially in France and Italy (Wilson 1972, pp. 183–261), the beginning of his influence on the Continent was linked to the reception given to his famed *Principles of Geology* (Lyell 1830–1833). It is well known that the first volume of this work, published in the summer of 1830 while Lyell was doing field work in France and Spain, was a great success with the reading public in Britain and was promptly reviewed in some British journals by distinguished authors such as George Poulett Scrope (1830) in the *Quarterly Review*, William Conybeare (1830) in the *Philosophical Magazine* and William Whewell (1831) in the *British Critic*.

Nevertheless, some copies also reached Paris, where Constant Prévost (1787–1856), the French geologist and co-founder with Ami Boué of the Société Géologique de France, already owned the proof sheets of the first volume of the *Principles*, which he had received directly from his friend Lyell in order to prepare a French translation (Wilson 1972, p. 301). Prévost, a former student of Georges Cuvier, had developed a firm actualistic view about the continuity of Earth history even before meeting Lyell, with whom he had shared geological ideas and several field trips from 1823 onwards (Bork 1990, Gohau 1995). However, in spite of his role as a staunch supporter of Lyell's uniformitarian theory – well shown, for example, in the 1835 debate on Buch's theory of craters of elevation (Dean 1980; Laudan 1987, pp. 192–193; Bork 1990, pp. 24–26) – Prévost failed to translate the *Principles* during the 1830s.

Instead, between 1831 and 1834, the three volumes of Lyell's work were reviewed for French readers by Ami Boué (1794–1881), who was at that time secretary for the foreign correspondence of the Société géologique de France, of which he became President in 1835. This geologist, born in Germany, had studied in Edinburgh, Paris, Berlin and Vienna: consequently he had a good knowledge of several languages, including English, and he made numerous field trips in France, Germany, Austria-Hungary and Italy (Birembaut 1970). Together with Constant Prévost, Boué has been considered 'the closest Lyell had come to an active supporter in France' (Lawrence 1978, p. 119). However, he did not embrace Lyell's most rigid uniformitarian conclusions on the constant intensity of the geological phenomena from the past to the present, just as he only partially accepted Elie de Beaumont's theory on mountain building and firmly refused its catastrophist historical framework (Laurent 1993). This position is well expressed in Boué's writings published in the *Bulletin de la Société géologique de France* and in his new periodical, the *Mémoires Géologiques et Paléontologiques*, which continued from the previous *Journal de Géologie*, also edited by Boué from 1830 to 1831.

In the first discussion of Lyell's geological work, Boué recognized its accuracy ('compilation judicieuse') and identified the author as 'un homme d'esprit'; but on the other hand the French geologist was not convinced by Lyell's 'philosophical' use of the present for explaining the past, which had led to the application of a 'marche philosophique du connu à l'inconnu' (the philosophical procedure from what is known to what is unknown) (Boué 1830–1831, pp. 182–183). Not by accident, the short review of the first two volumes of the *Principles* which was printed in the *Mémoires Géologiques et Paléontologique* identified Lyell's books as a very interesting 'traité philosophique de géogénie' (a philosophical treatise of geogeny) (Boué 1832, p. 317). Here a brief and neutral account of the content of the two volumes was significantly followed by a long French abstract taken from two critical papers on Lyell's theory published by William Conybeare in the *Philosophical Magazine* (Conybeare 1830–1831). According to Boué, Conybeare's attack on Lyell was an answer full of healthy criticism ('response pleine de saine critique') and deserved to be widely known among French scientists, although Boué himself clearly stated that he did not agree entirely with all Conybeare's ideas. Moreover, the publication of other critical contributions, in particular a paper by Charles Daubeny against Lyell, was also announced in the *Mémoires*.

Two years later, in the *Bulletin de la Société géologique de France*, Boué again reviewed Lyell's work (Boué 1834, pp. 170–176). Here the third volume of the *Principles*, published in 1833, was warmly welcomed because it was considered more geological (thus less 'philosophical') than the previous ones and because Boué found in its content a total correspondence with his own stratigraphical views, presented in Vienna in September 1832. According to the reviewer, Lyell's treatment of the Tertiary formations, based on a great amount of new data collected in the field, was by far the highlight of the *Principles* and deserved to be translated even on its own, because it introduced new ideas (including a new nomenclature of the periods, Eocene, Miocene and Pliocene) in the footsteps of Brongniart's research. The two chapters on the phenomena of denudation and erosion of valleys were also judged positively by Boué, who only regretted Lyell's indifference to the discussion of the facts supporting the theory of the parallel elevations in the same epoch, put forward by Léonce Elie de Beaumont (1798–1874). Lyell's denial of this interpretation of geological history in the *Principles* was based on theoretical arguments, which were considered by Boué insufficient for a convincing contrary explanation.

In spite of these and other elements of criticism expressed by French commentators on Lyell's work, it is a matter of fact that the group of geologists who established the French geological society of the early 1830s was essentially composed of the scientists who best knew Lyell's ideas. Apart from Prévost and Boué, the conchologist Gérard Paul Deshayes (1796–1875) was also one of the founding members of the Société géologique de France. He had provided Lyell with the indispensable series of data on the Tertiary molluscan faunas, and the two scientists had also examined some palaeontological collections in Paris together (Wilson 1972, pp. 259–260, 301–303). However, in spite of his cooperation with the British geologist, Deshayes remained substantially a catastrophist in his conception of the history of life (Laurent 1987, pp. 147–151).

At the beginning of his last review of the *Principles*, Boué (1834, p. 170) emphasized the lack of and need for a French translation of Lyell's work, pointing out that one of the German members of the Société géologique de France, the mineralogist Carl Friedrich Alexander Hartmann (1796–1863), had completed the publication of the first German translation of the *Principles*, based on its second edition. The three volumes entitled *Lehrbuch der Geologie* were printed from 1832 to 1834, immediately after the original publication in English (which included the first edition of the third volume), and can be regarded as the first translation published in Europe of Lyell's major work (Lyell 1832–1834). It must be noted that the German title was different from the original and the word 'Principles' was changed to 'Lehrbuch' (textbook). Now, we know that for Lyell the word 'principles' was not a synonym of 'textbook' and he did not intend to produce a summary or compilation of contemporary geological knowledge (Rudwick 1990, p. xiii). Consequently it is possible that he did not like the title of this first translation, because it contained a significant conceptual change. In fact in the second German translation, based on the sixth English edition of 1840, the title was changed to *Grundsätze der Geologie*, Fig. 1, the literal translation of *Principles of Geology* (Lyell 1841–1842).

The response to Lyell's concept of uniformitarianism

Carl Hartmann, the translator responsible for both these German editions, was the commissioner of mines in Brunswick and had an extensive knowledge of the English and French languages and of technical–scientific literature (Koch 1972). For this reason, during his career he translated, revised and edited about forty books on mining, mineralogy, metallurgy and geology, including Lyell's *Principles* and *Elements of Geology* (Lyell 1838).

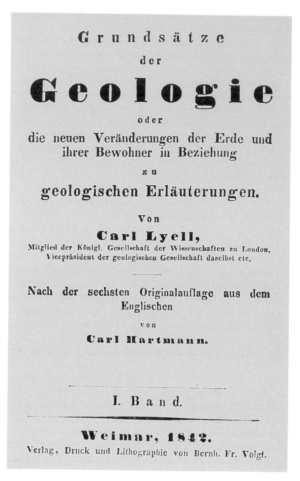

Fig. 1. Title page of Lyell 1841–1842 vol. 1.

The latter was promptly translated the year after the first English edition (Lyell 1839a), which demonstrates a continuity of interest among the German scientific community in Lyell's work. Not by accident, a note by the publisher Bernhard Friedrich Voigt in the preface to the *Principles'* second translation of 1841–1842 reveals that Lyell's work was positively reviewed in several German journals, such as *Helios* (no. 24, 1841), which named the British scientist as 'the best leading star in the field of geology' ('Es ist der beste Leitstern auf dem Gebiete der Geologie': Lyell 1841–1842, vol. 1, p. viii, footnote).

In spite of this, as Martin Guntau (1975) has clearly pointed out, the German scientists only partially accepted Lyell's ideas because on the one hand they were strongly influenced by the authority of Leopold von Buch and on the other hand they

generally 'understood the Earth's history as a process of development'. Among the most positive reactions, Karl Ernst Adolf von Hoff (1771–1837) warmly welcomed the publication of the *Principles*, which reinforced his own actualistic views (von Hoff 1834, pp. 3–8), while Carl Vogt (1817–1895) and Bernhard von Cotta (1808–1879) praised Lyell's anti-catastrophist and anti-diluvialist approach. Nevertheless, von Cotta, like other German geologists, criticized the rigidity of Lyell's views on the uniformity of the different stages in the Earth's past and underlined the qualitative peculiarities of the individual periods of geological history (Guntau 1975). In opposition to Lyell's uniformitarianism, von Cotta elaborated his 'law of development of the Earth' since 1839, which was explicitly formulated for the first time in 1858 and more widely discussed in 1866 (Cotta

1858, 1866). In short, according to von Cotta's 'natural law', the accumulation of all results of geological causes increased the degree of diversity far beyond its original magnitude (Wagenbreth 1992).

Significantly, in the same year, 1858, von Cotta also published the last volume of his translation of Lyell's *Elements of Geology*, the second German edition of this work, under the title *Geologie oder Entwickelungsgeschichte der Erde und ihrer Bewohner*, Fig. 2. At the end of the first volume von Cotta included an appendix on the chalk formations in Germany (Lyell 1857–1858, vol. 1, pp. 409–412). It was not the first time that a German translator added some new material to Lyell's original text: Carl Hartmann had already inserted a historical supplement ('Anhang. Neuere Geschicte der Geologie') at the end of the first volume of the *Grundsätze der Geologie* (Lyell 1841–1842, vol. 1, pp. 400–576). In this long appendix Hartmann praised the importance of Lyell's four chapters on the history of geology published in the first volume of the *Principles* and provided an enlargement based on the analysis of the works of several German, Swiss, English and French geologists of the nineteenth century. This 'modern' history of geology clearly emphasized the role of German scientists, in the light of the new historical accounts by Friedrich Hoffmann (1838) and Christian Keferstein (1840). However, Hartmann significantly concluded his supplement with a positive discussion of Elie de Beaumont's 'geognostische Systeme', considered impregnable in its essential principles and of great importance for the development of 'teoretische Geologie' (Lyell 1841–1842, vol. 1, pp. 548–576).

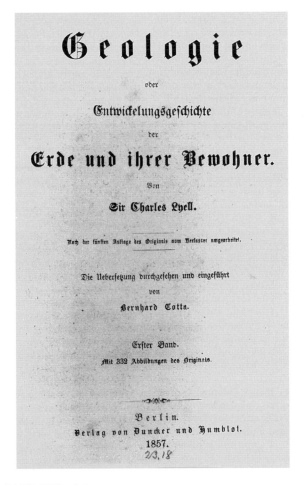

Fig. 2. Title page of Lyell 1857–1858 vol. 1.

Lyell's influence on geological thought in Europe

Other signs of Lyell's influence among German scholars around the middle of the nineteenth century can be found in authoritative geological textbooks such as Karl Cäsar von Leonhard's (1779–1862) *Lehrbuch der Geognosie und Geologie* (1835), Gustav Bischof's (1792–1870) *Lehrbuch der chemischen und physikalischen Geologie* (1847–1855), Carl Vogt's *Lehrbuch der Geologie und Petrefactenkunde* (1846–1847) and Karl Friedrich Naumann's (1797–1873) *Lehrbuch der Geognosie* (1850–1866). The *Principles* and the *Elements* are often quoted in these texts, sometimes even in the original English, but discussion of Lyell's theoretical positions remained undeveloped. As in the case of Alexander von Humboldt's (1769–1859) *Kosmos* (1845), the references to Lyell's work usually focused on his contribution to the definition of Tertiary formations.

Besides the Geological Society of London, Lyell was a member of other national geological societies in Europe: his name is included in the official list of the members of the Deutschen Geologischen Gesellschaft (the German Geological Society), published in 1856, although the library of the same institution did not contain a copy of the *Principles* or the *Elements* by June 1858 (Zeitschrift 1856–1858). The Société géologique de France also counted Lyell's name on the list of its early members (Boué 1834, p. x).

Nevertheless, as we have seen, this institution did not officially promote the first French edition of the *Principles,* which instead was translated between 1843 and 1848 under the auspices of Dominique François Jean Arago (1786–1853), distinguished scientist and influential member of the Chamber of Deputies in Paris (Lyell 1843–1848). This work was based on the sixth English edition of 1840 and the translation was made by Mme Tullia Meulien, an obscure figure compared with the scientific stature of the German translators Hartmann and von Cotta. According to the French translator, the main reasons of the publication of the *Principes de Géologie* were 'la grande célébrité que s'est acquise l'ouvrage de M. Lyell' (the great celebrity reached by Lyell's works) and its 'théorie ingénieuse, adoptée par l'auteur et soutenue de son brillant talent' (ingenious theory, adopted by the author and supported by his outstanding talent), that is to say 'le systeme si controversé des *causes actuelles*', the much debated uniformitarian system (Lyell 1843–1848, vol. 1, pp. vii–ix). Tullia Meulien had already translated into French the *Elements of Geology* (Lyell 1839b), regarded as a natural introduction to the *Principles*, again under

the auspices of Arago and in the same year as the first German edition (Lyell 1839a). Consequently, in spite of Ami Boué's stimulating reviews published in the early 1830s, French readers had to wait about fifteen years for the complete edition of the *Principles* in their own language. On the other hand, the prompt translation of the *Elements*, both in France and in Germany, indicates the appeal of Lyell's synthesis of his subject in a single volume.

French translations of the *Manual of Elementary Geology* (Lyell 1856–1857, 1863b: translated with the collaboration of Lyell himself), as well as other French editions of the *Principles* (Lyell 1873) and of the *Elements* (Lyell 1864a), see Fig. 3, the *Supplément au Manuel de Géologie* (Lyell 1857) and the *Abrégé des Éléments de Géologie* (Lyell 1875) were carried out by little known scholars such as M. Hugard – assistant of mineralogy at the Museum of Histoire Naturelle in Paris and vice-secretary of the Société geologique de France – and Jules Ginestou – librarian of the Société d'Encouragement pour l'Industrie Nationale. However, the French editions of Lyell's major works were widely known outside France during the nineteenth century, for example in Italy, where Lyell was never translated. Usually the French translator did not add new material to Lyell's text as the German ones did. In his translation of 1873, Fig. 4, based on the tenth English edition of the *Principles* (Lyell 1867–1868), Jules Ginestou inserted an appendix at the end of the second volume (Lyell 1873). However, this appendix contained only the amendments listed by Lyell himself in the preface of the new eleventh edition of the *Principles*, which had been published in London in 1872 while Ginestou was still completing his translation.

In the main French textbooks of geology and palaeontology published after the 1830s Lyell's ideas were generally ignored: the Belgian geologist Jean Baptiste Julien D'Omalius d'Halloy (1783–1875) could certainly not adopt them within the catastrophist framework of his *Éléments de Géologie,* where the mechanisms for explaining the upheaval of mountains were De Beaumont's concept of 'revolutions' and the action of the Deluge (Omalius d'Halloy 1835, 1843). On the other hand, François Sulpice Beudant (1787–1850), in his classic textbook *Géologie* (1851) – later officially adopted in French secondary and religious schools – clearly stated his views about the unacceptability of the uniform application of present geological facts to ancient phenomena: 'il est à croire en effet que les causes qui agissent aujourd'hui sous nous yeux sont aussi celles qui ont agi de tout temps, mais qui sans doute ont déployé une énergie supérieure à de certaines époques que

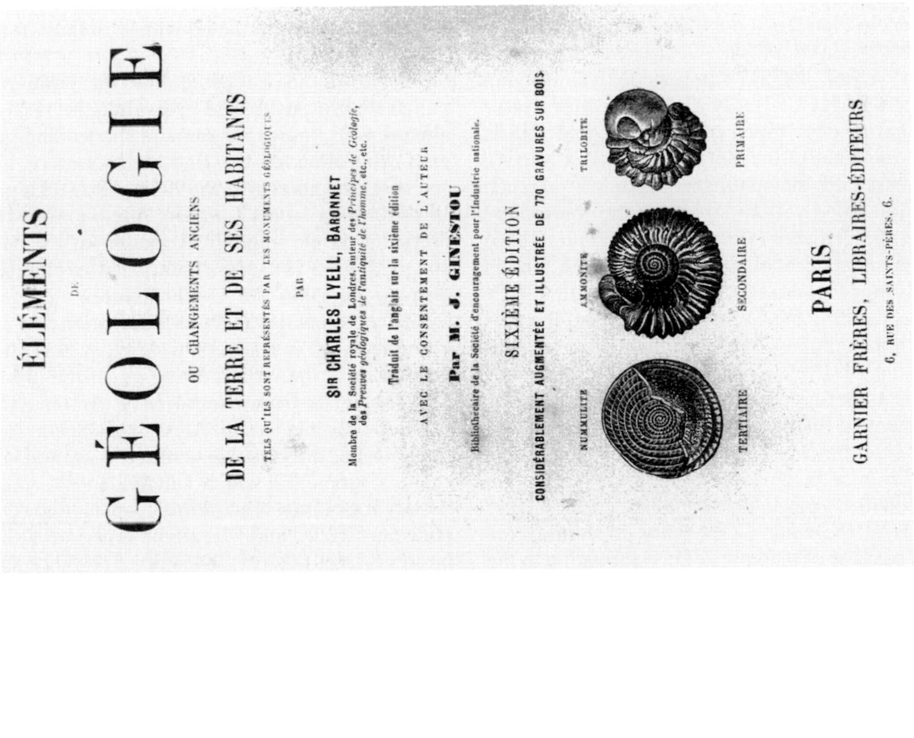
ÉLÉMENTS
DE
GÉOLOGIE
OU CHANGEMENTS ANCIENS
DE LA TERRE ET DE SES HABITANTS
TELS QU'ILS SONT REPRÉSENTÉS PAR LES MONUMENTS GÉOLOGIQUES
PAR
SIR CHARLES LYELL, BARONNET
Membre de la Société royale de Londres, auteur des *Principes de Géologie*, des *Preuves géologiques de l'antiquité de l'homme*, etc., etc.
Traduit de l'anglais sur la sixième édition
AVEC LE CONSENTEMENT DE L'AUTEUR
Par M. J. GINESTOU
Bibliothécaire de la Société d'encouragement pour l'Industrie nationale.
SIXIÈME ÉDITION
CONSIDÉRABLEMENT AUGMENTÉE ET ILLUSTRÉE DE 770 GRAVURES SUR BOIS
NUMMULITE AMMONITE TRILOBITE
TERTIAIRE SECONDAIRE PRIMAIRE
PARIS
GARNIER FRÈRES, LIBRAIRES-ÉDITEURS
6, RUE DES SAINTS-PÈRES, 6.

Fig. 3. Title page of Lyell 1864*a*.

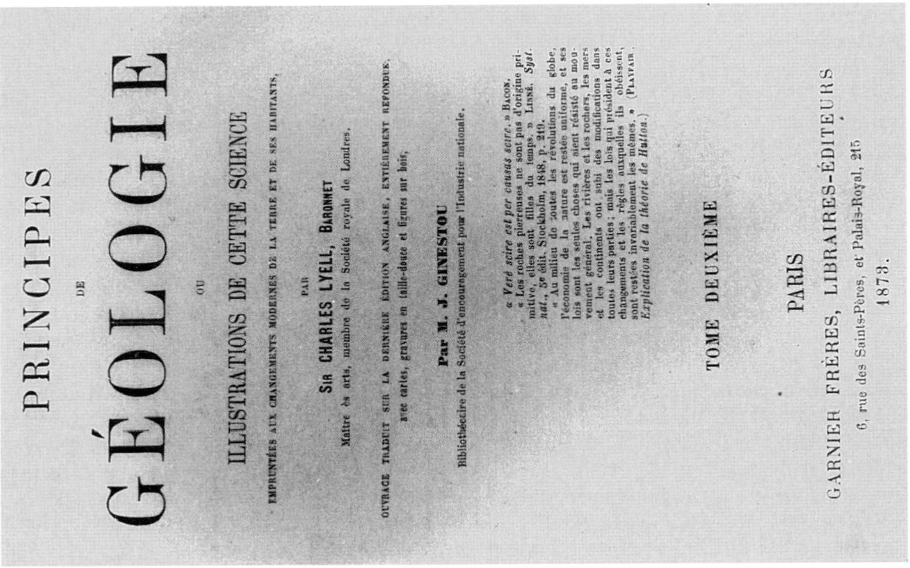
PRINCIPES
DE
GÉOLOGIE
OU
ILLUSTRATIONS DE CETTE SCIENCE
EMPRUNTÉES AUX CHANGEMENTS MODERNES DE LA TERRE ET DE SES HABITANTS,
PAR
SIR CHARLES LYELL, BARONNET
Maître ès arts, membre de la Société royale de Londres.
OUVRAGE TRADUIT SUR LA DERNIÈRE ÉDITION ANGLAISE, ENTIÈREMENT REFONDUE,
avec cartes, gravures en taille-douce et figures sur bois,
Par M. J. GINESTOU
Bibliothécaire de la Société d'encouragement pour l'Industrie nationale.

« *Verè scire est per causas scire.* » BACON.
« Les roches pierreuses ne sont pas d'origine primitive, elles sont filles du temps. » LINNÉ. *Syst. nat.*, 5e édit. Stockholm, 1848, p. 219.
« Au milieu de toutes les révolutions du globe, l'économie de la nature est restée uniforme, et ses lois sont les seules choses qui aient résisté au mouvement général. Les rivières et les rochers, les mers et les continents ont subi des modifications dans toutes leurs parties; mais les lois qui président à ces changements et les règles auxquelles ils obéissent, sont restées invariablement les mêmes. » (PLAYFAIR, *Explication de la théorie de Hutton.*)

TOME DEUXIÈME
PARIS
GARNIER FRÈRES, LIBRAIRES-ÉDITEURS
6, rue des Saints-Pères, et Palais-Royal, 215
1873.

Fig. 4. Title page of Lyell 1873.

l'observation va nous faire connaitre' (it is right to believe that the causes which act now under our eyes are also the same that have always acted, but they have certainly unfolded more energy in some particular epochs, as we learn from the observations) (Beudant 1851, p. 102). This does not mean that the French geologists neglected other aspects of Lyell's work, such as his lithological classification, which was in fact cited by Alcide d'Orbigny (1802–1857) in the *Cours Élémentaire de Paléontologie et de Géologie Stratigraphiques* (Orbigny 1852, pp. 262–263).

Lyell's reception in other countries, such as Switzerland is also of interest. The paper 'On the Proofs of a gradual rising of the land in certain parts of Sweden', read by Lyell at the Royal Society of London in 1834 and published in the *Philosophical Transactions* of the following year (Lyell 1835), was translated in its entirety into French by Louis Coulon, President of the Société des Sciences Naturelles de Neuchâtel (Lyell 1840). Although the *Bulletin Bibliographique* of the Swiss society published only abstracts of foreign papers already printed elsewhere, Lyell's paper was treated as an exception because of its great general interest. Moreover, numerous references to Lyell can be found in the geological textbook published by Bernhard Studer, professor in Bern (Studer 1844). Both Coulon and Studer knew Lyell personally, having met him during his travels in Switzerland in the summer of 1835 (Wilson 1972, p. 417).

The long paper 'On Tertiary strata of Belgium and French Flanders', published by Lyell in the *Quarterly Journal of the Geological Society of London* (Lyell 1852), interested two Belgian scholars: Charles Le Hardy de Beaulieu (professor at the École des Mines of Hainaut) and Albert Toillez (engineer of the Belgian Corps des Mines). Some years later they published a complete French translation of Lyell's paper in the periodical *Annales des Travaux Publique de Belgique* (1856) and as a separate booklet (Lyell 1856). Although the translators remarked on the utility of this work for understanding the geo-palaeontological features of Belgium, they also declared their disagreement with some of Lyell's conclusions: 'Nous devons déclarer, toutefois, que nous ne partageons pas entièrement le vues de l'auteur, surtout en ce qui concerne le passage graduel de la faune d'une formation à celle de la formation suivante' (we must declare, however, that we do not entirely share the ideas of the author, especially concerning the gradual passage of the fauna of one formation to the fauna of the following formation) (Lyell 1856, p. 3, footnote).

In Spain the *Elements of Geology* was translated in 1847 with the title *Elementos de Geologia* by the mining engineer Joaquín Ezquerra del Bayo

(1789–1859), one of the leading geologists of his country in the nineteenth century, who also added to his translation some pages on the geological features of Spain (Ordaz 1976). The *Elements* were chosen mainly to fill the absence of an exhaustive modern elementary treatise of geology in Spanish, but the translator also stated his agreement with Lyell's 'revolutionary' geological theories and remarked on the fact that Lyell's geological nomenclature had been accepted by all contemporary geologists: 'toda la nomenclatura que Lyell ha introducido en la ciencia ... ha merecido la aceptacion general de los geólogos, que la han recibido sin sequiera modificarla. Grande ha sido la revolucion que Lyell ha hecho en esta ciencia' (the full scientific nomenclature introduced by Lyell ... has deserved the general acceptance from all the geologists, who accepted it without even modifying it. Lyell has introduced a great revolution in this science) (Lyell 1847, pp. v–vi).

Finally, we can briefly examine Lyell's reception in Italy, which presents a singular situation. It is well known that Lyell studied several regions of the peninsula in 1828–1829 (Lyell 1881, vol. 1, pp. 200–246; Wilson 1972, pp. 218–261) and their geological features had an important role in the argument of the *Principles*. Nevertheless, this work and the *Elements of Geology* were not translated into Italian. The reason for this is not yet known, but the main consequence was that most Italian geologists started to read Lyell after the publication of the French editions of the *Principles* and the *Elements*, which circulated widely in Italy from the 1860s onwards.

But it should be noted that the stratigraphic nomenclature established by Lyell in the 1830s was already well known by some of the leading geologists who attended the first congresses of Italian scientists, beginning in 1839 (Morello 1983, p. 73). In 1840 the marquis Lorenzo Pareto (1800–1865), president of the geological section of the second congress held in Turin, mentioned explicitly the 'terreno miocene del Lyell' (miocenic terrain of Lyell) in referring to some alternations of strata near Siena in Tuscany (Atti 1841, p. 93). A few years later, at the sixth meeting organized in Milan, two geologists from central Italy, Antonio Orsini from Ascoli and count Alessandro Spada Lavini from Rome, presented a geological section of the central Apeninnes which subdivided the younger tertiary rocks into 'Terziario Subappennino (Older-pliocene Lyell)' and 'Terziario superiore (Newer-pliocene Lyell)' (Atti 1845, pp. 571–572). Orsini and Spada Lavini published the results of their research (1845, 1855) in the *Bulletin de la Société geologique de France* and included several geological sections which provided a detailed stratigraphy of central Italy, from lower Lias to Pleistocene (Parotto & Praturlon 1984, pp. 242–244). This did not represent the attitude of all Italian geologists towards Lyell's stratigraphic nomenclature. In the meeting of 1846 in Genoa, the division between Newer and Older Pliocene was discussed again by Lorenzo Pareto and the French geologist Henri-Jean-Baptiste Coquand (1813–1881). They both argued that in the sub-Apennines of Tuscany there was no field evidence of a precise boundary between the two formations (Atti 1847, p. 672).

Among the Italian geologists who critically considered Lyell's views, mention must be made of Leopoldo Pilla (1805–1848), an important exponent of the 'Pisa school of geology' in Tuscany (Corsi 1995; D'Argenio 1996). In 1847 Pilla published the first volume of the *Trattato di Geologia*, where he stated that most of the information and illustrations about the geology of northern Europe had been taken directly from the French translation of Lyell's *Elements of Geology* (Pilla 1847, p. x). Pilla had also adopted Lyell's nomenclature of the rocks from post-Pleistocene to Tertiary, but did not agree with the idea of the rigid uniformity of the geological causes. In his opinion the slow cooling of the Earth had determined a constant diminution in the intensity of geological phenomena, such as volcanic eruptions and earthquakes, from the past to the present (Pilla 1847, pp. 383–386).

It is important to point out that Lyell's geology was particularly well received and discussed by geologists who worked in the Italian Apenninic regions, such as Tuscany and Emilia Romagna. This happened because those terrains offered more possibilities for the application of the new Lyell nomenclature of Tertiary formations. In 1867 the Fisiocritici Academy of Science of Siena in Tuscany had acquired for its library the French translation of the *Manual of Elementary Geology*, because the book was widely regarded as a classic ('libro che si ritiene come classico') and its wide treatment of the Eocene, Miocene and Pliocene terrains was considered very useful for the geology of southern Tuscany (Pantanelli 1869, p. 56). In Emilia Romagna a group of geologists active in Bologna and Modena were strongly influenced by Lyell's geology. The young Giovanni Capellini (1833–1922), later professor of geology in Bologna, personally met Lyell in 1857 at La Spezia and two years later in London (Capellini 1914, pp. 136–141, 177). Giovanni Giuseppe Bianconi (1809–1878), professor of natural history at the University of Bologna, published in 1862 a *Catalogo della serie geognostica dei terreni bolognesi* which explicitly followed Lyell's stratigraphic subdivisions from the Recent back to the Eocene (Bianconi 1862, pp. 251–267). Finally,

the geologist Pietro Doderlein (1809–1895), who taught natural history in Modena and later in Palermo, published in 1870 a geological map of the provinces of Modena and Reggio Emilia entirely based on Lyell's stratigraphic classification, literally defined as 'la classica partizione di Lyell' (Doderlein 1871, p. 17).

The task of preparing a complete geological map of Italy was a fundamental desiderata in nineteenth century Italian geology and was often a major topic in the discussion of the geological sections of the annual congress of Italian scientists (*Riunione degli Scienziati Italiani*). This tendency towards a more descriptive geology based on detailed fieldwork, well synthesized by Igino Cocchi (1827–1913) in the introduction to the *Memorie per servire alla descrizione della carta geologica d'Italia* (Cocchi 1871–1893, vol. 1), probably determined the way Lyell's work was received by several Italian geologists: far from adopting or even discussing in detail the validity of the most rigid theoretical uniformitarian statements, the Italian geologists preferred instead to verify Lyell's new stratigraphic classification directly in the field. It is significant that probably the most violent theoretical rejection of Lyell's ideas was published in the book *Diluvio* by Alberto Cetta (1814–1890), a Jesuit scholar and professor of philosophy, but certainly not a field geologist (Cetta 1886).

Another aspect of Lyell's reception in Europe was linked to his role as a historian of geology (Porter 1976; Rudwick 1990). Chapters 1 to 4 of the *Principles* probably stimulated the publication of similar historical accounts, for example by Friedrich Hoffmann (1838), Christian Keferstein (1840) Carl Hartmann (Lyell 1841–1842, vol. 1, pp. 401–576) and Carl Vogt (1846–1847) in Germany and by Carlo Gemmellaro (1862), Antonio Stoppani (1881) and Giovanni Omboni (1894) in Italy.

The *Antiquity of Man* was also translated into both French (with the collaboration of Lyell himself) and German (Lyell 1864b, 1867), Fig. 5, and stimulated more studies, particularly in France and Italy, on the question of the local palaeontological evidence of 'fossil man' (Lyell 1864c; Cocchi 1866). However, this subject will not be treated here, as it deserves a detailed analysis on its own.

In conclusion, study of Lyell's reception in Continental Europe, especially related to the publication and diffusion of the *Principles of Geology* and *Elements of Geology*, presents some common themes throughout the continent. Generally, nineteenth-century geologists praised the systematic nature of Lyell's work and the internal organization of his treatises, but they often also criticized the dogmatic presentation of his

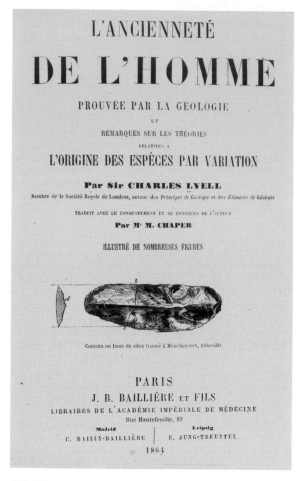

Fig. 5. Title page of Lyell 1864*b*.

uniformitarian theory. The theoretical aspects were considered and debated, but the numerous specific geographical–geological examples included in the pages of the *Principles* and the *Elements* were particularly appreciated by geologists on the continent. Sometimes quotations of certain parts of Lyell's works were accompanied by a request for additional data in future editions (for example, Boué 1834, p. 176).

As has been observed, it was not necessary to accept completely Lyell's methodology and theory in order to profit from the *Principles* or the *Elements*. These two works, especially the latter, were often taken as useful reference works, written in an enjoyable style and rich in nice illustrations. In spite of the great success of the several editions of the *Principles of Geology* in the English-speaking world, most of the contemporary geolo-

gists of Continental Europe remained doubtful about Lyell's uniformitarianism.

During the last twenty years some historical studies have tried to identify the cause or causes that hindered Lyell's reception in Europe. According to Mott Greene (1982, p. 76), Lyell 'never developed a satisfactory theory of mountain ranges' and consequently 'this lacuna was a great barrier to the acceptance of Lyell's geology in Europe', where the major geological question during the nineteenth century was 'precisely the origin of the great mountain ranges'. On the other hand, Greene continues, Lyell's contemporary, Léonce Elie de Beaumont, enormously influenced European geology with his catastrophist ideas on the causes of mountain formation, which firmly hindered Lyell's reception in Europe. Philip Lawrence (1978, p. 110) has stated that the general acceptance of the theory of central heat among nineteenth century geologists was reinforced by De Beaumont's synthesis which connected the cooling of the Earth's crust with periodic episodes of mountain building (orogenies), considered to be the main events of geological history. Consequently, according to Lawrence, 'the theory of central heat was potentially the most significant obstacle to the general acceptance of Lyell's theoretical system'. Rachel Laudan (1987, p. 221) has partially confirmed these conclusions on the opposition of two contemporary rival systems, in stating that Elie de Beaumont was probably more influential than Lyell in Europe, but in any case 'continental geologists read the French and German translations, appreciating Lyell's methodological ingenuity, but not changing their historical or causal theories'.

These historiographical judgements may be substantially confirmed by present research, at least in the case of France and Germany. Nevertheless, the case of Italy shows that, in spite of the lack of a translation, Lyell's stratigraphic nomenclature and classification of Recent and Tertiary formations were openly defined as 'classic' at the beginning of the 1870s, a few years before Lyell's death. Alongside the search for possible instances of acceptance or refusal of Lyell's complete geological theory, it may be important to investigate in detail how the works and the words of one of the most famous geologists of the nineteenth century were diffused, translated, read, discussed and used.

The research for this paper was made possible by my appointment as Resident Fellow for the academic year 1996–1997 at the Dibner Institute for the History of Science and Technology of Cambridge, MA. I would also like to thank Kenneth Taylor (Department of History of Science, University of Oklahoma), Marilyn Ogilvie (History of Science Collections, University of Oklahoma), Chiara Bratto (Accademia dei Fisiocritici di Siena, Italy), Martin Guntau (Rostock), Stefano Marabini (Faenza), the staff of the Kummel Library of Geological Sciences (Harvard University) and the staff of the Library of the Museum of Comparative Zoology (Harvard University). I am particularly grateful to the two anonymous referees whose comments and suggestions have greatly improved this paper.

References

ATTI 1841. *Atti della Seconda Riunione degli scienziati italiani tenuta in Torino nel settembre del 1840.* Cassone e Marzorati, Turin.
—— 1845. *Atti della Sesta Riunione degli scienziati italiani tenuta in Milano nel settembre del 1844.* Pirola, Milan.
—— 1847. *Atti della Ottava Riunione degli scienziati italiani tenuta in Genova dal 14 al 24 settembre 1846.* Ferrando, Genoa.
BARTHOLOMEW, M. 1976. The non-progress of non-progression: two responses to Lyell's doctrine. *British Journal for the History of Science*, **9**(2), 166–174.
—— 1979. The singularity of Lyell. *History of Science*, **17**, 276–293.
BEUDANT, F. S. 1851. *Géologie*. Ouvrage adopté par le Conseil d'Instruction Publique pour l'enseignement dans les Lycées et Colléges et approuvé par Monsigneur l'Archevêque de Paris pour l'enseignement dans les établissements religieux. 5th edn. Langlois & Leclerq / Masson, Paris.
BIANCONI, G. G. 1862. Sugli studj paleontologici e geologici in Bologna. Cenni storici. *Atti della Società Italiana di Scienze Naturali*, **4**, 241–267.
BIREMBAUT, A. 1970. Boué, Ami. *In*: GILLISPIE, C. C. (ed.) *Dictionary of Scientific Biography*. Charles Scribner's Sons, New York, vol. 2, 341–342.
BISCHOF, G. 1847–1855. *Lehrbuch der chemischen und physikalischen Geologie*. 2 vols. Marcus, Bonn.
BOLTON, H. C. 1885. *A Catalogue of Scientific and Technical Periodicals (1665–1882) Together with Chronological Tables and a Library Check-List.* Smithsonian Institution, Washington.
BORK, K. B. 1990. Constant Prévost (1787–1856). The life and contributions of a French uniformitarian. *Journal of Geological Education*, **38**, 21–27.
BOUÉ, A. 1830–1831. Lecture du compte rendu des "Progrès de la Géologie". *Bulletin de la Société géologique de France*, **1**, 105–124.
—— 1832. Les Principes de la Géologie … par M. Ch. Lyell. *Mémoires Géologiques et Paléontologiques*, **1**, 315–356.
—— 1834. Résumé des Progrès des Sciences Géologiques pendant l'année 1833. *Bulletin de la Société géologique de France*, **5**, 1–518.
CANNON, W. F. 1960. The uniformitarian – catastrophist debate. *Isis*, **51**, 38–55.
—— 1976. Charles Lyell, radical actualism, and theory. *British Journal for the History of Science*, **9**(2), 104–120.
CAPELLINI, G. 1914. *Ricordi. Volume I : 1833–1860.* Zanichelli, Bologna.
CETTA, A. 1886. *Diluvio*. Fratelli Speirani, Turin.
COCCHI, I. 1866. L'uomo fossile nell'Italia centrale. *Memorie della Società Italiana di Scienze Naturali*, **2**, 3–80.
COCCHI, I. (ed.) 1871–1893. *Memorie per servire alla descrizione della carta geologica d'Italia, pubblicate a cura del R. Comitato Geologico del Regno.* 5 vols. Barbera, Florence.
CONYBEARE, W. D. 1830. Letter on Mr. Lyell's "Principles of Geology". *Philosophical Magazine*, **8**, 215–219.
—— 1830–1831. An examination of those phenomena of geology which seem to bear most directly on theoretical speculation. / On the phenomena of geology. *Philosophical Magazine*, **8**, 359–362, 402–406; **9**, 19–23, 111–116, 118–197, 258–270.
CORSI, P. 1995. The "Pisa School of geology" of the 19th century: an exercise in interpretation. *Palaeontographia Italica*, **82**, III–VIII.

COTTA, B. VON 1858. *Geologische Fragen*. Engelhardt, Freiberg.

—— 1866. *Die Geologie der Gegenwart*. Weber, Leipzig.

D'ARGENIO, B. 1996. Gli anni della maturità di Leopoldo Pilla geologo. *In*: PILLA, L. *Notizie storiche della mia vita quotidiana a cominciare dal 1 gennaio 1830 in poi*. Edited by M. Discenza, Vitmar, Venafro, XIX–XXX.

DEAN, D. R. 1980. Graham Island, Charles Lyell and the craters of elevation controversy. *Isis*, **71**, 571–588.

DODERLEIN, P. 1871. Note illustrative della carta geologica del Modenese e del Reggiano, 1870. *Memorie della Regia Accademia di Scienze, Lettere ed Arti in Modena*, **12**, 5–114.

ELENA, A. 1988. The imaginary Lyellian revolution. *Earth Sciences History*, **7**(2), 126–133.

FRÄNGSMYR, T. 1976. The geological ideas of J. J. Berzelius. *British Journal for the History of Science*, **9**(2), 228–236.

GASCOIGNE, R. M. 1985. *A Historical Catalogue of Scientific Periodicals, 1665–1900, with a Survey of Their Development*. Garland Publishing, New York & London.

GEMMELLARO, C. 1862. Sommi capi di una storia della geologia sino a tutto il secolo XVIII pe' quali si detegge che le vere basi di questa scienza sono state fondate dagli italiani. *Atti dell'Accademia Gioenia di scienze naturali*, ser. 2, **18**, 5–40.

GILLISPIE, C. C. [1951] 1996. *Genesis and Geology. A Study in the Relations of Scientific Thought, Natural Theology, and Social Opinion in Great Britain, 1790–1850*. Harvard University Press, Cambridge, MA & London.

GOHAU, G. 1995. Constant Prévost (1787–1856), géologue critique. *Mémories de la Société géologique de France*, **168**, 77–82.

GREENE, M. T. 1982. *Geology in the Nineteenth Century. Changing Views of a Changing World*. Cornell University Press, Ithaca & London.

GUNTAU, M. 1975. *Charles Lyell's Influence on the Development of Geological Thought in Germany*. Paper read at the Charles Lyell Centenary Symposium, 6th INHIGEO Symposium, London, 1–5 September. Abstract in IUGS/INHIGEO *Charles Lyell Centenary Symposium. Programme & Abstracts*. Roberts, Cardiff, 24–25.

HALLAM, A. 1983. *Great Geological Controversies*. Oxford University Press, Oxford.

HOFF, K. E. A. VON 1834. *Geschichte der durch Ueberlieferung nachgewiesenen natürlichen Veränderungen der Erdoberfläche*. Perthes, Gotha, vol. 3.

HOFFMANN, F. 1838. *Geschichte der Geognosie und Schilderung der vulkanischen ercheinungen*. Nicolaischen Buchhandlung, Berlin.

HOOYKAAS, R. 1959. *Natural Law and Divine Miracle: A Historical-Critical Study of the Principle of Uniformity in Geology, Biology, and Theology*. Brill, Leiden.

HUMBOLDT, A. VON 1845. *Kosmos. Entwurf einer physischen Weltbeschreibung*. Cotta, Stuttgart, vol. 1.

KEFERSTEIN, C. 1840. *Geschichte und Literatur der Geognosie. Ein Versuch*. Lippert, Halle.

KOCH, M. 1972. Hartmann, Carl Friedrich Alexander. *In*: GILLISPIE, C. C. (ed.) *Dictionary of Scientific Biography*. Charles Scribener's Sons, New York, vol. 6, 142–143.

LAUDAN, R. 1987. *From Mineralogy to Geology. The Foundations of a Science 1650–1830*. University of Chicago Press, Chicago.

LAURENT, G. 1987. *Paléontologie et évolution en France, 1800–1860*. Éditions du CTHS, Paris.

—— 1993. Ami Boué (1794–1881): sa vie et son oeuvre. *Travaux du Comité Français d'Histoire de la Géologie*, ser. 3, **7**(3), 19–30.

LAWRENCE, P. 1978. Charles Lyell versus the theory of central heat: a reappraisal of Lyell's place in the history of geology. *Journal of the History of Biology*, **11**(1), 101–128.

LEONHARD, K. C. VON 1835. *Lehrbuch der Geognosie und Geologie*. Schweizerbart, Stuttgart.

LYELL, C. 1830–1833. *Principles of Geology, Being an Attempt to Explain the Former Changes of the Earth's Surface, by Reference to Causes Now in Operation*. 3 vols. Murray, London.

—— 1832–1834. *Lehrbuch der Geologie. Ein Versuch, die früheren Veränderungen der Erdoberfläche durch noch jetzt wirksame Ursachen zu erklären* (trans. C. Hartmann on 2nd English edn). 3 vols. Basse, Quedlinburg & Leipzig.

—— 1835. On the proofs of a gradual rising of the land in certain parts of Sweden. *Philosophical Transactions of the Royal Society*, **125**, 1–38.

—— 1838. *Elements of Geology*. Murray, London.

—— 1839a. *Elemente der Geologie* (trans. C. Hartmann). Voigt, Weimar.

—— 1839b. *Elements de géologie*. (Traduit de l'anglais sous les auspices de M. Arago, trans. T. Meulien). Langlois & Leclerq, Paris.

—— 1840. *Sur les preuves d'une élévation graduelle du sol dans certaines parties de la Suède* (trans. P. L. A. Coulon). Jent et Gassmann, Soleure (Bulletin Bibliographique des Mémoires de la Societé des Sciences Naturelles de Neuchâtel, Vol. 1).

—— 1841–1842. *Grundsätze der Geologie oder die neuen Veränderungen der Erde und ihrer Bewohner in Beziehung zu geologischen Erläuterungen* (trans. C. Hartmann on 6th English edn). Voigt, Weimar, 3 vols.

—— 1843–48. *Principes de géologie, ou illustrations de cette science empruntées aux changements modernes que la terre et ses habitants ont subis* (Traduit de l'anglais sous les auspices de M. Arago, trans. T. Meulien on 6th English edn). 4 vols. Langlois & Leclerq, Paris.

—— 1847. *Elementos de Geologia* (trans. D. Joaquin Ezquerra del Bayo). Yenes, Madrid.

—— 1852. On tertiary strata of Belgium and French flanders. *Quarterly Journal of the Geological Society of London*, **8**, 277–371.

—— 1856. *Mémoire sur les Terrains Tertiaires de la Belgique et de la Flandre Française* (trans. C. Le Hardy de Beaulieu & A. Toilliez). Van Dooren, Brussels.

—— 1856–1857. *Manuel de géologie élémentaire ou changements anciens de la terre et de ses habitants tels qu'ils sont représentés par les monuments géologiques* (trans. M. Hugard on 5th English edn). 5th edn, 2 vols. Langlois & Leclerq, Paris.

—— 1857. *Supplément au Manuel de géologie élémentaire* (trans. M. Hugard on 10th English edn). Langlois & Leclerq, Paris.

—— 1857–1858. *Geologie oder Entwickelungsgeschichte der Erde und ihrer Bewohner* (trans. B. Cotta on 5th English edn). 2 vols. Dunker & Humblot, Berlin.

—— 1863a. *The Geological Evidences of the Antiquity of Man, with Remarks on Theories of the Origin of Species by Variation.* Murray, London.

—— 1863b. *Manuel de Géologie Élémentaire ou changements anciens de la terre et de ses habitants tels qu'ils sont représentés par les monuments géologiques* (trans. M. Hugard on 5th English edn). 6th edn. Garnier Frères, Paris.

—— 1864a. *Éléments de Géologie ou changements anciens de la terre et de ses Habitants tels qu'ils sont représentés par les monuments géologiques* (trans. J. Ginestou on 6th English edn). 6th edn, 2 vols. Garnier Frères, Paris.

—— 1864b. *L'Ancienneté de l'homme prouvée par la géologie et remarques sur les théories relatives à l'origine des espèces par variation* (trans. M. Chaper). Baillière et fils, Paris.

—— 1864c. *L'ancienneté de l'homme. Appendice par sir Charles Lyell. L'homme fossile en France. Communications faites à l'Institut (Académie des Sciences) par MM. Boucher de Perhtes, Boutin, P. Cazalis de Fondouce, Christy. J. Desnoyers, H. et Alph, Milne Edwards, H. Filhol, A. Fontan, F. Garrigou, Paul Gervais, Scipion Gras, Ed. Hebert, Ed. Lartet, Martin, Pruner-Bey, De Quatrefages, Trutat, De Vibraye.* Baillière et Fils, Paris.

—— 1867. *Das Alter des Menschengeschlechts auf der Erde und der Ursprung der Arten durch Abänderung, nebst einer Beschreibung der Eiszeit in Europa und Amerika.* (trans. L. Büchner on 3rd English edn). Thomas, Leipzig.

—— 1867–1868. *Principles of Geology or the Modern Changes of the Earth and its Inhabitants Considered as Illustrative of Geology.* 10th edn, 2 vols. Murray, London.

—— 1873. *Principes de Géologie ou illustrations de cette science empruntées aux changements modernes de la terre et de ses habitants* (trans. J. Ginestou on 10th English edn). 2 vols. Garnier Frères, Paris.

—— 1875. *Abrégé des Éléments de Géologie* (trans. J. Ginestou). Garnier Frères, Paris.

LYELL, K. (ed.) 1881. *Life, Letters and Journals of Sir Charles Lyell.* 2 vols. Murray, London.

MASLANKIEWICZ, K. & WOJCIK, Z. 1975. Charles Lyell's ideas in Poland in the 19th century. *In*: IUGS/INHIGEO *Charles Lyell Centenary Symposium. Programme & Abstracts.* Roberts, Cardiff, 27–28.

MORRELL, J. B. 1976. London institutions and Lyell's career: 1820–1841. *The British Journal for the History of Science,* **9**(2), 132–146.

MORELLO, N. 1983. La geologia nei congressi degli scienziati italiani, 1839–1875. *In*: PANCALDI, G. (ed.) *I congressi degli scienziati italiani nell'età del positivismo.* CLUEB, Bologna, 67–81.

NAUMANN, C. F. 1850–1866. *Lehrbuch der Geognosie.* 3 vols. Engelmann, Leipzig.

OMALIUS D'HALLOY, J. J. D' 1835. *Élements de Géologie, ou, seconde partie des éléments d'histoire naturelle inorganique.* 2nd edn. Levrault, Paris.

—— 1843. *Précis élémentaire de Géologie.* Bertrand, Paris.

OMBONI, G. 1894. *Brevi cenni sulla storia della geologia.* Sacchetto, Padova.

ORBIGNY, A. D' 1852. *Cours Élémentaire de Paléontologie et de Géologie Stratigraphiques.* Masson, Paris, vol. 2.

ORDAZ, J. 1976. The first Spanish translation of Lyell's "Elements of Geology". *British Journal for the History of Science,* **9**(2), 237–240.

ORSINI, A. & SPADA LAVINI, A. 1845. Note sur la constitution géologique de l'Italie centrale. *Bulletin de la Société géologiques de France,* ser. 2, **2**, 408–411.

—— & —— 1855. Quelques observations géologiques sur les Apennins de l'Italie centrale. *Bulletin de la Société géologiques de France,* ser. 2, **12**, 1144–1172.

PAGE, L. E. 1976. The rivalry between Charles Lyell and Roderick Murchison. *British Journal for the History of Science,* **9**(2), 156–165.

PANTANELLI, A. 1869. Relazione annuale del Museo Mineralogico. *Atti dell'Accademia dei Fisiocritici di Siena,* ser. 2, **5**, 55–58.

PAROTTO, M. & PRATURLON, A. 1984. Duecento anni di ricerche geologiche nell'Italia centrale. *In*: Società Geologica Italiana *Cento anni di geologia italiana.* Pitagora, Bologna, 241–278.

PILLA, L. 1847. *Trattato di geologia. Diretto specialmente a fare un confronto tra la struttura fisica del settentrione e del mezzogiorno di Europa.* Vannucchi, Pisa., vol. 1.

PORTER, R. 1976. Charles Lyell and the principles of the history of geology. *British Journal for the History of Science,* **9**(2), 91–103.

RUDWICK, M. J. S. 1967. A critique to uniformitarian geology: a letter from W.D. Conybeare to Charles Lyell, 1841. *Proceedings of the American Philosophical Society,* **111**, 272–287.

—— 1975. Charles Lyell, F.R.S. (1797–1875) and his London lectures on geology, 1832–33. *Notes and Records of the Royal Society of London,* **29**(2), 231–263.

—— 1986. International arenas of geological debate in the early nineteenth century. *Earth Sciences History,* **5**(1), 152–158.

—— 1990. Introduction. *In*: LYELL, C. *Principles of Geology.* Facsimile reprint of 1st edn, vol. 1. University of Chicago Press, Chicago, v–lviii.

RUSE, M. 1976. Charles Lyell and the philosophers of science. *British Journal for the History of Science,* **9**(2), 121–131.

—— 1979. *The Darwinian Revolution.* University of Chicago Press, Chicago & London.

SCROPE, G. P. 1830. 'Principles of Geology' … by Charles Lyell. *Quarterly Review,* **43**, 411–469.

STOPPANI, A. 1881. Priorità e preminenza degli italiani negli studi geologici. *In*: STOPPANI, A. *Trovanti.* Agnelli, Milan, 87–125.

STUDER, B. 1844. *Lehrbuch der Physikalischen Geographie und Geologie*. Dalp, Bern, Chur & Leipzig.

VOGT, C. 1846–1847. *Lehrbuch der Geologie und Petrefactenkunde. Zum Gebrauche bei Vorlesungen und zum Selbstunterrichte. Theilweise nach L. Elie de Beaumont's Vorlesungen an der Ecole des Mines.* 2 vols. Vieweg und Sohn, Braunschweig.

WAGENBRETH, O. 1992. Bernhard von Cotta's "Law of development of the Earth". *In*: NAUMANN, B., PLANK, F. & HOFBAUER, G. (eds) *Language and Earth. Elective Affinities Between the Emerging Sciences of Linguistics and Geology.* John Benjamins, Amsterdam & Philadelphia, 357–367.

WHEWELL, W. 1831. "Principles of Geology" … by Charles Lyell. *British Critic, Quarterly Theological Review and Ecclesiastical Record*, **9**, 180–206.

WILSON, L. G. 1972. *Charles Lyell. The Years to 1841: A Revolution in Geology*. Yale University Press, New Haven & London.

—— 1996. Brixham Cave and Sir Charles Lyell's… the *Antiquity of Man*: the roots of Hugh Falconer's attack on Lyell. *Archives of Natural History*, **23**(1), 79–97.

ZEITSCHRIFT 1856–1858. *Zeitschrift der Deutschen geologischen Gesellschaft*. Hertz, Berlin, viii–ix.

Charles Lyell's debt to North America: his lectures and travels from 1841 to 1853

ROBERT H. DOTT, JR

Department of Geology and Geophysics, University of Wisconsin,
Madison, WI 53706, USA

Abstract: Charles and Mary Lyell visited North America four times from 1841 to 1853. During the first three visits, Lyell lectured and they travelled widely. The lectures were great successes with the American public in spite of Lyell's poor elocutionary skills. Comparisons of lecture topics covered over a decade provide insights into the rapid development of the young science of geology and of Lyell's changing preoccupations. Included were crustal movements; uniformity of causes through time; coral reefs and oceanic subsidence; Carboniferous conditions; the early appearance of reptiles; palaeoclimate; the submergence of land and the origin of drift; biogeography; and a uniform organic plan with arguments against transmutation and historical progression.

The Lyells saw more of the United States and Canada than had most citizens. Their conveyances included horseback, stagecoach, train, railroad handcar, and steamboat. Accommodation varied from elegant hostelries to dirt-floored shacks. Their travels took them from the Atlantic coast to the Ohio and Mississippi Rivers and from the St Lawrence valley to the Gulf Coast. America gained from Lyell the enhancement of the stature of geology, several original contributions to the understanding of its geology, and a positive impression of the New World conveyed back to Britain. Lyell's reward was greater, however, thanks to the generosity of American guidance in his field work. He gained many fresh geological examples for his books, published two travel journals and more than 30 journal articles and presented at least eight lectures in Britain about American geology, making him the local British authority on America. He especially gained new evidence for his opposition to biological progression and for his submergence theory for the drift. Besides enhancing his geological reputation, Lyell also made an incisive comparison of British and American education, which drew him into public reforms at home sponsored by Prince Albert. Finally, his exposure to American Unitarianism, coupled with disenchantment with the Anglican Church, caused Lyell to shift his religious allegiance.

For many years, a myth prevailed that Charles Lyell came in 1841 like a crusading white knight and rescued North American geology from the twin pagan heresies of neptunism and catastrophism. The reality is that geology was already well developed and becoming professionalized in America (Carozzi 1990; Newell 1993). At least half a dozen British émigré geologists had preceded Lyell and an equal number of American-born geologists had spent time in Europe, several of the latter having met Lyell before his first visit. Benjamin Silliman's *American Journal of Science*, established in 1818, carried accounts of important European developments and reprinted *in toto* numerous foreign articles, including some by Lyell. Several years had elapsed since the first publication of Lyell's *Principles of Geology* in 1830–1833 and Elements of Geology in 1838. A number of books on geology, including unauthorized editions of both *Principles* and *Elements*, had already been pub-

lished or republished in America. Consequently, a full spectrum of geological opinions was well entrenched, which was no more Wernerian or catastrophist than prevailed in Europe. Nonetheless, the visit of one of Britain's most prominent geologists was an important event. The status of American geology could only be enhanced by such a visit and local practitioners were anxious to show off their accomplishments, which they did by escorting Lyell to the field and by having him participate in the third annual meeting of the American Association of Geologists at Boston in April 1842 (American Association of Geologists 1843, p. 42–76). Lyell was duly impressed with the quality of American work, and both he and Mary Lyell were captivated by the great energy and friendliness of the egalitarian Americans. Lyell summarized their feelings nicely when he told the audience at the close of his 1842 New York lectures that 'I shall always look back to the time spent here

DOTT, R. H. JR 1998. Charles Lyell's debt to North America: his lectures and travels from 1841 to 1853. *In*: BLUNDELL, D. J. & SCOTT, A. C. (eds) *Lyell: the Past is the Key to the Present.* Geological Society, London, Special Publications, **143**, 53–69.

with a *home-feeling*, which will always make it difficult to regard America as a foreign land.' (Raymond 1842–1843, p. 52) (Note 1 Appendix).

It seems clear, however, that ultimately Lyell gained more from his four visits to North America between 1841 and 1853 than did his New World hosts. He came with a definite agenda and with London as his chief audience. First, with his chosen vocation of publishing for profit, he recognized the potential of the American market and the need to negotiate contracts for the publication of his books there in order to stop the plagiarism that had already occurred. Second, he could exploit the great popularity of exploration travel journals by publishing popular accounts of his experiences in America. Third, he could communicate the results of his American geological investigations through letters to the Geological Society of London, where they would be read orally and published promptly. Finally, upon returning home, he could publish full journal articles about American geology. That he succeeded spectacularly is attested by contemporary London newspaper and scientific journal accounts of his travels, the addition to his publication list of two travel books (Lyell 1845, 1849), and more than 33 published letters and articles (Skinner 1978). In addition, he obtained countless examples of American geological phenomena to embellish his books and to support his position on such important issues as the theory of marine submergence and floating ice for drift, his anti-progressionist and steady-state view of earth history, and his scheme for subdividing the Tertiary. Indeed, he was so successful that he quickly became the local London authority on North America, thus enhancing his already substantial reputation and increasing the intensity of his rivalry with Roderick Murchison (Page 1976). (Note 2 Appendix).

The present article provides a condensed account from my more extended treatment of Lyell's three series of lectures and associated travels in America over an 11 year period during a very important time in the history of geology (Dott 1996). Lyell's lectures and his American writings provide valuable insight into the development of his thinking; they also provide evidence of his interactions with the American community, which were not entirely amicable. After Lyell's first seven months in America, during which he had been guided by more than a dozen of the country's leading geologists and naturalists, his hosts became apprehensive that this guest would publish their data before they could do so (Newell 1993; Silliman 1995). Discussions of the American visits have appeared before in Lyell's own travel journals, his published correspondence (Lyell 1881), biographies (Bonney 1895; Bailey 1963; Wilson 1998) and other writings (Brice 1978,

1981; Dott 1996). Because there was much feedback between Lyell's American geological travels and the lectures, I felt it important to study them together. This exercise has illuminated the substantial debt to America that Lyell accumulated for important evidence to support many of his ideas, especially anti-progression and the origin of drift.

The lectures

The Boston Lowell Lectures were endowed in the will of John Lowell, Jr in 1836 to provide free lectures to benefit the people of Boston (Weeks 1966). Public lectures were then very popular in America, where self improvement enjoyed higher status than the passive entertainment of the theatre, especially in Boston. They were also a perfect vehicle for popularizing science. Benjamin Silliman had opened the first lecture season in 1840 in grand style and apparently suggested Lyell to inaugurate the second season. Accordingly, Charles and his charming wife, Mary Horner Lyell, arrived in Halifax on 31 July 1841 for a $12^{1}/_{2}$ month visit to the United States and Canada, the beginning of a great adventure in the New World which would span 12 years. The *American Journal of Science* in August 1841 carried an enthusiastic notice of the Lyell's arrival, the scheduled course of lectures at the Lowell Institute, and their subsequent itinerary. Three Boston newspapers also carried advertisements of the 12 lectures on geology, which were presented from 19 October to 27 November 1841 (Dott 1996). The audience averaged about 3000 enthusiastic and attentive 'persons of both sexes, of every station in society ... all well dressed and observing the utmost decorum' (Lyell 1845, p. 86). Lyell was delighted to receive US $2000 (more than US $30,000 in today's dollars), which was three times the going fee in London for the most reknowned lecturers. This, together with smaller fees for repeat offerings of the lectures in Philadelphia and New York, would finance the Lyells' extensive travels in America (Fig. 1).

Lyell's Lowell Lectures of 1841, 1845 and 1852 were all poorly reported by the Boston newspapers. By contrast, the 1842 repeat lectures in Philadelphia and New York were thoroughly reported, with the New York series (condensed to eight lectures) being published verbatim as a pamphlet with 16 illustrations taken from Lyell's *Principles of Geology* (1830–1833) and *Elements of Geology* (1838, 1841) (Raymond 1842–1843). Lyell's surviving handwritten notes for many lectures and Mary Lyell's correspondence from America provide additional evidence about the content of this first lecture series and they are the principal sources for the 1845 and 1852 series, for

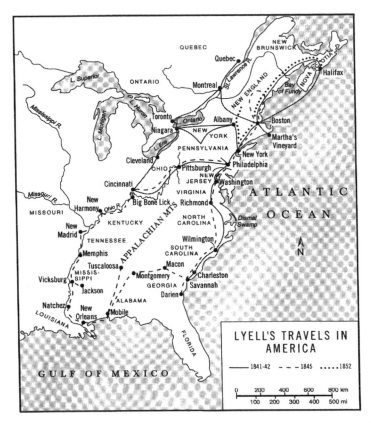

Fig. 1. Map of travels in America by Charles and Mary Lyell between 1841 and 1852, showing their routes and major locations visited.

which newspaper coverage was almost nil. Table 1 summarizes the topics of the three American lecture series and, for comparison, those of Lyell's 1832–1833 London lectures (Rudwick 1974–1975). Although specific topics changed from series to series, two things characterized all, namely the citation of many geological uniformities both in time and in space and an unrelenting stress upon the enormity of geologic time. The importance of the last item to Lyell is epitomized in the concluding sentence of a letter to his father-in-law in 1837: 'A generation must die off before geologists will know how to make use of an ample allowance of time' (Lyell 1881, p. 253).

Lyell's preparation for his American lectures was much the same as for his earlier, London ones (Rudwick 1974–1975; Dott 1996). Lyell wrote and rewrote notes for each lecture, had large diagrams painted on canvas by scene painters, and assembled specimens for display. His notes show a common pattern. First he made a numbered list of topics, then he wrote out a script for at least part of a

lecture, and, finally, he made a list of illustrations to be used; often he also made a sketch showing how he would display these. Line-outs, annotations and occasional complete rewritings of notes indicate that he experimented considerably with organization. Generally Lyell intended to begin with the description of specific, local phenomena using concrete field examples from his own wide experience, then he would extrapolate to a larger scale, and finish with an important methodological or philosophical conclusion. Accounts show, however, that he was a poor lecturer and rarely achieved this admirable goal. Instead, like many who followed later, he wandered, dwelt over long, hesitated, assumed distracting postures, often tried to cover too many points, and frequently had to carry over material to the following session. His subject matter and visual aids saved him. The 'very beautiful illustrations got up on quite a magnificent scale' (*New York Herald* 17 March 1842) consistently drew rave reviews. Young Josiah Whitney, who was soon to become a reputable geologist

Table 1. *Lyell's lecture topics*

1832–1833 London	1841–1842 USA (3 times)	1845 Boston	1852 Boston
General structure of the Earth's crust	Major rock types – Auvergne examples	Uniformity and extinctions	Changes now in progress
Types of rocks	Earth's strata	Extinct *Mastodon*	Geology of London area
Inference of stratigraphic order of chronology	Upheaval and subsidence	Fossil quadrupeds – mostly French examples	The chalk – entirely marine
Tertiary history vs. human history	Coral reefs – Darwin's theory	Cretaceous and Tertiary of US Coastal Plain	Young age of Alps – does not require catastrophe
Catastrophism vs. present causes	Coal – from plants; *in situ* origin	Coal – origin and evidence of subsidence	Mississippi River alluvium
Fossils as records of geological time	Fossil footprnts – *Cheirotherium*	Glaciers and glaciation	Carboniferous period – flora, fauna, climate
Tertiary divisions and marine vs. freshwater record in Auvergne	Silurian strata and fossils – superior, flat-laying record in America	Present and fossil land quadrupeds – geographic provinces and barriers	Reptiles with Coal Measures – Joggins bones and various footprints
Subterranean rocks – 'hypogene' coined	Recession of Niagara Falls – escarpment wave cut, but gorge cut by river	Present and fossil distributions support 'same causes' as now	Erratic boulder trains in Massachusetts
			Extinction of species
Strata of Sicily – thick but young	Glaciers and icebergs – diluvium vs. drift; glaciers form drift	Climate – factors include latitude of large lands and patterns of ocean currents	Introduction of new species
Mt Etna – still younger but ancient to humans	only in mountains, icebergs elswhere		Organic progression more apparent than real

himself, was inspired by the 1841 lectures to pursue geology as a career.

After the first presentation in Boston, the Lyells spent the winter travelling south through the Coastal Plain (Fig. 1). The *American Journal of Science* in January 1842 carried a notice of their winter itinerary and noted that the lectures would be repeated in Philadelphia. Lyell delivered 12 lectures in Philadelphia from 2 February to 12 March 1842 to an enthusiastic audience of 300–400 people, who paid US $4 each (about US $75 in today's dollars). During late March and early April, he presented the same material condensed into eight lectures for a New York audience of about 600 people, who paid US $3 each. It was the latter that were printed verbatim by the *New York Tribune* (Raymond 1842–1843), and which provide the greatest detail about any of Lyell's lectures.

In 1845 Lyell returned to present a series of 12 more Lowell Lectures from 21 October to 28 November. He had meanwhile repeated his first American lectures in London in the winter of 1843

and had lectured on American geology at York, England, in 1844. In 1852 Lyell gave a third and last series of 12 Lowell Lectures from 19 October to 26 November. The 1845 and 1852 lectures received even poorer coverage by the Boston newspapers than did those in 1841 despite Lyell's considerable increase in fame. By 1852, he had published his two American travel journals (1845 and 1849) and more than 15 journal articles and letters about American geology, and had presented four more lectures on America at the Royal Institution in London. Meanwhile, he had travelled several times to Europe, had been knighted, had been President of the Geological Society of London for a second term, and had served on royal commissions for the 1851 Great Exhibition and for the reform of Oxford and Cambridge. His praise of American education, as contrasted with the deplorable state of British education, in his 1845 travel journal caused Prince Albert to recruit him to these commissions. In 1853 the Lyells returned to America for the last time, but Charles did not lecture. This was a brief visit on the the the occasion of a New York Industrial Exhibition for which he was a Royal Commissioner representing Great Britain. Since the 1852 visit, Lyell had discussed American geology in two more journal articles (Lyell 1853; Lyell & Dawson 1853) and two more Royal Institution lectures.

Lyell's 1841–1842 lectures, like his 1832–1833 London series, used many examples from the Auvergne district of France and from southern Italy to introduce basic concepts for deciphering geologic history (Table 1). He showed connections between the present and remote past, but emphasized the profound changes that had occurred in those two regions and stressed the implications for the enormity of geologic time. Fossils as records of the past and the divisions of the Tertiary also received significant attention in both sets of lectures, as did evidence of upheaval and subsidence and their possible subterranean causes. New topics in 1841–1842 included Darwin's new theory of coral reefs (Darwin 1839), fossil footprints, the origin of coal, Silurian strata (especially in New York), Niagara Falls, as well as palaeoclimate and the origin of drift in terms of glaciers versus floating ice. Feedback from his field work in America provided much of the material for the last four topics (Dott 1996). For the 1845 and 1852 lectures, the Auvergne and Italy figured less prominently, and coral reefs, Silurian strata and Niagara Falls were dropped (Table 1). Carboniferous life, coal and paleoclimate, together with the origin of drift, continued to be highlights, and Lyell emphasized his floating ice hypothesis even more (Fig. 2). To these topics were added coverage of extinctions (especially of Cenozoic mammals), the appearance of new species, biogeography and more about palaeoclimate.

Throughout all of his lecture series, Lyell emphasized repeatedly the superiority of appeal to 'present causes' and gradualism over catastrophism for the analysis of geological phenomena. The closer the similarity between *any* ancient phenomenon and its modern counterpart the better. For example in 1841–1842 he noted the identity between the compound eyes of Silurian trilobites and those of modern arthropods, and concluded that the Silurian sea was as transparent and the solar light as brilliant as today. He also inferred from the similarity between ancient and modern ripple marks that ocean currents and waves had remained uniform. And he denied that ancient coralline limestones had formed on a 'grander scale' than their modern counterparts. Uniformity was also seen in the persistence of an organic plan, best illustrated among vertebrate animals, from earliest fish dating all the way back to the 'Silurian' (i.e. lower Palaeozoic) up to modern mammals,

"Berg with rocks...1840, lat. 66° S."

Fig. 2. Sketch of a 'Berg with rocks Jan. 1 1840 – Lat. 66° S' made by Joseph D. Hooker while a member of the James Ross Expedition to Antarctica, 1839–1841. This and some other drawings, together with written descriptions, were included with Lyell's 1845 lecture notes on glacial phenomena (Table 1) together with a list of illustrations for the lecture showing item 3 to be 'Hooker's ice berg carrying rocks'. The primary objective of the Ross expedition was to determine the position of the south magnetic pole. Young Hooker was the assistant surgeon naturalist, who later became a renowned botanist, described the flora of the Himalaya, and succeeded his father, William J. Hooker, as Director of Kew Gardens. He and Lyell were in frequent communication throughout their careers.

including even humans. In the fourth 1852 lecture topic (Table 1), Lyell clearly was rebutting Murchison's claim (1849) for catastrophic upheaval of the Alps. This was merely the latest round in his perpetual campaign against widely held views that the intensity of tectonism had changed directionally through time. His nearly life-long anti-progression stance is in evidence throughout, but it became more overt in the last two lecture series. America provided several new examples of the early appearance of fossil reptiles, which Lyell promptly enlisted in his anti-progression campaign, for the earlier their appearance the better. In his 1841–1842 lectures, he first invoked as evidence of the early appearance of reptiles some Triassic *Cheirotherium* ('hand animal') tracks, which had been discovered recently in Germany and Britain (Tresise 1993). These finds were soon reinforced by discoveries of even older evidence of Carboniferous reptiles in America (Fig. 3), which Lyell cited in his 1852 lectures (Table 1, topic 7). He also mentioned a more tenuous claim for turtle tracks reported by Logan and Owen (1851) from the Potsdam Sandstone (Cambrian) near Montreal – which are now considered to be of arthropod origin. It was not until the 1860s that Lyell begrudgingly abandoned his stubborn contention that any progress or directionality in the geological record is purely illusory, being the result of imperfections in that record.

All of Lyell's American visits pre-dated the Darwin–Wallace theory of evolution through the

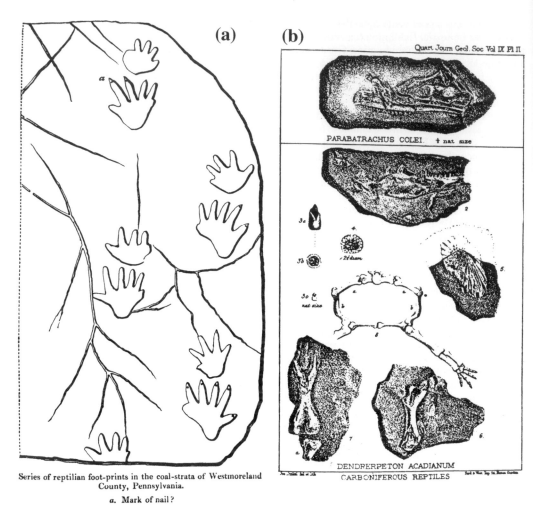

(a)

(b)

Series of reptilian foot-prints in the coal-strata of Westmoreland
County, Pennsylvania.

a. Mark of nail?

Fig. 3. (**a**) Fossil reptilian footprints from Carboniferous strata at Greensburg, Westmoreland County, in western
Pennsylvania (about 40 km east of Pittsburgh). These had been described by one 'Dr King' in 1844, so Lyell
determined to seek the locality *en rout*e from Pittsburgh to Philadelphia in April 1846, on the last leg of that year's
travels. 'The genuineness of these foot-marks was a point on which many doubts were still entertained both in Europe
and America ... [but] was of the highest theoretical importance.' (From Lyell 1849, vol. 2, p. 304 and figs 12–14, pp.
306–307; appeared later in *Elements of Geology*, e.g. Lyell 1852, 1868.) Yet another, slightly older Carboniferous
footprint locality was discovered in eastern Pennsylvania in 1849 and described by Lea (1853). (**b**) Plate from Lyell
and Dawson's 1853 paper describing the reptilian skeletal material and land mollusc fossils found in Carboniferous
coal measures at Joggins, Nova Scotia, in September 1852. For more than 100 years, vertebrates from Joggins
included the oldest known reptile fossil in the world (see Scott this volume). (From Lyell & Dawson 1853, plate II.)

process of natural selection, although other transmutational theories were already being debated. In
his lectures, Lyell dealt with the difficult question
of species by arguing vaguely that some 'unity of
plan' resulted in the continuous, piecemeal origin
of new species all over the Earth, always nicely
balanced by extinctions, thus maintaining a biological steady state. Environment, especially

climate, was seen by Lyell as the chief cause of
extinctions, extinct species having been adapted to
the environment of yesterday, whereas newly
created ones are adapted to that of today. Were a
change to bring back the environment of yesterday,
then quite possibly organisms now extinct, such as
Ichthyosaurus, would reappear. He believed that
the rate of extinction-speciation turnover differed

for different organisms, being faster for vertebrates than for invertebrates. Transmutation was difficult for Lyell to accept because of its anti-steady-state directionality, because there were no known examples of new species being formed today by 'present causes', but especially because of its implications for the human intellect.

Field work in America

As is shown in Fig. 1, during their first three visits, the Lyells travelled from the St Lawrence Valley and Nova Scotia in the north to Georgia and Louisiana in the south, and from the Mississippi and Ohio Rivers in the west to the Atlantic coast. As a result, they saw more of North America than had most people born there. Lyell had read widely and had consulted authorities for help in planning his travels. He especially wished to study Niagara Falls, to test his system for subdividing the Tertiary in the Coastal Plain, to examine the flat-lying and richly fossiliferous Lower Palaeozoic strata of New York, to compare the character and age of American coals with those of Britain, to see the structure of the Appalachian Mountains, to visit the famed Big Bone Lick, Kentucky, to see the Mississippi River and its delta, to study the earthquake effects around New Madrid, Missouri, and, wherever possible, to see drift phenomena and the remains of extinct mammals such as *Mastodon*. The St Lawrence Valley was also a must on his itinerary because Captain Henry Bayfield, who had charted the river valley and estuary for the Royal Navy, had published several articles about the region (e.g. 1840) and in 1835 had given Lyell some marine shells associated with drift near Quebec City. Besides seeing geological phenomena, however, Charles and Mary were also eager to experience a young nation that was still inventing itself. Always supportive, Mary, whom Charles' father characterized as a wife of gold, was as adventuresome as her husband, enduring cheerfully the most trying frontier accommodation and invasions of privacy. (Note 3 Appendix).

During their first visit of 1841–1842, which spanned 12½ months (Lyell 1845), Charles studied the Palaeozoic strata of New York, coal-bearing Carboniferous strata with *Stigmaria* rootlets and coal-roof floras in Pennsylvania and Nova Scotia, and multiple buried forests with upright tree stumps in Nova Scotia; he concluded that 80 per cent of the Carboniferous plants of North America were the same as European species (later reduced to about 35 per cent by Lesquereux (1879–1884)). He saw fossil footprints, ripple marks and raindrop impressions in New Red (Triassic) strata of the Connecticut Valley and New Jersey; was impressed by Appalachian structure in Pennsylvania, includ-

ing the eastward increase of intensity of both folding and metamorphism; studied Coastal Plain Cretaceous and Tertiary strata and fossils; and visited the Great Dismal Swamp, which he considered a fine analogue of Carboniferous coal swamps. Near Savannah, Georgia, Lyell saw elevated Recent shoreline deposits overlain by extinct *Mastodon* and other mammalian remains, and urged local naturalists to publish about them (Hodgson 1846); visited Ohio River terraces with their associated fossils, especially at the famous Big Bone Lick, Kentucky (Fig. 4), where a rich extinct mammalian fauna had been discovered 100 years earlier. (Note 4 Appendix). He studied Niagara Falls; visited *Mastodon* localities in New York and New Jersey, deposits of which seemed to overlie the drift, and he noted the southernmost extent of erratic boulders in Ohio and Pennsylvania; examined successive beach ridges above Lakes Erie and Ontario; and studied marine shells of Arctic species interstratified with drift along and near the St Lawrence valley (Fig. 5). Finally, he experienced the extreme tides of the Bay of Fundy region and observed on red intertidal muds such features as shrinkage cracks, raindrop impressions, bird tracks, and furrows gouged by winter shoreline ice.

During both Atlantic crossings for the second visit of 1845–1846, Lyell was elated to observe large icebergs, at least one of which carried stones, and he compared the southwesterly drift of the ice with the orientation of 'glacial furrows' in eastern

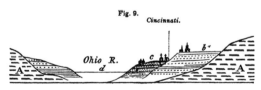

Fig. 9. *Cincinnati*.

A. Blue limestone (Lower Silurian).
b. Upper terrace.   c. Lower terrace.
e. Fossil wood and nuts, &c., found here in silt.

Fig. 4. Cross section across the Ohio River at Cincinnati, Ohio, prepared by Lyell after his 1842 visit for his first travel journal (1845, vol. 2, p. 51) and later used in *Elements of Geology* (e.g. Lyell 1852, 1868). The upper (older) river terrace deposits (b) had mammoth, *Mastodon*, and *Megalonyx* teeth and bones associated with Recent species of nonmarine molluscs. The famous Big Bone Lick (Kentucky) fossil mammal locality lies 37 km southwest of Cincinnati; Lyell inferred that it was associated with the terrace sediments shown on the left (south) side of the river (Lyell 1842–1843; 1845, vol. 2, pp. 50–60). A few erratic boulders were reported to Lyell on the upper terrace at Cincinnati, and he encountered continuous drift north of Cincinnati.

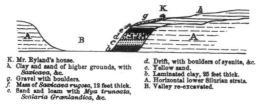

K. Mr. Eyland's house.
A. Clay and sand of higher grounds, with *Saxicava*, &c.
g. Gravel with boulders.
f. Mass of *Saxicava rugosa*, 12 feet thick.
c. Sand and loam with *Mya truncata, Scalaria Grœnlandica*, &c.

d. Drift, with boulders of syenite, &c.
c. Yellow sand.
b. Laminated clay, 25 feet thick.
A. Horizontal lower Silurian strata.
B. Valley re-excavated.

Fig. 5. Cross section across the St Lawrence River valley near Quebec City prepared by Lyell after his 1842 visit for his first travel journal (1845, vol. 2, p. 123) and later used in *Elements of Geology* (e.g. Lyell 1852, 1868). Bouldery drift (d) is interstratified with marine shelly layers containing Arctic molluscan species (c, f and h). This locality, which was first reported orally by Captain Bayfield in 1833 (published in 1840), was frequently cited by Lyell and compared to similar localities in Sweden in support of the submergence and floating ice theory for the drift. (From Lyell 1868, fig. 138, p. 164.)

Canada and New England (Lyell 1849). After landing at Halifax on 17 September 1845, the Lyells observed more marine shells interstratified with drift and 'glacial furrows' near the New England coast. After Charles had delivered the 12 Lowell Lectures, they left Boston on 3 December 1845 for warmer climes. At Richmond, Virginia, Lyell studied coal referred tentatively to the Oolitic (Jurassic) by R. C. Taylor (1835) and W. B. Rogers (1843). In 1857, the Zurich palaeobotanist Oswald Heer pronounced the plant fossils collected by

Lyell in Virginia to be Triassic, confirming a suspicion expressed by Rogers (see Heer letter in Skinner 1978). At Tuscaloosa, Alabama, Lyell determined from fossil plants that coal of uncertain age there must be Carboniferous rather than like that in Virginia. A return visit to the Georgia coast revealed two additional localities near Darien with Recent marine shell layers overlain by strata containing extinct mammalian fossils, all slightly above the present-day sea level. Crossing Georgia and Alabama by train, handcar, stage coach and steamboat, Lyell studied Cretaceous and Tertiary strata and was able to clarify their stratigraphy, but he also added a few new errors (Dott 1996). New Orleans provided further evidence of subsidence and the burial of multiple generations of forests as well as an instructive excursion to one mouth of the delta. Stopovers on the subsequent long steamboat trip up the Mississippi provided more examples of buried forests, floodplain swamps, and exposures of Tertiary strata in river bluffs capped by loess (Fig. 6). Although the origin of loess had not yet been established, Lyell recognized the similarity to loess along the Rhine Valley, and was especially interested in freshwater molluscs and extinct mammalian remains associated with the loess. He recognized that human bones assumed to have come from the loess near Natchez were probably eroded from a much younger Indian grave. At New Madrid, Missouri, earthquake effects from 1811–1812, such as sunken terrain and dead trees, were still clearly visible. *En route* across Pennsylvania, Lyell collected specimens of fossil reptilian

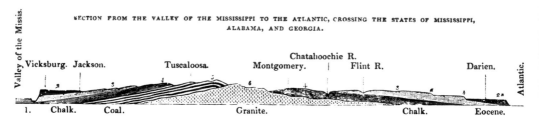

SECTION FROM THE VALLEY OF THE MISSISSIPPI TO THE ATLANTIC, CROSSING THE STATES OF MISSISSIPPI, ALABAMA, AND GEORGIA.

Length of Section 550 miles.

1.   Modern alluvium of the Mississippi.
2.   Ancient fluviatile deposit with recent shells and bones of extinct mammalia ; loess.
2*. Marine and freshwater deposit with recent sea shells and bones of extinct land animals.
3.   Eocene, or lower tertiary with Zeuglodon.   *a. b.* Terraces. Vol. i. p. 345.
4.   Cretaceous formation ; gravel, sand, and argillaceous limestone.
5.   Coal-measures of Alabama (Palæozoic).   See vol. ii. p. 80.
6.   Granite.

Fig. 6. Cross section from the Mississippi River valley at Vicksburg, Mississippi, to the Atlantic coast at Darien, Georgia. This diagram shows many of the features studied by Lyell during his 1846 travels through the southern United States; it was prepared for his second travel journal (volume and page references listed refer to that journal). A special purpose of the diagram was to illustrate the relationships of the 'Ancient fluviatile' terrace deposits with loess at Vicksburg to the elevated, interstratified 'Marine and freshwater deposit (2*)' on the Georgia coast. Both contain extinct mammalian remains and Lyell thought them to be of the same 'Post Pliocene' age. (From Lyell 1849, vol. 2, p. 262.)

footprints from Carboniferous strata (Fig. 3(a)), which were very important to him because they were older than the Triassic *Cheirotherium* footprints illustrated in his 1841–1842 lectures. Just before returning home, he was shown at Harvard University 'two magnificent skeletons, recently obtained, of the huge mastodon' (Lyell 1849, vol. 2, p. 343); one contained remains of its last supper of twigs and shoots of coniferous trees, some of which Lyell took to London for identification. After nine months in America, the Lyells sailed from Halifax for England on 1 June 1846.

During the Great Exhibition of 1851 in London, John A. Lowell engaged Lyell to return a third time in 1852 to present another set of 12 Lowell Lectures. So, on 31 August 1852, the Lyells arrived once again in Halifax and travelled in Nova Scotia and New England (Fig. 1). Lyell first went to Joggins, Nova Scotia, with J. William Dawson to restudy the remarkable succession of buried Carboniferous forests exposed there (see Scott this volume). In the stump of one fossil tree, they discovered reptilian bones (Fig. 3(b)) and terrestrial molluscs, both of which Lyell took on to Boston for confirmation of their identities. For many years, these were the oldest known fossils of their kinds. Before the Lowell Lectures, the Lyells travelled to Philadelphia, where Lyell saw fossil mammals from Nebraska. Then he was escorted by James Hall and Edward Hitchcock to revisit the famous Triassic footprints of the Connecticut Valley and to collect Tertiary fruits and seeds in Vermont. Next they studied spectacular but controversial trains of erratic boulders in southwestern Massachusetts (Lyell 1855; Silliman 1994). Lyell featured both these Richmond erratic boulder trains and the Joggins fossils prominently in his lectures a few weeks later. The Lyells returned home immediately after the lectures, arriving in Liverpool on 12 December. This concluded their geological travels in North America. During their last visit, in 1853, they ventured from New York only to pay brief social calls at Philadelphia and Boston. In later years, the Lyells continued to follow American developments through publications and correspondence. They were greatly depressed by the Civil War and feared for the fate of many of their friends in that most tragic event of American history.

Lyell's theory for the drift and a bold geomorphic synthesis for eastern North America

Contending explanations of the drift were being actively debated in Britain in the 1830s and 1840s, making this issue a high priority for Lyell when he came to America (see Lyell 1838, 1840), where he found much important evidence to support his theory for drift's origin from marine submergence and floating ice. The deluge-inspired diluvial theory had lost general favor during the 1830s, being too biblical, too catastrophic and also too neptunian, although the idea of some sort of aqueous deposition persisted. On the other hand, Agassiz's dramatic new glacial theory seemed too speculative and non-actualistic to most people. (Rudwick in 1969 suggested that Agassiz was initially motivated to explain the extinction of the diluvial fauna, and was influenced by Cuvier in seeking a revolutionary cause). A further problem was that an ice age followed by warming of climate ran counter to the prevailing directionalist paradigm. Gradual, secular cooling of the Earth had been an almost universal belief since Descartes, but Agassiz now proposed a reversal of that historical trend, so it should not be surprising that he had few supporters. Silliman (1994) has suggested that acceptance of Agassiz's glacial theory was delayed by Lyell's presumed greater professional stature. Although that social construction may be true, I believe that the greater popularity of the submergence theory is understandable without it.

As for the Americans, Timothy Conrad adopted Agassiz's glacial explanation almost immediately, while Edward Hitchcock briefly flirted with it, but then opted for an ambiguous 'glacio-aqueous' non-explanation (Carozzi 1973; Stiling 1991). Even the publication in the *American Journal of Science* of a detailed and sympathetic review of Agassiz's glacial theory by the Scotsman Charles MacLaren (1842), which contained the prescient, first-ever recognition of large-scale changes of global sea level implied by the theory, had no visible effect. Like their European counterparts, most American geologists vascillated in confusion, but leaned mostly towards submergence even after Agassiz emigrated to Boston in 1846 and catalogued abundant evidence of the widespread glaciation of North America. Not until the 1860s, thanks largely to the assimilation of new knowledge of the Greenland and Antarctic ice caps by Whittlesey (1860), did continental glaciation begin to become acceptable to Americans.

Why was the submergence theory for drift so widely accepted? First, it was virtually impossible during the 1840s to conceive of vast glaciers forming on flat terrain when nothing was yet known about the physics of ice flow. Moreover, geologists resident in the British Isles had no living glaciers at hand, but were intimately familiar with the potency of waves and ocean currents as geological agents. Indeed, they were already committed to the latter processes as the principal agents for sculpting present landscapes during a recent marine submergence, as exemplified most vividly by Lyell's theory for the denudation of the Weald by marine waves and currents (Lyell 1838; Davies 1969; Tinkler 1985). Fluvialism as yet had no place in the British school of geology. Dry valleys lacking any stream, and broad valleys with tiny or underfit streams in the Chalk Downs were cited as evidence against the efficacy of fluvial processes. Moreover, it was commonly believed (though not by Lyell) that soil and vegetation prevented rivers from accomplishing significant erosion. Finally, the presence of marine fossil shells of Arctic species in strata intimately associated with drift, and erratics in such widely scattered places as Scotland, Sweden, the St Lawrence Valley (Fig. 5) and in New England seemed to Lyell to demand that the origin of drift must be closely linked with the sea.

In 1840, Lyell had suggested the term 'drift' to replace 'diluvium' in order to divorce the debate about origin from catastrophic deluge speculations. Deluge overtones survived for some, however, with the notion of a great Arctic flood carrying countless icebergs south over Europe and North America. William Hopkins of Cambridge (1852) offered the speculation that, during submergence, the warm Gulf Stream shifted far westward over the central interior of North America and a compensatory Arctic current flowing south submerged northern Europe and brought the glacial climate there (see Smith 1989). Gradualists like Lyell argued instead for slow submergence and dispersal of drift by floating ice, which would gouge and deform sediments and scratch bedrock where icebergs ran aground (see Lyell 1840). Lyell even noted that the southwesterly flow of Atlantic icebergs parallels the direction of scratches seen on bedrock in New England and southeastern Canada. The cold climate, he explained, was caused by the concentration of continental land at high, northern latitudes during late Tertiary time.

The submergence theory seemed both more actualistic and more parsimonious than Agassiz's glacial theory because it was consistent with the accounts by Arctic and Antarctic explorers of icebergs carrying stones (Fig. 2) and because it explained both topographic features and the drift by a single, natural event resulting from observable causes now in operation (Rudwick 1969). Although Lyell may have been its most persistent advocate, submergence was also preached by most of his contemporaries (e.g. Conybeare, Greenough and Murchison). Lyell's father-in-law, Leonard Horner, discussed it at length in his presidential address to the London Geological Society (1846) and Robert Chambers (1848) published an entire book about past changes of the level of sea and land, which enthusiastically endorsed submergence. It was only

after multiple glaciations were recognized around 1890 that glacial eustatic changes of sea level began to be accepted (Meyer 1986; Dott 1992).

In 1834 Lyell had visited Sweden to assess the evidence cited by Celsius (1743) and Von Buch (1810) for the historical emergence of portions of Scandinavia from the sea. He viewed large rocks along the Gulf of Bothnia upon which Celsius had scratched sea level marks in 1731, and which had risen as much as 0.76 m above mean sea level in 103 years; anecdotal accounts by locals indicated a still greater change of about 3 m since 1500. Of equal interest to Lyell were stratified sand and gravel terraces on both sides of the Swedish peninsula now as much as 60 m above sea level and containing Arctic species of Recent marine molluscs (Lyell 1835). At most localities, these deposits were overlain by erratic boulders. This evidence from Sweden, coupled with the evidence of alternating uplift and subsidence of the crust beneath the famous Roman Temple of Serapis at Puzzuoli near Naples, which he had visited in 1828, as well as evidence recently presented by Darwin (1839, 1842, but presented orally to the Geological Society in 1837) for the subsidence of the Pacific basin and uplift of Patagonia, convinced Lyell of the importance of vertical movements of the crust. Because there is 'no human experience of a universal sea level change', he concluded that it was only the crust that could move up or down, and he never deviated from that conviction. Lyell also cited repeatedly in both lectures and publications the necessity for crustal subsidence required to explain thick successions of strata containing many levels with nonmarine and shallow marine features such as footprints, buried forests, raindrop impressions and ripple marks. 'Thus, then, do mountains rise – and oceans change their beds; and the "Atlantis" of Plato ceases to be [merely] a poetical creation' (from the 1842 Philadelphia lectures; Dott 1996, note 31).

During his first visit to North America, Lyell noted drift and erratic boulders as far south as the Susquehanna River in eastern Pennsylvania and to the Ohio River farther west. He observed marine Arctic molluscan species associated with the drift now as much as 150 m above sea level along the St Lawrence River at Quebec (Fig. 5), near Montreal, above Lake Champlain in New York and near the coast of New England. He crossed many concentric beach terraces above the present Lakes Erie and Ontario and saw widely distributed localities in northern areas that had yielded extinct *Mastodon* and other mammalian remains from bog and lake deposits overlying the drift. Along the Ohio River valley (Fig. 4), Lyell observed river terraces also containing extinct mammalian remains together with both freshwater and air-breathing molluscs.

Although unable to prove the relative age of these deposits and the drift by unambiguous field relationships, Lyell surmised that the terrace deposits were probably younger because of the presence of mammalian fossils like those that post-date the drift farther north (Lyell 1842–1843, 1845). The lower Mississippi Valley also contained terrace deposits, and the blufftops were capped with loess containing similar mammalian and terrestrial molluscan fossils (Fig. 6). In addition, the Georgia coastal terrace deposits now a few metres above sea level also contained the same mammalian fossils, but here they overlay strata containing marine molluscs like those still extant in the adjacent Atlantic Ocean (Fig. 6). Lyell had earlier noted an analogy of these Georgia fossils with similar ones reported by Darwin in Patagonia (Lyell 1845, vol. 1, p. 132; 1849, vol. 1, p. 346). As an aside, he concluded that the extinction of the large mammals had not been caused by a cold climate, as was commonly supposed, because they had continued to live after the deposition of the drift.

In Chapter 34 of his second travel journal, Lyell presented a bold geomorphic synthesis of the 'Post Pliocene' (Pleistocene and Recent) history of the entire region south of the St Lawrence and east of the Mississippi based upon the relationships summarized in the previous paragraph (1849, vol. 2, pp. 242–265). First, he attempted to compute the approximate age of the Mississippi delta and its lower alluvial plain. To accomplish this he calculated the approximate volume of alluvium using a few borings to estimate its thickness. He then determined the time necessary to deposit all of the alluvium using limited available data for the mean annual sediment load of the river. The result was 100 500 years, but Lyell acknowledged the uncertainty of the result. Next he turned his attention to the terrace deposits and capping loess of the Mississippi and Ohio Valleys. He inferred that they could be explained only by supposing, 'first, a gradual sinking down of the land after the original excavation of the valley ... and then an upheaval when the river cut deep channels through the freshwater beds [i.e. alluvium]' (Lyell 1849, vol. 2, p. 257) (Fig. 7). He then referred to the documented gradual rise of Scandinavia and to the sinking of western Greenland, and also noted the subsidence of the 'sunk country' around New Madrid, Missouri, during the great 1811–1812 earthquakes. Subsidence of a large region of the Ohio and Mississippi drainage basins would lower river gradients, causing the spilling of sediment on to the adjoining flood plains, thus raising their levels with successive layers of fluviatile silt containing air-breathing and freshwater mollusc shells and mammalian skeletons. If the region were subsequently uplifted, the rejuvenated rivers would now scour

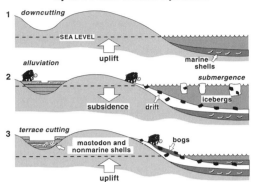

Fig. 7. Diagram illustrating Lyell's 'Post Pliocene synthesis for eastern North America as discussed in Chapter 34 of his second travel journal (1849, vol. 2, pp. 242–265). (1) Regional uplift of the land first caused downcutting throughout the Mississippi and Ohio river systems while marine deposition occurred along the Atlantic shore. (2) Subsidence of the crust then lowered river gradients, causing them to deposit much alluvium in their valleys; large mammals such as *Mastodon* and *Megalonyx* roamed over the alluvial plains, the uplands and the Atlantic coastal plain. Meanwhile, marine submergence flooded most northern areas and the Atlantic margin of southern areas (e.g. Georgia). During submergence, floating ice deposited the drift and erratic boulders over all of the region north of the Ohio and Susquehanna Rivers. 3. Renewed uplift caused the sea to retreat from the north and east and rivers to cut into their own alluvium, creating fluvial terraces. In the north, bogs developed upon the drift, and mammals such as *Mastodon* became mired in them.

out some alluvium and deepen their channels to produce the terraces seen fringing the valleys today. Moreover, Lyell argued, the same series of movements could account also for the present position of marine strata of comparatively modern date forming the Georgia coastal terraces containing extinct mammalian fossils overlying marine shells; similar circumstances had been reported to him for the Texas coast as well (Lyell 1849, vol. 2, pp. 261–262).

At the end of Chapter 34, Lyell extrapolated to the northern, drift-covered regions. The subsidence inferred for the more southerly United States could also account for the marine 'submergence of the great part of the continent drained by the headwaters of the Mississippi, Missouri, and their northern tributaries'. (Lyell 1849, vol. 2, p. 264), which he had discussed in greater detail in his first travel journal (Lyell 1845, Chapters 2 and 19). In the latter, he had postulated the cutting of escarpments such as at Niagara by marine waves and currents as well as the grooving and polishing of the bedrock and the deposition of the northern drift by floating ice. As the crust was raised and the sea retreated, marine beach terraces were formed around the lower two Great Lakes, along the St Lawrence Valley, and around Lake Champlain. He considered it 'a mere local accident' that no marine shells were associated with the terraces around Lakes Erie and Ontario in contrast with the latter localities (Lyell 1845, vol. 2, p. 81). During and after the subsequent uplift of eastern North America, roaming *Mastodon* and other mammals became mired in bogs that had formed upon the drift in the north, as well as at Big Bone Lick and on the southern river and coastal terraces (Fig. 7). Assuming a subsidence of at least 200 feet at a rate of 2.5 feet per century, based upon Scandinavia, Lyell calculated 16 000 years for subsidence and deposition, or 32 000 years for the complete subsidence–uplift cycle. This result implies further that the maximum areal deposition of the drift would have occurred approximately 15 000–16 000 years ago.

The discussion of the age of the Mississippi delta and alluvial plain from the first part of Chapter 34 was the subject of a lecture by Lyell to the British Association in 1846, but only an abstract was published (Lyell 1847a). This interesting synthesis of terraces, drift and fossils due to postulated subsidence and uplift was never published elsewhere. Curiously, it reflects an early idea for the origin of Rhine Valley loess, which Lyell had abandoned before 1841, in which he postulated that subsidence and submergence of the lower Rhine had caused alluviation followed by uplift and partial erosion of the loess (Lyell 1841, Chapter 11). Lyell rarely missed a chance to republish a good idea, so we must infer that he felt that his American synthesis was too speculative and too fraught with tenuous assumptions. In any case, it represents a fascinating attempt at explanation and illustrates Lyell's instinct and talent for synthesis. Inspired thoughts proven wrong are inspired nonetheless, and this example serves to remind us that every favorite paradigm – no matter how powerful or of whatever time, including our own – is susceptible to extinction.

Conclusions

The Lyells' visits to North America were very positive for both hosts and visitors. The Americans were stimulated in spite of Charles' inadequacies as a public lecturer, which were mostly offset by his authoritative command of the subject, his enthusiasm for geology and his 'quite magnificent' large illustrations. Although the public was more enthralled by Lyell's lectures than were American

scientists, the latter profited from the prominence that his visits gave to geology and from the insights he could share with them from his wide experiences in Europe.

Although at first flattered and excited by Lyell's visit in 1841, within six months his insatiable acquisitiveness for their data had led American geologists to regard him as a threat. Lyell was open about his intention to publish a journal about his travels and to include a geological map of the United States and Canada, which of necessity would have to be compiled from his hosts' maps. Also he intended to incorporate much American geology in later editions of his *Principles of Geology* and *Elements of Geology*. Assurances that none of these would appear before his hosts had already published their work provided little consolation. The perceived threat, coupled with resentment that Lyell had already mailed several communications to London without the expected American co-authorship, led to paranoia. The growing apprehensions reached the boiling point when an anonymous letter was published in Boston on 20 March 1842, charging Lyell with geological piracy (Clarke 1921; Newell 1993; Silliman 1995). Its author was none other than James Hall, who had been Lyell's first and most generous New World guide. Appearing just before the Boston meeting of the American Association of Geologists, the letter cast a tense atmosphere over the proceedings, and Hall was mortified at his own impetuous action. For his part, Lyell participated in the discussions as though nothing had happened, and the friendship with Hall recovered before his subsequent visits. In reality, this altercation was but one of several feuds over priority among various Americans during the 1830s and 1840s and may tell us more about his hosts than about Lyell (Dott 1996, note 43).

Lyell gained even more than the Americans from his four visits. To be sure, he made several significant original contributions to American geology, including the demonstration of the trans-Atlantic biogeographic homogeneity of Palaeozoic faunas and floras in contrast with the greater provinciality of their Cretaceous and Tertiary counterparts; the discovery with J. William Dawson of the oldest known terrestrial reptilian and molluscan fossils at Joggins, Nova Scotia; clarification of several stratigraphic uncertainties in the Coastal Plain; the discovery with Hall of a drift-filled river channel at Niagara; recognition of Arctic species of marine molluscs interstratified with drift north of Cape Cod, as in Scandinavia; and the suggested relationship between Ohio and Mississippi river terraces and Georgia coastal terraces, the northern drift and the extinct mammals associated with all of these. These contributions, however, seem overshadowed by the great treasure of American geological examples that embellished the later editions of Lyell's books. The seventh edition of the *Principles of Geology* (1847b) and the fourth edition of the *Elements*, now renamed *A Manual of Elementary Geology* (1852), prominently featured many North American examples; Figures 3(a), 4 and 5 of this paper are examples. North American phenomena also graced the *Antiquity of Man* (1863) and enabled him to add to his output the two travel journals, more than 33 journal articles and at least eight lectures in Britain about American geology. Of greatest importance for Lyell, however, was the evidence that America provided him against historical progression and in favor of the submergence theory for the drift.

The most important new evidences for Lyell's campaign against organic progression included the Carboniferous footprints in Pennsylvania and, especially, the actual skeletal remains of Carboniferous reptiles from Nova Scotia (Fig. 3). These greatly extended the known stratigraphic range of reptiles, which Lyell welcomed in support of his contention that imperfections in the geological record deluded one by making it only *seem* that different phyla appeared in a directional and progressive fashion rather than in a steady-state manner with new, piecemeal appearances always being balanced by extinctions. As for the drift question, the widespread marine Arctic molluscan fauna interstratified with drift in Canada and New England seemed to support Lyell's passion for marine submergence. His own observation of Atlantic icebergs containing large stones reinforced his commitment to floating ice as the mechanism for dispersal of most drift; this commitment was underscored by the parallelism of the southward drift of modern icebergs with the striations on bedrock beneath the drift across New England and Nova Scotia. His linkage of submergence and deposition of drift in the north with the formation of the Ohio and Mississippi river terraces and Georgia coastal terraces was a masterpiece of geomorphic synthesis when viewed in terms of knowledge then existing (Fig. 7). With the exception of James Dwight Dana, no American was attempting such grand-scale integration. (Note 5 Appendix).

Ironically, the New World probably helped Lyell hold out overlong against rapidly mounting evidence in favour of both the transmutation and glacial theories. By 1860, he acknowledged that he had lost the anti-progression campaign (Bartholomew 1976; Benton 1982), but his conversion was reluctant and qualified because of the apparent descent of humans from apes with dire implications for the human soul and intellect. As for the drift question, Lyell came to allow small 'encrustations of ice' upon Scotland, Wales and New England, but never gave up submergence and floating ice as the

principal agency for deposition of drift (see Lyell 1863, 1868). Clearly, as Leonard Wilson has observed, Lyell had an unusual 'capacity for suspended judgement in the face of seemingly overwhelming evidence' (Wilson 1971, p. 55).

More subtle, but no less important, was America's effect upon Lyell's metaphysical thinking. Charles Lyell was considered a liberal thinker for his time, and Charles Darwin assures us that Lyell was liberal in his religious belief but was nonetheless a strong theist (Darwin 1887, pp. 100–101), which contributed to his great difficulty in accepting his friend's theory of evolution. Many passages in his lectures reveal Lyell's theism and belief in design in nature; the following two examples from the 1842 New York lectures are typical:

We have been able to prove that beings lived, called by the Creator into existence on this planet–to display the beautiful and perfect harmony of the Universe–to show that all is modeled on one plan; that different as are the various genera that have lived...Geology shows that all things are the work of one Intelligence–One Mind–all links of one chain: that the Earth must have been admirably fitted for successive states which were to endure for ages. Thus do we learn to admire the variety and beauty of design displayed when we find traces and signs of the same design, the same unity of plan, the same harmony of wisdom through so vast a series as has been established by the Infinite and Eternal Creative power. (Raymond 1842–1843, lec. III, p. 25).

When I speak of the laws that govern these successive changes, I am aware that some persons look upon the idea, that such revolutions in the organic world are governed by laws, as favoring the doctrines of the Materialist. But we may regard these laws–when given to either the animate or inanimate world–as exponents fully to express the will of the Supreme Being. A law cannot be referred to any thing material; it belongs not to the material creation; it is higher than the world of matter–it is spiritual in its essence and leads us up to the comtemplation of one Immaterial and Spiritual Lawgiver. (Raymond 1842–1843, lec. VI, p. 42)

The biographer Bonney (1895) reported that Lyell became disenchanted with the authoritarianism of the Anglican Church, and this is borne out in Lyell's correspondence (1881). In America, Charles and Mary sampled a wide variety of denominations, including several African American parishes. He was particularly taken with the American Unitarians, and sought to hear their most distinguished ministers preach, including William Ellery Channing and Theodore Parker. He read Channing's writings and was delighted to meet him in 1841. After their American visits, the Lyells began attending a Unitarian chapel in London (Bonney 1895, pp. 212–213). Although they never officially renounced their Anglican roots, clearly they now found liberal Unitarianism more to their liking. So Charles Lyell was indebted to America for the consolidation of his theological as well as his geological views.

This article is a condensation from a larger study of Lyell's lectures and field work in America (Dott 1996), but this one presents a fuller discussion of the 'Post Pliocene' geomorphic synthesis. Unwittingly I was drawn into the Lyell history industry by Gordon Y. Craig, who in 1990 presented me with a copy of handwritten notes by Lyell for one of the 1842 lectures in Philadelphia. This piqued my curiosity, and subsequently Leonard Wilson, Hugh Torrens and Robert Silliman encouraged me to investigate all of Lyell's American lectures. Wilson was most generous in making available his extensive files of Lyell's lecture notes and Mary Lyell's unpublished correspondence from America. Earle Spamer and Carol Spawn of the Academy of Natural Sciences of Philadelphia discovered hitherto unknown newspaper accounts of the Philadelphia lectures. In addition to Wilson, Torrens and Silliman, many people have offered helpful criticisms and clues to additional reference materials. These include Martin Rudwick, Mott Greene, William Brice, Earle Spamer and others. I am deeply grateful to all of them, as well as to Derek Blundell and Andrew C. Scott, who invited my participation in the Bicentennial Hutton–Lyell Conference. Leonard Wilson and Andrew Scott provided helpful criticism of the manuscript.

References

AMERICAN ASSOCIATION OF GEOLOGISTS 1843. *Reports of the First, Second and Third Meeetings of the Association of American Geologists and Naturalists.* Gould, Kendall & Lincoln, Boston. (Proceedings of the 1842 meeting were also published in *American Journal of Science and Arts*, **43**, 146–184.)

BAILEY, E. 1963. *Charles Lyell.* Doubleday, Garden City.

BARTHOLOMEW, M. 1976. The non-progress of non-progression: two responses to Lyell's doctrine. *British Journal for the History of Science*, **9**, 166–174.

BAYFIELD, H. W. 1840. Notes on the geology of the north coast of the St Lawrence. *Transactions of the Geological Society of London*, ser. 2, **5**, 89–102.

BEDINI, S. A. 1985. Thomas Jefferson and American vertebrate paleontology. *Virginia Division of Mineral Resources Publication*, **61**.

BENTON, M. J. 1982. Progressionism in the 1850s: Lyell, Owen, Mantell and the Elgin fossil reptile *Leptopleuron (Telerpeton)*. *Archives of Natural History*, **11**, 123–136.

BONNEY, T. G. 1895. *Charles Lyell and Modern Geology.* Macmillan, London.

BRICE, W. R. 1978. Charles Lyell and Pennsylvania. *Pennsylvania Geology*, **9**, 6–11.

—— 1981. Charles Lyell and the geology of the Northeast. *Northeastern Geology*, **3**, 47–51.

CARROZI, A. V. 1973. Agassiz's influence on geological thinking in the Americas. *Archives des Sciences Geneve*, **27**, 5–38.

—— (ed.) 1990. Trans-Atlantic exchange of geological ideas in the 19th century (a symposium). *Earth Sciences History*, **9**, 95–162.

CELSIUS, A. 1743, Anmärkning om vatnets förminskande så i Östersion som Vesterhafvet. *Kongliga Svenska Wetenskaps Academiens Handlingar*, **4**.

CHAMBERS, R. 1848. *Ancient Sea Margins as Memorials of Changes in the Relative Level of the Sea and Land.* W. S. Orr, London.

CLARKE, J. M. 1921. *James Hall of Albany.* Privately published, Albany.

DANA, J. D. 1856. On American geological history: address before the American Association for the Advancement of Science, August, 1855. *American Journal of Science and Arts*, **22**, 305–334.

DARWIN, C. 1839. *Journal of Researches into the Natural History and Geology of Countries Visited During the Voyage of H.M.S. Beagle Round the World.* Colburn, London.

—— 1842. *The Structure and Distribution of Coral-Reefs.* Smith & Elder, London.

—— 1887. *The Autobiography of Charles Darwin, 1809–1882.* 1958 Barlow edition with omissions restored. Norton, New York.

DAVIES, G. L. 1969. *The Earth in Decay.* American Elsevier, New York.

DOTT, R. H., JR (ed.) 1992. *Eustasy: The Historical Ups and Downs of a Major Geological Concept.* Geological Society of America, Boulder, USA, Memoir 180.

—— 1996. Lyell in America – his lectures, field work, and mutual influences, 1841–1853. *Earth Sciences History*, **15**, 101–140.

—— 1997. James Dwight Dana's old tectonics – global contraction under divine direction. *American Journal of Science*, **297**, 283–311.

HAILE, N. S. 1997. The 'piddling school' of geology. *Nature*, **387**, 650.

HODGSON, W. B. 1846. *Memoir on the Megatherium and Other Extinct Gigantic Quadrupeds of the Coast of Georgia with Observations on its Geologic Features.* Bartlett & Welford, New York.

HOPKINS, W. 1852. On the causes which may have produced changes in the earth's superficial temperature. *Quarterly Journal of the Geological Society of London*, **8**, 52–92.

HORNER, L. 1846. Anniversary address. *Quarterly Journal of the Geological Society of London*, **2**, 145–221.

JEFFERSON, T. 1799. A memoir on the discovery of certain bones of a quadruped of the clawed kind in the western parts of Virginia. *Transactions of the American Philosophical Society*, **4**, 246–260.

LEA, I. 1853. On the fossil foot-marks in the red sandstone of Pottsville, Schuylkill County, Penna. *Transactions of the American Philosophical Society*, **10**, 307–317.

LEIDY, J. 1855. A memoir on the extinct sloth tribe of North America. *Smithsonian Contributions to Knowledge*, **7**.

LESQUEREAUX, L. 1879–1884. Description of the coal flora of the Carboniferous formation in Pennsylvania and throughout the United States. *Second Geological Survey of Pennsylvania*, Harrisburg.

LOGAN, W. E. & OWEN, R. 1851. On the occurrence of a track and foot-prints of an animal in the Potsdam Sandstone of Lower Canada, *Quarterly Journal of the London Geological Society*, **7**, 247–252.

LYELL, C. 1830–1833. *Principles of Geology, Being an Attempt to Explain the Former Changes of the Earth's Surface, by Reference to Causes Now in Operation*. Murray, London.

—— 1835. The Bakerian Lecture – On the proofs of a gradual rise of the land in certain parts of Sweden. *Philosophical Transactions of the Royal Society of London*, Part I, **125**, 1–38.

—— 1838. *Elements of Geology*. 1st edn. Murray, London.

—— 1840. On the boulder formation, or drift and associated freshwater deposits composing the mud-cliffs of eastern Norfolk. *London and Edinburgh Philosophical Magazine and Journal of Science*, ser. 3, **16**, 347–380.

—— 1841. *Elements of Geology*. 1st American edn. Murray, London; Hilliard & Gray, Boston.

—— 1842–1843. On the geological position of the *Mastodon giganteum* and associated fossil remains at Bigbone Lick, Kentucky, and other localities in the United States and Canada. *Proceedings of the Geological Society of London*, **4**, 36–39. (Reprinted 1844 in *American Journal of Science*, **46**, 320–323.)

—— 1845. *Travels in North America*. Wiley & Putnam, New York.

—— 1847a. On the delta and alluvial deposits of the Mississippi River, and other points in the geology of North Amerca, observed in the years 1845, 1846. *American Journal of Science*, **3**, 34–39.

—— 1847b. *Principles of Geology*. 7th edn. Murray, London.

—— 1849. *A Second Visit to the United States of North America*. Murray, London.

—— 1852. *A Manual of Elementary Geology*. Murray, London.

—— 1853. On the discovery of some fossil reptilian remains and a land-shell in the interior of an erect fossil-tree in the coal measures of Nova Scotia. *Proceedings of the Royal Institution of Great Britain*, **1**, 281–288.

—— 1855. On certain trains of erratic blocks on the western border of Massachusetts, United States. *Proceedings of the Royal Institution of Great Britain*, **2**, 86–97.

—— 1863. *Geological Evidences of the Antiquity of Man with Remarks on Theories of the Origin of Species by Variation*. Murray, London.

—— 1868, *Elements of Geology*. 6th edn. Appleton, New York.

—— & DAWSON, J. W. 1853. On the remains of a reptile (*Dendrerpeton acadianum*, Wyman and Owen) and of a land shell discovered in the interior of an erect

fossil tree in the coal measures of Nova Scotia. *Quarterly Journal of the Geological Society of London*, **9**, 58–62.

LYELL, K. M. (ed.) 1881. *Life, Letters and Journals of Sir Charles Lyell, Bart*. Murray, London.

MACLAREN, C. 1842. Review of *The Glacial Theory of Prof. Agassiz. The American Journal of Science*, **42**, 346–365.

MEYER, W. B. 1986. Delayed recognition of glacial eustasy in American science. *Journal of Geological Education*, **34**, 21–25.

MURCHISON, R. I. 1849. On the structure of the Alps, Apennines and Carpathians. *Quarterly Journal of the Geological Society of London*, **5**, 157–312.

NEWELL, J. R. 1993. *American Geologists and Their Geology: The Formation of the American Geological Community*. PhD dissertation, University of Wisconsin, Madison.

PAGE, L. E. 1976. The rivalry between Charles Lyell and Roderick Murchison. *British Journal for the History of Science*, **9**, 156–165.

RAYMOND, H. J. 1842–1843. *Lectures on Geology, Delivered at the Broadway Tabernacle in the City of New York by Charles Lyell, F.R.S. Reported for the New York Tribune*. Greeley & McElrath, New York.

ROGERS, W. B. 1843. On the age of the coal rocks of eastern Virginia. *In*: AMERICAN ASSOCIATION OF GEOLOGISTS. Reports of the First, Second and Third Meetings of the Association of American Geologists and Naturalists. Gould, Kendall & Lincoln, Boston, 298–316.

RUDWICK, M. J. S. 1969. Essay review of *Studies on Glaciers*, preceded by the *Discourse of Neuchatel* by Louis Agassiz, translated and edited by Albert V. Carozzi, New York: Hafner, 1967. *History of Science*, **8**, 136–157.

—— 1974–1975. Charles Lyell, F.R.S. (1797–1875) and his London lectures on geology, 1832–33. *Notes and Records of the Royal Society of London*, **29**, 231–263.

SCOTT, A. C. 1998. The legacy of Charles Lyell: advances in our knowledge of coal and coal-bearing strata. *This volume*.

SILLIMAN, R. H. 1994. Agassiz vs. Lyell: authority in the assessment of the diluvium-drift problem by North American geologists, with particular reference to Edward Hitchcock. *Earth Sciences History*, **13**, 180–186.

—— 1995. The hamlet affair, Charles Lyell and the North Americans. *Isis*, **86**, 541–561.

SKINNER, H. C. 1978. *Charles Lyell on North American Geology*. Arno, New York.

SMITH, C. 1989. William Hopkins and the shaping of dynamical geology: 1830–1860. *British Journal for the History of Science*, **22**, 27–52.

STILING, R. L. 1991. *The Diminishing Deluge: Noah's Flood in Nineteenth-Century American Thought*. PhD dissertation, University of Wisconsin, Madison.

TAYLOR, R. C. 1835. Review of geological phenomena, and the deductions derivable therefrom, in two hundred and fifty miles of sections in parts of Virginia and Maryland. Also notice of certain fossil acotyledonous plants in the secondary strata of

Fredericksburg. *Transactions of the Geological Society of Pennsylvania*, **1**, 314–325.

TINKLER, K. 1985. *A Short History of Geomorphology*. Barnes & Noble, Totawa.

TRESISE, G. 1993. Triassic footprints from Lymm quarry, Cheshire in the collections of Warrington Museum. *Archives of Natural History*, **20**, 307–332.

VON BUCH, L. 1810. *Riese durch Norwegen und Lappland*. Nauck, Berlin.

WEEKS, E. 1966. *The Lowells and Their Institute*. Little Brown, Boston.

WHITTLESEY, C. 1860. On the drift cavities or "potash kettles" of Wisconsin. *Proceedings of the American Association for the Advancement of Science*, 13th Meeting of 1859, 297–301.

WILSON, L. G. 1971. Sir Charles Lyell and the species question. *American Scientist*, **59**, 43–55.

—— 1998. *Lyell in America: a Transatlantic Geologist, 1841–1853*. Johns Hopkins, University Press, Baltimore (in press).

Appendix

Notes

1. The following passages from the travel journals illustrate further the impact of America upon the Lyells:

> One of the first peculiarities that must strike a foreigner in the United States is the deference paid universally to the [feminine] sex, without regard to [socio-economic] station. Women may travel alone here in stage-coaches, steam-boats, and railways, with less risk of encountering disagreeable behaviour, and of hearing coarse and unpleasant conversation, than in any country I have ever visited. ... We soon became tolerably reconciled to living so much in puiblic. Our fellow-passengers consisted for the most part of shopkeepers, artizans, and mechanics, with their families, all well-dressed, and ... polite and desirous to please. (Lyell 1845, vol. 1, p. 57)

> Instead of the ignorant wonder, very commonly expressed in out-of-the-way districts of England, France or Italy, at travellers who devote money and time to a search for fossil bones and shells, each [Alabama] planter seemed to vie with another in his anxiety to give me information as to the precise spots where organic remains had been discovered. Many were curious to learn my opinion as to the kind of animal [they had found]. (Lyell 1849, vol. 2, p. 74).

> The different stages of civilization to which families have attained, who live here [Alabama] on terms of strictest equality, is often amusing to a stranger ... sometimes, in the morning, my host would be of the humblest of 'crackers' [poor whites], or some low, illiterate German or Irish emigrants ... In the evening, I came to a neighbour whose library was well stored with works of French and English authors, and whose first question to me was, 'Pray tell me, who do you really think is the author of the *Vestiges of Creation*?' (Lyell 1849, vol. 2, p. 73).

> The factories at Lowell [Massachusetts] are not only on a great scale, but have been managed as to yield high profits, a fact which sould be impressed on the mind of every foreigner who visits them, lest, after admiring the gentility of manner and dress of the women and men employed, he should go away with the idea that he had been seeing a model mill ... (Lyell 1845, vol. 1, p. 94).

> There is no other region in Anglo-saxondom where national education has been carried so far ... [But] The fears entertained by the rich of the dangers of ignorance, is the only good result which I could discover tending to counterbalance the enormous preponderance of evil arising in the United States from so near an approach to universal suffrage. ... The political and social equality of all religious sects tends to remove the greatest stumbling block, [which is] still standing in the way of national instruction in Great Britain ... (Lyell 1845, vol. 1, pp. 95–96).

2. Although Lyell began his serious geological pursuits by accompanying Murchison to the Auvergne District of France and thence to Italy in 1828, the two came increasingly into competition thereafter. A measure of the tension between them is provided in a letter from Murchison to Henry De la Beche dated 3 April 1851, in which the writer expresses fear that Henry 'had become an inch by inch geologist and had gone over to the piddling school of Lyell, who had piddled so succesfully in England and in the United States'. (from Page 1976, p. 161). De la Beche's own long standing dislike of Lyell is evidenced by several of his clever cartoons. A recently discovered one drawn around 1828–1830 is relevant to the fluvial controversy (Haile 1997, see also Leeder this volume). It shows a small boy making a valley containing an underfit stream by piddling into it .

3. Examples include the following comments from their Mississippi River steamboat trip:

> As I was pacing the deck one passenger after another eyed my short-sight glass [monocle], suspended around my neck by a ribbon ... without leave or apology, brought their heads into close contact with mine, and looking through it, exclaimed in a disappointed and half

reproachful tone, that they could see nothing. [They must have expected a telescopic effect.] (Lyell 1849, vol. 2, p. 218).

... the wives and daughters of passengers of the same class were sitting idle in the ladies' cabin, occasionally taking my wife's embroidery out of her hand, without asking leave, and examining it with many comments, usually in a compli- mentary strain ... (Lyell 1849, vol. 2, p. 219).

We sometimes doubted how far an English party, travelling for mere amusement, would enjoy themselves. If they venture on the experiment, they had better not take with them an English maid-servant, unless they are prepared for her being transformed into an equal. (Lyell 1849, vol. 2, p. 219).

4. No less than Thomas Jefferson had described a new form from here, which he named *Megalonyx* ('great claw') and thought was carnivorous (Jefferson 1799). Just as he was submitting his manuscript for publication, however, he discovered in the British *Monthly Magazine* for September 1796 a brief account of a great clawed animal from Paraguay, which seemed nearly identical with his *Megalonyx*. That beast of Paraguay had just been studied by Georges Cuvier, who named it *Megatherium* and classed it as a relative of the sloth, thus a vegetarian. Jefferson quickly modified his manscript and called attention to the probable synonymy between the two fossils. (See Bedini 1985). After 1800, *Megalonyx* fossils were found widely throughout the southeastern United States in association with *Mastodon* and other extinct mammalian forms (Leidy 1855).

5. Dana was an early convert to the glacial theory, was friendly with Agassiz and was also a staunch fluvialist. He was sharply critical of Lyell's ultragradualism, for he considered the geologic record to be punctuated by profound tectonic and biological revolutions such as his 'Appalachian revolution' at the end of the Palaeozoic era. Dana's revolutions were by no means synonymous with the earlier catastrophist revolutions, however, being instead the results of global spasms of secular cooling and contraction of the earth (Dott 1997). Dana was also critical of Lyell's presentation of geological history 'back end foremost–like a history of England commencing with the reign of Victoria' (Dana 1856, p. 306).

Charles Lyell in New York State

GERALD M. FRIEDMAN*

Department of Geology, Brooklyn College and Graduate School of the City University of New York, Brooklyn, NY 11210, USA
*Correspondence address: Northeastern Science Foundation affiliated with Brooklyn College, Rensselaer Center of Applied Geology, 15 Third Street, PO Box 746, Troy, NY 12181-0746, USA

Abstract: In 1842 Amos Eaton (1776–1842), founder of the Rensselaer School in Troy, New York (later Rensselaer Polytechnic Institute), had become the most influential American geologist. The period between 1818 and 1836 is known as the 'Eatonian era'. In 1841, Lyell visited Eaton at Rensselaer. He states 'at Troy I visited Professor Eaton ... The mind of this pioneer in American geology was still in full activity, and his zeal unabated; but a few months after my visit he died.' In company with Eaton's former student James Hall (1811–1898), Lyell journeyed to Niagara Falls. After describing sediments and geological features near Niagara Falls, Lyell returned to his uniformitarian concepts and used the Niagara Falls as a point of departure to reiterate his message of the enormity of geological time but continuity of geological processes. Lyell was most impressed with the Helderberg Mountains. The 'Helderberg war', however, distracted him. According to a history, 'Sir Charles Lyell, ... geologizing on the Helderbergs that day, found the farmers "in a ferment." Sir Charles took time from his study of fossils embedded in the rocks of the Helderbergs to criticize (New York) Governor Seward.' In Troy Lyell reported on landslides in which people were killed. Much of Lyell's account relates to non-geological descriptions.

In his travels in North America, Charles Lyell spent much time in New York State. He visited New York on five separate occasions between 1841 and 1853. Three key areas were of interest to him: (1) the Troy–Albany Capital District, (2) Niagara Falls and (3) New York City. I shall discuss his visits to each of these areas and will begin this paper with a background on the state of geological pioneering in New York at the time of his visits.

The Troy-Albany Capital District: birthplace of geological science in America

Amos Eaton and the Van Rensselaer family

The Capital District, and Troy in particular, has the distinction of being the birthplace of the study of geological science in America during the early nineteenth century. The understanding of geology was in its infancy at that time and virtually nothing was published on the subject up until 1818. It was largely through the work of Amos Eaton (1776–1842) (Fig. 1), founder and first professor of the Rensselaer School in 1824, now the Rensselaer Polytechnic Institute, that the study of geological science in America took a giant leap forward. Indeed, Eaton was so influential during these early

years that in American geology the period 1818–1836 is known as the 'Eatonian era.' Merrill (1924) initially coined this term for the period 1820–1829, but Wells (1963) extended it to encompass the larger period. The term 'Eatonian era' pays tribute to the astonishingly effective public promotion of geology by Eaton. According to Johnson (1977), 'this rustic figure was nothing less than the one-man equivalent of an army of zealots with an all-consuming passion for science education. His greatest contribution to American geology was probably his training of an entire generation of geologists who staffed the earliest state geological surveys.' Hence Troy became known as the hallowed ground of geological pioneers. Eaton's legacy is still felt today. When the 28th International Geological Congress met in the United States in 1989 a field trip followed in Eaton's footsteps for several days. In fact part of the field trip's title was 'in the footsteps of Amos Eaton' (Rodgers *et al.* 1989).

Troy is located in Rensselaer County, New York, named after the distinguished Van Rensselaer family who established the only successful Dutch Patroonship, which thrived as a manorial estate from 1630 to the mid-1800s. One branch of the family produced Jeremiah Van Rensselaer

FRIEDMAN, G. M. 1998. Charles Lyell in New York State. *In*: BLUNDELL, D. J. & SCOTT, A. C. (eds) *Lyell: the Past is the Key to the Present*. Geological Society, London, Special Publications, **143**, 71–81.

71

Fig. 1. Amos Eaton. Courtesy Rensselaer Polytechnic Institute.

New England, and the transition and secondary ranges of Eastern and Western New York' (Barnard 1839, p. 75). He engaged Amos Eaton who completed this survey in 1823. His section extended from Boston to Lake Erie, a distance of about 550 miles.

'The crowning glory' (Barnard 1839, p. 76) resulted on 5 November 1824 in the founding of the Rensselaer School, now Rensselaer Polytechnic Institute, to which Van Rensselaer appointed two professors, a senior professor (Amos Eaton, pathfinder of North American stratigraphy) and a junior professor (Lewis C. Beck, later to be famous as State Mineralogist of New York). Beck was followed by Ebenezer Emmons (1799–1863), one of the giants of the nineteenth century American geology. By the time of and shortly after Lyell's visits, Rensselaer had furnished to the geological community more State Geologists than had been furnished in the same time by all the colleges of the Union (Johnson 1977; Friedman 1979a, b, 1981, 1983, 1989).

The importance of Stephen Van Rensselaer to the early study of geology cannot be overemphasized. Before 1830, the science was in its infancy and was being actively pursued in only few places, among them Troy, with a population under 11 000, and London, the largest city in the world. Troy's preeminence was due to Van Rensselaer's encouragement and generous sponsorship of the activities of Eaton, and his founding of the Rensselaer School

(1793–1871), a geologist who wrote one of the first geology textbooks published anywhere. Entitled *'Lectures on Geology; Being Outlines of the Science'* and published in 1825, this book preceded the textbooks of the other two 'giants of geology' from Troy, Ebenezer Emmons (1826) and Amos Eaton (1830). Stephen Van Rensselaer (1764–1839) (Fig. 2) was a twelfth-generation descendant of the original Dutch immigrant patroon. In 1819, the legislature of the State of New York elected Stephen Van Rensselaer as President of the Central Board of Agriculture. This board published two volumes on the geology of Albany and Rensselaer Counties authored by Amos Eaton. After republishing the studies on the geology of Albany and Rensselaer Counties at his own cost, he next turned his attention to a more extended scientific survey, to be carried through the entire length of the state along the line of the Erie Canal. Van Rensselaer considered the geological studies of these two counties and the Erie Canal route part of a grander scheme, a plan for a major contribution to the science of geology: 'this plan embraced a particular examination of the strata and formation of American rocks, by the survey of a transverse section, running across the great primitive ranges of

Fig. 2. Stephen Van Rensselaer. Courtesy Rensselaer Polytechnic Institute.

(now Rensselaer Polytechnic Institute), the first in America dedicated predominantly to the study of science.

Eaton guided his students on long excursions into New England, and more importantly along the newly built Erie Canal. One result of the 'Rensselaer School Flotilla' was an 1826 report titled *Van Rensselaer's Canal Survey*, which revolutionized regional geology through its introduction of precise nomenclature for the rocks of New York State (Rezneck 1957, 1959, 1965; Wells 1963; Grasso 1989; Spanagel 1996). This was the earliest account of the rock strata in the Niagara district, a subject whose study Charles Lyell continued later. Yet, despite changes in stratigraphic nomenclature by later investigators, most names of the rocks and groupings named by Eaton were adopted by New York geologists. Over a century would pass before geologists could view Eaton's achievements objectively enough to appreciate his accomplishments fully.

In 1841, Lyell (Fig. 3) visited Eaton at Rensselaer. He states (1845, pp. 66–67) 'at Troy I visited Professor Eaton...The mind of this pioneer in American geology was still in full activity, and his zeal unabated; but a few months after my visit he died.' In company with Eaton's former student James Hall (1811–1898) (Fig. 4), Lyell journeyed along the course of the Mohawk River and the Erie Canal to Niagara Falls – but more on this trip later.

To visit Eaton and the New York State Geological Survey, Lyell embarked from New York City on 16 August 1841. Lyell 'sailed in the splendid new steam-ship the *Troy*, in company with

Fig. 4. James Hall, *c.* 1832, as a faculty member of the Rensselaer School (Clarke 1923).

about 500 passengers'. Lyell states:

...when I was informed that 'seventeen of these vessels went to a mile,' it seemed incredible, but I found that in fact the deck measured 300 feet in length. To give a sufficient supply of oxygen to the anthracite, the machinery is made to work two bellows, which blow a strong current of air into the furnace. The Hudson is an arm of the sea or estuary, about twelve fathoms deep, above New York, and its waters are inhabited by a curious mixture of marine and freshwater plants and mollusca. At first on our left, or on the western bank, we had a lofty precipice of columnar basalt from 400 to 600 feet in height, called the Palisades, extremely picturesque. This basalt rests on sandstone...On arriving at the Highlands, the winding channel is closed in by steep hills of gneiss on both sides, and the vessel often holds her course as if bearing directly on the land. The stranger cannot guess in which direction he is to penetrate the rocky gorge, but he soon emerges again into a broad valley, the blue Catskill mountains appearing in the distance. The scenery deserves all the praise which has been lavished upon it, and when the passage is made in nine hours it is full of variety and contrast. (1845, pp. 15–16)

Fig. 3. Charles Lyell. A drawing by George Richmond hanging in the National Portrait Gallery in London.

James Hall, Lyell's field colleague in New York

Among the most influential alumni of Rensselaer was James Hall, the 'father of the geosyncline' and 'founder of American palaeontology'. Hall observed that the ancient marine rock strata of New York are flat lying and only a few hundred metres thick, whereas the rock strata of the same age found in the Appalachians to the east are not horizontal and are tens of thousands of meters thick. Hall hypothesized that the Appalachian rock strata had accumulated in a subsiding trough (later to be named a 'geosyncline' by James D. Dana, a student of Eaton's student Fay Edgerton), thus allowing a great thickness of strata to accumulate and, in the process, become folded (Fig. 5). Hall showed Lyell key outcrops in the field, particularly the New York Taconics and the deposits of the Appalachian Basin, on which his hypothesis was based. Lyell rejected Hall's conception of lateral compression. Hall's theory was not published until 1859 (Hall 1859), and Dana's term was introduced in 1873 for the origin of the Appalachian Mountains (Dana 1873). From the historical record it is not evident how far Hall's theory had crystallized in his mind in 1841 when Lyell rejected Hall's concepts.

Hall is alleged to have walked 220 miles from his home in Hingham, Massachusetts, to enroll and study under the great Eaton. Hall's first job at Rensselaer included white-washing one of its

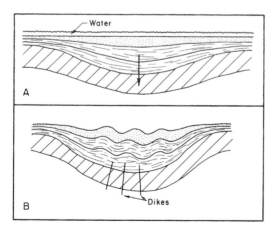

Fig. 5. Sketch of the origin of geosynclinal sediments and structures, according to the ideas of James Hall (1859, pp. 69–70). (A) The trough subsides; strata in the centre become thicker than on sides. (B) With great subsidence, the crust beneath the center of the trough fails, cracks open on the stretched part along the base of the trough, and dikes are intruded. Strata above are folded by the squeezing together of the limbs of the trough (Friedman *et al*. 1992, p. 644, fig. 17-9).

buildings and tidying up the school. Later he became librarian, and by 1835 he was listed as a full professor. Persuaded by Eaton, the New York State Legislature established the New York State Geological Survey in 1836. At the age of 26, James Hall was appointed State (Chief) Geologist of the western district of the state. He was said to be a pompous, aggressive workaholic. He browbeat (backed by the threat of an ever-present shotgun) his employees and apprentices and cajoled and threatened politicians (with his ever-present cane). Hall remained on the faculty of the Rensselaer School, and later Rensselaer Polytechnic Institute, on a part-time basis for almost 70 years.

Lyell in the field in the Capital District

In Albany, Lyell met Hall and proceeded with him into the field. Before starting his field programme, Lyell had visited the newly formed New York State Geological Survey:

> At Albany, a town finely situated on the Hudson, and the capital of the State of New York, I found several geologists employed in the Government survey, and busily engaged in forming a fine museum, to illustrate the organic remains and mineral products of the country. This state is divided into about the same number of counties as England, and is not very inferior to it in extent of territory. The legislature four years ago voted a considerable sum of money, more than 200,000 dollars, or 40,000 guineas, for exploring its Natural History and mineral structure.

In the field, Lyell noted:

> ...the resemblance of certain Silurian rocks on the banks of the Hudson river to the bituminous shales of the true Coal formation was the chief cause of the deception which misled the mining adventures of New York. I made an excursion southwards from Albany, with a party of geologists (principally James Hall), to Normanskill Creek, where there is a waterfall, to examine these black slates, containing graptolites, trilobites, and other Lower Silurian fossils. By persons ignorant of the order of superposition and of fossil remains, they might easily be mistaken for Coal Measures, especially as some small particles of anthracite, perhaps of animal origin, do actually occur in them.

Lyell wanted to see the Upper Palaeozoic section, including the:

> ...entire succession of mineral groups from the lowest Silurian up to the coal of Pennsylvania. Mr. James Hall, to whose hands the north-west division of the geological survey of New York

had been confided, kindly offered himself as my guide. Taking the railway to Schenectady, and along the Mohawk valley [Fig. 6], we first stopped at Little Falls, where we examined the gneiss and the lowest Silurian sandstone resting upon it. We then pursued our journey along the line of the Erie Canal and the Mohawk River, stopping here and there to examine quarries of limestone, and making a short detour through the beautiful valley of Cedarville in Herkimer County, where there is a fine section of the strata. (1845, pp. 18–19)

Among the most fascinating scenery and rocks of the Troy–Albany Capital District area are those of the Helderberg Mountains. So it was natural that Lyell selected this area for study (Figs 6 and 7). The Helderbergs have been called 'the key to the geology of North America' (Fig. 8).

Lyell visited the Helderberg Mountains in September 1841 and rejoiced that 'the precipitous cliffs of limestone, render this region more picturesque than is usual where the strata are undisturbed' (Lyell 1845 p. 67). The strata that are exposed illustrate the characteristics of marine sequences in Devonian carbonate rocks, including stromatoporoid reefs, spectacular karst-generated dissolution collapse features, stromatolites, erosion surfaces, and limestone as well as dolostone. These strata were formed when a depositional slope pro-graded seawards, and the surfaces bounding conformable successions are inferred to have resulted from rapid submergence (Friedman *et al.* 1992, p. 180). These examples serve now as classic case histories of modern sequence stratigraphy (Friedman 1990, 1995; Friedman *et al.* 1992, p. 180). Yet almost no geological information can be gleaned on the geology of the Helderberg Mountains from Lyell's travels or his other books that post-date his visits to North America: *Principles of Geology* (1847, 1850, 1853), *Manual of Elementary Geology* (1852, 1855), *The Student's Elements of Geology* (1871). Lyell must have hoped that an examination of the Helderberg Mountains might have given a boost to his broad views of Earth history. Yet Lyell's published information on the Helderberg Mountains relates almost entirely to the 'Helderberg war'.

At the time of his visit in 1841, the Helderberg war had recently begun. This was waged by the farmers against the Van Rensselaer family after the death of Stephen Van Rensselaer. Lyell observes:

...we found the country people in ferment, a sheriff's officer having been seriously wounded when in the act of distraining for rent, this being the third year of the 'Helderberg war', or a successful resistance by an armed tenantry to the legal demands of their landlord, Mr. Van

Rensselaer. It appears that a large amount of territory on both sides of the river Hudson, now supporting, according to some estimates, a population of 100,000 souls, had long been held in fee by the Van Rensselaer family, the tenants paying a small ground rent. This system of things is regarded by many as not only injurious, because it imposes grievous restraints upon alienation, but as unconstitutional, or contrary to the genius of their political institutions, and tending to create a sort of feudal perpetuity...the tenants...declared that they had paid rent long enough, and that it was high time that they should be owners of the land.

A few years ago, when the estates descended from the late General (Stephen) Van Rensselaer to his sons, the attempt to enforce the landlord's rights met with open opposition. The courts of law gave judgement, and the sheriff of Albany having failed to execute his process, at length took military force in 1839, but with no better success. The governor of New York was then compelled to back him with the military array of the state, about 700 men, who began the campaign at a day's notice in a severe snow storm. The tenants are said to have mustered against them 1500 strong, and the rents were still unpaid, when in the following year, 1840, the governor, courting popularity as it should seem, while condemning the recusants in his message, virtually encouraged them by recommending their case to the favourable consideration of the state, hinting at the same time at legislative remedies. The legislature, however, to their credit, refused to enact these, leaving the case to the ordinary courts of law.

The whole affair is curious, as demonstrating the impossibility of creating at present in this country a class of landed proprietors deriving their income from the letting of lands upon lease. (1845, pp. 68–70)

On his second visit to the United States in 1845, during his field work in the Helderberg Mountains, Lyell was still concerned with the Helderberg war:

I learned that 'the Helderberg war,' which I have alluded to in my former 'Travels,' is still going on, and seems as far from a termination as ever. The agricultural population throughout many populous counties have now been in arms for eight years, to resist payment of rents due to their landlords, in spite of the decisions of the courts, of law against them. Large contributions have been made toward an insurrectionary fund – one of its objects being to support a newspaper, edited by a Chartist refugee from England, in which the most dangerous anti-social doctrines are promulgated. The 'anti-renters' have not only

Fig. 6. The Helderberg escarpment where Lyell studied carbonate deposits and became witness to the Helderberg war. (**a**) Photograph of exposures of Middle Devonian (Manlius and Coeymans formations) limestone beds that Lyell studied. (**b**) Photograph of the same exposures, with the valley floor of the Mohawk River along which Lyell and Hall pursued their journey.

Fig. 7. Memorial plaque in the Helderberg Mountains recalling the research of Lyell (arrow), Hall, Eaton and other geologists, conducted between 1819 and 1850, which made this region classic ground.

set the whole militia of the state at defiance, in more than one campaign, but have actually killed a sheriff's officer, who was distraining for rent! If any thing could add to the disgrace which such proceedings reflect on the political administration of affairs in New York, it is the fact that the insurgents would probably have succumbed ere this, had they not been buoyed up by hopes of

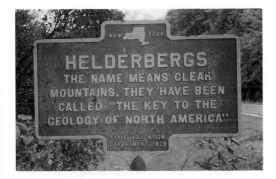

Fig. 8. Historical marker of the New York State Education Department explaining the meaning of the term 'Helderbergs'.

legislative interference in their favor, held out to them by popularity-hunting candidates for the governorship, and other official places. (1849, p. 260)

In contrast to Lyell's point of view, the American public considered the Helderberg war 'a decisive episode in the emergence of democracy' (Christman 1945) and 'the triumph of the democratic spirit over the unjustified pretensions of aristocracy'; further, it was 'the struggle of the people...against undemocratic and feudal practices' (Carmer 1945). Stephen Van Rensselaer was considered the 'good patron' and his descendants 'bastards' – giving rise to the title of 'the Bastard War'. None of the patroon's sons is said to have known humility and they had no sympathy for farmers and working people. Yet even philanthropic Stephen Van Rensselaer received an estimated US\$41 million from Rensselaerwyck (his estate), while he extended his holdings to include extensive tracts in northern New York State and considerable real estate in New York City. During his career, he was considered the richest man in the United States (Christman 1945). Stephen Van Rensselaer, whose generous support promoted the advancement of geology in the first half of the nineteenth century, became a villain in the eyes of those who supported the rights of the farmers and working people. Lyell's views expressed support for the Van Rensselaer family.

After the farmers, derisively referred to as 'so-called natives', captured a sheriff, Lyell interrupted his 'geologizing on the Helderbergs that day', and found the farmers 'in a ferment'. According to Christman (1945, p. 44), Lyell 'took time from his study of fossils embedded in the rocks of the Helderbergs to criticize Governor Seward for "courting popularity", and to praise the conservative legislature, who "to their credit" refused the tenants relief'.

Current interpretation of the events leaves Lyell far off the mark. As Christiansen (1992) notes of the Helderberg war,

...in the early years of the American nation, certain wealthy landowners lived like feudal lords in grand manor houses along the Hudson River Valley in New York State. After the American Revolution these landowners promised to sell some of their lands to the returning war veterans, on condition that the veterans-turned-farmers would clear the land and share their harvests with the landowners for seven years. For generation after generation the farmers worked the land, passing it along to their sons, and their sons' sons. Still, the landowners ignored their promises and refused to sell the land outright.

Avalanches in Troy

Lyell (1845) reported on landslides in the city of Troy, in which people were killed. Lyell does not provide his source of information, but it may be the 4 January 1837 edition of the *Albany Argus*, which under the headline 'Dreadful Calamity – Several Lives Lost' reported that in Troy

> ...an avalanche of clay came tumbling from an eminence of 500 feet, moving down the base of the hill to level land, and then continued from the impulse it received to the distance of about 800 feet, covering up acres of ground, accompanied with a cataract of water and sand, which kept up a terrible roar. The mass moved along with great rapidity, carrying with it two stables and three dwelling houses.

Lyell does not describe the geological setting for these slides, which may be noted as follows: at the end of the Pleistocene the slope between the Hudson River and the plateau on which South Troy is located formed the shelf and slope of a vanished glacial lake, now known as Lake Albany. Slippery varved lake clays mark its sediments. This kind of setting is an invitation for disaster. In a local section known as Mount Ida, an avalanche killed seven residents. At South Troy's Prospect Park the scars of avalanches are still visible today. Lyell probably inspected these same scars.

Trip to Niagra Falls

In August 1841, Lyell travelled with James Hall to Niagara Falls. On the way, he admired the mode of transport along the Mohawk River (Fig. 6(b)), where the railway 'often supported on piles' moved at the rate of 16 miles per hour. He was surprised to see, in the midst of so much unoccupied land, one flourishing town after another, including Utica, Syracuse, Auburn and Rochester. In Rochester, Hall and Lyell examined a recent find of a *Mastodon* with ivory and fossil teeth. They also noted fossil 'beach ridges', which were probably glacial features. Lyell also described tortoises and butterflies that he and Hall observed on the way.

Lyell's first near view of Niagara Falls is impressively described: 'we first came in sight of the Falls of Niagara when they were about three miles distant. The sun was shining full upon them – no building in view – nothing but the green wood, the falling water, and the white foam. At that moment they appeared to me more beautiful than I had expected' (Fig. 9). With this view in mind Lyell expounded his uniformitarian ideas: 'the Falls of Niagara teach us not merely to appreciate the power of moving water, but furnish us at the same time with data for estimating the enormous lapse of ages during which that force has operated. A deep and long ravine has been excavated, and the river has required ages to accomplish the task, yet the same region affords evidence that the sum of these ages is as nothing, and as the work of yesterday, when compared to the antecedent periods, of which there are monuments in the same district' (1845, p. 29).

A colour plate of the Falls, a coloured map of the Niagara District, a fold-out plate of the Niagara Falls from the original Utrecht Edition of 1697 as seen by Father Louis Hennepin, and a cross-section of the geology at Niagara Falls illustrate Lyell's

Fig. 9. Idealized birds-eye sketch of Niagara Falls showing the town of Queenston in the foreground, the Falls in the middle, and Lake Erie in the distance. Frontispiece in Lyell's first volume of *Travels in North America* (1845).

descriptions. Lyell and Hall confirmed the previous observations by Amos Eaton that the Niagara escarpment was not fault related.

In his discussion of the recession of the falls, Lyell takes care to credit Hall, thus: 'the geological representation of the rocks... has been introduced from the State Survey by Mr. Hall'; 'in discussing the theory of recession, I was assisted by Mr. Hall' (1845, pp. 29–30); and 'the water falls which I visited with Mr. Hall in New York' (1845, p. 34). As Sir Edward Bailey (1962, p. 146) phrased it, 'Lyell had the invaluable guidance of James Hall of New York.' Such a courtesy resulted from previous criticism. For example, a letter published in the *Boston Daily Advertizer* written by Hall under the pseudonym of 'Hamlet' had charged Lyell with geological piracy (Silliman 1995; see also Dott in this volume). In his first volume of Travels in North America (1845), Lyell inserted as a coloured frontpiece (Fig. 9) an idealized birds-eye sketch of the Falls of Niagara and adjacent country, in which he employed the names of the various geological formations Hall had named.

In his elucidation of how uniformitarianism applies to Niagara Falls, Lyell traced the approximate position of the 'first' cataract and how 'the river has been slowly eating its way backwards through the rocks for a distance of seven miles'. He concluded that the Falls had been originally nearly twice the present height and predicted correctly that they would diminish in the future. Since the 1950s, the Falls have retreated periodically with spec-

tacular collapse of the undermined carbonate strata of the Niagara escarpment (Fig. 10). These new observations are in line with those of Lyell. Lyell quotes Samuel Hooker, guide to Lyell and Hall, remarking that the 'sudden descent of huge rocky fragments...at the Horseshoe Fall, in 1828, and another at the American Fall, in 1818, are said to have shaken the adjacent country like an earthquake'. In June 1842, Lyell's interests in the Falls induced him to revisit Niagara Falls 'to re-examine diligently both sides of the river from the Falls to Lewiston and Queenston [Fig. 9], to ascertain if any other patches of the ancient river-bed had escaped destruction'. After describing and interpreting the sand, gravel, erratic blocks, ridges and terraces found at different elevations, he returned to his uniformitarian concepts and used the Niagara Falls as a point of departure to reiterate his message of the enormity of geological time but continuity of geological processes:

> But, however much we may enlarge our ideas of the time which has elapsed since the Niagara first began to drain the waters of the upper lakes, we have seen that this period was one only of a series, all belonging to the present zoological epoch; or that in which the living testaceous fauna, whether freshwater or marine, had already come into being. If such events can take place while the zoology of the earth remains almost stationary and unaltered, what ages may not be comprehended in those successive tertiary periods during which the Flora and Fauna of the globe have been almost entirely changed. Yet how subordinate a place in the long calendar of geological chronology do the successive tertiary periods themselves occupy! How much more enormous a duration must we assign to many antecedent revolutions of the earth and its inhabitants! No analogy can be found in the natural world to the immense scale of these divisions of past time, unless we contemplate the celestial spaces which have been measured by the astronomer. (1845, pp. 51–52)

New York City

Even in the urban jungle of New York City, Lyell made geological observations. However, most of his accounts of the city relate to non-geological descriptions, such as the Trinity Church, the social fabric of society, and local politics. In 1842, he gave a series of eight lectures in New York City to approximately 600 people at the Broadway Tabernacle. These lectures, which the New York Lyceum had organized, were published verbatim (see Dott in this volume). In July 1853, Lyell represented Britain at a New York Industrial Exhibition

Fig. 10. Niagara Falls before the collapse in the 1950s to 1970s of the uppermost beds of limestone. Note at the bottom of the Falls blocks from previous retreats of the escarpment. Photograph taken in June 1950.

(Dott 1996). During his travels to the United States in 1845–1846 he spent much of December 1845 and May 1846 in New York City.

The New York City's water-supply system amazed him on his visit in 1845. He noted 'we observe with pleasure the new fountains in the midst of the city supplied from the Croton water-works, finer than any which I remember to have seen in the center of a city since I was last in Rome' (1849, p. 180). The Croton reservoir in Westchester County to which Lyell refers is located about 40 miles north of New York City. Of the water pipe-line, he remarks that 'a work more akin in magnifi-cence to the ancient and modern aqueducts has not been achieved in our times...The health of the city is said to have...gained by greater cleanliness and more wholesome water for drinking...The water can be carried to the attics of houses.'

Features in New York City which Lyell admired included its fine churches. Trinity Church, now in its 300th year, was in Lyell's time already an imposing landmark (Fig. 11). A heavy snowstorm had weakened the church's structure in 1838–1839, and the building was torn down (Anon. 1996; Byard 1996). The present building, which Lyell saw on 3 December 1845, was consecrated on 21 May 1846 – five months after Lyell inspected it.

Lyell includes much data on New York – population statistics, children in school, private schools versus state public schools, libraries, budgets, etc. As a current faculty member of the City University of New York, I was interested to learn that in 1845 male teachers during the winter term earned US$14.16, whereas female teachers earned only US$7.37.

Comments

Although Lyell was fond of society and status, he impressed the local citizens as a scholar who enjoyed his visit to New York. He praised its scenery, its geology and its people. He enjoyed his field junkets with colleagues of the New York State Geological Survey, especially with James Hall. However, he had initially offended Hall, who under the pseudonym 'Hamlet' written in the *Boston Daily Advertiser* 'we raise our voice against (Lyell's) kind of piracy' (Silliman 1995). Lyell's career was in publishing his *Principles of Geology,* and Hall and others felt that he pumped state geologists in the field for information and then published as his own their unpublished work. Once the 'Hamlet' affair blew over, Hall, Ebenezer Emmons and the other members of the New York State Geological Survey enjoyed geologizing with Lyell. They looked forward to their joint field programmes and attended some of his lectures.

Fig. 11. Trinity Church, New York City, on 21 May 1846. This lithograph shows the clergy of the diocese of New York entering the church for the consecration. Lyell admired this church on his visit in December 1845.

References

ANON. 1996. Three centuries of service. *Trinity News,* **43**(1), 8–11.

BAILEY, E. 1962. *Charles Lyell.* Thomas Nelson & Sons, London.

BARNARD, D. D. 1839. *A Discourse on the Life, Services, and Character of Stephen Van Rensselaer Delivered Before the Albany Institute April 15, 1839 with a Historical Sketch of the Colony and Manor of Rensselaerwyck in an Appendix.* Hoffman & White, Albany.

BYARD, P. S. 1996. Appreciating Trinity's Church: an anniversary view of a landmark. *Trinity News,* **43**(1), 12–15.

CARMER, C. 1945. Introduction. *In*: CHRISTMAN, H. *Tin Horns and Calico: A Decisive Episode in the Emergence of Democracy.* Henry Holt & Company, New York, xv–xvii.

CHRISTIANSEN, C. 1992. *Calico and Tin Horns.* Dial Books, New York.

CHRISTMAN, H. 1945. *Tin Horns and Calico: A Decisive Episode in the Emergence of Democracy.* Henry Holt & Company, New York.

CLARKE, J. M. 1923. *James Hall of Albany: Geologist and Paleontologist, 1811–1898.* New York State Geological Survey, Albany, NY.

DANA, J. D. 1873. On some results of the Earth's contraction from cooling, including a discussion of the origin of mountains and the nature of the Earth's interior. *American Journal of Science*. ser. 3, **5**, 423–443.

DOTT, R. H., Jr 1996. Lyell in America – his lectures, field work, and mutual influences, 1841–1853. *Earth Sciences History*, **15**(2), 101–140.

—— 1998. Charles Lyell's debt to North America: his lectures and travels from 1841 to 1853. *This volume*.

FRIEDMAN, G. M. 1979a. Geology at Rensselaer: a historical perspective, *In:* FRIEDMAN, G. M. (ed.) *New York State Geological Association Guidebook, 51st Annual Meeting*. Rensselaer Polytechnic Institute, Troy, New York. 1–19.

—— 1979b. Geology at Rensselaer: a historical perspective. *Compass of Sigma Gamma Epsilon*, **57**, 1–15.

—— 1981. Geology at Rensselaer Polytechnic Institute: an American epitome. *Northeastern Geology*, **3**, 18–28.

—— 1983. "Gems" from Rensselaer. *Earth Sciences History*, **2**, 97–102.

FRIEDMAN, G. M. 1989. Troy, New York and the Van Rensselaers. *In:* RODGERS, J., GRASSO, T. & JORDAN, W. H. (eds) *Boston to Buffalo, in the footsteps of Amos Eaton and Edward Hitchcock, 28th International Geological Congress, Field trip Guide Book*, T169, 99. American Geophysical Union, Washington, DC, 39–43.

—— 1990. Vertical parasequences of Lower Devonian limestone, Helderberg Escarpment: the Indian ladder trail at the John Boyd Thatcher State Park near Albany, New York. *Northeastern Geology*, **12**, 14–18.

—— 1995. Sequence stratigraphy of platform carbonates: Devonian limestones of John Boyd Thatcher State Park, southwest of Albany, NY. *In:* GARVER, J. I. & SMITH, J. A. (eds) *Field Trips for the 67th Annual Meeting of the New York State Geological Association*. Union College, Schenectady, NY, 303–311.

——, SANDERS, J. E. & KOPASKA-MERKEL, D. C. 1992. *Principles of Sedimentary Deposits: Stratigraphy and Sedimentology*. Macmillan, New York.

GRASSO, T. X. 1989. Erie Canal history and geology. *In:* RODGERS, J., GRASSO, T. & JORDAN, W. M., (eds) *Boston to Buffalo, in the Footsteps of Amos Eaton and Edward Hitchcock: 28th International Geological Congress, Field Trip Guidebook*, T169, 99. American Geophysical Union, Washington, DC, 48–83.

HALL, J. 1859. *Paleontology. Vol. III, Containing Descriptions and Figures of the Organic Remains of the Lower Helderberg Group and the Oriskany Sandstone*. New York Geological Survey, Albany, NY, Natural History of New York, part 6. 532.

JOHNSON, M. E. 1977. Geology in American Education: 1825–1860. *Geological Society of America Bulletin*, **88**, 1192–1198.

LYELL, C. 1845. *Travels in North America; with Geological Observations on the United States, Canada, and Nova Scotia*. Murray, London.

—— 1847. *Principles of Geology*. 7th edn. Murray, London.

—— 1849. *A Second Visit to the United States of North America*. 2 vols. Harper & Brothers, New York; Murray, London.

—— 1850. *Principles of Geology*. 8th edn. Murray, London.

—— 1852. *A Manual of Elementary Geology*. Murray, London.

—— 1853. *Principles of Geology*. 9th edn. Murray, London.

—— 1855. *A Manual of Elementary Geology*. D. Appleton & Company, New York.

—— 1871. *The Student's Elements of Geology*. Murray, London.

MERRILL, G. P. 1924. *The First One Hundred Years of American Geology*. Yale University Press, New Haven.

REZNECK, S. 1957. Out of the Rensselaer past: a centennial in American science and James Hall, scientific scion of Amos Eaton and Rensselaer. *Rensselaer Review of Graduate Studies*, **12**, 1–4.

—— 1959. A traveling school of science on the Erie Canal in 1826. *New York History*, **40**, 255–261.

——1965. Amos Eaton: a pioneer teacher of science in early America. *Journal of Geological Education*, **13**, 131–134.

RODGERS, J., GRASSO, T. & JORDAN, W. M. 1989. *Boston to Buffalo, in the Footsteps of Amos Eaton and Edward Hitchcock. 28th International Geological Congress, Field Trip Guidebook*, T169, 99. American Geophysical Union, Washington, DC.

SILLIMAN, R. H. 1995. The Hamlet affair: Charles Lyell and the North Americans. *Isis*, **86**, 541–561.

SPANAGEL, D. I. 1996. *Chronicles of a Land Etched by God, Water, Fire, Time, and Ice*. PhD thesis, Harvard University, Cambridge, MA.

WELLS, J. W. 1963. *Early Investigations of the Devonian System in New York, 1656–1836*. Geological Society of America Special Paper, no. 74.

Charles Lyell and the evidences of the antiquity of man

CLAUDINE COHEN

Ecole des Hautes Etudes en Sciences Sociales (Centre Alexandre Koyré),
Muséum National d'Histoire Naturelle, 57, rue Cuvier,
75 231 Paris Cedex 05, France

Abstract: This paper examines the importance of Lyell's book, *The Geological Evidences of the Antiquity of Man* (1863), in Lyell's work and career, as well as its contribution to the founding of two scientific disciplines, Prehistoric Archaeology and Palaeoanthropology. It first focuses on Lyell's two 'conversions' (to the idea of the antiquity of man and to Darwin's theory of transmutation of species) both of which he acknowledged in his book ; it turns then to a study of the nature of the 'evidences' that Lyell advances. These evidences were of several kinds – archaeological, palaeontological, anthropological, and, most importantly, geological; at the time, the best way to prove the existence of fossil Man was to demonstrate geologically the contemporaneity of man-made objects and extinct animals.

This paper insists in particular on Lyell's relationship to French science of the time, and examines the circumstances and modalities of a collaboration which was essential to the international acceptance of the antiquity of man. The fact that Lyell was a major scientific authority, whose expertise could validate the authenticity of evidence, allowed him to play a central role in the founding of a new field of knowledge.

The years 1858–1863 have often been considered a key period for the development of Prehistoric Archaeology as a scientific discipline. These years saw the official recognition of the existence of man prior to history as laid down in written documents or biblical traditions – the existence of man in the age of the great extinct mammals. This recognition was a major event in the history of Western thought, as it thoroughly transformed the representations of man, of his place on Earth and his role in the living world. It was, in fact, the result of a long debate which goes back to the early nineteenth century, and which involved geological, palaeontological, archaeological and anthropological issues. It was also the result of the findings of researchers working independently of one another in several European countries, such as Denmark, France, Germany and England, whose work suddenly crystallized towards the middle of the century, especially with the collaboration of English and French men of science; Charles Lyell was to play a major role in all these events.

After Oakley's seminal work (1964), several studies have recently appeared on this important and interesting episode of the history of science which marked the acceptance of the antiquity of Man by the scientific community throughout Europe; Donald Grayson (1983) has examined this period in a perspective of 'longue durée', from the seventeenth century to the end of the nineteenth century, and Bowdoin Van Riper, in a less com-

prehensive monograph (1993) has limited his scope to the years 1857–1865, when this 'great and sudden revolution' (Murchison 1868, p. 486) occurred in England. Others studies have focused on the French context and discoveries (Laming-Emperaire 1964; Cohen & Hublin 1989; Coye 1993; Groenen 1994).

My purpose here is to underline the importance of Lyell's book, *The Geological Evidences of the Antiquity of Man*, which was published in 1863, in Lyell's work and career, as well as its contribution to the constitution of a new field of knowledge. This paper will first focus on Lyell's 'conversions' to two central ideas, which he had previously rejected (the existence of fossil Man and the 'transmutation of species'), then on the particular devices which he used as 'evidences' to prove the antiquity of man; I shall conclude by questioning the role Lyell played in the founding of two scientific disciplines, Prehistoric Archaeology and Palaeoanthropology. In this paper, I shall insist in particular on Lyell's relationship to French science of the time, and the circumstances and modalities of a collaboration that was essential to the constitution of the Prehistory of Man.

Lyell's 'conversions'

In 1863, four years after the publication of Darwin's book *On the Origin of Species* (1859), two major works, both published in England,

COHEN, C. 1998. Charles Lyell and the evidences of the antiquity of man. *In*: BLUNDELL, D. J. & SCOTT, A. C. (eds) *Lyell: the Past is the Key to the Present.* Geological Society, London, Special Publications, **143**, 83–93.

83

established the framework of knowledge on fossil man: T. H. Huxley's *Evidence as to Man's Place in Nature* and Lyell's *The Geological Evidences of the Antiquity of Man*. Huxley's work was the book of a young man, who was already an ardent follower of Darwin. Lyell's book was a work of maturity. When it appeared, he was 66 years of age. In fact this late book, which was described in the first reviews as his 'trilogy on the Antiquity of Man, Ice and Darwin' (K. Lyell 1881, p. 362) acknowledged two major 'conversions'(as he himself termed them): first, the conversion to the idea of the existence of Fossil Man, which Lyell had rejected until 1853, at least, and secondly, a conversion to Darwin's ideas on 'the transmutation of species'.

Throughout the first half of the century, the question of the antiquity of man had focused the interests of scientists and public alike, but the defenders of 'fossil' or 'antediluvian man' were confronted with strong religious opposition. In England, the first decades of the nineteenth century had seen the revival of natural theology. The Reverend William Buckland, the first lecturer on Geology at Oxford University, believed, like Cuvier, that man had only appeared after the last 'catastrophe', which was identified as the biblical flood.

In France, despite the refusal of the scientific establishment led by Cuvier (1812) and his followers, several authors successively put forward arguments and evidences as to the existence of what many at the time still called the 'Antediluvial Man'. During the 1830s, in France and Belgium, discoveries made by amateurs or marginal scientists such as Marcel de Serres in Montpellier (de Serres 1826, 1830) Tournal in the Bize Grotto of the Ariäge region (Tournal 1829, 1830) and Philip Schmerling in Belgium (Schmerling 1833, 1834), tended to prove the contemporaneity of man-made tools and antediluvial animal remains, such as *Elephas primigenius, Rhinoceros tichorhinus*.

But these discoveries had been made in caves, and therefore their stratigraphy was open to criticism. As early as 1832, the French geologist Jules Desnoyers, then president of the Geological Society of France, had rejected the validity of archaeological and palaeontological evidence of the antiquity of man found in cave sites as he considered they could have been disturbed by people, animals, or by the action of waters and other natural elements (Desnoyers 1831–1832). Charles Lyell followed this opinion. In the second volume of his *Principles of Geology* (Lyell 1832, pp. 226–227) he wrote:

More than ordinary caution is required in reasoning on the occurrence of human remains and works of art in alluvial deposits, since the chances of error are much greater than when we have the fossil bones of the inferior animals into consideration. For the floor of caves has usually been disturbed by the aborigenal inhabitants of each country, who have used such retreats for dwelling places, or for conealment, or for sepulture...

To decide whether certain relics have been introduced by man, or natural causes, into masses of transported materials, must almost always be of some difficulty, especially where all the substances, organic or inorganic, have been mixed together and consolidated into one breccia; a change soon effected by the percolation of water charged with carbonate of lime. It is not on such evidence that we shall readily be induced to admit either the high antiquity of the human race, or the recent date of certain lost species of quadrupeds

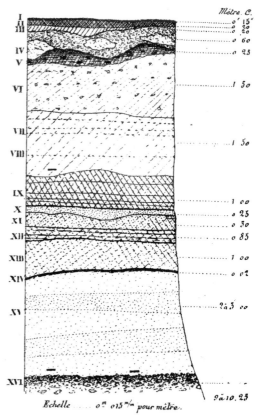

COUPE RÉDUITE DU TERRAIN DE MENCHECOURT PRÈS ABBEVILLE.
(Antiquités celtiques et antédiluviennes, tome 1er, page 234).

Fig. 1. Profile of the Menchecourt strata by Boucher de Perthes (1847). The black horizontal bars mark the location of stone artifacts.

He maintained his position on this subject at least until the ninth edition of his *Principles of Geology*, published in 1853. For Lyell as well as for Desnoyers, this criticism of evidence found in cave sites, made from the technical point of view of the geologist, also relied on a strong refusal of the idea that Man could be as ancient as what Cuvier (1812) had named 'antediluvial animals' or that humans could be of animal descent. At that time, for Lyell and for Desnoyers, man could only be a creation of God.

In 1837, Casimir Picard, a physician from Abbeville, conducted highly detailed and rigorous studies of the stratigraphy of the lower Somme

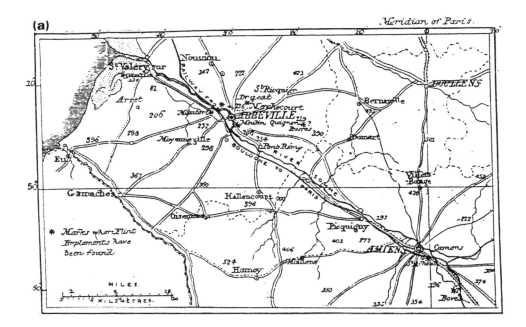

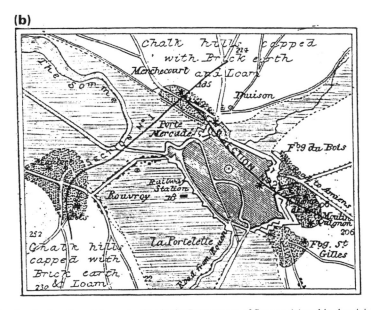

Fig. 2. Maps showing the location of the main sites in the Department of Somme (**a**) and in the vicinity of Abbeville (**b**) where flint implements had been found in association with remains of extinct animals (Prestwich 1861).

river valley (Picard 1838–1840); these sites of the region of Abbeville (Fig. 2) would become, after his premature death in 1844, the site of new and fruitful research. Another amateur from Abbeville, Jacques Boucher de Perthes, followed in Picard's footsteps and led the crusade for the official recognition of 'antediluvial Man' (Boucher de Perthes 1847, 1857, 1864). A customs officer of Abbeville and an enlightened amateur, Boucher de Perthes claimed that he had found evidence, as early as 1842, for the existence of 'Antediluvial Man' when he extracted from the clayey sands of the old riverbed of Menchecourt les Abbeville the jaw of a fossil elephant and a man-made primitive flint axe (Fig. 1). Boucher de Perthes was to have his victory, but in this the English were to play a major role. Whereas Boucher de Perthes had tried in vain to convince the French scientific establishment for over 20 years, English naturalists such as Falconer, Prestwich, Evans and Lyell, who came to Abbeville and Saint-Acheul (Fig. 3, 4), declared that Boucher de Perthes had been right; their authority was of great import among French academic authorities.

In England, important discoveries were made in this field from 1858. Brixham Cave, in Devonshire, had become the most important site of research in Prehistoric Archaeology (van Riper 1993, pp. 85–100). The excavations which were conducted between January and June 1858 led to the discovery of man-made stone tools which were shown geologically to be contemporary with the remains of large extinct animals. Although the excavations were made in a cave site, they seemed persuasive for several reasons: first, because the cave had been found untouched, with a stalagmite floor covering a thick layer of earth filled with bones of intact extinct fauna, such as *Elephas* and *Rhinoceros*, but also containing primitive stone tools which were found in the same strata; second, because the excavation itself was conducted for the first time using geological principles and relied on an 'horizontal method' which allowed archaeologists to obtain precise and rigorous data on the relative position of the vestiges found below the cave floor; and third, because it involved a close collaboration between trained amateurs, such as Pengelly, and professional geologists such as Falconer, Prestwich and Lyell.

These explorations at Brixham were to yield results that led the English scientists to accept the idea that man-made objects found buried in the earth could well be as ancient as deposits which Buckland had named the 'Diluvium' (Buckland 1837). But cave evidence was not sufficient, so the British decided to visit the French open air sites of the Somme valley, where Boucher de Perthes claimed to have found the same kind of associated remains.

In November 1858, Hugh Falconer was the first English geologist to visit Boucher de Perthes' sites and collections at Abbeville. On 27 April 1859,

Fig. 3. Flint tools from Saint-Acheul, near Amiens, collected by Lyell and illustrating his ... *Antiquity of Man* (1863, chap. VII, figs 10 and 11).

Fig. 4. The first use of a photography as evidence in prehistoric archaeology. On this picture (taken by Prestwich on 29th April 1859) one of the workers points to a stone implement found *in situ* in the quarry of Saint-Acheul (Musée d'Amiens).

Prestwich went to Amiens, just in time to witness the discovery of a flint tool *in situ* in the vicinity of Saint Acheul (Prestwich 1860). Lyell went to Abbeville at the end of July 1859. He came home with 65 flint tools, several of which had been found and extracted in his presence. After his journey to Abbeville, Lyell wrote to George Ticknor (K. Lyell 1881, p. 330)

...I have been much occupied with another geological subject, besides that which your niece, Ellen Twisleton, irreverently calls, the proving her to be the first cousin of a turnip (a violet she should have said). I mean the antiquity of Man, as implied by the flint hatchets of Amiens, undoubtedly contemporaneous with the mammoth, and also the human skeleton of certain caves near Liège which I believe to be of corresponding age. I regard the Pyramids of Egypt as things of yesterday in comparison of these relics. I obtained sixty five recently dug up, and Sir George Grey, of the Cape, and formerly governor of New-Zealand, recognises among them spear-heads like those of Australia, and hatchets and instruments such as the Papuans use for digging up roots, all so like as to confirm the saying you used to quote, of 'Man being a creature of few tricks'.

Like his predecessors at Abbeville, Lyell came back fully convinced, and 'converted' to the idea of fossil Man. In a memoir entitled *On the Occurrence of Works of Human Art in Post-Pliocene Deposits* (Lyell 1860), Lyell presented his defence of the idea of the antiquity of the stone implements found at Brixham and Abbeville at the 29th meeting of the British Society for the Advancement of Science, in Aberdeen in 1859. At the same session, he announced the upcoming publication of Darwin's book, *The Origin of Species*. The framework of Lyell's book on the antiquity of man was already laid out.

Indeed, Lyell's book contained a second, and even more spectacular conversion, to Darwin's ideas on 'the transmutation of species'. Against the traditional belief which he had previously adopted as his own, Lyell now accepted the idea of the animal descent of man.

Before asserting his acceptance of Darwin's ideas in the last four chapters of the book, Lyell pays hommage to Lamarck, whom he regrets he had previously rejected (Lyell 1863, pp. 388–391).

It is now 30 years since I gave an analysis in the first edition of my 'Principles of Geology' (vol. ii, 1832) of the views which had been put forth by Lamarck, in the begining of the century, on this subject... In that interval the progress made in zoology and botany... is so vast, that... what Lamarck then foretold has come to pass... Lamarck taught not only that species had been constantly undergoing changes from one geological period to another, but that there also had been a progressive advance of the organic world from the earliest to the latest times... from brute intelligence to the reasoning powers of Man.

The acceptance of Darwin's theory of transmutation rested, in Lyell's view, on several major arguments: first, it could explain the unity of morphological type throughout the whole organic world, (Lyell 1863, p. 413) which had been stressed by many naturalists who were proponents of the German school of transcendantal morphology since the end of the eighteenth century. But these naturalists, who included the great English palaeontologist Richard Owen, had tried to account for this 'unity of type' through metaphysical explanations. To these morphological features, Darwin's theory of 'common descent through phylogeny' offered a simpler, more materialistic elucidation. It explained not only 'the existence of 'rudimentary organs', which appeared then as 'being the remnants preserved by inheritance of organs which the present species once used' (Lyell 1863, p. 413), but also many biogeographical features, among which the reason 'why there are no mammals in islands far from continents, except bats, which can reach them by flying', and (Lyell 1863, p.414) 'a multitude of geological facts otherwise wholly unaccounted for, as, for example, why there is generally an intimate connection between the living animals and plants of each region of the globe and the extinct fauna and flora of the post-Tertiary or Tertiary formations of the same region'.

As he accepted at last Darwin's ideas, Lyell was conscious that this acceptance would necessarily lead him to admit that man was of animal descent, and to abandon 'old and long cherished ideas, which constituted the charm to [him] of the theoretical part of the science in [his] earlier days, when [he] believed with Pascal in the theory', as Hallam terms it (see Hallam this volume), of the 'archangel ruined' (K. Lyell 1881, pp. 361–362). Nevertheless, Lyell remained wary of the concept of natural selection, and tried to preserve something of a spiritualist view of human evolution. Only with great caution did he accept Darwin's doctrine, on which he maintained reservations and objections:

> ...I object in my *Antiquity of Man* to what I there called the deification of natural selection , which I consider as a law of force quite subordinate to that variety-making or creative power to which all the wonders of the organic world must be referred. I cannot believe that Darwin or Wallace mean to dispense with that mind of which you speak as directing the forces of nature. They in fact admit that we know nothing of the power which gives rise to variation in form, colour, structure, or instinct' (K. Lyell 1881, pp. 431–432).

Although Lyell's double conversion to the concepts of Fossil Man and of 'transmutation of species' opened the way to a new conceptual framework in scientific ideas, Lyell remained attached to a spiritualist view of the position of man in nature.The last pages of the book come back to natural theology.

> It may be said that, so far from having a materialistic tendency, the supposed introduction into the earth at successive geological periods of life – sensation, instinct, the intelligence of the higher mammalia bordering on reason, and lastly the improvable reason of Man himself, presents us with a picture of the ever-increasing dominion of mind over matter (Lyell 1863, pp. 505–506).

The nature of evidence

Although their perspectives were quite different, Huxley's and Lyell's works of 1863 had one important feature in common: at this early stage of the new science, they both stressed the need to produce evidence as to the antiquity and evolution of man. However, in these two works the nature of evidence was quite different. Huxley aimed to support the idea of a continuity between great apes and man, and his evidence drew mainly on comparative anatomy and embryology. Lyell's work relied on different types of evidence, such as archaeology, anthropology, palaeontology. But, the title of his work indicates, the main body of evidence was geological.

In Lyell's time, the best way to prove the antiquity of man was to demonstrate the contemporaneity of different objects found in the strata of the earth, such as flint tools and fossil bones of animals known to be extinct since 'antediluvial' times. Only a few human fossil remains were available at this time, and most of them were highly controversial. Even the remains of the Neandertal man were often considered 'too abnormal, too exceptional' to serve as evidence. On the other hand, archaeological finds, such as stone tools, objects of art, found in stratigraphical association with faunistic evidence, could provide clues for the antiquity of Man. In his book, Lyell described archaeological finds starting with more recent ages, Roman, Egyptian, etc., and reaching back to the most ancient times. For the most ancient periods of prehistory, he relied on the typology of flint instruments which had been studied in France. As early as 1837, Casimir Picard, in Abbeville, had clearly stressed the distinction between chipped and polished tools, and understood the principles of 'laminar debitage' which explained the method for the making of a stone 'knife' out of a nucleus (Picard 1836–1837). Subsequently, Boucher de Perthes (1847) placed these typological distinctions within a chronological framework and showed how the chipped handaxes of the Somme valley at Abbeville or Saint Acheul were more ancient than polished stone tools.

Vertebrate palaeontology also provided highly valuable evidence, when fossil bones had been found, as in Brixham, Menchecourt (Fig. 1) or Saint-Acheul, in association with man-made stone tools.

> The remains of elephants, ... purporting to come from the superficial deposits of Scotland have been referred to *Elephas primigenius*. ... the ocurrence of the mammoth and reindeer in the Scotch bulder clay, as both these quadrupeds are known to have been contemporary with man, favours the idea which I have already expressed, that the close of the glacial period in the Grampians may have coincided in time with the existence of man in those parts of Europe where the climate was less severe, for example in the basins of the Thames, Somme, and Seine, in which the bones of many extinct mammalia are

associated with flint implements of the antique type. (Lyell 1863, pp. 252–253).

Lyell relied in particular on the finds and studies of French scientists who had led several excavations in the valley of the Somme, referring for example to the remains of *Elephas primigenius*, *Rhinoceros tichorhinus*, *Equus fossilis* Owen, *Bos primigenius*, *Cervus somonensis* Cuvier, *C. Tarandus priscus* Cuvier, *Felis spelaea, Hyaena spelaea* as 'the most frequently cited as having been found in the deposits Nos. 2 and 3 at Menchecourt' (Lyell 1863, pp. 125–126). Through the study of these remains, even finer evidence of the contemporaneity of man and extinct animal could be put into light:

M. Lartet ... after a close scrutiny of the bones sent formerly to the Paris Museum from the valley of the Somme, observed that some of them bore the evident marks of an instrument, agreeing well with incisions such as a rude flint-saw would produce. Among other bones mentioned as having thus been artificially cut, are those of a *Rhinoceros tichorhinus*, and the antlers of *Cervus somonensis*'.

Lyell used this archaeological and palaeontological knowledge in order to argue the contemporaneity of fossil remains of extinct mammals and tools manufactured by man. The cover of the original edition of his book represents a primitive flint axe and mammoth molar, both in gold against a green background. The discovery of the two objects in the same ancient layer of earth was the very evidence of their contemporaneity. As a geologist, Lyell was able to make accurate reports on the precise location of these archaeological finds, and to go out into the field in order to confirm the observations of his predecessors.

Lyell relied on geological evidence to draw up a chronological and temporal framework for the existence of man through the changing climates of the last geological epochs. As an introduction to the book, he redefined terms used for divisions of Tertiary and 'Post-Tertiary' strata, outlining the framework for the existence of extinct mammals and early man. This chronology broke with the tradition of catastrophic events, and the ambiguous concepts of 'diluvial' and 'antediluvial' still adopted by Buckland in 1837 (Buckland 1837). The processes it threw into relief were not catastrophes, but gradual causes, among which Lyell stressed in particular the importance of the influence of glaciers, as he tried to draw a precise chronology of the relationships between the existence of fossil man and his geological environment, which he studied not only in Great Britain, but also in whole Europe:

The chronological relations of the human and glacial periods ... have taught us that the earliest signs of man's appearance in the British Isles, hitherto detected, are of post-glacial date, in the sense of being posterior to the grand submergence of England beneath the waters of the glacial sea. ... We may now therefore inquire whether the peopling of Europe by the human race and by the mammoth and other mammalia now extinct, was brought about during this concluding phase of the glacial epoch (introduction to chapter XIII of his book on the *Antiquity of Man*).

After examining in detail the succession of glacial deposits in Europe the glacial geology of Scandinavia, of the continental ice in Greenland and the glacial period of Scotland (Lyell 1863, chap. XIII), and dating the different positions of the earliest flint implements found in France and in England, he concludes:

If we reflect on the long series of events of the post-pliocene and recent periods contemplated in this chapter, it will be remarked that the time assigned to the first appearance of man, so far as our geological inquiries have yet gone, is extremely modern in relation to the age of the existing fauna and flora, and even to the time when most of the living species of animals and plants attained their actual geographical distribution. At the same time it will also be seen, that if the advent of man in Europe occurred before the close of the second continental period, and antecedently to the separation of Ireland from England and of England from the continent, the event would be sufficiently remote to cause the historical period to appear quite insignificant in duration, when compared to the antiquity of the human race (Lyell 1863, p. 289).

To archaeological, palaeontological and geological evidence, could be added anthropological inference. Eager to throw into relief the phylogenetic and cultural evolution of the human genus, Lyell also sought evidence for the common origin of modern humans by drawing a parallel between the origins and evolution of man and the development of human languages. Moreover, he underlined the fact that the stone tools found at Saint Acheul or at Brixham are similar to those of the Papuans, and may have belonged to similarly 'primitive' men. This heuristic parallel would be used systematically by John Lubbock in his book, significantly entitled *Pre-historic Times, as Illustrated by Ancient Remains, and the Manners and Customs of Modern Savages*, published in 1865; this work inaugurated a long tradition of the English and American school of 'cultural evolutionism' which flourished until the first decades of the twentieth century.

In his *Antiquity of Man*, Lyell sought to gather all evidence available at the time as to the 'establishment of human antiquity', this evidence being, as we have seen, of a mainly geological nature. Lyell was the champion of the principle of the 'uniformity' of causes in geology, according to which the same causes of natural processes acting in the past are still acting at present. Lyell stressed the possibility of comparing the activity of early man with that of present-day 'Savages', and in this respect, the research into the antiquity of man followed the same paths as geological research. Lyell also insisted on one of his favorite themes: that geology, as well as palaeontology, deal with scattered and incomplete materials. The archives of the earth have been partially destroyed, and Lyell wrote 'many of the witness of the past will never be found'. Like the palaeontologist, the archaeologist must seek the missing links in the history of human descent.

Lyell's authority and the foundation of a discipline

Lyell's book was not the first to give support to the notion of the antiquity of man. In France, the first two volumes of Boucher de Perthes' *Antiquités Celtiques et Antédiluviennes* had already appeared in 1847 and 1857, respectively. The third and last volume, published the year after Lyell's book (1864), brought together an impressive array of evidence, testimonies, drawings and maps.

Like Boucher de Perthes, Lyell accumulated in his book evidence of many varieties and natures drawn from geology, archaeology, palaeontology, anthropology and even linguistics. But whereas Boucher de Perthes never belonged to the scientific establishment and had to have the authenticity of his arguments confirmed by testimonies and recognized by the authority of other scientists, in Lyell's work, the empirical evidence and theoretical principles needed no such justifications. As Lyell was the major authority in geology, his own expertise validated the authenticity of the evidence.

Lyell's *Antiquity of Man* is as much a historiographic work as a scientific treatise, since it recorded for the first time in one book all the documented evidence on human antiquity since the eighteenth century: fossil and archaeological vestiges from Danish excavations which were the basis for their three-age chronology, German discoveries at Neandertal, Schmerling's research at Engis Cave in Belgium, Boucher de Perthes' activities in France, and excavations made by John Frere, Pengelly, Falconer, Prestwich and many others in England.

Not only did he use all these scientists as sources,

but he also recounted their careers, reread their writings and restudied their archaeological, palaeontological and geological materials. He retraced their steps in the field, went to Neandertal and Engis in 1857, travelled in 1859 to Abbeville (Fig. 2) and in 1860 to Hoxne (Suffolk) where John Frere, as early as 1799, had found flint tools associated with the remains of a great unknown animal, which he claimed to be of great antiquity (Frere 1800). Here, as in many seminal treatises, historiography is used as a rhetorical device to lay the foundation for a new science, and give it a legendary and noble past.

Lyell synthesized the whole spectrum of research conducted separately in different countries and disciplines since the end of the eighteenth century, and presented himself as the founder of a new science. He was accused of having appropriated the work of his collaborators (Wilson 1996), but one could argue in his defence that, reciprocally, his own authority and expertise came in support to their claims, and helped publicize their ideas .

Lyell's book could well be regarded, as Darwin wrote, as a 'compilation of the highest level'. But it can also be considered as a cornerstone in the study of the prehistory of man, as it defined the new science as a necessary synthesis of several disciplines (geology, palaeontology, palaeobotany, archaeology, evolutionary biology, anthropology, linguistics...) and as a necessarily collective and international enterprise. We understand from Lyell's book, and also from his correspondence, that the studies relating to the question of the antiquity of man led to the development of an important international network of scientists, such as Sedgwick, Darwin in England, Asa Gray, Louis Agassiz in the United States, Boucher de Perthes, Rigollot, the geologists Ravin and Buteux and the palaeontologist Edouard Lartet in France.

Moreover, Lyell's book made prehistoric knowledge accessible to a wide public (K. Lyell 1881, p. 376). During the previous years, the themes connected to the existence of fossil man and to the evolution of species had already promoted a great interest and, as Leonard Wilson wrote (Wilson 1996) Lyell's *Antiquity of Man* 'enjoyed an immediate and brillant success', 'within a week most of the 4000 copies of the first edition were sold'.

As he stressed the necessity for a change in the representations of the origin of man, and the necessity of a shift from religious to scientific knowledge, Lyell at age 66 appeared as an innovator who proclaimed the norms and criteria of the new knowledge, separating truth from fallacy, and establishing what Michel Foucault (1970) called a new 'état du vrai' (a new state of truth).

In fact, by the end of the century, the nature of

evidence in Prehistoric archaeology and in palaeo-anthropology had changed. What was now at stake was not only (as in Lyell's time) the proofs of the antiquity of man, but also the understanding of the details of his evolution and of his cultures. These disciplines had elaborated their own methods, typologies, intellectual and conceptual frameworks. The words Palaeolithic and Neolithic, coined in 1865 by John Lubbock (Lubbock 1865) to name the two most ancient periods of human Prehistory, described cultural features rather than geological layers. Similarly, the names of the different levels of the Palaeolithic coined by Gabriel de Mortillet (Mortillet 1880) – *Acheuléen, Chelléen, Moustiérien, Solutréen, Magdalénien* – were designed to identify mostly cultural assemblages.

On another hand, the 'Moulin Quignon Affair' (Boylan 1979; Cohen & Hublin 1989) revealed as early as the end of 1863 that geological criteria were not sufficient as evidence for the authenticity of prehistoric archaeological findings (Fig. 5). The recent human jaw, fraudulently planted in the most ancient ('diluvial') archaeological strata of the Somme valley, showed up the fragility of geological criteria as sole evidence to the antiquity of human remains, and the need for finer archaeological or anthropological evidence borrowed from comparative anatomy or stone tool typology and technology.

'Even in ten years, I expect, if I live, to hear of great progress made in regard to fossil man', Lyell wrote in 1863 (K. Lyell 1881, p. 373). Indeed, shortly after, new fossil and archaeological evidence came to light, especially in southwestern France (in the Dordogne and Vézäre river valleys) and in other regions of Europe, affording a more

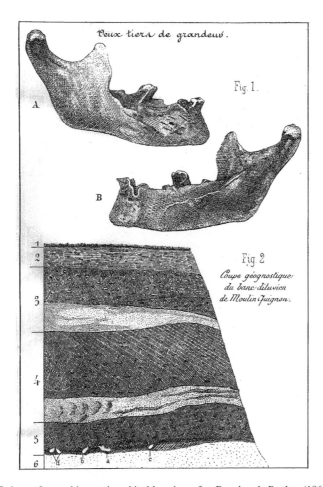

Fig. 5. The Moulin-Quignon Jaw and its stratigraphical location, after Boucher de Perthes (1864). In fact, this human mandible was a recent bone probably fraudulently planted into early Quaternary strata.

complex image of the anatomical features, evolution, culture and environment of fossil Hominids. Within a brief time-span, archaeological and palaeontological research into early man developed extensively to acquire immense popularity and found a place in scientific and academic institutions throughout Europe. But this would not have been possible without the events and collaborations of the years 1858–1863, which achieved what appeared to scientists and laymen alike as 'a great and sudden revolution'. Lyell's authority and the publication of his *Geological Evidences of the Antiquity of Man* had been essential for building the intellectual, institutional and international framework for the scientific researches into human origins.

References

BOUCHER DE PERTHES, J. 1847. *Antiquités Celtiques et Antédiluviennes*, vol. I. Treuttel et Wurtz, Paris, France.

—— 1857. *Antiquités Celtiques et Antédiluviennes*, vol. II. Treuttel et Wurtz, Paris, France.

—— 1864. *Antiquités Celtiques et Antédiluviennes*, vol. III. Jung-Treuttel, Paris, France.

BOYLAN, P. J. 1979. The controversy of the Moulin-Quignon Jaw : the role of Hugh Falconer. *In* : JORDANOVA, L. J. & PORTER, R. (eds) *Images of the Earth, Essays in the History of the Environmental Sciences*. London, 1979.

BUCKLAND, W. 1837. *Geology and Mineralogy Considered with Reference to Natural Theology.* 2 vols, London

COHEN, C. & HUBLIN, J.-J. 1989. *Boucher de Perthes, les Origines romantiques de la Préhistoire.* Belin, Paris, France.

COYE, N. 1993. *Des Mythes originels à la recherche archéologique. Sources, méthodes et discours de l'archéologie préhistorique en France avant 1950.* 2 vols, Thèse de Doctorat, Université de Provence Aix-Marseille I, France.

CUVIER, G. 1812. *Recherches sur les ossemens fossiles de Quadrupädes, où l'on rétablit les caractäres de plusieurs animaux que les révolutions du globe paroissent avoir détruites.* 4 vols, Déterville, Paris, France.

DARWIN, C. 1859. *On the Origin of Species by Means of Natural Selection, or the preservation of favoured races in the struggle for life.* Murray, London.

DESNOYERS, J. 1831–1832. Considérations sur les ossemens humains des cavernes du midi de la France. *Bulletin de la Société Géologique de France.* **2**, 126–133.

FOUCAULT, M. 1970. *L'Ordre du Discours.* Gallimard, Paris, France.

FRERE, J. 1800. Account of flint weapons discovered at Hoxne in Suffolk. *Archaeologia*, **13**, 204–205.

GRAYSON, D. K. 1983. *The establishment of Human Antiquity.* Academic Press, London, UK.

GROENEN, M. 1994. *Pour une Histoire de la Préhistoire.* Jerome Millon, Grenoble, France.

HUXLEY, T. H. 1863. *Evidence as to Man's Place in Nature.* William & Norgate, London, UK.

LAMING-EMPERAIRE, A. 1964. *Origines de L'archéologie Préhistorique en France.* Picard, Paris, France.

LUBBOCK, J. 1865. *Prehistoric Times, as Illustrated by Ancient Remains, and the Manners and Customs of Ancient Savages.* Williams & Norgate, London.

LYELL, C. 1832. *Principles of Geology, being an Attempt to Explain the Former Changes of the Earth's Surface by Reference to Causes now in Operation,* vol. II. Murray, London, UK.

—— 1860. *On the occurrence of works of human art in post-pliocene deposits.* Report of the 29th meeting of the British Society for the Advancement of Science, Notices and Abstracts, pp. 93–95.

—— 1863. *The Geological Evidences of the Antiquity of Man, with Remarks on Theories of the Origin of Species by Variation.* Murray, London, UK.

LYELL, K. 1881. *Life, Letters and Journals of Sir Charles Lyell.* 2 vols, Murray, London, UK.

MORTILLET, G. de, 1880. *Le Préhistorique.* Reinwald, Paris, France.

MURCHISON, C. 1868 *Paleontological memoirs and notes on the late Hugh Falconer (...) with a biographical sketch of the author. Vol. II. Mastodon, elephant, rhinoceros, ossiferous caves, primeval man and his contemporaries.* Hardwicke, London, UK.

OAKLEY, K. 1964. The problem of Man's antiquity. *Bulletin of the British Museum of Natural History,* Geological Series, **9**, 86–153.

PICARD, C. 1836–1837. Mémoire sur quelques instruments celtiques. *Mémoires de la Société Royale d'Emulation d'Abbeville*, pp. 221–272.

—— 1838–1840. Rapport de la Commission archéologique pour l'Arrondissment dAbbeville à Monsieur le Préfet du Département de la Somme ; en réponse à la circulaire de Monsieur le Ministre de l'Intérieur en date du 13 mars 1838. *Mémoires de la Société Royale d'Emulation d'Abbeville,* pp. 272–283.

PRESTWICH, J. 1860. On the occurrence of flint-implements, associated with the remains of extinct mammalia, in undisturbed beds of a late geological period. *Proceedings of the Royal Society of London,* **10**, 50–59.

SERRES, M. de 1826. Note sur les cavernes à ossemens et les brèches osseuses du midi de la France. *Annales des Sciences Naturelles,* **9**, 200–213.

—— 1830. Sur les ossemens humains découverts dans certaines cavernes du midi de la France, mêlés et confondus dans les mêmes limons où existent de nombreuses espèces de mammifères terrestres, considérées jusqu'à présent comme fossiles ou comme antédiluviennes. *Journal de Géologie,* **2**, 184–191.

SCHMERLING, P. 1833. *Recherches sur les Ossemens Fossiles découverts dans les Cavernes de la Province de Liège,* vol. I. Collardin, Liège, France.

—— 1834. *Recherches sur les ossemens fossiles découverts dans les cavernes de la province de Liège,* vol. II. Collardin, Liège, France.

TOURNAL, P. 1829. 'Considérations théoriques sur les caevrnes à ossemns de Bize, près Narbonne (Aude) et sur les ossemens humains confondus avec des restes d'animaux appartenant à des espèces perdues. *Annales des Sciences Naturelles*, **18**, 242–258.

—— 1830. Observations sur les ossemens humains et les objets de fabrication humaine confondus avec des restes d'animaux appartenant à des espèces perdues. *Bulletin de la Société Géologique de France*, **I**, 195–200.

VAN RIPER, B. 1993. *Men among the Mammoths, Victorian Science and the Discovery of Human Prehistory*. Chicago University Press, Chicago, USA.

WILSON, L. G. 1996. Brixham Cave and Sir Charles Lyell's ... *the Antiquity of Man*: the roots of Hugh Falconer's attack on Lyell. *Archives of Natural History*, **23**(1), 79–97.

Part 2. Lyell and the development of geological science

One could be forgiven for thinking that basin evolution and dynamics was a field of study developed in the later part of the twentieth century. The rapid rise in sedimentology and basin analysis would support this view. In attempting to understand, and give credit for, the origin of a particular discipline, there is often a need to identify individuals who may be regarded as the founder. Clearly with sedimentology such a man was Sorby. In his contribution, Mike Leeder argues that a case can be made that it was Lyell rather than Sorby who was the true originator of sedimentology and basin analysis. Leeder critically analyses Lyell's sedimentological descriptions and field observation as published in the *Principles*. Leeder demonstrates that Lyell grasped many concepts both of processes and products, indeed he recognized a number of sophisticated sedimentological concepts. Leeder further documents Lyell's willingness to abandon firmly held views. It is also clear that Lyell was not anticatastrophist as is often claimed. From Leeder's analysis it is firmly established that Lyell was engaged in and indeed contributed to establishing geology as a truly scientific discipline.

Of major interest to Lyell was the concept of geological time and its divisions. Lyell's studies on Tertiary molluscs and the establishment of geological periods based upon their percentage similarity to modern forms is well known. William Berggren discusses the development of Cenozoic stratigraphy including Lyell's terms Eocene, Miocene, Pliocene and Pleistocene and suggests changes to current nomenclature and concepts.

The establishment of time as a major element in geological ideas and the recognition of changing faunas and floras through time caused many difficulties to geologists in the nineteenth century. In his early work, as Tony Hallam points out, Lyell argued against organic progression. However, as data accumulated, Lyell eventually began to accept some kind of organic progression in the stratigraphic record whilst still believing in the imperfection of the fossil record. Hallam argues that it is not fair to criticize Lyell for his late and lukewarm conversion to evolution.

Joe Burchfield takes the concept of time a step further and gives a considered account of how, through the course of the nineteenth century, Lyell and his contemporaries developed ideas from a general impression of the vastness of geological time to the clear recognition of a geological history of the Earth as a succession of events in the stratigraphic record and attempts to determine a chronology for the age of the Earth. If the general concensus at the end of the century gave the Earth an age of *c.* 100 Ma, to be disproved within a decade, no matter, the concept of geological time had been established and with it the essential scientific basis to quantify the rates of geological processes.

Rapid climatic change and Quaternary glaciation provided a major challenge to Lyell. Patrick Boylan in his paper discusses Lyell's work in relation to the Glacial Theory of Louis Agassiz. Boylan documents the intensive field research of Lyell centred on his Scottish estate at Kinnordy. Lyell at first supported the Glacial Theory but hostility by a number in the Geological Society persuaded Lyell to revert to his earlier views of the importance of floating icebergs. However, by the end of his life he had begun to accept some highland glaciations but still continued to attribute deposits of the 'Glacial Period' to submergence.

Whilst Lyell continued to have problems with the Glacial Theory, he was still particularly interested in climate change. In his paper, James Fleming examines Lyell's position on climatic change in geological and historical times, and explores the mutual influence of Lyell with James Croll who was a proponent of an astronomical theory of Ice ages. Clearly the period of Charles Lyell's active geological life was one when numerous important astronomical discoveries were being made and many theories being advanced about climate change. Fleming argues that Lyell was slow in modifying his views on climate change but only because he tempered his judgements with solid evidence gathered from the record of the rocks – a lesson we might all learn!

It is clear from many of the contributions in this volume that labelling Lyell simply as a uniformitarianist who did not entertain catastrophes was not correct. Baker argues that working scientists are prone to misconceptions as to the relationship of philosophy to science. The 'new catastrophism' that Baker proposes is rooted in firm geological observation.

Finally, in this section, Lyell Professor John Mather relates the historical development of hydrogeology in Britain through the nineteenth century, a subject of passing interest to Charles Lyell which was advanced in his day by geologists who needed practical solutions to locating groundwater supplies to support the growing industrialization of the nation. Lyell recognized that rainwater percolated through the ground to issue forth as springs at the junction of permeable and impermeable strata, and was interested in this as an aspect of the Earth's surface processes. But it was really only in the latter half of the century that the science of hydrogeology was established and hydrogeological maps were published.

Derek J. Blundell
Andrew C. Scott

Lyell's *Principles of Geology*: foundations of sedimentology

M. R. LEEDER

School of Earth Sciences, University of Leeds, Leeds LS2 9JT, West Yorkshire, UK

Abstract: This chapter examines the extensive arguments Lyell brought to bear on the interpretation of sedimentary rocks through the operation of 'present causes' in the first Edition of *Principles of Geology* (1830–1833). A case is made *inter alia* for Lyell, rather than Sorby, being the true originator of sedimentology and basin analysis, amongst much else of course. Lyell had a special interest in Earth surface processes, and the effects of tectonics and climate on them, because he saw that the evidence is firmly written in the sedimentary products of observable events. His own explanations for the sedimentary and geomorphic processes of erosion and deposition were acutely sensible: he analysed (in today's parlance) river avulsions, controls on delta morphology, oceanic brine pools, cross stratification, confluence bars, boundary layers, hydraulic geometry, debris flows, sediment budgets, alluvial basin architecture, and clinoforms. Much of Sorby's work in some of these areas must have been inspired by his reading of the *Principles*. Lyell was occasionally too much guided by theory, as in the famous sophistry of his 1830 analysis (in the first edition) of the sedimentary and geomorphological evidence for Holocene Fenno-Scandinavian uplift. His willingness in subsequent editions of *Principles* to abandon such firmly held views in the light of empirical and personally collected field evidence to the contrary presages his momentous decision to throw his weight behind Darwin's natural selection theories some 30 years later. Lyell was not so rigidly an anti-catastrophist and inductionist as is commonly made out. His writings make very clear his ability to state bold theories. For example, his discussions of climatic change and his concept of the 'great year' were outstandingly incisive, holistic and original. They include the role of land–sea interactions, ocean currents and precessional orbital cycles. He even considered the probablity that continents and oceans changed position, albeit by vertical movements. The motive for all this came from observations he and others had made on the distribution of Cenozoic molluscs and the need for some general global cooling to explain these. He included in his actualism all manner of extreme (but not 'catastrophic') events, for example earthquakes and volcanic eruptions. He was also willing to consider (but then rejected) the possibility of catastrophes, most obviously in his discussion in the seventh edition of *Principles* (1846) of the possible 'lake-burst' of Lake Superior into the headwaters of the Mississippi. One feels he would have welcomed probability theory and the development of magnitude/frequency analysis, and that he would have laughed at any modern description of himself as the 'father of uniformitarianism'.

It is tempting to view Charles Lyell as a remote geological figure of only historical interest, with little relevance today. The great work of his early years, the first edition of *Principles of Geology* (Lyell 1830–1833), can thus be seen as a museum item in the pantheon of early geological literature, along with James Hutton's *Theory of the Earth* and William Smith's *Geological Map of England and Wales*. This view is wrong, because geology is a special science in that much depends upon the primacy of field observations. Lyell was an acute observer of Earth surface features and processes and the use of these in the interpretation of rocks. We can all relate to geologists of any age through their field observations; this stress on the empirical basis of geology is Lyell's hallmark, although, as we shall stress later, he was also a serious theorizer when the mood took him. To illustrate the arguments for relevance, consider the case of W. Q.

Kennedy, who, according to his biographer (Sutton 1980, p. 301), during his distinguished tenure as Chair at Leeds (1946–1967), used to read through *Principles* before the beginning of each new academic year to give himself the necessary inspiration for teaching. Nowadays the wide availability of the first edition of *Principles of Geology*, via the facsimile University of Chicago reprint, enables us all to confirm the essential modernity of much of Lyell's geological logic. *Principles* is an eternal stream from which we can all drink and refresh ourselves periodically. That the stream is a deep one, more than 1400 pages in total, is somewhat off-putting but the majority of Lyell's views on sedimentary topics are to be found concentrated in volume 1, published in 1830.

Nowhere is Lyell's own contribution to geology more apparent than in his discussions of sedimentary and stratigraphic issues (his observations

LEEDER, M. R. 1998. Lyell's *Principles of Geology*: foundations of sedimentology. *In*: BLUNDELL, D. J. & SCOTT, A. C. (eds) *Lyell: the Past is the Key to the Present*. Geological Society, London, Special Publications, **143**, 97–110.

on volcanics owed much to his friend Scrope). One often wonders how Lyell developed his abilities to read the rocks and the landscape so well. We cannot pretend that he invented the philosophy of actualism, but he certainly relentlessly pursued it. Case after case is presented, from personal observations and from a very wide variety of literature references (for a helpful analysis see Rudwick 1991) from explorers, encyclopaedias and classical authors concerning modern analogues useful in the interpretation of sedimentary rocks in the stratigraphic record. The example of his much-travelled grandfather, the early readings in the great library at Bartley Lodge (see Clasby 1997) and his Mediterranean-orientated classical education must all have played their part in setting up such an exceptionally broad-based approach. Although the analysis of physical processes is never technical in the sense of the use of mathematics, the approach is very modern in commonly using physical intuition and guideline calculations of rates of change (see the sections on fluid flow and sediment budgeting below)

A world view: reproduction versus destruction

It has been said (Rudwick 1991, p. xi) that Lyell's contemporaries were not particularly interested in a Huttonian worldview involving cyclic changes. But in the *Principles* they found exactly that, though more subtly expressed. Most importantly to field geologists, for on the whole they are not philosophically inclined, they could also find a piecemeal empirical approach to day-to-day geological investigations based on 'modern causes acting at their present intensities'. This non-theorizing aspect influenced (usually to their advantage, but sometimes to their detriment) subsequent generations of British and American geologists. But ignorance of the wider picture is a misreading of the global message of *Principles*. Nowhere is Lyell's intent more obvious than in Volume 1, where he lays out his basic philosophical position: that the outer Earth, the only realm of *geological* interest, is subject to the opposing tendencies of what he calls 'reproduction' (sediment deposition, lava eruption, uplift due to earthquakes) and 'destruction' (erosion). He saw all Earth surface processes as aimed towards one or the other of these ends, with the resulting balance being the state of the Earth as we see it now, or at any time in the the the past. Thus it was Lyell's philosophy that as we look back through the Earth's history we should look for a similar balance. The geological revolution of the past 30 years tells us emphatically that Lyell's basic geological intuition and logic were correct.

There can be little doubt that Lyell evolved his worldview largely through his experiences of field work in the southern European and Mediterranean regions. In his terminology these are areas of strong reproductive *and* destructive forces, contrasting mightily with the relatively staid and stable landscapes of his native Britain. Here in southern Europe he got most of his mollusc evidence for Cenozoic history and climate change, his earthquake and volcanic evidence for uplift and depression and his observations on the rates of infill of lakes and seas by deltas. We now know that the richness of this Mediterranean field laboratory arises from the complexities of its plate interactions and tectonics and the complications induced by Quaternary climatic changes. But at the same time it is interesting to note that the primacy there of earthquakes and magmatic flow in causing uplift and subsidence severely embarassed Lyell because it led to his earlier inability to understand how such processes operated *without* the aid of earthquakes and volcanoes. Hence his lofty dismissal (Lyell 1830–1833, vol. 1, pp. 227–232) of the published field evidence for the uplift of Fenno-Scandinavia. His subsequent visit there in 1834 overcame his theoretical opposition and he then accepted the field evidence. Without a theory to explain uplift, this volte-face showed Lyell's devotion to empiricism and his intellectual courage in rejecting a formerly strongly held conclusion. It presaged his more momentous decision to throw his weight behind Darwin's natural selection theories some 30 years later.

Lyell returns to the mass balance aspects of erosion and deposition in Chapter 17 of *Principles* Volume 1, where he questions the destination of sediment produced by coastal erosion. In line with his philosophy of reproduction/destruction he deduces that it must all go to make up the tidal sandbanks and coastal salt marshes that are so characteristic of areas around the southern North Sea. In a splendid re-emphasis of his 'present causes' philosophy (vol. 1, p. 311), he exorts geologists to go out and look for such effects in the geological record:

> Those geologists who are not averse to presume that the course of Nature has been uniform from the earliest ages, and that causes now in action have produced the former changes of the earth's surface, will consult the ancient strata for instruction in regard to the reproductive effects of tides and currents ... they will then search the ancient lacustrine and marine strata for manifestations of analogous effects in times past.

Perhaps the most stirring passage concerning the balances and checks between subsidence, uplift, deposition and erosion available in the natural world (note his reference to the Earth's surface as a 'system') comprises the last two sentences of *Principles* Volume 1:

> ...subterranean movements...the constant repair of the dry land... are secured by the elevating and depressing power of earthquakes...This cause ...is...a conservative principle in the highest degree, and, above all others, essential to the stability of the system.

Sedimentation rates and budgeting

In Chapter 13 of Volume 1', Lyell turns to the 'reproductive' effects of running water. In a series of striking phrases (p. 220), he engages us with his ideas of the mechanical transfer of energy and a useful definition of total sediment discharge: 'The aggregate amount of matter accumulated in a given time at the mouths of rivers ... affords clear data for estimating the energy of the excavating power of running water on the land...' Such approaches still underpin the estimation of past sediment yields from river catchments. Lyell then proceeds to outline some of the problems associated with such estimates, chiefly concerning the role of lakes in trapping sediment upstream from deltas. He returns to the theme of estimating sediment yields at the end of Chapter 14 ('Quantity of sediment in river water'), with a discussion of the likely annual sediment influx from the Yellow River, Ganges and other rivers. He draws on previous conceptual advances and researches by others, but it is clear that his ultimate goal is to estimate the total global sediment flux from the continental surface to the oceans (p. 246):

> Very few satisfactory experiments have as yet been made, to enable us to determine with any degree of accuracy, the mean quantity of earthy matter discharged annually into the sea by some one of the principal rivers of the earth'

Lyell computes estimates of the Ganges sediment yield and compares it to the volume of recent Etna lava flows. Here he is encouraging geologists to estimate total global fluxes for continental erosion and oceanic deposition. This is entirely in line with his global view of balance and change. We continue to stumble towards such goals today.

Fluid flow

Lyell frequently reveals an appreciation of the basic physics of sediment transport by fluid flow which was greatly in advance of his time. In discussing the 'transporting power of water' (vol. 1, p. 172 *et seq.*), he notes that one must always take into account the immersed weight of mineral grains

when estimating the mechanical power expended by running water. Quoting as reference the article on rivers in *Encyclopaedia Brittanica,* he carefully restates (pp. 172–173) the experimental fact that stream channels have three-dimensional zones of retardation (known, since Prandtl's work early this century, as boundary layers) due to retardation of flow by friction at their solid boundaries. He also notes from the same source the steadily increasing velocities needed to entrain and transport, respectively, clay, silt, sand, gravel and pebbles. Later, in a discussion on the transporting capacity of rivers (vol. 1, p. 247), he admits surprise at the tremendous variation in suspended sediment concentrations recorded by different workers in different rivers. He is particularly intrigued by the record of 25 per cent of suspended sediment by volume recorded by Rennell from the Ganges at flood stage. Although this particular figure has not been confirmed, we now know that such concentrations are commonly reached in the Yellow River, leading to the phenomena of hyperconcentrated underflows into the China Seas (Wright *et al.* 1986)

In a long letter written in September 1830 to his friend Scrope (Lyell 1881, pp. 296–299), in which he ruminates (quite wrongly it turns out) on the origin of subaqueous sand ripples, Lyell concludes by proposing a fluid mechanics experiment (Fig. 1) that they might do to further their understanding of ripple formation:

> A large and deep trough, with gently slanting sides, might enable us to experiment. Get a paddle-wheel which will turn with the hand, and make a ripple *ad libitum*, and sand and mud of different kinds to be deposited. Then we will

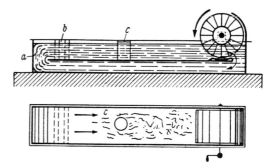

Fig. 1. Side (top) and plan (bottom) views of the hand-operated paddle wheel flume used by Ludwig Prandtl around 1900 (see Acheson 1990, pp. 260–299) to elucidate the nature of boundary layer separation. Lyell suggested to Scrope in 1830 that they might use a similar artifice to investigate ripple formation, perhaps the first planned (but never instigated) loose boundary hydraulics experiment in the history of sedimentology.

afterwards mix matter in chemical solution. After a due series of failures, blunders, wrong guesses etc, we will establish a firm theory.

Lyell ends with a typically acute remark, another product of his voracious reading and his predilection for lateral thinking: 'Have you read Dr. Young's experiments on the arrangement of sand upon boards vibrating by different notes of stringed insruments? The symmetrical forms obtained are wonderful.' Here he is hinting at the ability of sand piles to arrange themselves into well organized forms when stimulated by vibrations. Nowadays we are more familiar with the arrangement of self-similar forms and the intriguing properties of sand grains subject to shear and vibrations (Umbanhowar *et al.* 1996).

Sedimentary structures

Lyell had a special interest in sedimentary structures and endeavours at several points in *Principles* to give detailed explanations of their origins. He was clearly determined that geologists should use such features to improve their interpretations of sedimentary rocks and ancient environments. Inductive reasoning, from small to large, from present to past, epitomizes most of Lyell's geological philosophy. It is nevertheless clear that observations should be made in a deductive spirit, i.e. testing preformed hypotheses, as revealed in another extract from his long letter to Scrope of September 1830:

> I have for a long time been making minute drawings of the lamination and stratification of beds, in formations of very different ages, first with a view to prove to demonstration that at every epoch the same identical causes were in operation.

Close description is followed by incisive and general comparisons with recent processes like those first encountered at the end of Chapter 14 (section on 'Stratification of deposits in deltas'). Here (p. 254), he points out that deposited strata may not always be near-horizontal but may be 'disposed diagonally at a considerable angle'.

He suddenly presents a sketch (Fig. 2), taken at the confluence of the Rivers Arve and Rhone, of a stratified sequence he observed in the incised flood deposits of the former river in the latter. This is the first illustration and serious explanation of cross-bedding in geological literature. The sketch shows tangential thinning-downward foresets, truncated above by upper-phase plane beds, which are in turn succeeded by pebbly sand lenses. His description and the exact location of the deposit leave little doubt that we are looking at the deposits of a con-

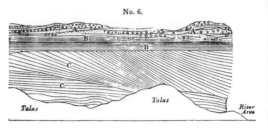

Fig. 2. A copy of the woodcut version (vol. 1, p. 254 fig. 6) of Lyell's field sketch of cross stratification taken at the confluence of the Rivers Rhône and Arve in 1829. The field of view is 3.66 m by 1.52 m. This is the first published illustration, description and correct interpretation of a major sedimentary structure. The deposits shown are the 1828 spring flood deposits of the Arve, as subsequently dissected by the Rhône, observed at low river stage by Lyell in January 1829. See Fig. 3 for a modern explanation.

fluence bar, specifically fronto-lateral avalanche and bar-top gravel-train strata:

> These layers must have accumulated one on the other by lateral apposition, probably when one of the rivers was very gradually increasing or

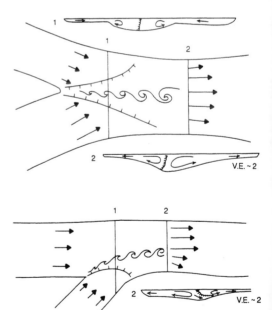

Fig. 3. The flow field and avalanche faces of bars formed at river confluences (from Bridge 1993; after Best (1986) and Best & Roy (1991)). Lyell's sketch in Fig. 1 pertains to the flow field at about section 1 of the lower example of an unequal channel confluence.

No. 31.

Section of shelly crag near Walton, Suffolk.

No. 32.

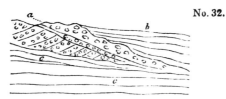

Section at the lighthouse near Happisborough.   Height sixteen feet.

a, Pebbles of chalk flint, and of rolled pieces of white chalk.

b, Loam overlying *a.*   *c, c,* Blue and brown clay.

No. 33.

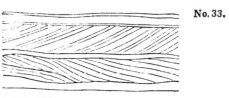

Section of part of Little Cat cliff, composed of quartzose sand, showing the inclination of the layers in opposite directions.

No. 34.

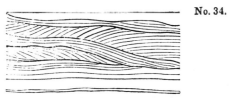

Lamination of shelly sand and loam, near the Signal-house, Walton.
Vertical height four feet.

Fig. 4. Copies of the woodcut versions (vol. 3, pp. 174–175; figs 31–34) of Lyell's field sketches of cross-stratification from the Plio-Pleistocene Crag deposits of East Anglia. Figure 31 is about 6.1 m high; fig. 32 4.88 m high; fig. 33 is 1.83 m high; fig. 34 is 1.22 m high. Lyell's fig. 33 was the first published illustration of 'herring-bone' cross-stratification.

diminishing in velocity, so that the point of greatest retardation caused by their conflicting currents shifted slowly, allowing the sediment to be thrown down in successive layers on a sloping bank.

A modern analysis of confluence bar formation (after Best 1986, Best & Roy 1991) is seen in Fig. 3.

During the past 170 years, the study of cross-stratification has revealed more about the nature of ancient environments than has any other sedimentary structure. Cross-stratification has revealed evidence for, among other things, palaeoflow vectors, depth of flowing water, types of bedforms, nature of lateral accretion and the way up of folded beds. It is usually assumed that Sorby (claimed by some as the 'father of sedimentology') was the first to recognize the true origins of cross-stratification, but clearly this honour (and appellation?) should go to the young Lyell. Sorby (1908) was certainly the first to relate the succession of sedimentary structures to the relative strength of a flow and the nature of cross lamination to the rate of deposition, but it was Lyell who first figured and correctly interpreted a variety of types of cross stratification (Fig. 4).

The main account is in the section on the East Anglian Plio-Pleistocene Crags, in *Principles* Volume 3, entitled 'Forms of stratification' (pp. 173–177). There is a revealing, though sometimes rather confusing, account in the letter to Scrope already noted. There is also a longer account in the surviving London lectures of 1832–1833 (Rudwick 1976), in which Lyell attributes the formation of thick deposits of rippled sandsones to 'the drifting of sand grains at the bottom', presumably over ripple forms. Climbing ripple cross-laminations were termed 'ripple-drift' well into the mid twentieth century, usually associated with Sorby's 1908 remarks but ultimately derived from this usage by Lyell. It is interesting that Lyell's interpretation of cross-stratification came from detailed observations of tiny wind-blown ballistic ripples (Fig. 5) forming on the coastal sand dunes at Calais. His arguments are too long to repeat here but we may note (Fig. 6) his recognition of the asymmetric form of both wind-blown and subaqeous ripples, steep leeward faces, lee-side avalanching and gradual downcurrent ripple migration. His impressive ability to scale up and his recognition that the fluid physics of air and water are comparable (Bagnold's detailed physical contrasts and comparisons (Bagnold 1951) lay far in the future) are impressive (vol. 3, p. 177):

> We think that we shall not strain analogy too far if we suppose the same laws to govern the subaqueous and subaerial phenomena; and if so,

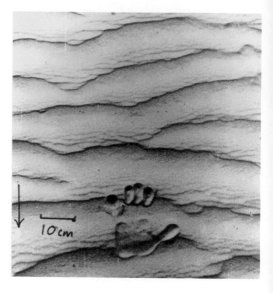

Fig. 5. Wind-blown 'ballistic' ripples of the type recognized and described by Lyell from the Calais coastal dunes. Photo by the late Ian Wilson (his handprint) taken *c.* 1970 in the Algerian Erg Oriental (Wilson 1972). The arrow indicates wind direction.

> we may imagine a submarine bank to be nothing more than one of the ridges of ripple on a larger scale, which may increase...by successive additions to the steep scarps.

This is a profound advance, with Lyell deducing that a heirarchy of bedforms exists–from the lowly current ripple up to larger dunes and channel-sized sand waves.

He notes that in the Crags (and also in analogous forms in sandstones of the Old Red Sandstone, Lower Palaeozoic and other ages) shelly calcareous sands have layers oblique or diagonal to the general true dip direction of the strata. He had obviously measured the directions of dip of the steep foresets over large areas because he notes (p. 175) that all along the Suffolk coast these were generally to the south, with the current responsible therefore from the north. These observations confirm Lyell, not Sorby (whose observations were made in the late 1840s and 1850s), as the first geological recorder and interpreter of cross stratification and palaeocurrents. He further notes (see Fig. 3), that certain successive cross sets were orientated in opposing directions. This was the first published description of a key structure, which more than a century later (Reineck 1963) became known as 'herring-bone' cross-stratification (Fig. 7). Lyell then goes on (p. 177) to make a masterly analysis of the probable

No. 35. •

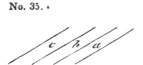

No. 36.

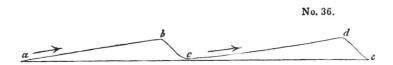

Fig. 6. Copies of the engraved versions (vol. 3, p. 176; figs 35–36) of Lyell's sketches of asymmetric (wind-blown) ripples (fig. 36) and the origin of successive avalanche deposits of cross-stratification (fig. 35).

tidal origins of such structures and prepares the ground for geologists to measure the structures in other parts of the geological record. Sorby began this mission some 20 years later.

Modern sedimentary environments

Rivers

In his discussions on the alluvial valley of the River Po (Chapter 11 of Volume 1), Lyell clearly refers to the infill of flood-plain lakes and marshes, and documents a number of 'deserted river courses'. This introduces the concept of river channel switches, or 'avulsions' as we now know them. Perhaps the most definite statement comes in Volume 1 (pp. 432–433), accompanied by his naming of the depositional river system as an 'alluvial plain':

> When we read of the drying up and desertion of the channels of rivers, the accounts most frequently refer to their deflection into some other part of the same alluvial plain, perhaps several miles distant. Under certain circumstances a change of level may undoubtedly force the water to flow over into some distinct hydrographical basin; but even then it will fall immediately into valleys already formed.

Such events are now best known and documented from areas like the Saskatchewan River (Smith *et al.* 1989; Smith & Perez-Arlucea 1994; Fig. 8) and from the active rift of the Rio Grande in New Mexico (e.g. Gile *et al.* 1981).

Lyell's discussion of the Mississippi River uses material published by early explorers and navigators. He presents descriptions of channel meanders, point bars, raised levees, the famous log jam at the

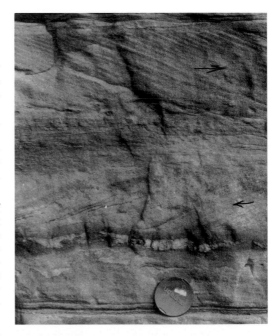

Fig. 7. Illustration of opposed large-scale sets of 'herring-bone' cross-stratification of the kind recognized by Lyell (see Fig. 4). Lebberston Member of the Middle Jurassic Cloughton Formation, Cloughton Wyke, Yorkshire. Arrows indicate flow directions.

Fig. 8. Oblique aerial photograph looking NNW to show the famous avulsion node of the New Channel (NC) of the Saskatchewan River, Canada (Smith *et al.* 1989; Smith & Pèrez-Arlucea 1994). The avulsion from the course of the Old Channel (OC) occurred in the early 1870s, during Lyell's lifetime. Photo courtesy of M. Pèrez-Arlucea.

confluence with the Atchfalaya and the general tendency of the Red River tributary junctions to be the site of periodic major flooding and lake formation (alternating deposits formed by flooding at tributary junctions are further discussed in an insightful manner in Volume 1, pp. 245–246). Such considerations are rarely found in modern discussions of alluvial facies and environments.

Lyell's most original and penetrating comments on river behaviour come when he integrates knowledge of river channel shifting, lake formation and the occurrence of earthquakes like that of the large 1812 New Madrid event in the Mississippi valley (vol. 1, p. 191):

> The frequent fluctuations in the direction of river-courses, and the activity exerted by running water in various parts of the basin of the Mississippi, are partly, perhaps, to be ascribed to the co-operation of subterranean movements, which alter from time to time the relative levels of various parts of the surface.

This is strikingly modern in highlighting tectonic tilting as a factor in river channel migrations and avulsions (see Alexander & Leeder 1987).

The short discussion of major historic river floods and their sediment transporting power (pp.

192–197) is marked by great attention to detail. Both the New Hampshire and Swiss examples cited involve descriptions of floods that we would now refer to as hyperconcentrated flows: 'with as large a quantitiy of earthy matter as the fluid could hold in suspension'.

Lyell is again thoroughly modern in frequently quoting flood speeds and sediment/water discharge magnitudes. Most important, in the light of subsequent mocking cartoons by De La Beche (Haile 1997; Fig. 9), are his comments on the difficulties that misfit streams provide for 'present cause' philosophy (p. 196, with author's emphasis):

> It is evident, therefore, that when we are speculating on the excavating force which running water may have exerted in any particular valley, the most important question is not the volume of the existing stream, nor the present levels of the river channel, nor the size of the gravel, *but the probability of a succession of floods*, at some period since the time when some of the land in question may have been first elevated above the bosom of the sea.

Lyell would surely have delighted in the development of probablity analysis of floods and their magnitudes (Nash 1994).

Fig. 9. De la Beche's satirical cartoon (brought to light by Haile 1997) mocking the effects of present causes. The child is peeing into the huge valley and a caption has his nurse exclaiming, 'Bless the baby! What a walley he have a-made.!!!' The cartoon is entitled 'Cause and Effect'. Reproduced courtesy of Professor W. J. Kennedy, University Museum, Oxford.

A further neglected example of Lyell's propensity for speculative deductive analysis is his direct contemplation of a truly catastrophic scenario painted in the seventh edition of *Principles* in 1846 (p. 152). He is considering the vexed question of 'deluges' and states:

> ...it would seem that the two principal sources of extraordinary inundations are, first, the escape of the waters of a large lake raised far above the sea; and secondly, the pouring down of a marine current into lands depressed below the mean level of the ocean.

Lyell specifically mentions, as a possible example of the former, the escape of the waters of Lake Superior into the headwaters of the Mississippi. He rejects the possibility of such an event occurring suddenly and considers that, even if it did, catastrophic effects downstream would not necessarily follow. This is a 'catastrophist' thought experiment (albeit an 'Aunt Sally') of considerable daring and ingenuity that we must now see in the light of the twentieth-century discovery of the 'lake-burst' of glacial Lake Missoula. Examples of the marine flooding scenarios that were cited include the possible inflow of Mediterranean or Black Sea waters into the depressed Caspian–again an 'Aunt Sally' but made more interesting to the historian of ideas by the discovery of the desiccated Miocene Mediterranean and its subsequent filling by the Atlantic (Hsü *et al.* 1977), and by the glacial isolation of Lake Euxine and the subsequent inflow of Mediterranean waters to form the present Black Sea (Ross *et al.* 1970).

Deltas

In Chapter 13 of Volume 1 Lyell divides deltas into three types according to the nature of their outfall: freshwater, inland sea (e.g. the Mediterranean) and oceanic. He clearly defines and separates the influences of tidal current and river influences on delta form and processes, and in Chapters 13 and 14 discusses in detail the lake and marine deltas of the Rhône, Po, Nile, Ganges and Mississippi.

His discussion of the impact of the River Rhône upon the shoreline of Lake Geneva draws upon depth soundings made by De la Beche and is distinguished by a clear account of what we would now term delta progradation and the production of low-angle delta foreset clinoforms (vol. 1, p. 221):

> We may state, therefore, that the strata annually produced are about two miles in length: so that, notwithstanding the great depth of the lake (about 160 fathoms), the new deposits are not inclined at a high angle; the dip of the beds, indeed, is so slight, that they would be termed, in ordinary geological language, horizontal.

After noting the likely coarse/fine alternations (i.e. 'varves' in modern parlance) to be expected in the foresets due to alternating yearly snowmelt and low-stage summer discharges, Lyell states: 'If then, we could obtain a section of the accumulation formed, in the last eight centuries, we should see a great series of strata, probably from six to nine hundred feet thick, and nearly two miles in length, inclined at a very slight angle.' He then goes on to contrast these low angle clinoforms with the increased dips expected from other smaller but steeper torrents entering the lake margins. He here anticipates G. K. Gilbert's account (1885) of 'angle-of-repose' deltas in the Pleistocene lakes of the Great Basin; deposits we now know as Gilbert deltas.

Lyell's careful analysis of delta and shoreline advance around Lake Geneva and his reliance on earthquakes as the sole means of crustal uplift led to the famous sophistry of his weak attempt to demolish Celsius' and subsequent workers' evidence for Fenno-Scandinavian uplift (vol. 1, pp. 227–232). Lyell simply and dogmatically ends his discussion by saying: 'No earthquakes, no uplift; all relative changes in shoreline position here are due to shoreline deposition'.

Lyell's discussions of the marine deltas of the Po, Rhône, Ganges and Mississippi are perhaps his most sustained and successful analysis of 'present causes' in the whole of *Principles*. His descriptions of freshwater, marine and hypersaline interactions and their effects upon faunas at the mouth of the Rhône anticipate by 150 years our concepts of schizohaline (Folk's term of 1974) environments. He clearly separates estuaries from deltas and, for the Ganges, notes in telling prose (vol. 1, pp. 244–245) that the correspondence of river flood with strong gales and high spring tides generates catastrophic (his word!) flooding and thus sedimentation or erosion. Noting the low tides affecting the Mississippi delta in the Gulf of Mexico, he remarks (p. 245) that 'the delta of the Mississippi has somewhat of an intermediate character between an oceanic and Mediterranean delta.'

He returns to discuss the effects of tidal currents on North Sea estuaries and deltas, chiefly the Rhine, in Chapter 16 (pp. 285–286), highlighting the antagonistic forces of the tide and river: 'the one striving to shape out an estuary, the other to form a delta'. Earlier comments on river channel avulsions are reinforced by this statement, 'It is common, in all great deltas, that the principal channels of discharge should shift from time to time.' This presages the influential concept of delta lobe switching introduced by Gulf Coast workers (e.g. Coleman & Gagliano 1964) in the 1960s.

Estuarine shorelines and shelf seas

In the surviving notes for the 1832–1833 London lectures (see Rudwick 1976), we hear Lyell analysing his evidence for estuarine shoreline conditions pertaining to deposition of the Lower Cretaceous Horsham Sands (see Allen 1975, pp. 413–415). He describes extensive thin bioturbated sandstones, the tops extensively rippled and the bases preserving desiccation cracks penetrating the interbedded clays. Noting the presence of freshwater shells and large reptiles in the sequence he deduces that the desiccation took place between tides.

As noted, Lyell's studies of the sedimentary structures in the East Anglian Crags were epoch making. Taking this together with faunal evidence, he proposed an insightful environmental analogue for the conditions of Crag deposition (vol. 3, p. 182): 'the formations which may now be in progress in the sea between the British and Dutch coasts, – a sea for the most part shallow, yet having here and there a depth of 50–60 fathoms, and where strong tides and currents prevail'.

Basin analysis and ancient sedimentary environments

The penultimate section of Chapter 14 of Volume 1 (pp. 249–251) contains an inspiring account of 'some general laws of arrangement which must evidently hold good in almost all the lakes and seas now filling up'. We would now call this 'arrangement' the stratigraphic architecture of an infilling basin. It is worth quoting Lyell in full here to reveal the depth of his thoughts on the matter and their relevance for the geologist interpreting ancient rocks. Basically he is describing the gradual infill of any tectonic depression with transverse and axial river and delta systems:

If a lake, for example, be encircled on two sides by lofty mountains, receiving from them many rivers and torrents of different sizes, and if it be bounded on the other sides, where the surplus waters issue, by a comparatively low country, it is not difficult to define some of the leading geological features which will characterize the lacustrine formation when this basin shall have been gradually converted into dry land by influx of fluviatile sediment. The strata would be divisible into two principal groups; the older comprising those deposits which originated on the side adjoining the mountains, where numerous deltas first began to form; and the newer group consisting of beds deposited in the more central parts of the basin, and towards the side farthest from the mountains. The following characters would form the principal marks of distinction between the strata in each series. The more ancient system would be composed, for the most part, of coarser materials, containing many beds of pebbles and sand often of great thickness, and sometimes dipping at a considerable angle. These, with associated beds of finer ingredients, would, if traced round the borders of the basin, be seen to vary greatly in colour and mineral composition, and would also be very irregular in thickness. The beds, on the contrary, in the newer group, would consist of finer particles, and would be horizontal, or very slightly inclined. Their colour and mineral composition would be very homogeneous throughout large areas, and would differ from almost all the separate beds in the older series.

The following are the causes of the diversity here alluded to between the two great members of the lacustrine formation. When the rivers and torrents first reach the edge of the lake, the detritus washed down by them from the adjoining heights sinks at once into deep water, all the heavier pebbles and sand subsiding near the shore. The finer mud is carried somewhat farther out, but not to the distance of many miles, for the greater part may be seen, where the Rhone enters the Lake of Geneva, to fall down in clouds to the bottom not far from the river's mouth. Certain alluvial tracts are soon formed at the mouths of every torrent and river, and many of these, in the course of ages, become several miles in length. Pebbles and sand are then transported farther from the mountains, but in their passage they decrease in size by attrition, and are in part converted into mud and sand. At length some of the numerous deltas, which are all directed towards a common centre, approach near to each other – those of adjoining torrents become united, and are merged, in their turn, in the delta of the largest river, which advances most rapidly into the lake, and renders all the minor streams, one after the other, its tributaries. The various mineral ingredients of each are thus blended together into one homogeneous mixture, and the sediment is poured out from a common channel into the lake.

Today's models for rift or foreland basin infill (see Busby & Ingersoll 1995) are essentially similar, though with 150 years' worth of additional accumulated jargon.

Sea level, ocean currents and deep brine pools

Lyell was concerned about the possibility the world's oceans do not share a common absolute elevation of mean surface (vol. 1, p. 293 *et seq.*), a possibility we now know as fact. He cites several measurements suggesting that mean levels of different seas and oceans did in fact differ, although we must stress (following Dott 1996) that Lyell never advances ideas on world wide (eustatic) sea level change. Lyell also thought correctly that the action of tides caused the mean sea level at any one time to vary. In a comment directly relevant to modern sequence stratigraphy he states, '*It* is scarcely necessary to remark how much all points relating to the permanence of the mean level of the sea must affect our reasoning on the phenomena of estuary deposits.' He draws attention to the nature of currents formed by strong onshore winds and correctly identifies the phenomena of leeshore superelevation. He deduces (p. 294) that, on cessation of any storm, strong offshore currents would result which could transport sediment far offshore. Nowadays we call these a storm surge, superelevation causing a steady seaward surge (gradient current) during the storm, due to the balance of hydrostatic and dynamic forces.

On p. 296 of *Principles* Volume 1 there is a particularly insightful discussion based upon naval surveys and Wollaston's analysis of the water balance and currents between the Mediterranean and the Atlantic via the Straits of Gibraltar. This leads to the recognition of saline deep water currents and what was probably the first proposal for the origin of ocean floor brine pools. Examples of these have been discovered in the Mediterranean in the last two years. Lyell clearly gives geologists a deep-brine model for the origin of ancient evaporites:

> the heavier fluid does not merely fall to the bottom, but flows on till it reaches the lowest part of one of these submarine basins into which we must suppose the bottom of this inland sea to be divided. By a continuance of this process, additional supplies of brine are annually carried to deep repositories, until the lower strata of water are fully saturated, and precipitation takes place ... on the grandest scale – continuous masses of rock salt ... like those in the mountains of Poland, Hungary, Transylvania and Spain.

Climate and ocean currents

Lyell was passionately interested in climatic and oceanographic influences on the distribution and succession of faunas, floras and sediments. It is noteworthy that he devotes the first three scientific chapters of *Principles* to these themes after lengthy, and polemical, introductory philosophical and historical material. These chapters mark him out most emphatically as a holistic Earth scientist in the modern sense. After considering the geological evidence for climatic change in Chapter 6, he constructs possible scenarios that could account for it on a global scale. He first points out the nature of continentality and proximity to oceans via a discussion of the Gulf Stream and its effects. He sees the changing distribution of land and sea as the main immediate cause for climate change and he is ready to invoke uplift and submergence on massive scales in order to effect the necessary changes in ocean current circulations. Here (vol. 1, p. 105) is his main point:

> But if, instead of vague conjectures as to what might have been the state of the planet at the era of its creation, we fix our thoughts steadily on the connection at present between climate and the distribution of land and sea; and if we then consider what influence former fluctuations in the physical geography of the earth must have had on superficial temperature, we may perhaps approximate to a true theory.

By 1847 and the seventh edition these notions had hardened into maps showing the likely disposition of continental masses that might lead to temperature extremes in the geological past. (see Fleming Fig. 1, in this volume).

Lyell's willingness to undertake bold and startling thought experiments in his quest for scientific explanation of geological evidence is best seen in this statement (vol. 1, p. 121):

> That some part of the vast ocean which forms the Atlantic and Pacific, should at certain periods occupy entirely one or both of the polar regions, and should extend, interspersed with islands, only to the parallells of 40°, and even 30°, is an event that may be supposed in the highest degree probable, *in the course of many great geological revolutions*.

I have emphasized the last phrase to draw attention to Lyell's use of the word 'revolution', clearly meant literally, in the context of cyclical change.

More shocks are in store. On p. 110 of Volume 1, Lyell draws attention, a little hesitantly, to the precession of the equinoxes and their role in a possible 20 ka climatic cycle. After noting the existence of such long period cycles of climatic influence, he develops his concept of a 'great year', a profound proposal somewhat ruined by the following fanciful remarks (much publicized and ridiculed by his opponents) on the reappearance of ichthyosaurs and other clearly extinct beasts during its course. Lyell's concept of a 'great year' is analogous to, but clearly on a longer time scale than, the astronomical cycle of the approximately 20 ka precessional orbital modulation (itself discovered by the French physicist d'Alambert in the eighteenth century). Lyell held its effects jointly responsible (though as junior partner) for climate change, in conjunction with the changing distributions of land and sea and the albedo effects of land and sea, ice and snow. Historians of both

Lyell's science (e.g. Dott 1996 p. 126) and of theories for climate change (Imbrie & Imbrie 1979) have generally neglected that these proposals were made as early as the 1830s. Here is the relevant passage in *Principles* (vol. 1, p. 110):

> It is, however, of importance to the geologist to bear in mind, that in consequence of the precession of the equinoxes the two hemispheres receive alternately, each for a period of upwards of 10 000 years, a greater share of solar light and heat. This cause may sometimes tend to counterbalance inequalities resulting from other circumstances of a far more influential nature; but, on the other hand, it must sometimes tend to increase the extreme of deviation which certain combinations of causes produce at distant epochs.

In later editions, from the third (1834) onwards, Lyell also discussed the possible influence upon climate of eccentricity variations in the Earth's orbit around the Sun. It was the astronomer John Herschel who originally suggested that eccentricity may have a role in climate change, albeit with little direct effect on the magnitude of incoming solar radiation, in a paper read to the Geological Society (but not published) in 1830. Herschel acknowledged (letters of February 1865 to James Croll; in Irons 1896, pp. 121–122, 126–127) that Lyell was the first to suggest a role (albeit subordinate) for the influence of orbital variations on ancient climates. Such serious investigations by Lyell and Herschel into orbital theories of climate change considerably pre-date the efforts of the Frenchman Adhémar in 1842, the person wholly responsible according to one influential account (Imbrie & Imbrie 1979, pp. 80–83). I am not claiming here that Herschel and Lyell pre-empted glacial theories but there can be little doubt that the emergence of Croll's glacial theory after Agassiz's epoch-making field observations owed much to them as progenitors of the concept of climate change (Croll had carefully read Lyell's *Principles*) (see also Fleming in this volume).

Lyell's sedimentary legacy

It seems to the author that there are four major trends in modern sedimentary geology:

(1) an analytical one seeking explanations of processes in terms of fundamental physics and chemistry;
(2) an empirical one pursuing explanations of sedimentary rocks in terms of a descriptive comparison with modern processes and products;

Fig. 10. Henry Clifton Sorby (hand lens at the ready) as a young man in the 1840s. Sorby and Lyell are recorded as having met on some of the former's rare visits to London from Yorkshire. Original in the City of Sheffield Museum.

(3) a regional and integrative one that relates erosion and sedimentation to tectonic uplift and igneous processes in sedimentary basins;

(4) a visionary global one that sees Earth surface processes uniting to provide a global budget of sediment and chemicals into the oceans, where they are recycled and renewed by plate margin processes.

Lyell's *Principles* can be seen as the fertile beginning of all of these interrelated themes. It becomes clear that the young Sorby (Fig. 10) had stood upon the shoulders of a giant when we read that he was able to recollect, in old-age (Kendall & Wroot 1924, p. 110), the 'discovery' he had made during a walk in about 1847:

> when walking from Woodbourne to Orgreave, I was caught in a shower of rain, and whilst sheltering in a quarry near Handsworth my attention was attracted by what I afterwards called 'current structures', namely structures produced in stratified rocks by the action of currents present during the time of deposition.

This is not the place for more general comments on Lyell's legacy to geological posterity but it seems safe to conclude that his enthusiasm for field observations, his devotion to the pure sciences and, at the same time, for speculative deduction have had much to do with the healthy growth of what we might call (dangerously) hypothetico-empiricism in geology. Darwin, to take the most famous example, was certainly fired up by *Principles* because of this heady mixture of theory and observations; doubtless countless others were too, including, no doubt, Sorby. Exactly what made Lyell himself think and act this way must await more systematic study, but we can imagine the active adolescent, alive with ideas and energy, rushing home with some natural specimen (dead rock or living thing) to Bartley Lodge, tipping off his boots and sliding along to his father's library in stocking feet to consult this or that reference among the many books in that marvellous place. The author believes that Lyell's heritage continues today, for, even in the depths of the economic depression from the mid-1980s to mid-1990s, British students still flocked into universities in their thousands to study our subject. This was in stark contrast, for example, to the USA, where economic factors prevailed and doubtless Lyell's original profession, that of lawyer, seemed more lucrative to many.

I thank Derek Blundell and Andrew Scott for their invitation to contribute this paper as part of the Lyell bicentenary birthday celebrations. Thanks to them I have spent scores of happy hours poring over Principles and ruminating on the particular genius of the young Charles Lyell; a humbling but also inspiring experience. Thanks to my Leeds colleagues Jim Best and Jane Francis for reading and commenting on the manuscript and for their encouragement. I thank Philip England for some revealing discussions on Lyell and for correcting an error in my interpretation of Lyell's concept of the 'great year'. It is a particular pleasure to thank Perce Allen for the tremendous job he did in improving the clarity and grammar of my text, and also Paul Clasby for finding textural errors. I am also grateful to Andrew Scott for his editorial help, to W. J. Kennedy of the University Museum, Oxford, for permission to reproduce Fig. 9 and to Marta Pèrez-Arlucea for providing Fig. 8.

References

ACHESON, D. J. 1990. *Elementary Fluid Dynamics*. Oxford University Press, Oxford.

ALEXANDER, J. A. & LEEDER, M. R. 1987. Active tectonic control of alluvial architecture. *In*: ETHERIDGE, F. G. & FLORES, R. M. (eds) *Recent Developments in Fluvial Sedimentology*. Special Publication of the Society of Economic Palaeontologists and Mineralogists, Tulsa, OK, 243–252.

ALLEN, P. 1975. Wealden of the Weald: a new model. *Proceedings of the Geologists' Association*, **86**, 389–437.

BAGNOLD, R. A. 1951. The movement of a cohesionless granular bed by fluid flow over it. *British Journal of Applied Physics,* **2**, 29–34.

BEST, J. L. 1986. The morphology of river channel confluences. *Progress in Physical Geography,* **10**, 157174.

—— & ROY, A. G. 1991. Mixing-layer distortion at the confluence of channels of different depth. *Nature,* **350**, 411–413.

BRIDGE, J. S. 1993. The interaction between channel geometry, water flow, sediment transport and deposition in braided rivers. *In*: BEST, J. L. & BRISTOW, C. S. (eds) *Braided Rivers.* Geological Society, London, Special Publications, **75**, 13–71.

BUSBY, C. & INGERSOLL, R. V. (eds) 1995. *Tectonics of Sedimentary Basins.* Blackwell, Boston.

CLASBY, P. S. 1997. Bartley Lodge and Sir Charles Lyell. *The Hatcher Review,* Winchester, 1–11.

COLEMAN, J. M. & GAGLIANO, S. M. 1964. Cyclic sedimentation in the Mississippi river delta plain. *Transactions of the Gulf Coast Association of Geological Societies,* **14**, 67–80.

DOTT, R. H. 1996. Lyell in America – his lectures, field work, and mutual influences, 1841–1853. *Earth Sciences History,* **15**(2), 101–140.

FOLK, R. L. 1974. The natural history of crystalline calcium carbonate: effect of magnesium content and salinity. *Journal of Sedimentary Petrology,* **44**, 40–53.

GILBERT, G. K. 1885. The topographic features of lake shores. *Annual Report of the United States Geological Survey,* **5**, 69–123.

GILE, L. H., HAWLEY, J. W. & GROSSMAN, R. B. 1981. Soils and geomorphology in the Basin and Range area of southern New Mexico. *Guidebook to the Desert Project.* New Mexico Bureau of Mines and Mineral Resources Memoir, **39**, Socorro.

HAILE, N. S. 1997. The 'piddling school' of geology. *Nature,* **387**, 650.

HSÜ, K. J. ET AL. 1977. History of the Mediterranean salinity crisis. *Nature,* **267**, 399–403.

IMBRIE, J. & IMBRIE, K. P. 1979. *Ice Ages: Solving the Mystery.* Macmillan, NY.

IRONS, J. C. 1896. *Dr. Croll's Life and Work.* Stanford, London.

KENDALL, P. F. & WROOT, H. E. 1924. *Geology of Yorkshire.* Privately printed, Vienna.

LYELL, C. 1830–1833. *Principles of Geology.* 3 vols. Murray, London. (Facsimile reprint, University of Chicago Press, Chicago, 1991)

LYELL, K. (ed.) 1881. *Life, Letters and Journals of Sir Charles Lyell, Bart.* Murray, London.

NASH, D. B. 1994. Effective sediment transporting discharge from magnitude–frequency analysis. *Journal of Geology,* **102**, 79–95.

REINECK, H.-E. 1963. Sedimentgefüge im Bereich der südlichen Nordsee. *Abheilung senckenbergische naturforschung Gesselschaft,* **505**.

ROSS, D. A., DEGENS, E. T. & MACILVAINE, J. 1970. Black Sea: recent sedimentary history. *Science,* **170**, 163–165.

RUDWICK, M. J. S. 1976. Charles Lyell speaks in the lecture theatre. *Journal of the History of Science,* **9**, 147–155.

—— 1991. Introduction. *In*: LYELL, C. *Principles of Geology.* Facsimile reprint, 3 vols. University of Chicago Press, Chicago.

SMITH, N. D. & PÈREZ-ARLUCEA, M. 1994. Fine-grained splay deposition in the avulsion belt of the Lower Saskatchewan River, Canada. *Journal of Sedimentary Research,* **B64**, 159–168.

——, CROSS, T. A., DUFFICY, J. P. & CLOUGH, S. R. 1989. Anatomy of an avulsion. *Sedimentology,* **36**, 1–23.

SORBY, H. C. 1908. On the application of quantitative methods to the study of the structure and history of rocks. *Quarterly Journal of the Geological Society of London,* **64**, 171–233.

SUTTON, J. 1980. William Quarrier Kennedy. *Biographical Memoirs of Fellows of the Royal Society,* **26**, 275–303.

UMBANHOWER, P. B., MELO, F. & SWINNEY, H. L. 1996. Localised excitations in a vertically vibrated granular layer. *Nature,* **382**, 793–796.

WILSON, I. G. 1972. Aeolian bedforms – their development and origins. *Sedimentology,* **19**, 173–210.

WRIGHT, L. D. ET AL. 1986. Hyperpycnal plumes and plume fronts over the Huanghe (Yellow River) delta front. *Geo-Marine Letters* , **6**, 97-105.

The Cenozoic Era: Lyellian (chrono)stratigraphy and nomenclatural reform at the millennium

WILLIAM A. BERGGREN

Department of Geology and Geophysics, Woods Hole Oceanographic Institution, Woods Hole, MA 02543, USA

Abstract: The historical and intellectual framework and contemporary influences and sources leading to, and associated with, Charles Lyell's studies of Tertiary stratigraphy are reviewed. The nomenclatural hierarchy into which some of Lyell's stratigraphic terms have been subsequently grouped is then analysed.

Lyell's terms 'Eocene', 'Miocene', 'Pliocene' (1833) and 'Pleistocene' (1839) were essentially biostratigraphic/biochronologic in nature. Together with the subsequently defined 'Oligocene' (1854) and 'Paleocene' (1874), they have gradually aquired a chronostratigraphic connotation over the past 100 years. The term 'Neogene' (Hörnes 1853, 1856, 1864), when transferred from its original biostratigraphic/biochronologic (Lyellian) to a chronostratigraphic connotation, includes the stratigraphic record of the Miocene, Pliocene, Pleistocene and Recent (subsequently termed Holocene) epochs (equivalent to the 'periods' of Lyell). Thus a Neogene/Quaternary boundary is wholly inappropriate as a standard chronostratigraphic boundary. The 'Palaeogene' (Naumann 1866) now comfortably accomodates the stratigraphic record of the Paleocene, Eocene and Oligocene epochs

The terms 'Palaeogene' and 'Neogene' should be used as period/system subdivisions of the Cenozoic Era. 'Tertiary' and 'Quaternary' should be abandoned, relics of a now outmoded, and inappropriate, pre-Lyellian stratigraphy. Retention of the term 'Quaternary' for geopolitical purposes (institutional affiliation or scientific/professional identification), as opposed to its retention as part of the standard stratigraphic hagiography, is recommended and would reduce, if not eventually eliminate, continuing but unnecessary, and unscientific, rancorous debate over standard Cenozoic chronostratigraphic terminology.

Events in human history may be characterized in terms of the totally unrelated phenomena of serendipity and viticulture. 1769 was a good vintage year: it saw the birth in France of Leopold Christian Friedrich Dagobert Georges Baron de Cuvier, Jean Francois d'Aubuisson de Voisins, Napoleon Bonaparte and Francois René de Chateaubriand; in Germany of Freiherr Friedrich Wilelm Heinrich Alexander von Humboldt; in Great Britain of William Smith, Sir Walter Scott, Arthur Wellsley (the Duke of Wellington), Jane Haldimand Marcet (whose *Conversations on Chemistry*, published in 1806 (*fide* Wilson 1972, p. 548), momentarily tempted Charles Lyell to publish his *Principles* in a popular, rather than academic, version a generation later) and Charles Lyell II, father of Sir Charles Lyell III, the pre-eminent geologist of the Victorian era.

In similar manner the turn of the eighteenth/ nineteenth century witnessed a concatenation of births and deaths (which was to be pivotal in the geological sciences and related events under con- sideration here (and a few of an unrelated nature; see below): 1796 saw the birth of Lyell's close colleague and collaborator in France, Gérard Paul Deshayes (1796–1875); 1797 saw the birth of Charles Lyell III and the death of James Hutton – whose bicentennial we celebrate in this book – and his friend George Poulett Scrope (né Thompson) (1797–1876), with whom he shared an almost identical life span; and of Carl Friedrich Naumann (1797–1873; professor of geognosy at the University of Leipzig and author of the term 'Palaeogene' which is discussed in greater detail below. Two years earlier, Giovanni Arduino (1713– 1795), whose term 'Tertiary' was to form the focus of much of Charles Lyell III's scientific career, died, while one year earlier saw the birth of Sir Henry De la Beche (1796–1855) and the death of Lyell's grandfather, and one year later the birth of Jean Baptiste Armand Louis Léonce Élie de Beaumont (1798–1874), Lyell's cordial friend and adversary (in matters of the origin and age of mountain systems) in Paris. The turn of the century saw the death of Horace Benedict de Saussure and Lazzaro Spallanzani (1799); the births of Heinrich Georg Bronn (1800–1862), whose early work (1829–1838) paralleled to a surprising extent

BERGGREN, W. A. 1998. The Cenozoic Era: Lyellian (chrono)stratigraphy and nomenclatural reform at the millennium. *In*: BLUNDELL, D. J. & SCOTT, A. C. (eds) *Lyell: the Past is the Key to the Present.* Geological Society, London, Special Publications, **143**, 111–132.

that of Lyell; John Phillips (1800–1874) who gave us the terms 'Palaeozoic', 'Mesozoic' and 'Caenozoic' in the mid-nineteenth century; Barthélemy de Basterot (1800–1887), French naturalist who made the first thorough description of the Aquitaine Basin in the vicinity of Bordeaux and provided a quantitative decription/catalogue of Miocene molluscan faunas; Jules Pierre Desnoyers (1800–1887), author of the term 'Quaternary'; Adolphe Théodore Brongniart (1801–1876); the death of Dieudonné Sylvain Guy Tancrède, called Déodat, gratet (count) de Dolomieu (1801); and the birth of Alcide Charles Victor Dessalines d'Orbigny (1802–1857), who formulated the first systematic subdivision of the Mesozoic and Cenozoic into 'stages' (1848–1852), characterized as rock units with temporal connotation characterized by (r)evolutionary (catastrophic) faunal changes at their boundaries. His deistic geostratigraphic framework represented the final demise of the school of Cuvierian 'catastrophism' that had informed the French school of geology for more than half a century.

Among unrelated but interesting, events that occured at about the same time: the French landscape painter Jean-Baptiste Camille Corot shared an almost identical life span with Charles Lyell III (1796–1875); 1797 saw the birth of Mary Wollstonecraft Shelley, the English novelist (who, in 1818, at the age of 21, published her most famous work, *Frankenstein, or the Modern Prometheus*, in the same year that Lyell made the first of his many extensive tours of the Continent, and immersed himself in its geology), and the births of Heinrich Heine, the German poet, Franz Schubert, the German composer, the Italian composer Gaetano Donizetti and the French poet and playright Comte Alfred Victor de Vigny; 1826 saw the deaths of Juan Chrisostomo, the 'Italian Mozart', at the premature age of 20, and the German composer Carl Maria von Weber; in 1827 Ludwig van Beethoven died and in 1828 so did Franz Schubert, the same year that Lyell set out with Sir Roderick Impey Murchison on the trip to southern France, Italy and Sicily that was to determine the course of his life. It would appear that the birth of the Romantic period in music (with the compositions of von Weber) and that of 'modern' stratigraphy (with the publications of William Smith) were serendipitously contemporaneous at the turn of the eighteenth/nineteenth century, although they were hardly inspired by the same muse.

This chapter deals with the historical intellectual setting in which 'Tertiary' stratigraphy developed during the course of the nineteenth century and a review of the terminology Lyell applied to the subdivisions of the 'Tertiary', the contemporary and subsequent additions to the hierarchy of stratigraphic terms and the need for reform in Cenozoic chronostratigraphic nomenclature at the approach of the millennium. While I have examined a broad selection of the literature dealing with Lyelliana, as well as the historical sources on which Lyell drew in developing his views of 'Tertiary' stratigraphy, I have tried to avoid the polemics that have marked recent scholarship on the subject of Lyell and his contemporaries and the issues with which they were dealing. I can instead recommend to the reader the relevant studies by Albritton (1963, 1980), Bartholomew (1976, 1979), Berry (1968), Conkin & Conkin (1984), Geikie (1897,1905), Gould (1987), Oldroyd (1996), Porter (1976, 1982), Rudwick (1963, 1969, 1970, 1971, 1974, 1975, 1978), Rupke (1983), Schneer (1969), Secord (1987), Zittel (1901) and Wilson (1969, 1970, 1972, 1980), amongst others.

Historical background

Abraham Gottlob Werner – founder of the 'plutonist' school of geology – died in 1817, the same year that Lyell's interest in geology was aroused by attendance at William Buckland's lectures in mineralogy at Oxford University. John Playfair (1748–1819), who had brought the important writings of James Hutton – founder of the 'neptunist' school of geology – to the attention of British (and Continental) geologists, died two years later, in the same year that Queen Victoria was born (1819). It was the passing of an era and the next two decades saw the successive decadal rise of the influence of the 'English school' of geology (as personified by the academic quartet of William Buckland (Oxford), William Daniel Conybeare (Oxford; later he worked independently), Adam Sedgwick (Cambridge) and William Whewell (Cambridge) and the 'Lyellian school'. The English school was devoted in the 1820s to diluvialism and catastrophism and in the 1830s to geological progressivism). The 'Lyellian school' of geology was latterly descended from the 'Huttonian/Scottish school' of Edinburgh, devoted to uniformitarianism, and was personified by Lyell and his closest colleagues and sometimes defenders, Gideon Algernon Mantell, Dr John Fleming, Hugh Falconer (with whom Lyell broke following publication of his *Antiquity of Man* in 1863; see Falconer 1863), George Poulett Scrope, William Henry Fitton, Charles Darwin and, somewhat later in his career, Herbert Spencer (the philosopher) and Thomas Henry Huxley. Lyell did not so much found a *school* of geology as a methodology of interpreting Earth history, which, in turn, established historical geology as a multifaceted discipline and an integral part of the Earth sciences.

Indeed, being outside the 'mainstream' of the British academic establishment throughout his life it was several decades before the full impact of his 'revolutionary' studies was fully integrated into what may be appropriately labelled 'modern geology' in Britain as well as on the Continent, where the 'catastrophist' schools still reigned (as for instance in France in the form of the legacy of Baron de Cuvier, Élie de Beaumont and Alcide d'Orbigny, and in Germany under the influence of Alexander von Humboldt and Leopold von Buch). It is an interesting fact that Lyell had been accorded no significant position or recognition in the hagiography of geological giants 1900, the term 'outstanding' (as opposed to 'unique') being the sobriquet most often used (Woodward 1908, 1911).

The events of Lyell's postgraduate years have been well documented in several biographies and will be touched upon only lightly here. His extensive travels in southern France, Italy and Sicily with Sir Roderick Impey Murchison and Mrs Murchison in 1828–1829 confirmed in his mind the idea that geological history could be interpreted in terms of processes currently operating on Earth and laid the groundwork for his subsequent life work. He had been led to Gérard Paul Deshayes by a chance meeting with Jules Desnoyers in Paris (February 1829). Desnoyers was about to introduce the term *Quaternaire* or *Tertiaire récent* for marine deposits younger than those in the Loire–Seine Basin and informed Lyell that Deshayes had deduced that a threefold sudivision of the Tertiary strata of the Continent could be effectuated by means of the successive associations of molluscan faunas. There began a close collaboration between Lyell and Deshayes which was to last the remainder of their professional lives and was to serve as the linchpin in Lyell's forthcoming subdivision of the Tertiary strata into 'formal' stratigraphic units in the third volume of the *Principles of Geology* (1830–1833).

The period 1830–1833 was one of fervent activity for Lyell. The first two volumes of the *Principles* appeared in 1830 and 1832, respectively. In 1831 he sought, and was appointed to, a professorship in geology at Kings College, London (see p. vii, this volume), a position that he appears to have enjoyed for its academic prestige, if not its financial compensation, but from which he resigned in October 1833, once the financial security issuing from his writings became assured. He travelled briefly to the Continent during the summer of 1831, ostensibly to examine the extinct volcanoes of the Eifel region, but opportunely for the purpose of courting and proposing (on 12 July) to Mary Horner, eldest daughter of Leonard Horner, the recently resigned (1831) warden of London University, at Bad Godesberg (near Bonn). A year

later , while wrestling with the classification and terminology of the Tertiary stratigraphic units which were to become the hallmark of Volume 3 of the *Principles*, Lyell interrupted his writing to return to Bonn to get married (exactly a year to the day from his proposal). The two-month honeymoon trip took Mary and Charles Lyell to the major tourist sites and academic inner sanctums of the Rhine Valley in Germany, and to Switzerland, northern Italy and France (during which time Cuvier died in Paris). The winter of 1832/1833 saw Lyell busily at work back in London, completing Volume 3 of the *Principles,* which appeared in April 1833 and in which his (bio)stratigraphic subdivision of the Tertiary appeared.

Meanwhile Volume 1 had been well received and reviewed by several of Lyell's colleagues (e.g. Scrope 1830). William Whewell (1831) and Adam Sedgwick (1831), clergymen-professors of geology at Cambridge, wrote generally critical, sympathetic and polite reviews. Whewell, however, concluded that much of Lyell's theorizing about the 'long' chronology required to account for faunal and tectonic observations and events in the young strata of southern Italy and Sicily, rather than the 'short' orthodox mosaic chronology, was geofantasy; whereas Sedgwick used the opportunity to deride Lyell's uniformitarianism and champion Élie de Beaumont's tectonic catastrophism. Lyell's friend George Poulett Scrope, although predisposed to fault Lyell for his failure to place finite limits on Earth history, reviewed the volume in favorable and admiring terms. In reviewing the second volume (1832) in similarly favorable terms, Whewell coined the terms 'uniformitarianism' and 'catastrophism', which have remained fixtures in geologic parlance (see the discussion in Gould 1987). Lyell was so pleased with Whewell's review, and so impressed with his linguistic facilities, that he asked his advice in formulating the appropriate terminology for his forthcoming Tertiary units in Volume 3 of the *Principles*. Whewell obliged and it is to him that we owe the derivation of the epoch/series terms introduced by Lyell (see Fig. 1) and that have become the standard terms of Cenozoic chronostratigraphic hagiography (Whewell 1831).

In an irony of history Lyell's friend, the Irishman William Henry Fitton (1780–1861), in reviewing the *Elements of Geology* (which appeared in 1838 and whose method consisted of extracting from the *Principles* the basic aspects of geology that would be comprehensible to a more general, lay audience and adding to it Book IV of the *Principles,* in which the Tertiary succession of strata and its terminology were discussed), questioned Lyell's introduction of the term 'metamorphic' objected to the introduction of the terms 'Eocene', 'Miocene' and 'Pliocene' and questioned the value of the percentage system

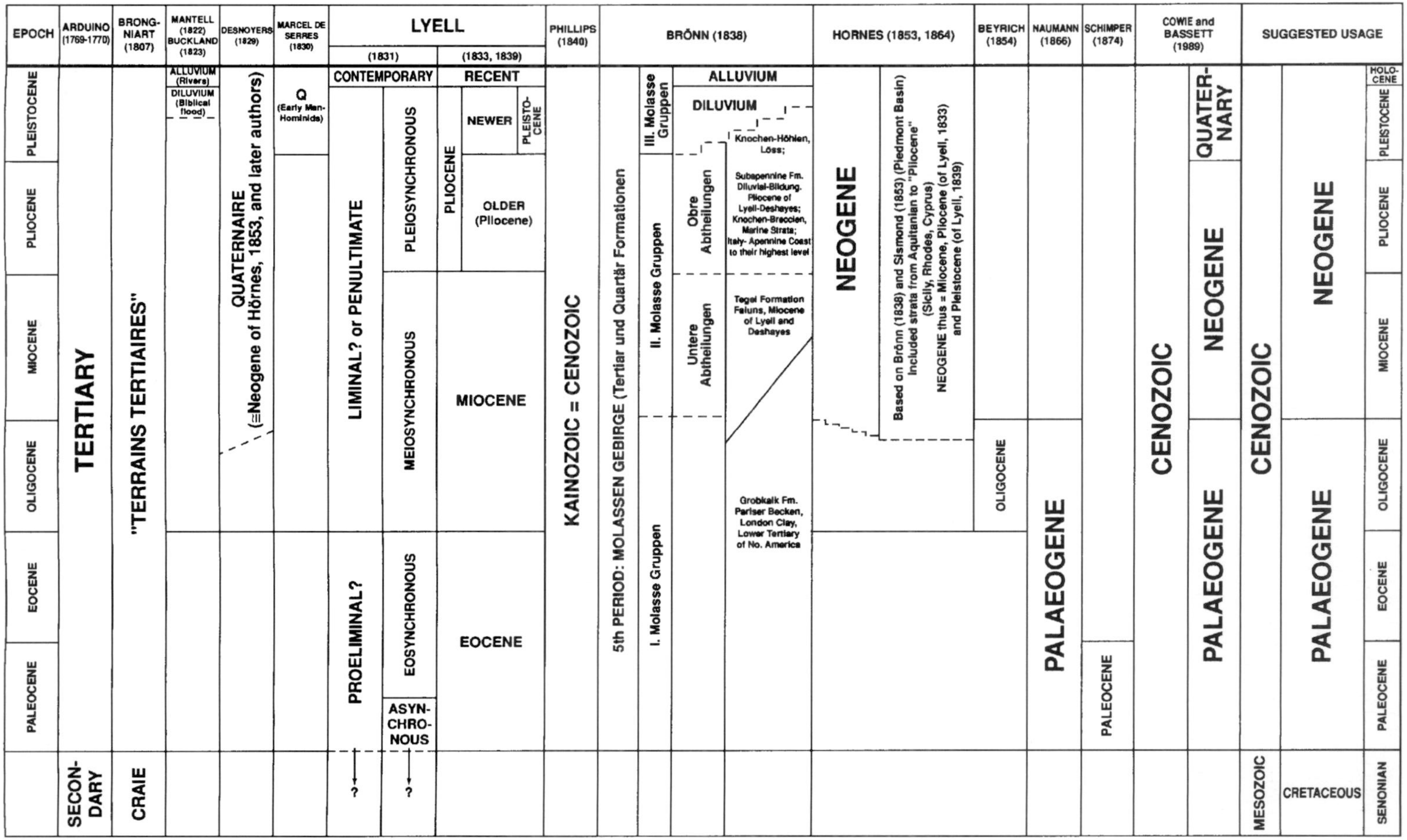

Fig. 1. History of terminology applied to Cenozoic chronostratigraphic units. Note that Tertiary nomenclature ascribed to Lyell (1831) represents an informal attempt to derive a suitable classification for what subsequently (1833, 1839b) became the standard, classic terms still in use today. In a letter dated 21 January 1831, the informal terms were submitted to William Whewell, who replied ten days later suggesting to Lyell some etymologically more appropriate terms which Lyell enthusiastically accepted and published in his first edition.

for placing the succession of Tertiary strata in relative order, saying that he doubted that these terms would be 'permanently expedient'. *Sic transit gloria!* The *Elements of Geology* is considered by many the first textbook on historical Geology.

The reputation of Lyell was securely established by 1841 when he left on the first of his lecture tours of the United States. Lyell never looked back, and there are some who might say he did not look very far forward (or sideways) either in making revisions to his *Principles* that would have enhanced the work in terms of contemporary advances in areas ancillary to historical geology. Nor did he enhance his reputation with the publication of his *Antiquity of Man* (1863), but these are issues beyond the scope of this chapter (see Falconer 1863, Rupke 1983, Bynum 1984).

A fourfold subdivision of mountains and their associated rock types was introduced by Giovanni Arduino (1760, or 1759?) based on field observations in Tuscany and Vincente Province (the valley of the Agno, northwest of Vicenza): Primary, Secondary, Tertiary, and modern alluvial deposits. Field work during the second half of the eighteenth century on the Continent and in Britain led to the recognition that these terms were more appropriate as descriptive rock terms than they were applicable to any chronological succession. As a result, in the course of the first half of the nineteenth century the meanings of the terms were transformed to include groups of units based on an observed succession of rock types and associated fossils. The Tertiary was given its essentially modern connotation by Brongniart during the period 1807–1810 (Fig. 2) in being applied to the succession of rocks that occur above the Chalk in the Paris Basin (which essentially encompasses the Palaeogene in modern terms) and which was the focus of the major study by Cuvier and Brongniart (1811) a year later. Its chronologic/stratigraphic extension to include rocks subsequently encompassed by most of the Cenozoic (Phillips 1840) was quickly established. However, the Quaternary (Desnoyers 1829) was carved out of the upper part of the Tertiary (stratigraphically equivalent to the Neogene of Hörnes (1853) and this paper), but its limits were immediately modified, and temporally shortened to include the diluvium deposits in accordance with the biblical deluge (De Serres 1830). It was in its recently modified, truncated form that Lyell inherited the term 'Tertiary' when he set about subdividing it into his 'periods' in Volume 3 of the *Principles* (1833).

It should be remembered that Lyell's subdivison of the 'Tertiary' into 'periods' was based on an assumption of uniformity of rate and state using a decay constant of species formation in molluscs

(Gould 1987). While d'Orbigny's definition of 'stages' (1849–1852) was of rock units deposited during finite temporal intervals whose boundaries were characterized/delimited by (r)evolutionary faunal changes/upheavals – 28 of them! – , Lyell used a statistical methodology in subdividing the 'Tertiary' into constituent units. Inasmuch as Lyell believed that species were evenly spaced in time – like beans being put into and taken out of a jar at regularly spaced intervals, to use Gould's (1987) effective metaphor – his 'periods' were not capable of capturing more than a fleeting moment, a window, as it were, in the continuous flow of time, and were not bounded by distinct, precisely defined moments in time. In short, Lyell's 'periods', as originally defined, were biostratigraphic, not chronostratigraphic, in concept. The distinctions between these categories were not recognized and defined for another half century until the deliberations of the Sixth Geological Congress in Bologna in 1881. Lyell was at repeated pains to point out that his 'zoological periods in geology should not acquire too much importance' (1833, p. 57), fully expecting missing parts of the (Tertiary) geological record to be recognized in the course of future studies. But he remained steadfast in his belief in the primacy of his uniformitarian approach to time's cycle (although he left his geological timescale open ended), which he continued to defend for another 30 years, until his capitulation to geological progressivism and time's arrow in 1862 in the face of the mounting evidence for evolution as promulgated by his friend Charles Darwin. Perhaps his capitulation served as a means of cutting his losses in surrendering his allegiance to 'time's cycle' while retaining the other cherished linchpins in his model: unformity of law, rate and process (Gould 1987, pp. 172, 173).

Lyell invested a significant amount of time on ascertaining the reliability of the percentage method in his assessment of the relative stratigraphic position of Tertiary rocks. When Lyell met Deshayes the latter told him that he could divide the Tertiary into three periods and he published a tabulation of his fossil collections a year later (Deshayes 1830). There began a collaboration between the two that was to last for several decades. Initially Lyell asked (and paid) Deshayes to classify the extensive collections of molluscs he had made on his tour of Sicily and Italy in 1828–1829. Comparison with Deshayes' extensive collections from the Paris, Aquitaine, London, Vienna, Touraine and other continental basins formed the framework for Lyell's (1833) fundamental subdivision of the Tertiary into the Eocene (<3 per cent extant species), Miocene (>8 per cent) and Older (35–50 per cent) and Younger (90–95 per cent) Pliocene periods.

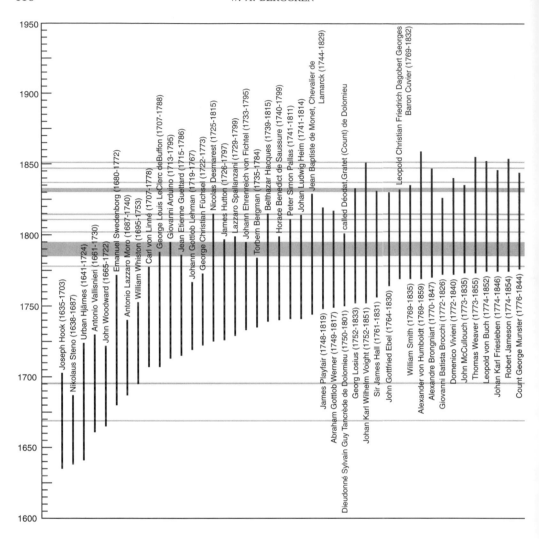

Fig. 2. (Bio)chronology of some of the major figures and significant events in the history of geology and stratigraphy (from the mid-seventeenth to mid-nineteenth century). The events shown are keyed as follows. (1) 1669: Nicolaus Steno describes the law of superposition. (2) 1695: John Woodward notes the marine origin of some fossils and that 'different strata have different fossils'. (3) 1696: William Whiston suggests the role of extraterrestrial phenomena as agents of geological change. Dismissed by Lyell as having 'retarded the progress of truth', the role of asteroids in global catastrophes was revived nearly three centuries later by Alvarez *et al.* (1979, 1980) and is now accepted by most Earth scientists. (4) 1778: First use of the term *géologie* by Jean André de Luc. (5) 1785: James Hutton's theory of the cyclic history of the Earth and of an Earth with a long chronology (published in abstract form). (6) 1795: Hutton's *Theory of the Earth* is published, postulating an Earth with 'no vestige of a beginning, no prospect of an end'. (7) 1795: The use of fossils to correlate geologic strata by William Smith. (8) 1802: Publication of William Playfair's *Illustration of the Huttonian Theory*. (9) 1811: Baron de Cuvier and Alexandre Brongniart describe the upper Cretaceous (Chalk) and lower Tertiary succession in the Paris Basin and interpret alternating marine and brackish water fossils in terms of successional, catastrophic faunal replacements due to the changing position of sea and land. (10) 1814: Thomas Webster demonstrates continuity (across the English Channel) of the Chalk and overlying lower Tertiary strata on the Isle of Wight and in the Paris Basin and names the London Basin. (11) 1814: Giovanni Battista Brocchi recognizes the 'Subapennine Beds' (the Older Pliocene of Lyell 1833), describes the Tertiary fossil shells of Italy and ascribes the difference of these late Neogene faunas from those of the Palaeogene of the Paris Basin to geography not age. (12) 1815: Publication of the first geological map of England by William Smith, the 'father of stratigraphy'. (13) 1830–1833: Charles Lyell publishes his groundbreaking *Principles of Geology*. Subdivision of the Tertiary into informal biostratigraphic units based on quantitative assessment of the succession of

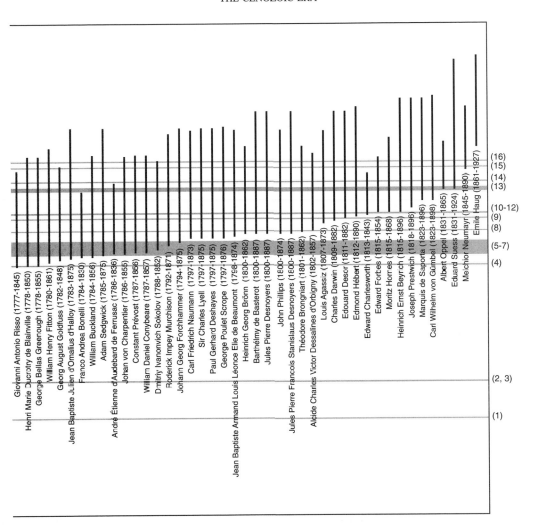

molluscan faunas by Gérard Deshayes. Comparable, contemporary faunal analysis (but no formal application to stratigraphic subdivision of the Tertiary) by Hans Georg Bronn. (14) 1837: Louis Agassiz promulgates his glacial theory, which helps bury diluvianism. (15) 1848: Alcide d'Orbigny introduce the deistic concept of 'stage' as a rock unit with a temporal connotation characterized by (r)evolutionary (catastrophic) faunal changes at its boundaries. (16) 1851: The term *stratigraphie* is introduced by Alcide d'Orbigny (although it had already been practised for over a half century).

Lyell's subdivision of the Tertiary based on this 'biochronologic' methodology was not long unchallenged. In the course of the famous incident of the Crag (see the detailed analysis in Wilson 1970), which Lyell (1833) had considered to be of Older Pliocene age, he became embroiled in a dispute between two worthies of the Geological Society, Edward Charlesworth and Samuel Woodward, who had differing opinions on the identity and number of Crags. Charlesworth (1835a, b, 1837a, b) proved a worthy opponent, observing that there were three Crags (Norwich, Red Crag and Coralline Crag) rather than one (Woodward's view) or two (as Lyell and Charlesworth had originally thought), and pointing out that analysis of the molluscan faunas of these units by the contemporary experts had led to significantly different age determinations. He mounted a frontal attack on Lyell by calling into question the accuracy of the latter's quantitative methods in assigning ages to these rocks units. The question of the age of the Crag was to occupy Lyell for four years and it was not until 1839 that he was able to reply to Charlesworth and others concerned with this problem (Lyell, 1839a). (The Coralline Crag has been the subject of several recent studies on calcareous plankton and dinoflagellates, which, when integrated, appear to indicate that the upper Ramsholt Member is of late early Pliocene (Zanclean; > 3.6 Ma) age, whereas the upper Sudbourne Member may be of middle late Pliocene (Reuverian C) age based on pollen analysis; see Head 1997.)

Paul Gérard Deshayes (a taxonomic 'splitter', in modern parlance) and Heinrich Beck (curator of the Museum of Natural History of Prince Christian of Denmark, and a 'lumper') had differing taxonomic/species concepts regarding Tertiary molluscs. Lyell returned to Copenhagen in 1837 to determine whether it was possible to reconcile the differences in interpretation by these specialists as well as by George Sowerby in Britain, thus to justify his subdivisions of the younger Tertiary. Charlesworth had observed that the Crag could be identified as Eocene (in Denmark based on Beck's analyses), Miocene (in Britain based on Sowerby's studies) and Pliocene (in France according to Deshayes). On the basis of consultation with Beck, Lyell realized the problems in taxonomic bias between specialists, and asserted that it was also possible to estimate the general degree of resemblance between fossil (in this case the Crag) and Recent assemblages as a test of the age of Tertiary deposits. Following subsequent field work Lyell was able to resolve to his own satisfaction (Lyell 1839a) that Charlesworth had been correct in recognizing three distinct crag units. The discrepancies between Beck's and Deshayes' taxonomy confirmed his

earlier assignment of the Norwich Crag to the Older Pliocene (equivalent to the Subapennine Series), while vindicating at the same time Charlesworth's assertion that the other two crags were of greater (Miocene) age.

Lyell breathed a sigh of relief that his 'statistical' method of subdividing the 'Tertiary' was (more or less) still intact. However, we see that Lyell did not necessarily believe exclusively in a rigorous application of his statistical methodology in determining relative stratigraphic and/or chronologic placement of Tertiary strata. As a result of differing taxonomic concepts held by different specialists, Lyell recognized that an equally important approach to the problem was the overall character of fossil assemblages and their degree of resemblance to modern ones as a means of determining the relative antiquity of Tertiary strata. This approach is seen to have its parallel in the contemporaneous work of Heinrich Georg Bronn (Heidelberg) and in the subsequent work of Moritz Hörnes (Vienna) a generation later (see below).

The subdivisions of the Tertiary remained relatively stable for the remainder of Lyell's professional life. Lyell never recognized the Oligocene of Beyrich (1854), although he discussed it at length (Lyell 1857a). While agreeing that there were significant differences between the molluscan faunas of the *Faluns de Touraine* type of his Miocene, and those of the Fontainebleau Sands (the *Cerithium plicatum* 'zone') of the Paris Basin, the older (Eocene) affinities of the latter formation were used to justify rejection of the term 'Oligocene'. At the same time Lyell recognized that the Eocene was becoming excessively unwieldy with the recent addition of the Thanet Sands and Lower Landenian of Belgium below its originally conceived lower limit. Consequently he agreed to delimit the Eocene and Miocene at the same level as adopted in France 'provided that we always regard it as an arbitrary and purely conventional line, – one which has no pretensions to be founded on any great change of species, still less on any general revolution in the earth's physical geography assumed to have happened at the era referred to' (Lyell 1857a, p. 11).

Lyell's compromise solution can, I think (from the advantageous, if perhaps comfortable and defensible, position of hindsight) be seen as an egregious affront to Beyrich. Rather than accept the inclusion of a new (Teutonic, no less) term into the exclusive Lyellian hagiography of standard Tertiary periods, Lyell chose the more convoluted solution of substituting the term 'Lower Miocene' for what had been heretofore called 'Upper Eocene', extracting several of the units formerly included in his Upper Eocene and transferring them to the Lower Miocene (e.g. the Hempstead Beds of England and the Fontainebleau Sands and *Calcaire de Beauce* of the Paris Basin, and 'Oligocene' strata of Northern Germany, i al.). He then subdivided the Middle Eocene in the following manner:

Middle Eocene

Bagshot and Bracklesham Beds of England (and foreign equivalents listed in stratigraphic order from yougest to oldest)
Calcaire Grossier of the Paris Basin,
Upper Soissonais, Sands of Cuise-Lamotte of France and
Nummulitic Formation of Europe, Asia etc.

The Lower Eocene remained the same as in the Table in the first edition (Lyell 1833, p. 106). In this way Lyell retained sole authorship of Tertiary stratigraphic terminology. (Dare one think what he might have said about the Palaeocene following its introduction by Schimper (1874), especially once some of his favorite Lower Eocene units were placed therein? In fairness to Lyell, however, it should be pointed out that in his major study of the Tertiary strata of Belgium and French Flanders (1852, p. 279, table 1) he had assigned the *Tuffeau de Lincent* (lower Landenian Dumont 1849) and the marls and glauconite of Heers (Heersian stage of Dumont) to a level 'intermediate between the lower Eocene [Plastic Clay–Lignite Soissonnais and London Clay] and the Cretaceous'.)

Lyell (1833) may be forgiven for his inconsistent use of stratigraphic terminology a half century before the stratigraphic nomenclatural hierarchy was formally established. He referred to the 'Recent' as 'period', 'era' and 'epoch' (p. 106) and divided the Tertiary 'epoch' into four 'periods'– the Eocene, Miocene and Older and Younger Pliocene – the reverse of the hierarchical ranking ultimately adopted when stratigraphic terminology was formalized at the sixth International Geological Congress in Bologna (1881). His original definition of the Tertiary system was based on biostratigraphic affinities of (predominantly molluscan) fossil faunas contained in disjunct rock sections with Recent (extant) taxa.

The Neogene

A historical review of the Tertiary period/system and of its consituent epochs/series has been presented by Berggren (1971), and of the Quaternary period/system by Hays & Berggren (1971) and Berggren & Van Couvering (1982). At this point I shall consider briefly some contemporaneous investigations (1830–1870) on the Continent that had an impact on Lyellian stratigraphic terminology.

Heinrich Georg Bronn (1800–1862), professor of zoology and technology at the University of Heidelberg made extensive trips between 1824 and 1827 in southern France and Italy. In 1831 Bronn published a major study on the fossils in the Italian Tertiary (Bronn 1831) which also included a comprehensive catalogue of the relative and absolute numbers of invertebrates, vertebrates, worms and 'zoophagous' (animal-eating) and 'phytophagous' (plant-eating) forms from the formations of various basins in Europe. He listed the numbers of definitely recognizable species in Italy (770), the Paris Basin (546), Bordeaux Basin (296), Montpellier Basin (529) and Vienna Basin (113) in terms of Lamarckian classification, and remarked on the degree of similarity of the Paris and London Basin faunas. He recognized a division of the Tertiary into lower and upper parts based on the presence or absence, respectively, of the large foraminiferid *Nummulites* in the Italian, Paris, London and Vienna Basins. Finally, he recognized the need for extensive additional studies on the faunas of European basins before a fundamental subdivision of the Tertiary could be made (Bronn 1831, p.174). Lyell made this subdivision two years later based on his extensive investigations with Deshayes on molluscan faunas of the European Tertiary basins, but there is no evidence that Lyell was aware of Bronn's concurrent studies or that they played any role in his own subdvision of the Tertiary. Indeed, Lyell lamented the fact that many important treatises were in German, in which he possessed 'only perfunctory skills'. He met Bronn finally, and briefly, during his honeymoon visit to Heidelberg in 1832. Bronn appears not to have suffered from the chronic Anglo-Saxon illness of linguaphobia; he referred copiously and knowledgeably to Lyell, Deshayes and Charlesworth, among others, in his later discussions on faunal affinities among 'Tertiary' faunas of Europe (see below).

Of greater importance in the context of this paper is the role Bronn played in subsequent subdivisions of the Tertiary into older (Palaeogene) and younger (Neogene) components. His major work, the *Lethaea Geognostica*, published between 1834 and 1838 (in several volumes), was followed by two later editions, the third of which was completed with the aid of Ferdinand Roemer, and appeared between 1851 and 1856 (in several volumes) and won for Bronn the 'reputation of being the most distinguished palaeontologist in Germany' (von Zittel 1901, p. 364). Here Bronn reviewed the stratigraphic and palaeontological content of all known subdivisions of the geological record. He provided meticulous illustrations and detailed decriptions of the most noteworthy of fossil forms from various formations in a set of 47 folio plates (1838).

Bronn divided Earth history into five periods and denominated as his fifth period of geological history, the *Molasse Gebirge* (Fig. 2), which included the *Tertiar und Quartär Formationen* and which corresponded essentially to what Lyell, following the historical progression in geological terminology since Arduino's introduction of the term in 1760, referred to as the 'Tertiary', i.e. all the rocks above the Chalk. The minor difference was that Lyell terminated his 'Tertiary' conceptually, if not actually, within the Pleistocene, in (unhappily for posterity) restricting the term 'Recent' to the 'interval since the earth has been tennanted by man'. The *Molasse Gebirge,* in turn, was subdivided into three groups, termed the *Molasse Gruppen.* The first (lowest) unit was correlated with Lyell's recently defined Eocene 'Period'; the second, or *Molassen Gruppe (sensu stricto),* included the Miocene and Pliocene 'Periods' *sensu* Lyell and were treated as subunits and referred to as the *untre Abtheilung* (Miocene) and *obre Abtheilung* (Pliocene), respectively. The upper part of this unit was correlated with the Subapennine Formation of Italy, the *Diluvial Bildungen* of Continental, predominantly Austrian and German, authors, the Pliocene of Lyell and Deshayes and the Middle Tertiary of North America (*partim*). The third *Molassen Gruppe* contained the *Alluvial und Quartär-Gebilde zum Theille* (corresponding essentially to what is termed Recent or Holocene at present), the Pleistocene proper *Knochen Breccien, Knochen-Hohlen und der Loss,* the latter of which Bronn interpreted, as did Lyell, as marine deposits that were already incorporated within the upper part of his second *Molassen Gruppe* (see Fig. 2). Bronn noted that he was unable to distinguish the faunas from the Vienna and Hungarian Basins, Poland and the Siebenburg from those of the Subapennine strata of Italy and thus was unable to consider them as older, contrary to the opinion of Deshayes. He also compared the percentage of fossil taxa relative to living forms in a manner similar to Deshayes but does not appear to have applied it rigorously as a biostratigraphic tool.

Twenty-five years later, in the course of preparing a monograph on the molluscan faunas of the Vienna Basin, Moritz Hörnes (1815–1868), Director of the Museum of Natural History in Vienna (1836–1868), noted that the Miocene and Pliocene faunas were more similar to each other than to those of the Eocene. In creating the term Neogene for these upper, younger faunas, Hörnes (1853, 1864) referred specifically to the biostratigraphic subdivision of the Tertiary and Quaternary made by his friend Bronn in 1838. Indeed, Hörnes drew attention to his observation that so similar were the typical Mio-Pliocene faunas of the Vienna Basin to the 'Pliocene' faunas of Sicily, Rhodes and Cyprus that no clear line of demarcation was possible between the two units; he favoured combining the two into a single unit. He further remarked that rigorous adherence to quantitative methodology was of little worth. Rather he preferred to use the overall character of the fauna and, in particular, the more or less common occurrence of index forms as indicators of relative stratigraphic position in preference to what he saw as the shortcomings of current practice. These included (a) unique occurrences, (b) poorly determined forms and (c) immature individuals, all of which yield unreliable results. The original Neogene concept thus includes *a priori* all post-Eocene molluscan faunas. Hörnes appears to have understood the distinction between typical Eocene, Miocene and the term his friend Beyrich was about to create (1854) and insert between the two, namely the Oligocene (see Hörnes 1853, p. 808). Hörnes included in his term Neogene the strata in the Vienna Basin up to and including those in glacial loess and diluvial deposits, as well as correlative Mediterranean faunas in Sicily, Rhodes and Cyprus which would now be included in the Pleistocene. It will be recalled that Lyell coined the term 'Pliocene' in 1833 and subsequently (1839*b*, 1857*a,b*) subdivided it into an Older Pliocene and Younger Pliocene (the latter equivalent to the Pleistocene). At the same time it should be remembered that Hörnes' Neogene specifically referred to, and incorporated inclusively, the subdivision of the Tertiary and Quaternary of Bronn (1838) and not that of Lyell (1833).

The Neogene was originally proposed to characterize the faunal (and particularly molluscan) and floral changes that denote the middle part of the Cenozoic (beginning, but not clearly delimited originally, at the present day Oligocene/Miocene boundary) and continue to the present day. While its original connotation was that of a biostratigraphic term (as were Lyell's Tertiary 'periods'), in its modern resurrection as a chronostratigraphic unit it corresponds to the interval since the beginning of the Miocene epoch/series.

The terms 'Neogene' and 'Palaeogene' found little favour in British geological circles until the 1960s. Indeed, there was widespread resistance to the whole concept of 'stages' in British geological circles until after World War II. In contrast to this, in continental Europe and to a lesser extent in North America the term 'Neogene' was widely acknowledged, and generally accepted, in subdivisions of the upper Cenozoic during the latter half of the nineteenth century (cf. Van Couvering 1997, p. xii). Renevier (1897), for example, recognized it as a fundamental unit in the subdivison of the upper part of the Cenozoic era. By way of background a few notes are listed here on Renevier's use of terminology:

(1) In conforming to the rules established at the International Geological Congress at Bologna in 1881, Renevier adopted a nested hierarchy of four orders of subdivision with decreasing importance in regional extent:
  (i)   eras = groups; of global/mondial extent/ value;
  (ii) periods = systems; of very general extent/value;
  (iii) epochs = series; primarily of European extent/value;
  (iv) ages = Stages; of only regional extent/ value.
A fifth order, substages/beds, is viewed as strictly of local value.

(2) The term *Cenozoaire* (fr.) = 'Cenozoary' (engl. [*sic*!]) is used to signify the ordinal value of the division. *Neogenique* (fr.) = 'Neogenic' (engl. [*sic*!]) is used for the second order.

(3) 'Ceno-' is preferred to 'Kaino-' in conformity with the construction of the words 'Pliocene', 'Miocene', 'Eocene', etc. Inasmuch as the roots are identical, if one acccepts 'Kainozaire', one must also accept 'Pliokaine'.

(4) 'Plistocene' is used in preference to 'Pleistocene', which is consistent with 'Pliocene'. Inasmuch as the root of the former (Plistocene) is the superlative of that of the second (Pliocene), they must be interpreted in the same manner. Whereas it may be etymologically more correct to say 'Pleistocene', it would then be necessary to use 'Pleiocene' and 'Meiocene'; 'any other usage is illogical!' observed Renevier.

(5) Consistent with earlier editions of his work, Renevier included in the Tertiary Era the Plistocene and Recent ('*actuelle*') and equated the Tertiary and Cenozoic. By way of justification he referred to his earlier (1874) explanation that there are no important organic changes between the Tertiary (older usage) and the Quaternary. Few types disappeared and few new forms appeared except humans, whose existence during the Pliocene in Italy and Belgium has been indicated by several contemporary authors. Compare the major differences between the Primary, Secondary and Tertiary eras on the other hand, he declared. He observed that his (deceased) teacher F.-J. Pictet had expressed the same views in 1857, and repeated from his earlier publication (Renevier 1874, p. 233) the contemporary quote of Gervais:

L'époque que l'on continue, on ne sait trop pourquoi à appeler Quaternaire, comme si elle constituait une nouvelle grande série de faunes et de flores... ('the epoch which we continue to call, for some reason which is not all clear, Quaternary, as if it constituted a new and important series of faunas and floras...')

(6) Renevier pointed to the diverse usages of 'Tertiary' and 'Quaternary'. D'Orbigny and Mayer set the Recent epoch apart but combined the Tertiary and Quaternary. Lyell and Gaudry terminated the Tertiary above the Plistocene with *Elephas meridionalis*, i.e. in the middle of the Quaternary of most authors. Finally Naumann subdivided the Cenozoic into Quaternary and Tertiary. Renevier (1897, p. 558) concluded, 'Ces divergences confirment ma thèse qu'il n'y a point là de division primordiel naturelle. (These differences confrm my thesis that this is not a natural and fundamental division.)

Renevier's (1897) subdivision of the upper part of the Cenozoic era is shown below in Fig. 3. Renevier's use of the term 'Prepliocene' was unfortunate in hindsight, but fully understandable in light of the knowledge available at the time on the stratigraphic position of the Pontian Beds of eastern Europe. The Pontian stage now rests relatively secure in the bosom of the late Miocene (Steininger *et al.* 1989, 1996). It is readily seen that Renevier's (1897) denotation of the term 'Neogene' is essentially identical to that endorsed here a hundred years later.

De Lapparent (1885, 1906), in the second and

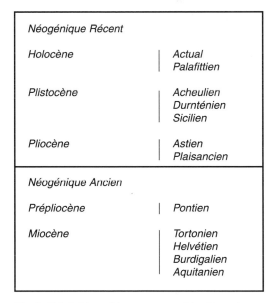

Fig. 3. Subdivision of the upper part of the Cenozoic Era by Renevier (1897).

third editions of his monumental *Traité de Géologie*, equated the Cenozoic with the *Neozoique* (fr.) = 'Neozoic'. He followed Lyell's threefold subdivision of the Tertiary, included the Miocene and Pliocene in the Neogene and equated the '*Ere moderne ou Quaternaire*' with the '*époque Pleistocène*'. The latter was considered to be characterized by the appearance of humans and of glacial climates on Earth; De Lapparent indicated a preference for the term 'Pleistocene' (rather than 'Quaternary') because of its unequivocal denotation of continuity with the term Pliocene. While admitting that it was still impossible for the science of geology to provide a precise chronometry for the Pleistocene, he believed this interval was relatively short, on the order of several hundreds of thousands of years. It is an irony of history that De Lapparent's second edition (1906) was published in the same year that the first use of radiometry was made in the service of a geological timescale, following the discovery of radioactivity a decade earlier by Becquerel (1896). In this context it is also important to remember that De Lapparent followed contemporary opinion in including the Sicilian 'stage' in the 'upper Pliocene' (*contra* Lyell 1833); he equated it with the Norwich Crag, the Forest Bed of Cromer, the Amstelian of Belgium–Holland and the appearance of *Elephas meridionalis* in southern France (Languedoc-Roussillon) and considered the base of the Pleistocene essentially contemporaneous with the first appearance of *Elephas antiquus, Rhinoceros mercki, Hippopotamus major, Ursus spelaeus, Cervus megaceros* and (in neritic environments) *Mya truncata*. This demonstrates that the base of the Pleistocene (or Quaternary) at the turn of the nineteenth century (in the sense shown here) was located at a level corresponding to what is currently considered close to the early/middle Pleistocene boundary near the Brunhes/Matuyama boundary at a little over 0.8 Ma (Jenkins *et al.*, 1985; Berggren *et al.* 1985; Van Couvering 1997).

In his *Manual of Geology*, Dana (1894) referred to the use by some contemporary authors (e.g. William H. Dall and Gilbert D. Harris), and even by the US Geological Survey map of 1884, of 'Neocene' for the combined Miocene and Pliocene. He noted (Dana 1894, p. 880) the curious, if not iconoclastic subdivision by Heilprin in 1887 of the Cenozoic into the Eogene (Eocene plus Oligocene), Metagene (Miocene) and Neogene (Pliocene plus Quaternary). Finally Dana observed that the Tertiary 'is a relic of early geological science which retains its place simply because of the convenience of continuing an accepted name' and notes that the term 'Neozoic' (originally described by Edward Forbes (1846) to denote the Mesozoic and Cenozoic) is sometimes used in a more restricted

sense to denote the Tertiary and, occasionally, as a substitute for the entire Cenozoic.

The culprit behind the current misunderstanding of the correct denotation of the term 'Neogene' appears to be Maurice Gignoux (1913), who used the term in his doctoral thesis on the Plio-Pleistocene succession in Italy without providing a discussion of the historical framework of the term. He simply excluded the Pleistocene (or Quaternary) from the Neogene. This viewpoint was promulgated in his subsequent textbook (Gignoux 1955). The term appears to have served as little more than 'a casual literary convenience', as Van Couvering (1997a, p. xii) so aptly put it. It should be recalled that whereas Hörnes (1853) did not recognize a separation of Pliocene and Pleistocene in his original distinction of the Neogene, Gignoux (1913, 1950, 1955) took for granted such a separation. It may be of interest to note that one of the few instances in which the Neogene has been used correctly in its original sense (incorporating the Miocene through the Holocene) is in a recent text book on historical geology (Cooper *et al.* 1996, fig. 3.1, p. 66, fig. 3.2, p. 67).

This failure to recognize the original intent for the term 'Neogene' has had unfortunate consequences in attempts to modernize the standard chronstratigraphic subdivision of the Cenozoic. In overlooking the fundamental historical background to the term 'Neogene', the INQUA Subcommission on the Neogene/Quaternary (N/Q) boundary inherited a 'dead cat' when it, and eventually the International Commission on Stratigraphy (ICS), wrestled with the boundary problems in the late 1970s and early 1980s. The Neogene period/system, in its original sense, was a biochronologic unit (Berggren & Van Couvering, 1974, 1979), overlapping chronologically the Quaternary era/erathem in its modern sense, and quite irrelevant to questions of a lower boundary for the Quaternary. Perhaps most damning to proponents of a distinction between Neogene and Quaternary is the simple fact that no upper limit to the Neogene was specified either by Hörnes (1853) or by Gignoux (1913, 1950, 1955). There is, simply, no post-Neogene! We are still living in the Neogene period!

It is ironic that Gignoux's revised concept of the Neogene has been accepted by those who advocate eliminating the term 'Tertiary' (Harland *et al.* 1982, 1990), whereas the IUGS (Cowie & Bassett 1989) simply omitted the Tertiary in their 'official' timescale and juxtaposed the Neogene and Quaternary, the latter a rather inappropriate denomination for a third, rather than fourth, subdivision of the Cenozic, as Van Couvering (1997a, p. xii) has ruefully observed. The original IUGS commission declaration (King & Oakley, 1950) specified that it was the Tertiary/Quaternary boundary that was to be identified with the Pliocene/Pleistocene boundary. This has finally been achieved in a satisfactory manner (Aguirre & Pasini 1985, Pasini & Colalongo 1997). Current attempts in some quarters to lower the boundary of the Pleistocene to a level around 2.6–2.75 Ma, based on climatic criteria, are viewed here as misguided and ill-informed. They will confuse a situation that has only recently seen the agreement of both the Neogene and Quaternary Subcommissions of the IUGS in selecting a Global Stratotype Section and Point (GSSP) for the base of the Pleistocene (and Quaternary).

The Palaeogene

If the Neogene has suffered from neglect and connotational and denotational misinterpretation, the Palaeogene may be said to have suffered a crisis of author identity. The term was attributed to Hörnes by no less than Dana (1894, p. 880), Harland *et al.* (1990, p. 61), Schoch (1989, p. 25) and seemingly by Papp (1959, p. 5; 1981, p. A499), among others, a classic example of Norman Watkins' famous dictum on the reinforcement syndrome: 'repeat it often enough and it becomes the accepted truth'. Fortunately, authorship was correctly ascribed by Denizot (1962, p.151) in his contribution to the *Lexique Stratigraphique International*.

The term 'Palaeogene' owes its origin to Carl Friedrich Naumann, professor of mineralogy and geology at the University of Leipzig (beginning in 1842). Before this, Naumann had been successor to Werner (who had been professor of crystallography in Freiburg/Saxony) as professor of geognosy. The term 'Palaeogene' was introduced in the third volume of Naumann's monumental *Lehrbuch der Geognosie* (1866, p. 8). In his second volume (1854, pp. 1029–1033) Naumann had reviewed the role played by Deshayes in establishing the quantitative methodology as a biochronologic tool in Tertiary stratigraphy, and Lyell's subsequent application of this to establish formal subdivisions of the Tertiary. He drew attention to his friend Hörnes' warning that caution should be used in applying Deshayes' quantitative methodology in a rigorous manner, owing to the problem of identifying rare species, etc. (an echo of what Hörnes expressed the year before in his 1853 letter to Naumann about the term 'Neogene'; see above). He noted his indebtedness to Hörnes, who had recently (1853) confided to him his observations on various faunas of the Tertiary and his intent and reasons (i.e. the major difference between Eocene and Mio-Pliocene molluscan faunas) for proposing the new term 'Neogene'. He noted that none of this was particularly new, for Bronn had focused

attention on these facts in the first edition of his *Lethaea Geognostica* (1838), as well in the third edition of the work, when dividing the Tertiary into a lower and an upper part. He observes that, according to Hörnes, the Tertiary formations can be subdivided in the following manner:

A. Eocene
   1. Older Eocene; Paris Basin, London
   2. Younger Eocene; Lesbaritz, Tongrian and Rupelian System in Belgium, Westeregeln near Magdeburg (which was to become part of Beyrich's Oligocene in 1854)

B. Neogene
   3. Older Neogene; Touraine, Bordeaux, Vienna, Turin, Poland
   4. Younger Neogene; Asti, Castell'Arquato, Sicily, Rhodes, *i.al.*

We see that Naumann was not ready to subdivide the lower part of the Tertiary at this stage (1854). At this time he used the terms *Nummuliten und Flysch Formation* for what he subsequently termed 'Palaeogene'. However, in the third volume of his *Lethaea,* published a decade later (1866, p. 8), Naumann states,

Da sich nun Hörnes noch neurdings dahin erklärt hat, dass er durch den Collectiv-Namen neogen nur auf die scharfe Gränze zwischen den eocänen und den neueren Tertiärbildungen hinweisen wolllte, ohne deshalb die Möglichkeit einer weiteren Eintheilung dieser beiden Haupt-Etagen in Abrede zu stellen, so lässt sich die viergliederige Eintheilung er Tertärformationen auch in folgender Weise darstellung machen: (Inasmuch as Hörnes has recently explained that by means of the collective name of Neogene he is referring only to the sharp boundary between the Eocene and the younger Tertiary strata, without the possibility of any further subdivision of these two main 'stages' [periods/systems in modern parlance], this allows further subdivision of the Tertiary formations in the following manner):

A. Palaeogene Tertiary
   1. Eocene Formations
   2. Oligocene Formations

B. Neogene Tertiary
   3. Miocene Formations
   4. Pliocene Formations

In other words, it seems that Naumann felt that he was obliged by default to name the Palaeogene since Hörnes had shown no inclination to do so following his introduction of the term 'Neogene' a decade earlier. In an interesting comment, Naumann (1866, p. 9) drew attention to the recent

(1857) classification of the Tertiary of Europe by Mayer-Eymar into 'Stufen' (or stages), based on his belief in the superiority of this approach as opposed to the classification of Deshayes and the nomenclature of Lyell. Naumann then listed the six Lower Tertiary and six Upper Tertiary stages of Mayer, and observed that the upper two stages (Astian and Saharan, which included the Norwich Crag, as well as the Sands of the Subapennine Formation and many younger units) would be included under the term Pliocene. This further illuminates the original intent of Hörnes' definition of the term 'Neogene' and contemporary understanding (by Naumann) of this. The stages of Mayer (and those of d'Orbigny 1849–1852) and their role in Cenozoic stage terminology have been discussed in greater detail by Berggren (1971).

The denotation of the Palaeogene has not suffered the same convoluted history as its younger counterpart, the Neogene. With the addition of the term 'Palaeocene' by Schimper (1874), a threefold subdivision of the Cenozoic was generally accepted on the Continent); elsewhere adoption of the term 'Paleocene' was delayed for various reasons (it was not acknowledged in Britain, nor was it formally accepted in the United States until 1939). Acceptance of the term 'Palaeogene', however, has been uneven. For instance, our hero of the Neogene, Renevier (1897) balked at accepting it on rather flimsy grounds (seen from this vantage point), saying it was too similar to 'Palaeocene' or 'Palaeozoic', while acknowledging current usage of 'Eogene' as a synonym. 'Je comprends Néogène (nouvellement formé), mais Eogène (aurore formée) c'est ridicule!' (I understand the term Neogene (recently formed), but Eogene, that is ridiculous!) (Renevier 1897, p. 562). Instead he fell back on the familar and popular French term *Nummulitique*, which remains popular in French texts to this day but is inappropriate as a chronostratigraphic term on stratigraphic and etymologic grounds (*Nummulites* ranged from the late Palaeocene – Thanetian age – until the end of the early Oligocene – Rupelian Age – and the term defined in this manner scarcely encompasses the time between the Cretaceous/Palaeogene (Senonian/Paleocene) boundary and the Palaeogene/Neogene (Oligocene/Miocene = Chattian/Aquitanian) boundary.

The Pleistocene and the Quaternary

Lyell (1839b) created the term 'Pleistocene' for his Younger or Newer Pliocene (1833), almost as an afterthought, in an appendix to a French translation of his *Elements of Geology* (1838). Its boundary (and that of the Tertiary) with the Recent was said to conform with the appearance of humans, a definition that, in hindsight, was quite unfortunate and led to varying degrees of confusion in determining the limits of upper Cenozoic (chrono)stratigraphic terms. Lyell used the term 'Post-Tertiary' for the formations above/younger than the Tertiary and retained the subdivision into post-Pliocene and Recent through many editions of his *Principles*.

The Pleistocene had been given a 'glacial' connotation by Forbes (1846) in referring to what he thought Lyell (1833) had termed Newer Pliocene (or Pleistocene). But Lyell pointed out that Forbes' use referred to post-Pliocene and Lyell withdrew the term Pleistocene. This inappropriate use of the term resulted from the fact that Lyell retained glacial and associated topics under the Newer Pliocene through to the fifth edition of the *Manual of Elementary Geology* (1855). Lyell (1857a,b) modified his Tertiary terminology in distinguishing the following units (in descending order): Recent (deposits with human remains, alluvial deposits of the Thames Valley with buried ships); post-Pliocene (in which he grouped deposits with fossil shells of living species in which no human remains have been found and shell-marls of Scottish and Irish lakes); Newer Pliocene, or Pleistocene (with which he equated glacial deposits as well as preglacial deposits of the Thames Valley and the Norwich Crag and the *Terrain quaternaire, diluvium* and *Terrains tertiaires supérieurs* of the Continent). Below these units followed the Older Pliocene, Miocene (subdivided into an upper and lower part, the latter consisting of what had previously been included in his upper Eocene), and an upper and lower Eocene. Above the post-Pliocene was the post-Tertiary. Lyell clearly considered the Tertiary to encompass all but the youngest, superficial deposits of the present day.

Lyell (1873) later adopted Forbes' usage and incorporated his post-Pliocene into Newer Pliocene or Pleistocene (the latter term he finally accepted and substituted for 'Newer Pliocene'). Modern usage of the term 'Pleistocene' encompasses Lyell's (1833) Newer Pliocene or Pleistocene and his vaguely defined 'Post-Pliocene'. The term 'Holocene' has generally been used for the present interglacial with a defined base at about 10 ka (radiocarbon), but is little more than an interglacial interval of the Pleistocene (Berggren & Van Couvering 1974, pp. 51, 55).

The base of the Pleistocene has recently been stratotypified at the Vrica section in Calabria (southeastern Italy), at a level near the top of the Olduvai Magnetic Polarity Chrononozone (C2n) with an estimated age of 1.81 Ma (Berggren *et al.* 1995; Pasini & Colalongo 1997; Van Couvering 1997). With the recent addition of a Milankovich-based astronomical timescale, we now have a precise chronology for the events associated with this boundary. We can link the bounding limits of the Olduvai Subchron with isotope stages 64 (younger) and 72 (older), respectively, and can effectuate global correlations via integrated magnetobiochronologic studies. The Pliocene remains strongly anchored to proper Lyellian stratigraphic procedures with a threefold chrono-stratigraphic subdivision (Zanclean, Piacenzian and the recently proposed Gelasian stage).

The Quaternary (*Quaternaire ou Tertiaire récent*) was introduced by Jules Desnoyers (1829) as the fourth, and final, subdivision of the then threefold subdivision of the geological record (Primary, Secondary and Tertiary), for the rocks in the Loire–Touraine Basin and Languedoc that were demonstrably younger than those of the Seine–Paris Basin. He subdivided the rock units into three parts (from younger to older): 3) *Récent*; 2) *Diluvium*; 1) *Faluns de Touraine, la Molasse suisse, le Pliocène marine de Languedoc*. In retrospect we see that this original definition essentially corresponded to, and included, what Lyell was to include in his Miocene, Pliocene and Recent (1833, 1839b) and Hörnes (1853) in his Neogene. Marcel de Serres (1783–1862), professor at the University of Montpellier, used the term *Quaternaire* in 1830, in considering it synonymous with the term *Diluvium* (as proposed by Buckland in 1822 for deposits of the biblical deluge), and observed that humans were contemporaneous with these Quaternary deposits, thereby restricting its stratigraphic and chronologic extent closer to its modern denotation (see also De Serres 1824). This apparently led him later (1855) to claim, with questionable justification, priority for creation of the term 'Quaternary'.

The term 'Quaternary' was given a faunal connotation by Reboul (1833), who distinguished it as containing living species of animals and plants (*Période anthropéenne*) as opposed to the Tertiary, which was believed to contain mostly, if not exclusively, extinct species. This work and the classic treatise of d'Archiac (1849) on the *Terrain Quaternaire* or *diluvien* gave added weight to acceptance of the term 'Quaternary' in geological literature. The Swiss archeologist C. A. Morlot (1854/1856) introduced the word *Quaternaren* into the German language and subsequently (1858) modified it to *Quartären*, a translation of the term *Quartaire* which he had proposed that same year and restricted to the post-Pliocene. The term *Diluvium* was used in earlier German literature as a synonym for the Pleistocene and *Alluvium* for the Recent (in sharp contrast to early English usage following Mantell's (1822) original designation of superficial sediments in two categories: 'Diluvium', sediments formed during, and by, the biblical Flood; and 'Alluvium', post-Flood sedi-

ments formed by modern rivers, streams, etc.). The two terms equated, respectively, with glacial and post-glacial time. They have historically been combined to represent the Quaternary period/system in both German and Anglo-Saxon literature, and also, but with a predominantly mammalian faunal denotation, in French literature. Modern studies have shown that there is no relationship between the initiation of polar glaciation and the base of the Quaternary, and that the placement of the boundary of the Quaternary (the base of the Pleistocene) should be based upon changes in marine faunas, as with all other Phanerozoic period/system boundaries (see reviews in Hays & Berggren 1971; Berggren & Van Couvering 1982).

Lyell (1833) introduced the term 'Recent' (see above) for the time 'which has elapsed since the earth has been tennanted by man' at the same time as Reboul (1833) was busy redefining the Quaternary. As such the Recent included both the Pleistocene and Holocene of modern usage (although the term 'Recent' has since been modified to correspond to the Recent, or inter-glacial, epoch of some geologists). To add further to the confusion, Gaudry, with the tacit approval of Prestwich and De Lapparent, introduced at the International Geological Congress in London (1888) the proposal that humans (represented by their artefacts in particular) were the characteristic element of the Quaternary (*fide* Hays & Berggren 1971, p. 670).

In presenting a review of the terminology of post-Pliocene stratigraphic terminology, Richard Foster Flint (1947, p. 281), patron saint of glaci-ologists and Quaternary geologists and professor of Geology at Yale University, pointed to the transitional nature of Pliocene–Pleistocene strata in terms of both lithostratigraphy and biostratigraphy and concluded that

> to consider such a boundary as separating time units of so major an order as periods, equivalent in rank to the Cretaceous period, is to over-emphasize its importance. The Pliocene–Pleistocene boundary cannot justifiably have 'system' value, but only 'series' value. It follows that the ancient concept of a 'Tertiary period' and a 'Quaternary period', though fully adequate for the time it was first used, does not now rest on firm ground and should be abandoned. In other words we should think of the present as a part of the initial period of the Cenozoic era.

He urged recognition of the Pleistocene as a time-stratigraphic unit based on fossils and not climate (although climatic perturbations would be expected to have left a characteristic imprint upon the stratigraphic record of the Pleistocene) and remained cool about use of 'Holocene' as a formal time-stratigraphic term, suggesting, instead, informal use of the terms 'recent' or 'postglacial' in geographically restricted areas.

Flint (1965, 1971, p. 384) reiterated these views a couple of decades later in updated discussions of the terms 'Pleistocene' and 'Quaternary' and concluded by saying

we dissent from common practice in that we favor the dropping of Tertiary System and Quaternary System from stratigraphic nomenclature. If this were done, the Pleistocene Series would include all post-Pliocene strata, as implied by Lyell in 1839. Further, we believe, in view of the long term span of the succession of late-Cenozoic cold climates that the Pliocene/Pleistocene boundary should not be based on climatic indications. Finally we think the terms recent and postglacial should be used only informally, and applied only within geographically restricted areas. We have little hope that these changes will come about soon, but we think they are soundly based. (Flint 1971, p. 384)

The recommendations by Schuchert and Dunbar (1941), Dunbar and Rogers (1957) and Flint (1947, 1965), among others, to discard the terms 'Tertiary' and 'Quaternary' have been echoed more recently in the publications of Berggren and Van Couvering (1974), Berggren *et al.* (1995) and Steininger (1996), and (at least for 'Quaternary') Harland and colleagues (1990). Pomerol (1973, p. 12) expressed the situation well in stating 'l'ére quaternaire' ...n'est paléogéographiquement parlant, que la projection du Néogène dans les temps actuels'. Unfortunately, after making this observation he proceeded to separate the Neogene from the Quaternary at the Pliocene/Pleistocene boundary thus adding to the proliferation of misconceptions about the denotation of the Neogene period/system.

Nomenclatural reform

The discovery of radioactivity by Becquerel in 1896 provided one of the great ironies in history. It squashed once and for all the intellectual conservatism and arrogance of the physicist William Thomson (Lord Kelvin) (Emiliani 1982) who had denied Charles Darwin, the 'deep time' he required to explain his observations on evolution as reflected in the geological record. Kelvin gave the geologists up to 100 Ma but more likely 20 Ma based on his estimates of cooling rates in the Earth's interior. Darwin died in 1892, believing there had been inadequate time for evolution to have occurred as he believed via natural selection. At the same time the discovery vindicated the 'deep time' expounded by James Hutton a century earlier based only on his

interpretation of the relatively long time necessary to account for the tectonic upheavals observed in the form of major Palaeozoic unconformities in Scotland.

In the intervening century, geologists had no means of independently calibrating to a standard scale the geological events that they – 'uniformitarians' and 'catastrophists' alike – gradually came to realize were spread out over a considerably longer interval of time than that represented in the 'orthodox' Mosaic chronology of six millennia allotted the Earth by Bishop Ussher. Lyell wrestled with these problems in the course of a long and productive career that spanned about 40 years. He developed an essentially steady-state, non-directional view of history in building upon Hutton's theory of recurring geological cycles. But in contrast to his intellectual mentor, Lyell championed the idea that a long period of time was required to record the Earth's history as manifested in the stratigraphic record, and thus effected a major breakthrough, allowing his contemporaries to come to grip with the concept of 'deep time'. The passage of time as recorded in the rocks was placed in a conceptual framework by Lyell; this remains one of his major legacies to modern stratigraphy. His stratigraphic terms were subsequently placed in a more rigorously based hierarchical system over the century following his death. We can do no greater honour to Lyell on this bicentenary of his birth than to place his now standard subdivisions of the Cenozoic erathem into a modern chronostratigraphic framework and to eliminate the last vestiges of antiquated stratigraphic nomenclature.

Modern concepts and principles of stratigraphic classification and procedure were formulated by the International Subcommission on Stratigraphc Classification (Hedberg 1976) of the International Geological Union of Geological Sciences (IUGS) International Commision on Stratigraphy (ICS), and recently updated by Salvador (1994). A corollary of this work has been the delineation of guidelines for establishing boundary stratotypes between the stage, series and systems of the Cenozoic erathem (Cowie 1986, recently updated in Remane *et al.* 1996). Working groups have been established under the aegis of the Palaeogene and Neogene Subcommissions of the ICS and several boundary stratotypes have now been submitted to, and ratified by, the ICS; for example, Cretaceous/Paleogene (= Danian/Maestrichtian); Eocene/Oligocene (= Priabonian/Rupelian); Palaeogene/Neogene (= Oligocene/Miocene = Chattian/Rupelian); Pliocene/Pleistocene (= Gelasian/Calabrian). Working groups are currently involved in studies on boundary stratotypes for the Palaeocene/Eocene and Miocene/Pliocene boundaries, as well as the intra-series stage boundaries with a view to completing most of this work by the next International Geological Congress at the turn of the century.

The proper stratigraphic procedure for defining upper Cenozoic GSSPs was debated at length during the late 1960s, the 1970s and early 1980s at numerous international conferences of the International Geological Correlation Program (IGCP) 41 (Pliocene/Pleistocene Boundary), the Neogene Subcommission and the INQUA Subcommission on the N/Q Boundary (particularly in the former Soviet Union). At these meetings most Soviet and other geologists specializing in Quaternary studies were finally convinced that chronostratigraphic boundaries are typified in marine stratigraphic sections and that the definitions are lithostratigraphic (GSSP), whereas the means of correlation were/are heterogeneous (biostratigraphic, palcomagnetic, radioisotopic, stable isotopic, etc.). More recently, astronomical periodicity in the stratigraphic record has played a major role in chronostratigraphy. INQUA geologists have often resorted to special pleading for definition of boundaries in the Quaternary (Hays & Berggren 1971) in terms of vague and unscientific criteria (such as climatic changes, evidence of glaciation, hominid evolution, mammalian evolutionary or immigration events). These special pleadings have now been silenced and it is time to move on in recognizing an appropriate hierarchical terminology for the Cenozoic erathem. We have more important issues to which we should direct our attention as we approach the millennium than sterile arguments about moving a well defined Pliocene/Pleistocene boundary stratigraphically downwards into what is for the vast majority of Earth scientists the Pliocene epoch (at 2.6–2.8 or 3 Ma). The time seems right to complete the updating of Cenozoic terminological hierarchy in accordance with the data reviewed in this paper.

The basic principles of stratigraphy and a terminological hierarchy to accommodate the primary subdivisions of the geological record were established only after Lyell's death, so that to speak of a Lyellian 'chronostratigraphy' is, strictly speaking, incorrect. Nevertheless, in order to recognize the fact that Lyell's fourfold subdivision (into epochs/series, in modern parlance) of the Cenozoic era has withstood the test of time and subsequently been given a chronostratigraphic connotation, I have chosen to place the 'chrono-' component of 'chronostratigraphy' in the title of this paper in parenthesis.

Lyell died in 1875, the same year as his longtime collaborator Paul Gérard Dehayes, his-long time colleague (and antagonist) Adam Sedgwick and the Belgian palaeontologist Jean Baptiste

Julien d'Omalius d'Halloy, a year after (1874) Léonce Élie de Beaumont (his French antagonist, with whom he retained cordial relations throughout his life) and a year before (1876) his long-time friend George Poullet Scrope. It was a 'rapid turnover'.

Conclusions

Chronostratigraphic terminology should reflect the current conceptual and operational framework of stratigraphy. While stability is generally accepted as having precedence over priority in stratigraphic names (as opposed to taxonomic priority in palaeontology), the elimination of demonstrably inappropriate (i.e. antiquated) terms from the nomenclatural hierarchy can be defended in certain circumstances. The terms 'Tertiary' and 'Quaternary' are prime examples, relics of a now outdated and terminologically inappropriate fourfold subdivision of the Phanerozoic dating back nearly 250 years (which also included the now abandoned terms 'Primary' and 'Secondary').

This study has shown that:

(1) The Neogene period/system includes – on first principles harking back to Hörnes' (1853) definition based partially on Bronn's (1838) subdivision of the *Molasse Bildungen* of the Vienna Basin, plus his own reference to the younger (i.e. Pleistocene or Newer Pliocene) outcrops of Malta, Sicily and the Italian coast – the time and corresponding rocks from the revised, post-Lyellian base of the Miocene up to and including the Pleistocene and Holocene.

(2) The term 'Tertiary' should be suppressed (as it was most judiciously by Cowie and Bassett (1989) in their IUGS Time Scale) along with the 'Quaternary' as remnants of a now outmoded and inappropriate terminology, as suggested by Flint half a century ago (1947) and reiterated a quarter of a century later (Flint 1971; see also Berggren *et al.* 1995). (I favour retention of the term 'Quaternary' for geopolitical purposes but not as a hierarchical part of our canonical chronostratigraphic hagiography.)

(3) The Palaeogene (Palaeocene, Eocene and Oligocene, together with the Neogene (Miocene, Pliocene, Pleistocene and (reluctantly) Holocene) represent appropriate period/system subdivisions of the Cenozoic era/erathem.

Adoption of the procedures recommended here will finally rid Cenozoic stratigraphy of the last vestiges of terminological conservatism, and diminish, if not eventually eliminate, unnecessary and unscientific rancorous debate and allow our science to flourish with renewed vigor at the turn of the millennium . This will serve as a fitting tribute to Sir Charles Lyell, whose pioneering studies in (bio)stratigraphy established the standard epoch/series units of the modern Cenozoic era/erathem, and whose eclectic approach led to the first great synthesis of geology, which established historical geology as the fundamental basis for a modern natural philosophy of life.

Epilogue

'La République n'a pas besoin de savants.' The remark has been attributed to Coffinhal, revolutionary tribunal judge during the Reign of Terror, in sentencing Lavoisier to the guillotine in 1794. Although the attribution is apocryphal (Gould 1989), it expressed the views clearly felt at the time by the disenfranchised.

Two hundred years later we are witnessing a disturbing resurrection of this attitude in modern society at the political level (drastic reduction or elimination of whole areas of science by governmental agencies in the name of fiscal accountability, and concomitant resurgence of pragmatic, goal-oriented applied research) and the level of popular culture (in the form of media attention to pseudoscience in prime-time TV, New Age crystals and associated accoutrements of 'alternative lifestyles' – flying saucers and alien abductions, astrology and the like). Now, more than ever, it is incumbent upon scientists to play an active role in providing the framework of connections between past, present and future knowledge of the natural world and thus furnish constraints upon theoretical speculations. Sir Charles Lyell, one of the leading geologists of the nineteenth century played such a role in his time by providing the conceptual framework for the transition from a Eurocentric worldview dominated by an ahistoric, deistic tradition to a modern, global view of the history of the Earth based on the evidence of the geological record, while establishing historical geology as the fundamental component of a natural philosophy of life. This essay is dedicated to his memory on the occasion of the bicentennial of his birth and to those geologists who still, regretfully, fail to place their own research in a historical context by seeking out original sources in their work, thereby failing to communicate our communal debt to antiquity and historical continuity in the advancement of knowledge.

I thank Derek Blundell and Andrew Scott (University of London) for their invitation to speak at the Lyell Meeting of the Lyell–Hutton Bicentennial Conference (July 1997) at Burlington House. The ideas expressed in this paper have been clarified through discussions with my colleagues John A. Van Couvering, Fritz Steininger, Robert Knox, Brian McGowran and my wife, Marie-Pierre Aubry. Fritz Steininger has been particularly helpful in clarifying the ideas of Moritz Hörnes. My special thanks go to Ms D. Fisher for her kindness in providing access to primary sources housed in the Rare Books section of the Museum of Comparative Zoology and to Debbie Rich for comparable kindness in connection with original literature consulted in the Bernard Kummel library of the Department of Geology, Harvard University Geology Department. Thorough reviews by two anonymous readers for the Geological Society aided in improving the organization of the final manuscript. Finally, I thank Michel and Monique Bonnemaison (Madrid) for providing the peace and serenity in which the first draft of this paper was completed.

This is Woods Hole Oceanographic Institution Contribution No. 9598.

References

AGUIRRE, E. & PASINI, G. 1985. The Pliocene–Plesitocene boundary. Episodes, **8**, 116–120.

ALBRITTON, C. C., JR. (ed.) 1963. *The Fabric of Geology*. Addison-Wesley, Reading, MA.

—— 1975. *Philosophy of Geohistory: 1785–1970*. Benchmark Papers in Geology, vol. 13, Dowden, Hutchinson & Ross, Stroudsburg, PA.

ALVAREZ, L. W., ALVAREZ, W., ASARO, F. & MICHEL, H. V. 1979. Extraterrestrial cause for the Cretaeous–Tertiary extinction: experiment and theory. *Lawrence Berkeley Report*, LBL-9666.

——, ——, —— & —— 1980. Extraterrestrial cause for the Cretaeous–Tertiary extinction. *Science*, **208**(4448), 1095–1109.

ARDUINO, G. 1760 (or 1759?). Letter by Arduino to Antonio Valisnieri, Professor of Natural History, University of Padua. Nuova raccolta di opuscoli scientifici e filologici del padre abate Angiola Calogierà, Venice, **6**, 142–143.

BARTHOLOMEW, M. 1976. The non-progress of non-progression: two responses to Lyell's doctrine. *British Journal for the History of Science*, **9**, 166–174.

—— 1979. The singularity of Lyell. *British Journal for the History of Science*, **17**, 276–293.

BECQUEREL, H. 1896. Sur les radiations invisibles émises par les sels d'uranium. *Comptes Rendus*, **122**, 689–694.

BERGGREN, W. A. 1971. Tertiary boundaries and correlations. *In*: FUNNELL, B. M. & RIEDEL, W. R. (eds) *Micropalaeontology of the Oceans*. Cambridge University Press, Cambridge, 693–809.

—— & VAN COUVERING, J. A. 1974. The Late Neogene. *Paleogeography, Palaeoclimatology, Palaeoecology*, **16**(1–2), 1–216.

—— & —— 1979. Biochronology. *In*: COHEE, G. V., GLAESSNER, M. F. & HEDBERG, H. D. (eds) *Contributions to the Geologic Time Scale*, American Association of Petroleum Geologists, Studies in Geology, **6**, 39–53.

—— & —— 1982. Quaternary. *In*: ROBISON, R. A. &D

TEICHERT, C. (eds) *Treatise on Invertebrate Paleontology*, Part A: *Introduction, Fossilization (Taphonomy), Biogeography and Biostratigraphy*. Geological Society of America, Boulder; University of Kansas Press, Laurence, KA, A505–A543.

——, HILGEN, F. J., LANGEREIS, C. G. *ET AL*. 1995. Late Neogene chronology: new perspectives in high-resolution stratigraphy. *Geological Society of America Bulletin*, **107**(11), 1272–1287.

—— KENT, D. V. & VAN COUVERING, J. A. 1985. The Neogene: Part 2 – Neogene geochronology and chronostratigraphy. *In*: SNELLING, N. J. (ed) T*he Geochronology of the Geological Record*. Memoir 10, Geological Society, London, 211–260.

BERRY, W. B. N. 1968. *Growth of a Prehistoric Time Scale*. W. H. Freeman, San Francisco.

BEYRICH, E. 1854. Über die Stellung der Hessischcen Tertiärbildungen. *Verhandlungen Köngliche Preussischen Akademie Wissenschaft Berlin, Monatsberichtungen*, (November), 640–666.

BRONGNIART, A. 1810. Sur les terrains qui paroissent avoir été formés sous l'eau douce. *Annales Musée Historire Naturelle Paris*, **15**, 357–405.

BRONN, H G. 1831. *Italiens Tertiär-Gebilde und deren organische Einschlusse*. Heidelberg.

—— 1834–1838. *Lethaea Geognostica, oder Abbildungen und Beschreibungen der fur Gebirgs-Formationen bezeichnendsten Verstinerungen*, vol. 1, 1–96 (1834); 97–192 (1835); 193–288, 384 and 480 (1836); 481–768 (1837); vol.2, 769–1346 (1838), E. Schweizerbart's Verlag, Stuttgart.

—— 1838. *Lethaea Geognostica, oder Abbildungen und Beschreibungen der für Gebirgs-Formationen bezeichnendsten Versteinerungen*. E. Schweizerbart's Verlagshandlung, Stuttgart, vol. 2, 769–1346.

BYNUM, W. F. 1984. Charles Lyell's *Antiquity of Man* and its critics. *Journal of the History of Biology*, **17**(2), 153–187.

CHARLESWORTH, E. 1835a. Observations on the Crag formation and its organic remains; with a view to establish a division of the Tertiary strata overlying the London Clay in Suffolk. *London and Edinburgh Philosophical Magazine*, **7**, 8–94.

—— 1835b. Reply to Mr. Woodward's remarks on the Coralline Crag; with observations on certain errors which may affect the determination of the age of Tertiary deposits. *London and Edinburgh Philosophical Magazine*, **7**, 464–470.

—— 1837a. Observations on the Crag and on the fallacies involved in the present system of classification of Tertiary deposits. *London and Edinburgh Philosophical Magazine*, **10**, 1–9.

—— 1837b. On some fallacies involved in the results relating to the comparative age of the Tertiary deposits obtained from the application of the test recently introduced by Mr. Lyell and M. Deshayes. *Edinburgh and London New Philosophical Journal*, **22**, 110–116.

CONKIN, B. M. & CONKIN, J. E. (eds) 1984. *Stratigraphy: Foundations and Concepts*. Van Nostrand Reinhold, New York.

COOPER, J. D., MILLER, R. H. & PATTERSON, J. 1996. *A Trip Through Time: Principles of Historical Geology*, Merrill Publishing Company, Columbus, OH.

COWIE, J. W. 1986. Guidelines for boundary stratotypes. *Episodes*, **9**, 78–82.

—— & BASSETT, M. G. 1989. International Union of Geological Sciences 1989 stratigraphic chart. *Episodes*, **12**(2) (insert).

CUVIER, G. & BRONGNIART, A. 1811. Essai sur la géographie minéralogique des environs de Paris, avec une carte géognostique et des coupes de terrain, Paris, Baudouin. *Annales Museum Histoire Naturelle Paris*, **1**, 293–326.

DANA, J. D. 1894. *Manual of Geology; Treating of the Principles of of the Science with Special Reference to American Geological History*. 4th edn, American Book Company, New York.

D'ARCHIAC, A. 1849. Histoire des progrès de la Géologie de 1834 à 1845, II, 2ème part, Tertiare. *Societé géologique de France*, Paris, 441–1100.

DE LAPPARENT, A. 1885. *Traité de Geologie*. 2nd edn. Masson, Paris.

—— 1906. *Traité de Géologie*. 5th edn. Masson, Paris, vol. 3, 1289–2015.

DENIZOT, G. 1962. Paléogène. *In*: DENIZOT, G. (ed.) *Lexique Stratigraphique International*. Commission de Stratigraphie, CNRS, Paris, vol. 1, 151–152.

DE SERRES, M. 1824. Observations sur des terrains d'eau douce récemment découverts dans les environs de Sète, à très peu de distance de la Mediterranée, et inférieurs au niveau de cette mer. *Memoire Musée Histoire Naturelle Paris*, **2**, 372–419.

—— 1830. De la simultancité dcs tcrrains de sédiment supérieurs. *In*: *La Géographie Physique de l'Encyclopédie Methodique*, vol. 5.

—— 1855. Des caractères et de l'importance de la période Quaternaire. *Societé géologique France, Bulletin*, series 2, **12**, 257–263.

DESHAYES, G. P. 1830. Tableau comparatif des espèces de coquilles vivantes avec les espèces de coquilles fossiles des terraines tertiaires de l'Europe, et des espèces de fossiles de ces terrains entr'eux. *Societé Géologique France Bulletin*, **1**, 185–187.

DESNOYERS, J. 1829. Observations sur un ensemble de dépots marins plus récents que les terrains tertiaires du bassin de la Seine, et constituant une formation géologique distincte: précédées d'un apercu de la nonsimultanéité des bassins tertiaires. *Annales scientifiques naturelles*, **16**, 171–214, 402–419.

D'ORBIGNY, A. 1849–1852. *Cours élementaire de paléontologie et de géologie stratigraphique*. Victor Massson, Paris.

DUMONT, A. 1849. Rapport dur la carte géologique du Royaume. *Bulletin Academie royale de Belge*, **16**(2), 351–373.

DUNBAR, C. O. & ROGERS, J. 1957. *Principles of Stratigraphy*. John Wiley & Sons, New York.

EMILIANI, C. 1982. A new global geology. *In*: EMILIANI, C. (ed.) *The Oceanic Lithosphere, The Sea*. John Wiley & Sons, New York & Chichester, vol. 7, 1687–1728.

FALCONER, H. 1863. 'Letter', *Athenaeum*, (4 April), 459–460.

FLINT, R. F. 1947. *Glacial Geology and the Pleistocene Epoch*. John Wiley & Sons, New York.

—— 1965. The Pliocene–Pleistocene boundary. In: WRIGHT, H. E. & FREY, D. G. (eds) International Studies on the the Quaternary. Special Paper, Geological Society of America, New York, **84**, 497–533.

—— 1971. Glacial and Quaternary Geology. John Wiley & Sons, New York.

FORBES, E. 1846. On the connexion between the distribution of the existing fauna and flora of the British Isles and the geographical changes which have affected their area, especially during the epoch of the Northern Drift. Great Britain Geological Survey Memoir, **1**, 336–432.

GAUDRY, A. 1878. Les enchainements du monde dans les temps géologiques: Mammifères Tertiares. Librarie F. Savy, Paris.

—— 1896. Essaie de Paléontologie philosophique, ouvrage faisant aux enchainments du monde animal dans les temps géologiques. Masson, Paris.

GEIKIE, A. 1897. The Founders of Geology. Macmillan, London.

—— 1905. The Founders of Geology. Macmillan, London.

GIGNOUX, M. 1913. Les formations marines pliocènes et quaternaires de l'Italie du Sud et de la Sicile. Université Lyon, Annales, **1**(36), 1–693.

—— 1950. Géologie stratigraphique. 4th edn. Masson, Paris.

—— 1955. Stratigraphic Geology (trans. G. G. Woodford). Freeman and Co., San Francisco.

GOULD, S. J. 1987. Time's Arrow, Time's Cycle: Myth and Metaphor in the Discovery of Geological Time. Harvard University Press, Cambridge, MA & London.

—— 1989. The passion of Antoine Lavoisier. Natural History, **6**, 16–25.

HARLAND, W. B., ARMSTRONG, R. L., COX, A. V., CRAIG, L. E., SMITH, A. G. & SMITH, D. G. 1990. A Geologic Time Scale 1989. Cambridge University Press, Cambridge,.

——, COX, A. V., LLEWELLYN, P. G., PICKTON, C. A. G., SMITH, A. G. & WALTERS. R. 1982. A Geologic Time Scale, Cambridge University Press, Cambridge.

HAYS, J. D. & BERGGREN, W. A. 1971. Quaternary boundaries and correlations. In : FUNNELL, B. F. & RIEDEL, W. R. (eds) Micropaleontology of the Oceans. Cambridge University Press, Cambridge, 669–691.

HEAD, M. J. 1997. Thermophillic dinoflagellate assemblages from the mid-Pliocene of eastern England. Journal of Paleontology, **7**(2), 165–193.

HEDBERG, H. D. (ed.) 1976. International Stratigraphic Guide – A Guide to Stratigraphic Classification, Terminology and Procedure. John Wiley & Sons, New York.

HÖRNES, M. 1853. Mittheilung an Professor BRONN gerichtet, Wien. Neues Jahrbuch fur Mineralogie, Geologie, Geognosie und Petrefaktenkunde, 806–810.

—— 1864. Die fossilen Mollusken des Tertiaerbeckens von Wien. Jahrbuch der geologischen Reichssanstalt, **14**, 509–514.

HUTTON, J. 1788. Theory of the Earth; or an investigation of the laws observable in the composition,

dissolution and restoration of the land upon the globe. Transactions of the Royal Society of Edinburgh, **1**, 209–305.

—— 1795. Theory of the Earth with Proofs and Illustrations. 2 vols. William Creech, Edinburgh.

JENKINS, D. G., BOWEN, D. Q., ADAMS, C. G., SHACKLETON, N. J. & BRASSELL, S. C. 1985. The Neogene: part 1. In: SNELLING, N. J. (ed.) The Chronology of the Geological Record. Memoir 10, Geological Society, London, 199–210.

KING, W. B. R. & OAKLEY, K. P. 1950. Report of the Temporary Commission on the Pliocene–Plesitocene Boundary, appointed 16th August 1948. In: BUTLER, A. J. (ed.) International Geological Congress, Report of the Eighteenth Session, Great Britain, 1948, Part I: General Proceedings, Geological Society, London, 2134–2214.

LYELL, C. 1830–1833. Principles of Geology, Being an Attempt to Explain the Former Changes of the Earth's Surface by Reference to Causes Now in Operation. Murray, London, vol. 1 (1830), vol.2 (1832), vol. 3 (1833).

—— 1831. Letter to W. Whewell, 21 January 1831 (cited in Wilson 1972, 305, 306).

—— 1838. Elements of Geology. Murray, London

—— 1839a. On the relative ages of the Tertiary deposits commonly called 'Crag' in the counties of Norfolk and Suffolk. Magazine of Natural History, **3**, 313–330.

—— 1839b. Nouveaux éléménts de géologie. Pitois-Levranet, Paris.

—— 1852. On the Tertiary strata of Belgium and French Flanders, part II. The lower tertiaries of Belgium. Quarterly Journal of the Geological Society of London, Proceedings of the Geological Society, **8**, 278–368.

—— 1855. A Manual of Elementary Geology, 5th edn, Appleton, London.

—— 1857a. Supplement to the fifth edition of a Manual of Elementary Geology. Murray, London.

—— 1857b. A Manual of Geology. Reprint of the sixth edition. Appleton, London.

—— 1863. On the Geological Evidences of the Antiquity of Man. Murray, London.

—— 1873. On the Geological Evidences of the Antiquity of Man. 4th edn. Murray, London.

MANTELL, G. A. 1822. The Fossils of the South Downs; or Illustrations of the Geology of Sussex. L. Relfe, London.

MAYER-EYMAR, K. 1857. Versuch einer synchronistischen Tabelle der Tertiär-Gebilde Europas. Verhandlungen Schweizerische naturhistorische Gesellschaft, **17–19**, 164–199.

MORLOT, A. 1854. Uber die quaternaren Gebilde des Rhonegebiets. Verhandlungen Schweizerische Gesellschaft Naturwissenschaften, **39**, 161–164.

—— 1856. Sur le terrain quaternaire du bassin du Léman. Bulletin Societé vaudoise Science naturelle, **6**, 101–108.

—— 1858. Uber die quaternaren Gebilde des Rhonegebiets. Verhandlungen Schweizerische Gesellschaft Naturwissenschaften, **43**, 144–150.

NAUMANN, C. F. 1854. *Lehrbuch der Geognosie.* Engelmann, Leipzig, vol. 2.

—— 1866–1872. *Lehrbuch der Geognosie.* 2nd edn. Engelmann, Leipzig, vol. 3, 1–192 (1866), 193–352 (1868), 353–576 (1872).

OLDROYD, D. R. 1996. *Thinking About the Earth: A History of Ideas in Geology.* Harvard University Press, Cambridge, MA.

PAPP, A. 1979. Tertiary. *In*: ROBISON, R. A. & TECHIERT, C. (eds) *Treatise on Invertebrate Paleontology.* Geological Society of America, Boulder, A488–A504.

PASINI, G. & COLALONGO, M. L. 1997. The Pliocene–Plesitocene boundary-stratotype at Vrica, Italy. *In*: VAN COUVERING, J. A. (ed.) *The Pleistocene Boundary and the Beginning of the Quaternary,* Cambridge University Press, Cambridge, 15–45.

PHILLIPS, J. 1840. 'Palaeozoic Series'. *In*: LONG, G. (ed.) *The Penny Cyclopaedia of the Society for the Diffusion of Useful Knowledge.* Charles Knight, London, vol. 17, 153–154.

PLAYFAIR, J. 1802. *Illustrations of the Huttonian Theory of the Earth.* Cadell & Davis, London; William Creech, Edinburgh.

POMEROL, C. 1973. *Ere Cénozoique (Tertiare et Quaternaire).* Doin, Paris.

PORTER, R. 1976. Charles Lyell and the principles of the history of geology. *British Journal of the History of Science,* **9**, 91–103.

—— 1982. Charles Lyell: the public and private faces of science. *Janus,* **69**, 29–50.

REBOUL, H. 1833. *Géologie de la périod Quaternaire et introduction à l'histoire ancienne.* F. G Levrault, Paris.

REMANE, J., BASSETT, M. G., COWIE, J. W., GOHRBANDT, K. H., LANE, H. R., MICHELSEN, O. & NAIWEN, W. 1996. Revised guidelines for the establishment of global chronostratigraphic standards by the International Commission on Stratigraphy (ICS). *Episodes,* **19**(3), 77–81.

RENEVIER, E. 1874. Tableau des terrains sédimentaires formés pendant les époques de la phase organique du globe terrestre. *Bulletin Societé Vaudoise Sciences Naturelles,* **12**(70), table 3; **13**(72), 218–252.

—— 1897. Chronographe géologique; seconde édition du Tableau des terraines sédimentaires. *In*: *VI Congrés géologique international, Compte-Rendu.* Georges Bridel, Lausanne, 522–695.

REYMENT, R. A. 1996. Charles Lyell (1797–1875). *Terra Nova,* **8**(4), 390–393.

RUDWICK, M. J. S. 1963. The foundation of the Geological Society of London: its scheme for co-operative research and its struggle for independence. *British Journal for the History of Science,* **1**(4), 325–355.

—— 1969. Lyell on Etna and the antiquity of the earth. *In*: SCHNEER, C. J. (ed.) *Toward a History of Geology.* MIT Press, Cambridge, MA, 288-304.

—— 1970. The strategy of Lyell's *Principles of Geology. Isis,* **61**, 4–33.

—— 1971. Unformity and progression: reflections on the structure of geological theory in the age of Lyell. *In*: ROLLER, D. H. D. (ed.) *Perspectives in the History of Science and Technology.* Norman, OK, 209–227.

—— 1974. Poulett Scrope on the volcanoes of Auvergne: Lyellian time and political economy. *British Journal of the History of Science,* **7**, 205–242.

—— 1975. Caricature as a source for the history of science: De la Beche's anti-Lyellian sketches of 1831. *Isis,* **66**, 534–560.

—— 1978. Charles Lyell's dream of a statistical palaeontology. *Palaeontology,* **21**(2), 225–244.

RUPKE, N. A. 1983. *The Great Chain of History,* Clarendon Press, Oxford.

SALVADOR, A. (ed.) 1994. *International Stratigraphic Guide: A Guide to Stratigraphic Classification, Terminology, and Procedure.* 2nd edn. Geological Society of America, Boulder.

SCHIMPER, W. P. 1874. *Traité de Paléontologie végétale,* J. B. Baillière et fils, Paris, vol. 3.

SCHNEER, C. J. (ed.) 1969, *Toward a History of Geology.* MIT Press, Cambridge, MA.

SCHOCH, R. M. 1989. *Stratigraphy: Principles and Methods.* Van Nostrand Reinhold, New York.

SCHUCHERT, C. & DUNBAR, C. O. 1941. *A Textbook of Geology; Part II – Historical Geology.* John Wiley & Sons, New York.

SCROPE, G. P. 1830. [Review of] *Principles of Geology...by Charles Lyell. Quarterly Review,* **43**, 411–469.

SECORD, J. A. 1987. *Controversy in Victorian Geology: The Cambrian–Silurian Dispute.* Princeton University Press, Princeton.

SEDGWICK, A. 1831. Address to the Geological Society, delivered on the evening of the 18th of February, 1831. *Proceedings of the Geological Society, of London,* **1**, 281–316.

SHELLEY, M. W. 1818. *Frankenstein, or the Modern Prometheus.* London.

SMITH, W. 1815. *A Memoir to the Map and Delineation of the Strata of England and Wales, with Part of Scotland.* John Cary, London.

—— 1817. *Stratigraphical System of Organized Fossils, with Reference to the Specimens of the Original Collection in the British Museum: Explaining Their State of Preservation and Their Use in Identifying the British Strata.* E. Williams, London.

STEININGER, F. F., BERGGREN, W. A., KENT, D. V., BERNOR, R. L., SEN, S. & AGUSTI, J. 1996. Circum-Mediterranean Neogene (Miocene and Pliocene) marine-continental chronologic correlations of European mammal units. *In*: BERNOR, R. L., FAHLBUSCH, V. & MITTMAN, H.-W. (eds) *The Evolution of Western Eurasian Neogene Mammal Faunas.* Columbia University Press, New York, 7–46

——, BERNOR, R. L. & FAHLBUSCH, V. 1989. European Neogene marine/continental chronologic correlations. *In*: LINDSAY, E. H., FAHLBUSCH, V. & MEIN, P. (eds) *European Neogene Mammal Chronology.* Plenum Press, New York, 15–46.

VAN COUVERING (ed.) 1997. *The Pleistocene Boundary and the Beginning of the Quaternary.* Cambridge University Press, New York.

VON ZITTEL, K. 1901. *History of Geology and Palaeontology to the End of the Nineteenth Century.* Walter Scott, London.

WHEWELL, W. 1831. [Review of] *Principles of*

*Geology...*by Charles Lyell. *Quarterly Theological Review and Ecclesiastical Record*, **9**, 180–206.

WILSON, L. G. 1969. Charles Lyell's *Principles of Geology*, 1830–1833. *In*: SCHNEER, C. J. (ed.) *Toward a History of Geology*. MIT Press, Cambridge, MA, 426–443.

—— (ed.) 1970. *Sir Charles Lyell's Scientific Journals on the Species Question*. Yale University Press, New Haven.

—— 1972. *Charles Lyell: The Years to 1841: The Revolution in Geology*. Yale University Press, New Haven & London.

—— 1980. Geology on the eve of Charles Lyell's first visit to America, 1841. *Proceedings of the American Philosophical Society*, **124**(3), 168–202.

WOODWARD, H. B. 1908. *The History of the Geological Society of London*. Longman, London.

—— 1911. *History of Geology*. Watts, London.

Lyell's views on organic progression, evolution and extinction

A. HALLAM

School of Earth Sciences, University of Birmingham, Birmingham B15 2TT, UK

Abstract: Following Cuvier, Lyell was readily prepared to accept a succession of species extinctions in the past but he rejected Cuvier's ideas on episodes of catastrophic mass extinction. In the *Principles of Geology* he argued against organic progression in the fossil record, believing instead that organic traces had been substantially removed from older rocks by metamorphism and other factors, and that groups such as land mammals should not be expected, except very rarely, in the marine strata of the Palaeozoic and Mesozoic. He did, however, accept that the human species was of modern origin, but emphasized that the most significant feature of humanity emergence was in the moral rather than the physical sphere. Lamarck's theory of evolution involving species transmutation was vehemently rejected; like Cuvier, Lyell believed in the stability of species. As evidence continued to accumulate in favour of some kind of organic progression in the stratigraphic record, Lyell's resistance eventually crumbled, but he continued to believe in the extreme imperfection of the fossil record. Like Darwin, he could find no direct evidence from it for evolution. In the 1860s he eventually came to accept some form of evolution, while adhering to the rest of his uniformitarinism, but continued to reject Darwin's theory of natural selection.

The first written record of Lyell's opinions on the subject of this article comes in his review (1827) of Scrope's monograph on the geology and geomorphology of central France (Scrope 1827). Lyell had read at least the English translation of Cuvier's celebrated *Discours sur les révolutions de la surface du globe* (Cuvier 1813), based primarily on his researches on the Tertiary strata of the Paris Basin. Cuvier was the first to establish clearly that strata contained the remains of fossil species that had subsequently gone extinct, and Lyell had no trouble in accepting this. He soon recognized that the succession of species through Tertiary strata, marking a series of extinctions and replacements, could be used as a natural chronometer for Tertiary time. He was, however, firmly opposed to Cuvier's postulation of a succession of catastrophic mass extinctions separated by longer interludes of comparative quiescence.

In his review, Lyell pointed out that the Massif Central had not experienced any Tertiary marine incursions, unlike the Paris Basin. Between Cuvier's Tertiary faunas and those of the present, Lyell suggested, was a period in the history of life 'over which the greatest obscurity still hangs' (Lyell 1827, p. 452). However the fauna had changed during this period, it could not have been by Cuvier-type revolutions: central France had evidently remained free from whatever physical changes had occurred in the lower-lying Paris

Basin. Lyell foresaw a piecemeal replacement of species as they went extinct, due no doubt to environmental causes. His extended visit to Sicily in 1828 confirmed for him that both biological and geological changes in the past could have occurred by processes as gradual in their action as those still in operation (Rudwick 1972).

Lyell's *Principles of Geology*

In establishing his uniformitarian doctrine Lyell (1830–1833) faced a potentially serious problem. Sufficient fossils had been collected to suggest to many a kind of organic progression through time, from simple plants and invertebrates to progressively more complicated or biologically sophisticated vertebrates, culminating in our own species. Chapter 9 of his first volume is devoted to an attempted refutation of this 'objection urged against assumption of uniformity in the order of nature' (1830–1833, vol. 1, p. 144).

Lyell argued that in the older rocks of what was subsequently called the Palaeozoic there must have been many agents – including metamorphism, tectonic deformation and solution of fossil material by the 'percolation of acidulous waters' (p. 147) – that would have acted either separately or in concert to obliterate to a greater or lesser degree the traces of organic remains. Thus it was hardly surprising that so few vertebrates had yet been

HALLAM, A. 1998. Lyell's views on organic progression, evolution and extinction. *In*: BLUNDELL, D. J. & SCOTT, A. C. (eds) *Lyell: the Past is the Key to the Present*. Geological Society, London, Special Publications, **143**, 133–136.

found in rocks of 'Palaeozoic' age. Nevertheless, fish were known from some of the oldest strata, which 'entirely destroys the theory of precedence of the simplest forms of animals' (p. 148). Furthermore, land mammals could not be expected in the marine strata of the Carboniferous and 'Mesozoic' of the well explored rocks of Great Britain. Vertebrates are abundant in the marine Mesozoic but consist almost entirely of fish and reptiles. Lyell noted, however, that mammalian jaws related to opossums (determined by Cuvier as belonging to the genus *Didelphis*) had been found in the Middle Jurassic Stonesfield Slate of the southern English Midlands. The Tertiary strata of the Paris Basin were partly of nonmarine origin and contained many land mammals, but fossils of this group were extremely rare in marine deposits of the same age. The London Clay, for example, contains only marine fish and reptiles, notably turtles and crocodiles. In contrast, the Jurassic Stonesfield Slate contains mammals, indicating a reversal in the presumed order of progression.

Lyell was prepared to accept that the human species is of modern origin. Was it then the culmination of organic progression? His reply to this query is that 'the superiority of man depends not on those faculties and attributes which he shares in common with the inferior animals, but on his reason by which he is distinguished from them' (p. 155). Indeed, 'the animal nature of man, even considered apart from the intellectual, is of higher dignity than that of any other species'.

> We may easily conceive that there was a considerable departure from the succession of phenomena previously exhibited in the organic world, when so new and extraordinary a circumstance arose, as the union, for the first time, of moral and intellectual faculties capable of indefinite improvement, with the animal nature. (p. 155)

In biological comparisons – for example, in competition for food and space, predation, and vulnerability to such natural phenomena as earthquakes and volcanoes – humans are indistinct from other animals. There remains, however, a crucial difference:

> But he would soon perceive that no one of the fixed laws of the animate or inanimate world was subverted by human agency, and that the modifications produced were on the occurrence of new and extraordinary circumstances, and those not of a *physical*, but a *moral* nature. (p. 164).

Thus did Lyell attempt to preserve the dignity of *Homo sapiens*, prior to Darwin. His uniformitarian views with regard to what he considered to be a very incomplete fossil record are perhaps best epitomized by the following well known passage:

> Then might those genera of animals return, of which the memorials are preserved in the ancient rocks of our continents. The huge iguanodon might reappear in the woods, and the ichthyosaur in the sea, while the pterodactyl might flit again through the umbrageous groves of tree ferns.

This passage provoked a cartoon by his colleague Henry de la Beche (see Rudwick this volume, p.11), portraying the future Professor Ichthyosaurus, lecturing to fellow reptilian students on the skull of a strange (human) creature of the last creation.

Lyell's first encounter with evolutionary theory concerned the transmutational ideas of Lamarck, whose *Zoological Philosophy* had just been reissued following Geoffroy de St Hilaire's revival of the subject. Volume 2 of the *Principles* contained an extensive critique of Lamarck's ideas. To have admitted that species were unreal and that organisms were in a state of constant flux would have undermined the validity of his natural chronometer. Lyell shared Cuvier's belief in the stability of species. A continually changing physical environment would lead to shifting patterns of ecological conditions, which could lead to species extinctions. The 'creation' of new species remained an unsolved puzzle, both to Cuvier and to Lyell. Neither invoked divine intervention.

The firm establishment of organic progression

Lyell's dismissal of the fossil evidence for 'progression' failed to convince most of his contemporaries, and almost all other publications on the subject, whether specialist or popular, emphasized that the fossil record was indeed broadly progressive. This became increasingly apparent as a result of further collecting, most notably in the undeformed Palaeozoic strata of cratonic regions of eastern Europe and North America. Enough nonmarine Palaeozoic and Mesozoic rocks were examined to rule out with confidence Lyell's supposition that 'higher' vertebrates arose much earlier in geological history than indicated by the pioneer studies of Cuvier and Brongniart. By the mid 1840s the major outlines of the fossil record were firmly established on lines that have survived with only minor modifications into the twentieth century.

Rudwick (1972) has emphasized the important role of the German palaeontologist Heinrich-Georg Bronn in establishing with authority what might be termed the consensus view of informed opinion in the middle of the nineteenth century. Earlier in his career, Bronn attempted to use numerical techniques to assess degrees of affinity between faunas of different formations. All his later compilative works illustrate the same belief that organic change is gradual, with the piecemeal extinction of old species and production of new ones. However, he never made Lyell's mistake of assuming that uniformity of gradual change necessarily entailed a steady-state picture of the history of life. The nature of the creative force for producing new species was emphatically a *natural* agency, albeit mysterious. (This was indeed a common view, with divine intervention never being seriously invoked; Darwin (1859) was unduly polemical in putting up such a straw man.)

Bronn saw evidence of a continual replacement of extinct species by more diverse and 'higher' forms of life, while there was maintained at all times a well balanced ecological assemblage. The organic world had come gradually to approximate more and more its present state, with no periods of massive extinction or of wholesale production of new species. Increased taxonomic diversity was related to the progressive physical diversification of the Earth's surface and the increased complexity of ecological relations between organisms. A true 'progressive' trend towards more complex forms of life could also be discerned within each of the major groups.

Bronn and Lyell were sharply divided about the *degree* of imperfection of the fossil record. Because Lyell clung desperately to his original belief in the steady-state history of the organic world, he was obliged to maintain with equal tenacity his view that the fossil record was extremely imperfect, such that any appearance of 'progress' was an illusion. In support, however, of Lyell, and opposed to Lamarck's evolutionary theory, there was no fossil evidence of species changing gradually into others when traced through successive strata, nor of major groups, with distinctive anatomy, being traceable to a common ancestor.

Lyell's changing position from mid-century

Lyell's last defence of non-progression was his Anniversary Address to the Geological Society of London in 1851 (Lyell 1851). As has been elucidated from study of his private journals, compiled between 1855 and 1861 (Wilson 1970), he subsequently began to waver and eventually surrendered with as good a grace as he could muster. The tenth edition of *Principles,* published in 1868, was the first to announce his retreat. The eleventh edition, appearing in 1872, just three years before his death, contained only minor revisions; Chapter 9 of this edition assents to 'progressive development of organic life'.

Lyell's relationship with Darwin during these two decades is a subject of critical importance to historians of science. From the time of the *Beagle* voyage, Darwin was an ardent disciple of Lyell's uniformitarian doctrine, with its emphasis on gradualism (Desmond & Moore 1991). Following Lyell, Darwin believed that discontinuities in the fossil record were the result of its imperfections. His view in the *Origin of Species* (Darwin 1859) was that dramatic faunal turnovers in the stratigraphic record reflected major stratigraphic hiatuses. 'The old notion [i.e. Cuvier's] of all the inhabitants of the earth having been swept away by catastrophes at successive periods is very generally given up.'

Lyell encouraged Darwin to publish his revolutionary theory of evolution, recommending his own publisher, John Murray, and arranging with Hooker the famous joint presentation of the papers of Darwin and Wallace at the Linnean Society's meeting rooms in London in 1858 (Desmond & Moore 1991). It is widely believed that Darwin was instrumental in provoking Lyell's assent to evolution. Gould (1987) considers that examination of Lyell's journals (Wilson 1970) forces us to reverse this conventional argument towards a new interpretation.

The journals record that Darwin first broached his theory to Lyell during his visit to Darwin's home in 1856, and reveal that Lyell was already obsessed with doubt about non-progressionism. Before Darwin's revelation, he had already come to the conclusion that he would probably have to abandon this anchor of his central vision. Gould argues that Lyell did not accept evolution because Darwin persuaded him. In particular, he never accepted natural selection, much to Darwin's disappointment.

If we take the three attributes of the deity of the Hindoo Triad, the Creator, Brahma, the preserver or sustainer, Vishnu, and the destroyer, Siva, Natural Selection will be a combination of the two last but without the first, or the creative power, we cannot conceive the others having any function. (Quoted in Wilson 1970, p. 369)

The journals reveal Lyell's great reluctance to place human origins into nature's ordinary course.

There is but little difference between the out and out progressionist and Lamarck, for in the one case some unknown *modus operandi* called creation is introduced and admitted to be govered by a law causing progressive development and by the other an extension of multiplication of the variety-making power is adopted of the unknown process called Creation. It is the theory of a regular series of progressively improved beings ending with Man as part of the same, which is the truly startling conclusion destined, if established, to overturn and subvert received theological dogmas and philosophical reveries quite as much as Transmutation. ... There seems less to choose between the rival hypotheses [evolution and progressionism] than is usually imagined. (Quoted in Wilson 1970, p. 222).

Gould (1987) regards this statement as the key to Lyell's conversion. He does not accept evolution because the facts proclaim it, since he finds little to choose between evolution and progressionism. Why then should one prefer evolution? Lyell's answer seems clear in the journals. Evolution is the fallback position of minimal retreat from the rest of uniformity. He could continue to hold firmly to uniformity of rate, uniformity of law, and actualism. Thus he could retain as much of his uniformitarianism as possible, when the facts of the fossil record finally compelled his reluctant allegiance to organic progression.

Lyell's first public statement of his altered views was in his semi-popular book the *Antiquity of Man* (1863). While he was able to sketch out an evolutionary history of the human species, he could not present any positive evidence of its origin, but nevertheless suggested, with some reluctance, that evolutionary theory might ultimately be applicable to human beings. The key issue at stake, as much earlier in his life, was the status of mind and consciousness, and he refers to 'new and powerful causes to explain the spiritual part of human nature'. To Darwin's evident disappointment (Desmond & Moore 1991), he could not accept humanity's place 'among the brutes'.

Lyell found himself in the curious position of believing so strongly that, because of the imperfection of the fossil record, palaeontology could not provide direct evidence for evolution, that he failed even to consider what fragmentary evidence could show. A good example of this is his failure to recognise the evolutionary significance of the

discovery in 1861 of the first specimens of *Archaeopteryx* as a kind of 'missing link' between birds and reptiles. Huxley made no such mistake (Rudwick 1972).

It would be wrong, however, to be unduly critical of Lyell's somewhat reluctant and lukewarm conversion to evolution. As an old man he could hardly be expected to embrace with enthusiasm a new doctrine which appeared to challenge in such a fundamental way his lifelong beliefs, especially about the place of humankind in nature. In the event he made a noble effort to make the best of a bad job.

References

CUVIER, G. 1813. *Essay on the Theory of the Earth. With Geological Illustrations by Professor Jameson.* Blackwood, Edinburgh.

DARWIN, C. R. 1859. *On the Origin of Species by Means of Natural Selection, or the Preservation of Favoured Races in the Struggle for Life.* Murray, London.

DESMOND, A. & MOORE, J. 1991. *Darwin.* Joseph, London.

GOULD, S. J. 1987. *Time's Arrow, Time's Cycle.* Harvard University Press, Cambridge, MA.

LYELL, C. 1827. "Memoir on the geology of Central France..." by G. P. Scrope. *Quarterley Review,* **36,** 437–483.

—— 1830–1833. *Principles of Geology, Being an Attempt to Explain the Former Changes of the Earth's Surface, by Reference to Causes Now in Operation.* 3 vols. Murray, London.

—— 1851. Anniversary address of the President. *Proceedings of the Geological Society,* 7, xxv–lxxvi.

—— 1863. *On the Geological Evidences of the Antiquity of Man.* Murray, London.

RUDWICK, M. J. S. 1972. *The Meaning of Fossils.* Macdonald, London.

—— 1998. Lyell and the *Principles of Geology.* This volume.

SCROPE, G. P. 1827. *Memoir on the Geology of Central France, Including the Volcanic Formations of Auvergne, the Velay and the Vivarais.* Murray, London.

WILSON, L. G. (ed.) 1970. *Sir Charles Lyell's Scientific Journals on the Species Question.* Yale University

The age of the Earth and the invention of geological time

JOE D. BURCHFIELD

Department of History, Northern Illinois University, DeKalb, IL 60115, USA

Abstract: During Charles Lyell's lifetime the concept of geological time was gradually rendered meaningful through the construction of a heuristic geological timescale, the development of quantitative methods to calculate the duration of that scale, and the acceptance of a quantitatively determinable limit to the age of the Earth. This paper describes a few episodes in the process of 'inventing' the concept of geological time, episodes chosen primarily for their relevance to the work of Lyell.

Reflecting on the panoramic record of the Earth's past exposed to view as he travelled across the North American continent on US interstate highway 80, John McPhee observed, 'The human mind may not have evolved enough to be able to comprehend deep time. It may only be able to measure it' (McPhee 1980, p. 127). Travelling from east to west in a horizontal plane, McPhee passed through the assemblies of rock that contained the record of the Earth's past in no apparent chronological order. Rocks of different geological ages were exposed seemingly at random. From time to time, in cuts and tunnels, a more orderly but limited chronological sequence of rocks was exposed in a vertical plane. To produce McPhee's sense of 'deep time', the record of the rocks had to be assembled and interpreted. In this paper I suggest that the 'invention' (or perhaps more accurately, the 'construction') of geological time was the historical process by which the concept that McPhee calls 'deep time' was first recognized and then rendered conceptually meaningful, if not truly comprehensible. For a working definition of 'conceptually meaningful', I will adopt Lord Kelvin's variation on McPhee's cogent distinction, namely his famous dictum that 'when you can measure what you are speaking about and express it in numbers you know something about it; but when you cannot measure it, when you cannot express it in numbers, your knowledge is of a meagre and unsatisfactory kind: it may be the beginning of knowledge, but you have scarcely, in your thoughts, advanced to the stage of *science*' (Thomson 1891, pp. 80–81).

Time has always presented unique conceptual difficulties for scientists and philosophers alike (see Kitts 1966, 1989). We experience time only in the present. Our sense of the duration of time is subjective, an inference based on our transitory perceptions of experienced events and our memory of the past. We create a concept of external, non-

personal time by analogy, by imposing an order, a constructed memory, on the evidence of external events. We make our concept of 'duration' more or less objective by reference to some repetitive quantifiable standard such as a day or a year, but our ability to comprehend duration remains subjective, both shaped and limited by the conditions of individual experience. Since geological time, like historical time, lies forever outside the scope of our direct experience, our concept of geological time is an artefact. It had to be created or 'invented'.

The invention of geological time

In this paper, I will argue that the invention of geological time involved at least five essential steps: the recognition of the evidence of a succession of past events in the static record of the rocks, the acceptance of a terrestrial age significantly greater than the historical record of humankind (the notion, however vague, of 'deep time'); the development of a historical sense of the Earth's past through the construction of a heuristic geological timescale; the creation of quantitative methods to calculate the duration of that scale; and the acceptance of a quantitatively determinable limit to the age of the Earth. Also important was the creation of a new concept of 'historical time' embracing what Stephen Toulmin has called 'the twin ideas of "*periods*" and "*development*" ' (Toulmin 1962–1963, p. 105). In the sense of step one, the process of inventing geological time began in the seventeenth century with the work of Nicolaus Steno and Robert Hooke: in the sense of steps three and four, it is still going on and has been particularly vigorous during the past two or three decades (see Berggren *et al.* 1995). By limiting this paper largely to British geologists in the age of Lyell, I can deal with only a few episodes in the

BURCHFIELD, J. D. 1998. The age of the Earth and the invention of geological time. *In*: BLUNDELL, D. J. & SCOTT, A. C. (eds) *Lyell: the Past is the Key to the Present*. Geological Society, London, Special Publications, **143**, 137–143.

137

early stages of invention, a few of the steps in sketching the preliminary design.

By the time of Charles Lyell's birth in 1797, the first hints of a notion of geological time had been appearing for several decades. The pioneering local stratigraphies published by Johann Gottlob Lehmann, Christian Füchsel and Giovanni Arduino between 1756 and 1761, for example, all pointed to an Earth formed by a succession of events rather than a single act of creation (Albritton 1980, pp. 117–119; Berry 1968, pp. 29–34). None of the three suggested that the duration of these events would require a terrestrial age significantly greater than that implied by the Mosaic narrative. Nonetheless, the notion of 'deep time' had already been suggested a decade earlier by the Cartesian speculations of Benoît de Maillet and the Newtonian cosmology of Georges Buffon, and by 1774 it had been quantified by Buffon's experimental extrapolation from Leibniz's hypothesis of a primordial molten Earth (Haber 1959, pp. 115–136). Despite the controversies aroused by de Maillet's countless ages for the diminution of the seas and Buffon's more modest 75 000 years for the cooling of the Earth, by the final quarter of the eighteenth century, a growing body of field evidence had begun to create an intellectual climate where some notion of deep time was, at least in principle, acceptable (Porter 1977, pp. 157–160). No single event, even the Noachian deluge, could account for the complexity of the strata, nor could the succession of life revealed by the fossils be accommodated in six literal days of creation. By 1785, this new intellectual climate was sufficiently widespread to inspire William Cowper's acerbic lines:

...Some drill and bore
The solid Earth, and from the strata there
Extract a register, by which we learn
That he who made it, and reveal'd its date
To Moses, was mistaken in its age (Cowper 1785, pp. 166–167).

Although a long span of time before the advent of humans seemed necessary to account both for the record of ancient life preserved in the fossils and for the thickness and complexity of strata exposed by the new field studies, there was no agreement about the magnitude of the time spans involved. Buffon's quantitative chronology was hardly less controversial than de Maillet's eternalism, and, though speculations on the nature of geological causes were sometimes hotly debated, scant attention was paid to their rates of action. Thus, when James Hutton, an Enlightenment rationalist, published the *Abstract* of his now famous 1785 address, the opening lines read: 'The purpose of this dissertation is to form some estimate with regard to the time the globe of this

Earth has existed, as a world maintaining plants and animals' (Hutton 1785, p. 3). But, after systematically reviewing the natural causes that he believed had shaped and continued to shape the Earth's crust, he was faced with the conclusion that

...as there is not in human observation proper means for measuring the waste of land upon the globe, it is hence inferred, that we cannot estimate the duration of what we see at present, nor calculate the period at which it had begun; so that, with respect to human observation, this world has neither a beginning nor an end. (Hutton 1785, p. 28)

Hutton had perhaps clarified the problem, but the concept of geological time remained formless.

Much has been written about the 'birth of geology' in the years that coincide closely with Charles Lyell's youth. I will not attempt to rehash the many ways in which time was invoked in the conflicts over 'Genesis and geology' and the disputes between the neptunists and plutonists. I will suggest only the obvious: that no meaningful conception of geological time had yet been formulated. Nonetheless, the foundations for such a conception were being laid (see Rudwick 1996). The palaeontological researches of Georges Cuvier, J. B. Lamarck, William Buckland and others made the strange inhabitants of the remote past vividly visual and thus sensibly real. They pointed unmistakably to a temporal succession of living forms. Fossils also provided William Smith with one of the three principles by which he solved the puzzle of the strata and showed how to correlate its widely scattered parts. The way had been opened for the record of the rocks to be organized into a single chronological column.

By the second decade of the nineteenth century, the identification and systematic classification of the fossiliferous strata were underway in earnest. In the two decades between J. B. J. d'Omalius d'Halloy's grouping of the Chalk and underlying sandstones and marl into a single *Terrain Crétacé* in 1822 and Roderick Murchison's identification of the *Permian* in 1841, most of the subdivisions now designated as systems were identified and arranged chronologically according to the principle of superposition. Also by 1840, Adam Sedgwick and John Phillips had suggested new names reflecting the temporal progression of life revealed by the fossil record – *Paleozoic, Mesozoic* and *Cenozoic* – for larger chronological groups of systems (Albritton 1980, pp. 123–130; Berry 1968, pp. 64–95). Agreement over the new systems was far from general. There was little consensus about the relative importance of lithology and fossils as the basis for the classification of strata, and heated controversies over disputed system boundaries sometimes dragged on for years (see Rudwick 1985; Secord 1986a). One conclusion, however, was inescapable. Placed in ordered sequence, the strata revealed unmistakable evidence of the temporal succession of past geological events. The Earth had been given a history. The rocks themselves gave no indication of the duration of the events they recorded, but the sheer magnitude of the reconstructed record together with the diversity of the successive periods of life revealed by the fossils it contained produced an inescapable impression of vistas of time stretching far beyond the scope of human history.

The concept of geological time

An impression of time, even 'deep time,' is not the same thing as a concept of geological time. Nevertheless, a brief examination of four geological treatises from the 1830s may give some insight into how such a concept began to form. The first and most influential of the four was Lyell's *Principles of Geology*, published between 1830 and 1833. As is well known, the time of the stratigraphers played little role in the first two volumes of the *Principles*. Lyell's concept of time was shaped to serve the ends of his concept of gradual, actualistic geological processes operating in a dynamically balanced, steady–state terrestrial mechanism (Lyell 1830–1833; see also Rudwick 1969a,b, 1971, 1974). Time, indefinite drafts of time, was necessary for Lyell's gradualism, but in its fervent anti-directionalism, his notion of the Earth's dynamics was curiously atemporal. When Lyell did turn to what he later referred to as 'Geology proper' in the third volume of the *Principles* (Wilson 1972, p. 503), he still gave only a brief account of the stratigraphy of his peers. Nonetheless, he did address both temporal progression and the quantification of the evidence for geological time in what Martin Rudwick has termed his 'statistical palaeontology' (Rudwick 1978). He thus contributed to the temporal classification of the stratigraphic record by introducing his four part subdivision of the Tertiary and coining the terms Newer and Older 'Pliocene', 'Miocene', and 'Eocene'. It was unquestionably Lyell's influence that gave prominence to a sense of 'deep time,' but his changeless, temporally indefinite Earth conveyed, at best, an obscure notion of 'geological time.'

Henry De la Beche presented a very different vision of terrestrial dynamics in his *Researches in Theoretical Geology*, published in 1834, the year after the third volume of the *Principles*. Chemistry and physics were essential to De la Beche's geology, and he devoted considerable attention to the theory of the Earth's central heat and the hypothesis of its igneous fluid origin. Thus, De la Beche's idea of finite, directional geological time owed as much to the physics of heat flow as to the record of the rocks. But it was the rocks that conveyed a more substantial sense of duration, and he referred to the 'millions of years' needed to produce the layers of strata he had described. The number meant little more than 'a whole lot', and in his concluding remarks, De la Beche foreshadowed McPhee's musing with which I began:

> Measuring time as man does in the minute manner suited to his wants and conveniences, a few thousand revolutions of the Earth in its orbit appear to him to comprise a period so considerable, that he feels it difficult to conceive that great lapse of time which geology teaches us has been necessary to produce the present condition of the Earth's surface. (De la Beche 1834, quotations 371, 397)

For De la Beche, geological time was vast but finite, difficult but perhaps not impossible to comprehend.

The first use that I have found of the term 'geological time' appears in volume 1 of John Phillips' *A Treatise on Geology*. Published in 1837 as part of Dionysius Lardner's *Cabinet Cyclopaedia*, Phillips' intended audience was broader and no doubt a bit down the scale from the public that had made Lyell's *Principles* a bestseller. He started with fundamentals. He started with time. 'The very first inquiry to be answered,' he wrote in his opening chapter, 'is what are the limits within which it is possible to determine the relative dates of geological phenomena? For if no scale of geological time be known, the problem of the history of the successive conditions of the globe becomes almost desperate' (Phillips 1837, pp. 8–9). Chronology provided the guide to explaining the record of the Earth's crust, and with its focus on examining the stratigraphic record methodically from the oldest to the most recent formations, volume 1 of Phillips' *Treatise* might be considered perhaps the first text in historical geology. In his opening chapter, Phillips briefly remarked that the total length of the stratigraphic scale might be an important 'element for direct computation of the total time elapsed in the formation of the crust of the globe.' He made no such calculation. As he had observed a few pages earlier, the periods involved were 'too great to comprehend' (pp. 16, 18).

Time and geological history

When Lyell's *Elements of Geology* appeared in 1838, it was much more than just an expanded treatment of all that was 'Geology proper' in the *Principles*. Part 2 of the *Elements* was essentially a treatise on historical geology. Lyell maintained his allegiance to his steady-state uniformitarian principles by declaring that the chronological sequences of the four classes of rocks – aqueous, plutonic, volcanic and metamorphic – would be 'considered as four sets of monuments relating to four contemporaneous, or nearly contemporaneous, series of events' (Lyell 1838, p. 266). In the *Elements*, Lyell treated the whole of the known European stratigraphic record, not just the Tertiary, as a chronological sequence and made constant reference to the relative ages of the rocks. Lyell's chronological grouping of the fossiliferous rocks was somewhat out of date. Most of the 18 groups into which he divided the stratigraphic column were taken from the older, lithologically based, divisions, supplemented by his own four-part classification of the Tertiary and by the Silurian and Cambrian systems of Murchison and Sedgwick. More interesting for my purpose, however, was Lyell's caution that he could not assert that the 18 groups represented equal periods of time, but he continued:

> If we were disposed, on Palaeontological grounds, to divide the entire fossiliferous series into a few groups, less numerous than the above table, and more nearly co-ordinate in value than the sections called primary, secondary, and tertiary, we might, perhaps, adopt the six following groups or periods. (Lyell 1838, pp. 280–281)

This is a very tentative statement, and Lyell's meaning is by no means explicit. But if we assume that 'more nearly co-ordinate in value' refers to time, Lyell's next grouping gives an insight into his idea of the relative duration of the periods of geological time. His groups would give a ratio of 1:4:1 for the duration of the Tertiary, Secondary and Primary. Lyell still ventured no guess as to the magnitude of the time involved, but in giving a history to that part of the Earth's crust that is most visible to study, he employed at least a utilitarian notion of geological time. A letter written to Andrew Ramsay in 1846 makes clear, however, that for Lyell 'an indefinite lapse of geological time' was still required for the deposition of even that part of the strata laid down since the last coal measures of South Wales (Geikie 1895, p. 86).

By 1840 at least the chapter titles for a 'history' of the Earth had been established, and the new historical categories of period and development had become part of the conceptual vocabulary of geology. Over the next several decades the content of that history was enormously refined and expanded as the level of geological activity and geological knowledge grew exponentially. The various geological surveys added volumes to the stratigraphic record (see Secord 1986b). Systematic study of the 'causes' emphasized in Lyell's *Principles* was begun in earnest, and a few attempts were made to measure them (Davies 1969, pp. 225–226). Perhaps most of all, palaeontology served to reinforce an ever more vivid sense of 'deep time' with a seemingly endless succession of dramatic discoveries. As Rudwick has shown, the popularity of visual images of the remote past spread the notion of deep time to an ever wider public (Rudwick 1992). But though the concept of geological time was perhaps given greater detail during those years, it was not qualitatively changed; it was vast, perhaps beyond imagination, but it had no definable magnitude. It lacked measure. Charles Darwin's brief lapse of caution in 1859 changed all that.

Time was as important for Darwin's theory as it had been for Lyell's, and he drew on it liberally. But when Darwin decided to illustrate the magnitude of the time he envisioned with a quick and dirty calculation of the time needed for the denudation of the Weald, his result – more than 300 million years for a small portion of recent geological time – drew a deluge of protest (Darwin 1859, p. 287; see Burchfield 1974). John Phillips again provides a pertinent example. In his presidential address to the Geological Society in early 1860, Phillips offered his own quick and dirty estimate of the denudation of the Weald. Assuming fluvial denudation rather than the marine denudation assumed by Darwin, Phillips calculated that Darwin's 300 Ma could be reduced to a mere 1.3 Ma years (Phillips 1860a, p. lii). Perhaps more significant in the long run, in his Read Lecture a few months later, Phillips followed up on the idea he had alluded to in 1837, and described his calculation of the time needed to erode (and by analogy, to form) the whole sedimentary column. Using one of the few quantitative measures of erosion available, Robert Everest's 1832 measurements of the sedimentary load of the Ganges, Phillips arrived at a figure of 96 Ma for the total age of the sedimentary rocks. Phillips was well aware that his calculation was as much a 'guesstimate' as Darwin's; the periods involved, he declared, were 'too vast and we must add too vague for conception' (Phillips 1860b, pp. 126–127). The method he employed, however, was in broad outline to become the method employed by geologists for nearly half a century in their effort to measure geological time. Meanwhile, the contrast between Darwin's result and Phillips' showed just how different the notion of incomprehensibly vast time could appear when expressed in numbers. Darwin quietly dropped the age of the Weald calculation from the later editions of the *Origin of Species*. Phillips, however, had already written to William Thomson.

Thomson, the future Lord Kelvin, had a long-standing interest in the cosmic evolution implicit in LaPlace's nebular hypothesis and in the physics of the interior of the gradually cooling and solidifying Earth which would be one of its consequences. He had little interest in what Lyell had called 'Geology proper'. Possibly stirred by Phillips' letter to respond to Darwin, or perhaps simply by the relative leisure afforded by a temporarily disabling accident, Thomson, in 1862, produced two short papers containing calculations of the ages of the Sun and Earth, respectively. The methods introduced in the two calculations were necessarily different, but both involved applying the newly developed principles of thermodynamics, on which Thomson was the recognized British authority, to conditions inferred from LaPlace's hypothesis. Both calculations pointed to a terrestrial age of about 100 Ma – possibly as little as 10 Ma, but certainly no more than 400 Ma (Thomson 1862, 1863; Burchfield 1975, pp. 21–56).

'Deep time' had been measured, or at least quantified, but only by invoking the cosmology that Lyell had rejected, and it was hardly deep enough to suit either Lyell's or Darwin's requirements. Some of Thomson's data, to be sure, were as uncertain as those of Darwin and Phillips, but they were also more remote and their faults harder to pinpoint. The methods themselves had an elegant simplicity, and the principles involved seemed irrefutable. The initial response from geology, however, was negligible until Thomson fired off a short broadside against the 'doctrine of uniformity' late in 1865 (Thomson 1865). Whether British popular geology was as uniformly 'uniformitarian' as Thomson claimed is debatable, but the soundness of his attack on Lyell's conception of a dynamically steady-state Earth as being contrary to the known laws of physics could hardly be ignored. Geology began to respond. One of the first responses was indirect but significant. In 1864, James Croll proposed a hypothesis linking the onset of the ice ages to changes in the eccentricity of the Earth's orbit (Croll 1864). By 1867, Croll had developed his hypothesis sufficiently to suggest that the calculated dates of periods of high orbital eccentricity might be used to determine the date of the most recent glacial epoch (Croll 1867). Stimulated by Croll's theory, Lyell made his own attempt to date the last ice age, choosing as the most likely date, a period of very pronounced eccentricity which, according to astronomical calculations, occurred between 750 000 and 850 000 years ago. Lyell then combined this date with his statistical palaeontology of the tertiary molluscs to suggest that a complete *revolution* of species would require about 20 Ma. Assuming that approximately equal periods of time were neces-

sary for each of the twelve revolutions that he estimated had occurred since the beginning of the Cambrian, he concluded that the total time required would be 240 Ma. Lyell published this result in volume one of the tenth edition of the *Principles* in 1867, thus finally giving a quantitative scale to his conception of 'indefinite time' (Lyell 1867, pp. 271–301). Lyell's quantitative scale was an order of magnitude smaller than that implied by Darwin's quantitative guess, but it was still too large for Croll.

Responding a year later, Croll questioned Lyell's choice of an early rather than a later period of high orbital eccentricity. Croll doubted that any trace of glaciation would remain if the forces of denudation had been operating at their present rates had the period of glaciation ended 700 000 years ago, as Lyell's assumption would suggest. If, instead, one chose the more recent period of high orbital eccentricity which occurred about 80 000 years ago, Croll showed that Lyell's method of calculation would yield an age of only 60 Ma since the beginning of the Cambrian. This result, Croll emphasized, was in good agreement both with what was known about the current rates of denudation and with the 100 Ma that William Thomson had shown to be the time available for the entire history of the Earth (Croll 1868, pp. 363–368). Like Darwin, Lyell quietly removed his one quantitative estimate of time from the later editions of his book.

During the 1850s and 1860s the number of efforts to measure some of Lyell's 'causes now in action' increased. Estimates of the composite maximum thickness of the stratigraphic systems were regularly revised, and measurements of the rates of denudation in several major river basins promised a possible way to estimate the time elapsed in the simultaneous destruction and formation of the Earth's crust by denudation and sedimentation (see Davies 1969, pp. 317–355). But when Archibald Geikie reviewed the quantitative results of these efforts in 1868, he concluded that the data available were not sufficient for geologists to make an independent calculation of geological time. What the results did indicate to him, however, was that modern denudation was a far more rapid process than geologists had tended to believe, and hence their demands for time had been exaggerated. Geologists, he asserted, had been 'drawing recklessly upon a bank in which it appears that there are no further funds at our disposal'. Referring finally to Thomson and Croll, Geikie implicitly accepted that 'the time assigned within which all geological history must be comprised' was about 100 Ma. (Geikie 1871, pp. 188–189). A year later, T. H. Huxley, then President of the Geological Society of London, also conceded that 'Biology takes her time from geology. ... If the geological clock is wrong, all the naturalist will have to do is to modify his notions of the rapidity of change accordingly' (Huxley 1869, p. 331).

When Lyell died in 1875, the Earth had both a history and an age. To be sure, the precise age remained in dispute, but there was broad general agreement as to its magnitude. 'Deep time' had been numbered, and to that extent, at least, geological time had been made comprehensible. The Earth's history, however, still lacked intermediate dates. Some older geologists, such as Lyell's slightly younger contemporaries A. C. Ramsay and Joseph Prestwich, doubted that geological measurements could ever provide adequate data for calculating the duration of geological time in years (Ramsay 1873–1874; Prestwich 1895). A few younger geologist made the attempt. The attitude of some of this new generation was perhaps best summarized by T. Mellard Reade's challenge in 1876 that 'to make Geology essentially a Science, the mathematical method must step in to measure, balance, and accurately estimate' (Reade 1878, p. 211).

By the century's end, both geology and physics had produced a number of methods for calculating the Earth's age and even a few estimates of the duration of the larger divisions of the geological record (Burchfield 1975, pp. 90–156). Most of the geological calculations employed some variation on the method used by Phillips in 1860, and they reflected the constant flux of data and opinion concerning the composite depth of the stratigraphic column, the rates of past and present denudation, and the size and shape of ancient sedimentary basins. Geological time had been measured, but it still required two chronological scales: a biostratigraphic scale of the relative ages of the successive divisions of the stratigraphic record and a far more fragmentary quantitative scale. But if there was little unanimity among geologists on the details of the latter, there was a broad general agreement as to the order of magnitude of the Earth's age and some consensus as to the relative duration of the three major eras of geological time. Perhaps equally suggestive of what I have called a 'meaningful' conception of geological time was the geologists' confidence in their quantitative methods and results. At the century's end, geologists could confidently marshal their own methods, calculations and measurements of geological time against the ever more restrictive results from the physical methods of Thomson (by then Lord Kelvin) and his followers. A decade later, they were equally confident in opposing the inconceivably vast new timescale introduced by the proponents of radioactivity (Burchfield 1975, pp. 121–205). For most geologists in 1900, geological time was vast but finite; it could be measured and comprehended.

I hardly need to add in conclusion that the geologists' confidence was misplaced. The implications of radioactivity did push back the boundaries of 'deep time,' perhaps beyond comprehension. But radioactivity also provided new methods for a more precise determination of the Earth's age, and, ultimately, methods for dating the separate divisions of the biostratigraphic scale. The invention of geological time has now proceeded for yet another century, but it still reflects much of the design sketched in the age of Lyell.

References

ALBRITTON, C. C., Jr 1980. *The Abyss of Time*. Farrar & Strauss, New York.
BERGGREN, W., KENT, D. V., AUBRY, M. P. & HARDENBOL, J. (eds) 1995. *Geochronology, Time Scales and Global Stratigraphic Correlation*. Society for Sedimentary Geology Special Publication, Tulsa, OK, **54**.
BERRY, W. B. N. 1968. *Growth of a Prehistoric Time Scale*. Freeman, San Francisco.
BURCHFIELD, J. D. 1974. Darwin and the dilemmas of geological time. *Isis*, **65**, 300–321.
—— 1975. *Lord Kelvin and the Age of the Earth*. Science History Publications, New York.
COWPER, W. 1785. 'The task.' *In*: BAIRD, J. D. & RYSKAMP, C. (eds) *The Poems of William Cowper*, vol. 2. Clarendon Press, Oxford.
CROLL, J. 1864. On the physical cause of the change of climate during geological epochs. *Philosophical Magazine*, ser. 4, **28**, 121–137.
—— 1867. On the excentricity of the earth's orbit and its physical relations to the glacial epoch. *Philosophical Magazine*, ser. 4, **33**, 119–131.
—— 1868. On geological time, and the probable date of the glacial and upper Miocene period. *Philosophical Magazine*, ser. 4, **35**, 363–384 & **36**, 141–154, 362–386.
DARWIN, C. 1859. *On the Origin of Species*, 1st edn. Murray, London.
DAVIES, G. L. 1969. *The Earth in Decay*. Macdonald, London.
DE LA BECHE, H. C. 1834. *Researches in Theoretical Geology*. Charles Knight, London.
GEIKIE, A. 1871. On modern denudation. *Transactions of the Geological Society of Glasgow*, **3**, 153–190.
—— 1895. *Memoir of Sir Andrew Crombie Ramsay*. Macmillan, London.
HABER, F. C. 1959. *The Age of the World: Moses to Darwin*. Johns Hopkins Press, Baltimore.
HUTTON, J. 1785. *Abstract of a Dissertation Read Before the Royal Society of Edinburgh*. Edinburgh University Library, Edinburgh. Reprinted 1997.
HUXLEY, T. H. 1869. Geological reform. *Quarterly Journal of the Geological Society*, **25**, xxxviii–liii.
KITTS, D. B. 1966. Geologic time. *Journal of Geology*, **74**, 127–146. Reprinted *In*: ALBRITTON, C. C. (ed.) 1975. *The Philosophy of Geohistory*. Dowden, Hutchinson & Ross, Stroudsburgh, PA.
—— 1989. Geological time and psychological time. *Earth Sciences History*, **8**, 190–191.

LYELL, C. 1830–1833. *Principles of Geology*. 1st edn, 3 vols. Murray, London.

—— 1838. *Elements of Geology*. 1st edn. Murray, London.

—— 1867. *Principles of Geology*. 10th edn, vol. 1. Murray, London.

MCPHEE, J. 1980. *Basin and Range*. Farrar, Strauss & Giroux, New York.

PHILLIPS, J. 1837. *A Treatise on Geology*, vol. 1. Longman, Rees, Orme, Brown, Green & Longman, London.

—— 1860a. Presidential address. *Quarterly Journal of the Geological Society*, **16**, xxx–lv.

—— 1860b. *Life on the Earth: Its Origin and Succession*. Macmillan, London.

PORTER, R. 1977. *The Making of Geology*. Cambridge University Press, Cambridge.

PRESTWICH, J. 1895. *Collected Papers on Some Controverted Questions of Geology*. Macmillan, London.

RAMSAY, A. C. 1873–1874. On the comparative value of certain geological ages (or groups of formations) considered as items of geological time. *Proceedings of the Royal Society of London*, **22**, 145–148.

READE, T. M. 1878. Geological time. *Proceedings of the Liverpool Geological Society*, **3**, 211–235.

RUDWICK, M. J. S. 1969a. The strategy of Lyell's *Principles of Geology*. *Isis*, **61**, 5–33.

—— 1969b. Lyell on Etna and the antiquity of the Earth. *In*: SCHNEER, C. J. (ed.) *Toward a History of Geology*. MIT Press, Cambridge, MA, 288–304.

—— 1971. Uniformity and progression. *In*: ROLLER, D. H. D., (ed.) *Perspectives in the History of Science*. University of Oklahoma Press, Norman, OK, 5–33.

—— 1974. Poulett Scrope on the volcanoes of Auvergne:

Lyellian time and political economy. *British Journal for the History of Science*, **7**, 205–242.

—— 1978. Charles Lyell's dream of a statistical palaeontology. *Paleaeontology*, **21**(2), 225–244.

—— 1985. *The Great Devonian Controversy*. University of Chicago Press, Chicago.

—— 1992. *Scenes from Deep Time*. University of Chicago Press, Chicago.

—— 1996. Cuvier and Brongniart, William Smith, and the reconstruction of geohistory. *Earth Sciences History*, **15**, 25–36.

SECORD, J. A. 1986a. *Controversy in Victorian Geology: The Cambrian–Silurian Dispute*. Princeton University Press, Princton.

—— 1986b. The Geological Survey of Great Britain as a research school, 1839–1855. *History of Science*, **24**, 223–275.

THOMSON, W. 1862. On the age of the Sun's heat. *Macmillan's Magazine*, 5 March, 288–293.

—— 1863. On the secular cooling of the earth. *Philosophical Magazine*, ser. 4, **25**, 1–14.

THOMSON, W. 1865. The doctrine of uniformity in geology briefly refuted. *Proceedings of the Royal Society of Edinburgh*, **5**, 512–513.

—— 1891. *Popular Lectures and Addresses*. vol 1. Macmillan, London.

TOULMIN, S. 1962–1963. The discovery of time. *Manchester Literary and Philosophical Society Memoirs and Proceedings*, **105**, 100–112. Reprinted in: ALBRITTON, C. C. (ed.) 1975. *The Philosophy of Geohistory*. Dowden, Hutchinson & Ross, Stroudsburgh, PA.

WILSON, L. G. 1972. *Charles Lyell, the Years to 1841: The Revolution in Geology*. Yale University Press, New Haven.

Lyell and the dilemma of Quaternary glaciation

PATRICK J. BOYLAN

City University, Frobisher Crescent, London EC2Y 8HB, UK

Abstract: The glacial theory as proposed by Louis Agassiz in 1837 was introduced to the British Isles in the autumn of 1840 by Agassiz and his Oxford mentor, William Buckland. Charles Lyell was quickly converted in the course of a short period of intensive fieldwork with Buckland in and around Forfarshire, Scotland, centred on the Lyell family's estate at Kinnordy. Agassiz, Buckland and Lyell presented substantial interrelated papers demonstrating that there had been a recent land-based glaciation of large areas of Scotland, Ireland and northern England – at three successive fortnightly meetings of the Geological Society of London, of which Buckland was then President, in November and December 1840. However, the response of the leading figures of British geology was overwhelmingly hostile. Within six months Lyell had withdrawn his paper and it had become clear that the Council of the Society was unwilling to publish the papers, even though they were by three of the Society's most distinguished figures. Lyell reverted to his earlier interpretation of attributing deposits such as tills, gravels and sands and the transport of erratics to a very recent deep submergence with floating icebergs, maintaining this essentially 'catastrophist' interpretation through to his death a quarter of a century later.

Many paradigm shifts in the development of science (Kuhn 1960) have depended less on a major, unexpected intellectual leap on the part of some heroic scientific figure than on the sudden recognition of a new interpretation of well established evidence: re-examining and reinterpreting perhaps very well established facts and observations within a new theoretical framework. In the context of the Pleistocene this is now widely recognized to be the case, particularly since the publication of the wide-ranging *Ice Ages: Solving the Mystery* by Imbrie and Imbrie (1979). The central theme of the present paper is the response of Charles Lyell (1797–1875) to the reinterpretation in the autumn of 1840 by Louis Agassiz (1807–1873) and Lyell's former teacher and mentor William Buckland (1784–1856) that much of the well known 'superficial' deposits of northern Britain were evidence of a Recent land-based ice age. In Lyell's case, following his conversion to the glacial theory, he drew some of his strongest evidence for a Recent glaciation from field work that had featured in his first communication to the Geological Society 14 years earlier (Lyell 1826), while Buckland similarly reinterpreted a wide range of observations made up to 29 years earlier (Buckland 1841, p. 332).

The discoveries of Agassiz

In regard to the recognition of glaciation, the old 'heroic' view of geological progress focused on the brilliant young Swiss palaeontologist, Louis

Agassiz and his *'Discours de Neuchâtel'* presidential address (1837) presenting the theory of a very recent major ice age to the Société Hélvétique des Sciences Naturelles (Agassiz 1837, 1838). However, though Agassiz was certainly of great importance, this is a major oversimplification. From at least the late 1820s there had been growing speculation about the possible role of a recent major land glaciation in processes of erosion, transport and deposition across much of the now temperate regions of the northern hemisphere. As is well known, more than a generation earlier both Hutton and Playfair had speculated on this possibility.

In 1826 Robert Jameson published in his *Edinburgh New Philosophical Journal* a translation from the original Norwegian of a paper by the Dane, Jens Esmark, arguing that both Norway and Denmark had been recently glaciated (Esmark 1826). Furthermore, Herries Davies has shown from both the recollections of J. D. Forbes and notes in the Jameson Papers in Edinburgh that by 1827 Jameson was discussing the former existence of glaciers in Scotland in his university lectures (Davies 1969, pp. 267–270.)

However, the 1837 'Discours' of Agassiz was certainly a key event, not least because of its indirect impact in Britain. A number of leading British scientists had been amongst the first to recognize the remarkable abilities of the 29-year-old Agassiz as a vertebrate palaeontologist, and British patrons and organizations (particularly the Geological Society) were important sources of

BOYLAN, P. J. 1998. Lyell and the dilemma of Quaternary glaciation. *In*: BLUNDELL, D. J. & SCOTT, A. C. (eds) *Lyell: the Past is the Key to the Present*. Geological Society, London, Special Publications, **143**, 145–159.

Agassiz's income from 1834 onwards through their funding of his work on British fossil fish collections. It was therefore with much alarm that his leading British patrons and supporters, above all Buckland, once regarded as an arch-catastrophist himself, learned of Agassiz's sudden espousal and exposition of the glacial theory in July 1837 (Agassiz 1886, pp. 248–251).

As early as 1831 Buckland himself had argued that the 'northern region of the earth seems to have undergone successive changes from heat to cold' (Buckland 1831). However, geological deposits of this last period of intense cold continued to be interpreted in terms of aqueous deposition, whether through catastrophic flooding events (in Buckland's case) or the gentle marine submergence of Lyell. Buckland saw the heterodox views of the 'Discours' as representing a potentially serious threat to further British support for Agassiz's highly important vertebrate palaeontology work, and resolved to visit Switzerland himself in order to dissuade Agassiz from pursuing his glacial theory any further.

Another important patron, Alexander von Humboldt, gave the same advice in a letter of 2 December 1837:

> I am afraid you work too much, and (shall I tell you frankly?) that you spread your intellect over too many subjects at once. I think that you should concentrate your moral and also your pecuniary strength upon this beautiful work on fossil fishes. ... In accepting considerable sums from England, you have, so to speak, contracted obligations to be met only by completing a work which will be at once a monument to your own glory and a landmark in the history of science. .... No more ice, not much of echinoderms, plenty of fish... (Agassiz 1886, pp. 267–272).

Because of Buckland's heavy teaching and religious duties at Oxford, his first opportunity to visit Agassiz in Neuchatel was the 1838 summer vacation. He remained at first 'an uncompromising opponent' of the glacial theory (Gordon 1894, pp. 140–141) and pleaded with Agassiz to recant. However, by the end of his Swiss tour Buckland appears to have been not only converted to Agassiz's argument that the Alps and adjacent lowlands had been extensively glaciated in geologically recent times, but also realized that he was observing unmistakable parallels to localities in Scotland and northern England which had in some cases puzzled Buckland for more than a quarter of a century (Buckland 1841, p. 332).

In those pre-railway days travel between Switzerland and Britain was a lengthy and expensive undertaking which Agassiz could not undertake lightly, so an early visit to Britain was not possible. However, interest in the new glacial theory was growing. Perhaps most significantly of all, Agassiz's 'Discours de Neuchâtel' was within a matter of months plucked from the relative obscurity and limited circulation of its original publication in the *Actes de la Société Hélvétique* (Agassiz 1837) and given worldwide circulation in English translation by Robert Jameson of the University of Edinburgh in Jameson's influential *Edinburgh New Philosophical Journal* (Agassiz 1838). (Significantly, Jameson translated and republished four other substantial Continental publications on the glacial theory between 1836 and 1839.)

Lyell was by this time well aware that there had been one or more geologically recent cold phases, as demonstrated by the 'Arctic' molluscan faunas that were being widely recognized in the Newer Pliocene (soon to be renamed Pleistocene) of temperate latitudes, and hence he was moving towards the acceptance of a glacial phase, at least in climatic terms. However, he was fundamentally opposed to Agassiz's inferred mechanism of a continental scale glaciation to explain the characteristic features of erosion, transportation and deposition, and therefore prepared a major paper intended to refute the emerging glacial theory. He avoided a frontal attack: within the accepted traditions of the Geological Society this would have been regarded as just as 'unphilosophical' as the glacial heresy itself (see, for example, Rudwick 1963; Morrell 1976; Thackray this volume). Instead, Lyell carefully followed the Geological Society's tradition, and in 1839 presented to the Society an apparently innocuous descriptive paper on the thick and extensive superficial deposits that are so abundant in the glaciated areas of eastern England, under the title 'On the Boulder Formation or drift and associated freshwater mud cliffs of eastern Norfolk' (Lyell 1840). In this, Lyell described in some detail a wide range of features including the very typical boulder clays, erratic boulders transported from long distances including some almost certainly transported all the way from Scandinavia, freshwater Arctic shell horizons and apparently near-contemporaneous contortions in deposits. Lyell accepted the evidence for a recent very cold phase, but attributed the actual deposition of the boulder clay and erratic blocks and boulders to a phase of relatively deep submergence, with long-distance transport by icebergs, the grounding of which produced contortions in many of the deposits described.

The Norfolk paper is of additional historic interest in that it introduced to the standard geological vocabulary two words that were to become very widely used in Pleistocene geology through to the present day: the Lowland Scots farmers' name of 'till' for boulder clay in the narrow sense, and Lyell's own term of 'drift' for glacial deposits as a whole. The latter term had major implications, as it very clearly and unambiguously indicated Lyell's argument for a marine origin of the material, i.e. that this had been 'drifted' to its present location by floating icebergs. Ironically, though it was to be almost a century before the very useful and relatively neutral term 'till' came into general use, the term 'drift' (arguably very tendentious because of its implications about the presumed origin of the deposits described) was quickly adopted across much of British geology, particularly through its adoption by the Geological Survey. (It is also interesting to note the recent revival of interest in a possible glacio-marine origin of coastal 'shelly drifts' around Britain: see, for example, Eyles & McCabe 1989).

Buckland, seen as the arch-diluvialist in the early 1820s (though he had long since abandoned this position), must have realized that most of the leading figures of the British geological establishment could all too easily easily portray his adoption of the glacial theory as a return to old-style catastrophism. He therefore seems to have said nothing about what he had seen in Switzerland, but instead concentrated his efforts on persuading Agassiz to return to Britain as soon as possible.

The British Association meeting of 1840

The planned September 1840 annual meeting in Glasgow of Britain's 'Parliament of Science', the British Association, offered an ideal opportunity. Agassiz agreed to come to report progress on his fossil fish studies at the Glasgow meeting, and afterwards to undertake further cataloguing of British and Irish collections of fossil fish. With Buckland making the arrangements, it was probably no accident that the first new collection to be visited after the Glasgow meeting was the Gordon Collection of Old Red Sandstone fish at Elgin in northeast Scotland. Travelling there would inevitably take Agassiz through areas of the Scottish Highlands where, Buckland was convinced, Agassiz would recognize much evidence of a recent glaciation.

In the meantime Buckland was serving a second two year term as President of the Geological Society. Near the end of the 1839–1840 winter session of the Society a paper from Agassiz 'On the polished and striated surfaces of the rocks which form the beds of glaciers in the Alps' was read to the Society on his behalf, presumably by one of the secretaries, and was later reported briefly in the *Proceedings*, (Agassiz 1841*a*) but seems to have attracted little attention or comment. However, Buckland also used the obituaries section of his

presidential Anniversary Address (Buckland 1840) to praise very warmly the recently deceased Foreign Member, Jens Esmark, who (as noted above) had argued as long ago as 1826 that there had been a recent regional-scale glaciation of much of Scandinavia (Esmark 1826). In relation to this, it is significant that the annual presidential address was the only thing in the Society's programme that by tradition was published in full without the risk of what in some cases amounted to direct or indirect censorship (or even total rewriting) by the Society's officers (see Thackray in this volume).

The events relevant to the glacial theory around the time of the September 1840 meeting of the British Association in Glasgow have been reconstructed in detail by Herries Davies (1968, pp. 271–283), and White (1970) added to this. My own subsequent documentary and field work has identified and re-examined on the ground the field evidence at each of the 117 localities across Scotland and northern England used by Agassiz, Buckland and Lyell in the autumn of 1840 (Boylan 1978, 1981, 1984, pp. 471–511, and the deposited supplementary publication to this paper – see below).

As planned, Agassiz was met by Buckland in Glasgow, probably on the second day of the British Association meeting (20 September 1840), having spent much of the summer virtually living on the Aar Glacier itself, where he had set up the long-term scientific monitoring station, dubbed the 'Hôtel des Neuchâtelois'. No doubt Buckland reported on his own identification of the Crickhope Linn, Dumfriesshire moraine *en route* to Glasgow (Boylan 1981), while a visit to the Bell's Park area of the city, where building work was in progress, convinced Agassiz that the boulder clay and moulded and striated rock surfaces of central Glasgow were themselves of glacial origin (Herries Davies 1969, pp. 274–276).

During the formal programme of the meeting, Agassiz reported progress on his work on fossil fish as expected, but also gave a paper with the innocuous title 'On glaciers and boulders in Switzerland', of which only a very brief abstract was published in the meeting's *Report and Transactions* (Agassiz 1841*b*). However, contemporary press reports and correspondence show that Agassiz publicly claimed that he expected to find evidence of the former existence of glaciers in Scotland during his stay, but he was strongly opposed by at least some of the geological establishment present, including Murchison (Herries Davies 1968, 1969, p. 275). In view of what was to happen only three weeks later it is interesting to speculate on the views of Lyell (who was certainly present at the Glasgow meeting: a portrait sketch of him painted during the meeting is now in the collection of Stuart A. Baldwin and was displayed during the Lyell Bicentenary Conference – see cover of this volume). However, so far no record seems to have been found of his participation in any of the public or informal discussion of Agassiz's claims.

Field work in Scotland

The British Association meeting ended on 23 September 1840 and Agassiz left Glasgow immediately, accompanied by Buckland and his wife Mary. It seems certain that they travelled northwards by the main west-coast road (originally an eighteenth-century 'General Wade' road, now the A82), via Loch Lomond, Inveraray and Ballachulish, to Fort William for the Ben Nevis region and Glen Roy, noting clear evidence of glaciation all along the route. For example, Agassiz later recalled the stagecoach approaching the Duke of Argyll's Inveraray Castle (where the party stayed at least one night): 'as the stage entered the valley, we actually drove over an ancient terminal moraine, which spanned the opening of the valley' (Agassiz 1886, p. 307); while Buckland, in a letter dated 4 October 1840 sent ahead to Professor John Fleming at Aberdeen, reported, 'We have found abundant Traces of Glaciers round Ben Nevis' (Gordon 1894, p. 141; White 1970).

After continuing along the Great Glen to Inverness and thence to Nairn, Forres and Elgin to see Old Red Sandstone fish fossil localities and collections, Agassiz and the Bucklands continued to Aberdeen, where they parted company on 9 October 1840. Agassiz went via Perth to Glasgow and continued immediately to northern Ireland, again to see fossil fish collections, though he found much evidence of glaciation there also (White 1970; Agassiz 1841c), while the Bucklands travelled southwards at a more leisurely pace, seeking out evidence of glaciation in northeast Scotland (Boylan 1984, pp. 487–489).

Eight years earlier, an Oxford student, Edward Jackson, recorded in his notebook (now in the British Geological Survey Library, MS ref. 1/635) covering Buckland's 1832 geological lecture series that Buckland told his students: 'advice – never try & persuade ye world of a new theory – persuade 2 or 3 of ye tip top men – & ye rest will go with ye stream, as Dr B. did with Sir H. Davy and Dr. Wollaston in case of Kirkdale Cave' (Boylan 1984, p. 648).

Following this precept, Buckland appears to have focused on the arch-uniformitarian Lyell as the least obvious but most telling third 'tip top' man (with Agassiz and himself) to involve in presenting the newly recognized abundant evidence of glaci-

ation in the Scottish Highlands. The Bucklands arrived on or around 10 October 1840 at Lyell's Scottish home, Kinnordy House near Forfar, and Buckland lost no time in presenting to Lyell both the conclusions of his tour with Agassiz of the Highlands and also his own conclusion that there was abundant evidence of a recent glaciation on the Kinnordy estate itself.

By 15 October 1840 Buckland was able to write to Agassiz in Ireland, 'Lyell has adopted your theory in toto !!! [sic] On my showing him a beautiful cluster of moraines within two miles of his father's house, he instantly accepted it, as solving a host of difficulties that have all his life embarrassed him' (Agassiz 1886, p. 309). It is, however, very interesting to note that perhaps because of that 'embarrassment' neither Buckland nor Lyell ever explicitly referred to evidence of glaciation at Kinnordy itself (at least by name). Nevertheless there are prominent morainic features within a few tens of metres of the house, while Kinnordy Loch adjacent to the house is clearly a typical example of the many water-filled kettle holes and other small 'glacial' lakes found on the extensive tills and 'hummocky moraine' of the area, which had in fact been the subject of Lyell's first scientific paper (Lyell 1826).

Lyell (as always) recorded his detailed observations in his notebooks. The two covering the three weeks 13 October to 6 November 1840 (Notebooks nos 84 & 85, Lyell Papers, Kinnordy) give some indication of the pace of Lyell's field research on the glacial question. Notebook no. 84, covering just one week, 13–20 October, includes 58 field drawings, sections and profiles plus detailed notes on around twenty localities or areas, ranging from the high corries near the head of Glen Clova down to the sea at Lunan Bay. Lyell travelled for much of the time with the Bucklands, who were both over 50 years old and in weak health following a serious coaching accident in Germany four years earlier. By way of comparison, my own field work in attempting to locate on the ground and re-evaluate Lyell's published localities in and around Forfarshire alone took me the equivalent of $2\frac{1}{2}$ weeks work over a three-year period, and I had the use of motor vehicles, metalled roads, modern mapping and Lyell's and Buckland's detailed notes to work from. For Lyell and his companions to have covered so much ground by carriage or on horseback in just a week was a remarkable achievement simply in terms of distances and travelling times, even if in many cases they were visiting familiar places which they now wished to reinterpret through, if not rose-tinted, then at least decidedly glacial, spectacles.

The speed and comprehensiveness of Lyell's conversion to the glacial theory is demonstrated by a detailed notebook entry dated 'Oct. 13. Kinnordy – 1840', at the most three or four days after the arrival of the Bucklands, in which he sets out why he considered a continental-scale land glaciation to be the only explanation, in contrast with his theory of marine deposition from icebergs etc., presented in his Norfolk paper of the previous year. These pages read (verbatim):

Moraines: The great valleys which descend from the Grampians into the Strath have been the chief vomitories of unstratified boulder formation capped with stratified loam, gravel & sand.

The proofs of the glacier origin of these remarkable deposits are:

1st. the entire absence of organic remains even in the clays & mud

2ndly. the absence of stratification in matter transported to great distances

3rdly The contortion of alternating layers of perfectly level gravel sand and mud in distinct layers resting on beds in which there has been little or no disturbance & sometimes ... [crossed? covered?] with horizontal layers.

4thly. The striated & polished surface of many of the boulders

5thly. The same on the surface of rocks in situ.

6thly. Local character of the boulders brought down different valleys

7thly. The frequent mounds or narrow ridges

transverse to the valley or general direction of the till resembling terminal moraines [*sic*] & which uniting with longitudinal ridges form numerous land locked hollows, lakes & peat mosses. (Lyell Notebooks, no. 84, pp. 3–4)

The first page of the following notebook, dated 21 October 1840, Kinnordy, addresses the marine submergence theory explicitly: 'If large errats. and boulds. [erratics and boulders] always or generally over small gravel then marine theory for stratd. [stratified] folld [followed] by glacial period for till will not do' (Lyell Notebooks, no. 85, p. 1).

Lyell continued to develop further tests which supported the glacial theory in both lowland and highland areas around Forfar. He spent much time examining two corries, Loch Brandy and Loch Whorrall, high up near the head of Glen Clova. He recognized and sketched clear retreat moraines in each corrie, and also other evidence of glaciation, and explicitly compared some of the evidence with his own observations in Switzerland during his earlier travels (Lyell Notebooks, no. 84, pp. 64–80; for examples see Figs 1–2 in this chapter). As the Geological Society's secretaries were to report in their abstract for the *Proceedings* of Lyell's November 1840 paper to the Society:

The distribution of an enormous mass of boulders on the southern side of Loch Brandy, and clearly derived from the precipices which overhang the Loch on the three other sides, is advanced as another proof in favour of the glacial theory. It is impossible to conjecture, Mr.

Fig. 1. Lyell's October 1840 sketch of the Glen Clova showing the moraine downstream of Loch Brandy (Lyell Notebooks, Kinnordy, no. 84, p. 68).

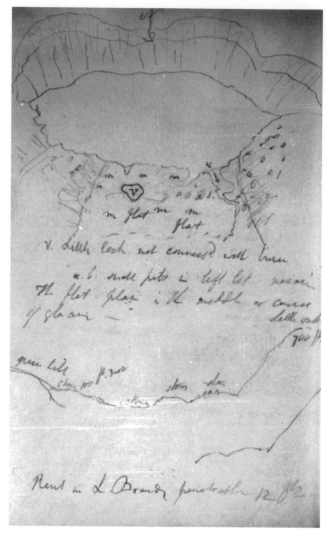

Fig. 2. *Main drawing:* October 1840 sketch of Loch Whorral, Glen Clova, showing that large blocks of rock matched with back-wall outcrops could have reached their present position only if the corrie basin had been filled with ice. *Middle sketch:* Sketch section across Glen Clova through Loch Whorral. *Bottom sketch:* Apparently a similar valley profile downstream of Loch Whorral, through Loch Brandy (Lyell Notebooks, Kinnordy, no. 84, p. 74).

Lyell says, how these blocks could have been transported half a mile over a deep lake; but let it be imagined that the Loch was once occupied by a glacier, and the difficulty is removed. Loch Whorral, about a mile to the east of Loch Brandy, is also surrounded on its north, east and western sides by precipices of gneiss, and presents on its southern [*sic*] an immense accumulation of boulders with some other detritus, strewed over with angular blocks of gneiss, in some instances twenty feet in diameter. The moraine is several hundred yards wide, and exceeds twenty feet in depth, terminating on the plain of Clova in a multitude of hillocks and ridges much resembling in shape some terminal moraines examined by Mr. Lyell in Switzerland. (Lyell 1841, pp. 339–340)

Lyell's notes also contain a number of sketch diagrams clearly intended to clarify in his own mind in the explanations and processes relating to his observations. For example, a sketch, apparently based on an active Swiss glacier with which he was familiar, indicates how the fluctuating ice margin of a valley glacier could create a complex of morainic ridges, hummocks and small lakes, of a form that has very clear parallels with those just re-examined in Glen Clova (Lyell Notebooks, no. 85, p. 53; Fig. 3 in this chapter). Another small sketch on a page of rough notes and reminders (Lyell Notebooks, no. 85, p. 60; Fig. 4 in this chapter) shows how three quite different glacial deposits – horizontally bedded deposits, steeply dipping or false-bedded mid-margin coarser deposits, and ground moraine tills – could all be formed more or less simultaneously in a steep-sided glaciated valley.

Lyell readily agreed with Buckland that he would contribute a major paper to the Geological Society on his findings in Forfarshire, to be presented along with papers from Agassiz on the comparisons between the Swiss evidence and his new observations in Britain and Ireland, and a more detailed paper by Buckland on the evidence of glaciation in Scotland (other than Forfarshire) and northern England. Buckland, as President, was able to ensure that the first three fortnightly meetings of the Geological Society in Somerset House from the start of its winter season on 4 November 1840 would be devoted entirely to these interrelated papers written by his 'three tip top men'. Lyell, always a methodical and careful worker, appears to have stayed at Kinnordy for most of the next fortnight carrying out further detailed research and writing (recorded in the notebooks) before going to London, arriving just in time for the opening of the Society's season.

The Bucklands, however, made a further rapid tour of several areas of the Highlands familiar to them from much earlier field work, including Upper Tayside, Schiehallion and the Trossachs, before travelling to Edinburgh. The planned rendezvous in Edinburgh with Agassiz on his return from Ireland probably did not take place, as the Bucklands apparently left for London on about 24 October, while Agassiz did not arrive until 27 October, just a week before the first Geological Society meeting. Buckland meanwhile continued to record observations of glacial phenomena, presumably from the London stagecoach, along much of the route of the Great North Road south from Edinburgh via Dunbar, Berwick-on-Tweed and Northumberland, referring to a number of these in his Geological Society paper the following month, including the Bradford Kames near North Charlton, Northumberland, which he had first examined in 1821 and which was one of the first British localities that he had reassigned to glacial deposition in the course of his Swiss studies in 1838 (Buckland 1841, p. 346).

In these few weeks Agassiz, Buckland and Lyell between them identified 117 localities or areas in Britain that they considered to present clear evidence of a recent glaciation, of which 113 have now been identified on the ground and re-evaluated (Boylan 1981, 1984, pp. 682–767). Agassiz reported on 21 of these, 12 were recorded by Lyell and the remaining 80 (including all 36 localities in northern England) were seen and cited by Buckland. Despite the apparent imbalance in the number of localities studied, there is no doubt that Buckland saw the powerful support of Agassiz and Lyell as crucial to the presentation and hoped-for acceptance of the glacial theory. The range of erosional and depositional features recorded by one or more of the three men at these 117 localities as evidence of a recent glaciation was very wide indeed, and included (using twentieth century

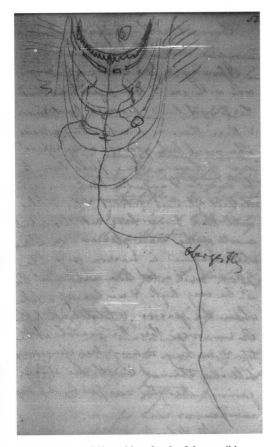

Fig. 3. October 1840 working sketch of the possible effect of ice margin fluctuations on the formation of moraines and perched lakes (*cf.* that on the Loch Whorral moraine – see Fig. 2), apparently based on recollection of a Swiss glacier (Lyell Notebooks, Kinnordy, no. 85, p. 53).

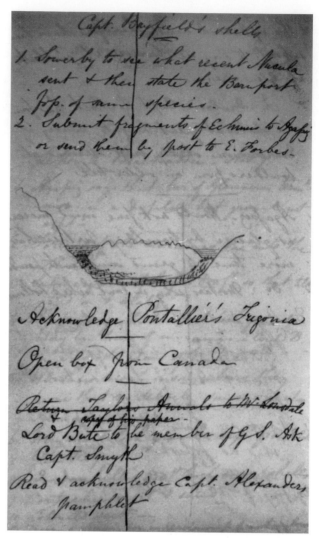

Fig. 4. Page of October 1840 notebook with various notes and reminders. The sketch in the middle of the page explores the way that quite different kinds of glacial deposits can be formed simultaneously by a valley glacier: ground moraine tills or boulder clays below the ice, steeply dipping or false-bedded mid-margin deposits along the side, and fine-grade horizontally bedded deposits at or near the surface. (Lyell Notebooks, Kinnordy, no. 85, p. 60).

terminology, and in order of frequency, and with percentages of the total number of localities):

Fluvioglacial/outwash deposits	33 sites (13.0%)
Kames/kame terraces	33 sites (13.0%)
Terminal or lateral moraines	28 sites (11.0%)
Glacial polishing and rounding	26 sites (10.2%)
Till	24 sites (9.4%)
Erratics	22 sites (8.7%)
Striations	21 sites (8.3%)
'Hummocky moraine'	20 sites (7.9%)
Ice damming and drainage diversion	11 sites (4.3%)
Drumlins	10 sites (3.9%)
Grooving of bare rock surfaces	9 sites (3.5%)
Rôches moutonées	8 sites (3.1%)
Eskers and related features	8 sites (3.1%)
Glacial re-advance within a moraine	1 site (0.4%)

Two further significant points can be made in the light of the identification and re-examination in the field of these localities. First, the range of features and sites is quite remarkable and could, for example, even today form a sound basis for university-level teaching of the subject, quite apart from the historic interest of such localities. Accordingly, a field guide to the 113 localities that have been identified (*English and Scottish Glacial Localities of Agassiz, Buckland and Lyell, 1840*) is available as both a Supplementary Publication No. SUP 18124 (30 pp) from the Geological Society Library and from the British Library Document Supply Centre, Boston Spa,Wetherby, W. Yorkshire LS23 7BQ, and as a World Wide Web document (http://www.city.ac.uk/artspol/glaclocs.html).

Second, all three pioneers of the 1840 were satisfied that what has for more than a century been termed 'hummocky moraine' was morainic in the strict sense of the word – on the basis of their knowledge of and comparisons with both present-day and Quaternary glacial margins in the Alps. However, since the work of Sharp (1949) and particularly of Hoppe (1952), the generally accepted view has been that this interpretation is incorrect, and that 'hummocky moraine' is deposition produced by the down-wasting of stagnant ice *in situ*. In Scotland, this was the interpretation of Sissons (1967) in his extensive studies of glacial landforms including his review of several of the classic sites and areas investigated in 1840 (e.g. Sissons 1974). However, recent work on active glacier margins in Svalbard now indicates that typical 'hummocky moraine' can indeed be morainic, formed by thrusting and ice-margin fluctuation in surge-type glaciers (Hambrey *et al.* 1997). It is perhaps appropriate that in Lyell's bicentenary year detailed contemporary research has at last confirmed the original interpretation of Agassiz, Buckland and Lyell, drawing on their Alpine observations of Alpine processes, 157 years ago.

Geological Society presentations

On 4 November 1840 large numbers of Geological Society members gathered in Somerset House, London, for the opening of the winter programme of fortnightly meetings. With Buckland as President in the chair the season opened with the announced commencement of the presentation by Agassiz, Lyell and Buckland of their recent findings. (It is not clear whether the truly sensational conversion of Lyell, the arch-opponent of old-style 'catastrophism', was announced or even hinted at that opening evening.) The meeting began with the reading by the author of Agassiz's paper, 'On glaciers, and the evidence of their having once existed in Scotland, Ireland and England' (Agassiz 1841c). Agassiz stressed the importance of

... investigating a country in which glaciers no longer exist, but in which traces of them might be found. This opportunity he had recently enjoyed, by examining a considerable part of Scotland,

the north of England, and the north, centre, west and south-west of Ireland; and he has arrived at the conclusion, that great masses of ice, and subsequently glaciers, existed in these portions of the United Kingdom at a period immediately preceding the present condition of the globe, founding his belief upon the characters of the superficial gravels and erratic blocks, and on the polished and striated appearance of the rocks *in situ*. (Agassiz 1841c, p. 328)

He argued that all the evidence pointed to an ice sheet of Greenland proportions, not merely Alpine-style valley glaciers. His glacial theory had thus expanded very greatly: both geologically and psychologically there is a great difference between arguing that sometime in the recent past existing Alpine glaciers may have extended a few tens of kilometres further down their valleys and perhaps spilled out on to the Alpine foreland, and contending that most if not all of Scotland, together with much of northern England and Ireland, had been under continental-scale ice sheets hundreds if not thousands of metres thick (Boylan 1981).

When Agassiz finished his paper, Buckland took over and began to read what was much the longest of the three communications, 'On the evidences of glaciers in Scotland and the North of England' (Buckland 1841), citing evidence from over 90 localities and areas which, he argued, demonstrated the reality of a recent glaciation. Because of its length, Buckland's paper was in fact spread over all three fortnightly meetings. The first part, dealing with Scotland, was commenced on 4 November following Agassiz's paper, and was continued on 18 November, sharing that evening with the first half of Lyell's paper, 'On the geological evidences of the former existence of glaciers in Forfarshire' (Lyell 1841). Then on 4 December the second half of Lyell's Forfarshire paper was read, and Buckland presented the third and final part of his paper, on his evidence for glaciation across large areas of northern England.

As explained below, only abstracts prepared by the Geological Society's secretaries were published by the Society (Agassiz 1841c; Buckland 1841; Lyell 1841), and neither Agassiz's nor Lyell's appear to have survived in manuscript form either. However, there is a fairly detailed draft of Buckland's paper in the Oxford University Museum collection of Buckland's lecture notes and working papers, and this shows just how significantly, though subtly, the published summary was edited (or perhaps more accurately, censored). Even Buckland's redefinition of 'till' was changed from the manuscript 'argillaceous detritus of Glaciers interspersed with pebbles' (Buckland Papers, Oxford University Museum, BuP Glacial

File) to the published 'unstratified glacier-mud containing pebbles' (Buckland 1841, p. 345). Similarly, writing of contortions within a small terminal moraine in the valley of the College Burn, near Kirknewton, Northumberland, Buckland's manuscript states that 'the laminae are variously & violently contorted, in a manner only explicable on the theory of a bed of laminated sand having been severed into fragments, which had subsequently been moved, &, contorted by the slow pressure of a Glacier descending the deep trough of the College Burn' (Buckland Papers, Oxford University Museum, BuP Glacial File).

However, the editors of the *Proceedings* clearly would not accept 'only explicable' and changed this to 'He is of the opinion...' (Buckland 1841, p. 346). Even less acceptable were claims about the revolutionary significance of what Agassiz, Buckland and Lyell were proposing. The published account does not contain the slightest hint of Buckland's ambitious claim, 'For some time to come the Glacial Theory must occupy a prominent place in Geological Investigation. The Subject appears to me the most important that has been put forth since the propounding of the Huttonian Theory & the surface of the whole Globe must be examined afresh...' (Buckland Papers, Oxford University Museum, BuP Glacial File).

From Lyell's contribution only the summary in the *Proceedings* survives, and presumably this would have been treated no better that that of the President. During the two evenings Lyell must have spoken for a total of at least $2^1/_2$ hours, and quite possibly for more than four hours. It therefore seems likely that even though only 12 localities or areas are specifically cited in the brief printed summary of Lyell's paper, he must have presented much more than this. It seems likely that many if not all of the detailed notes on sites and arguments in support of the glacial theory which he recorded so meticulously in Notebooks nos 84 and 85 between 12 October and the start of the Geological Society season on 4 November 1840 formed the core of his actual extended address. Certainly it was Lyell's style to present such abundant detail at great length when lecturing (a practice his detractors attributed to his training as a barrister in the English law courts).

However, Lyell is reported to have begun taking a much wider overview of his subject, quite unambiguously retracting his long-held views on the origin of many 'superficial' deposits, most recently and most explicitly in his substantial Norfolk 'drift' paper read to the Society just one year earlier:

Three classes of phænomena connected with the transported superficial detritus of Forfarshire, Mr. Lyell had referred, for several years, to the action of drifting ice; namely, 1st, the occurrence of erratics or vast boulders on the tops and sides of hill at various heights as well as the bottom of the valleys, and far from the parent rocks; 2ndly, the want of stratification in the larger proportion of the boulder formation or till; and 3rdly, the curvatures and contortions of many of the incoherent strata of gravel or of clay resting upon the unstratified till. [Footnote: 'See Mr. Lyell's paper on the Norfolk Drift, Phil. Mag., May 1840, and the Abstract of the paper, antè, p.171.'] When, however he attempted to apply the theory of drifting ice over a submerged country to facts with which he had been long acquainted in Forfarshire, he found great difficulty in accounting for the constant superposition of the till with boulders to the stratified deposits of loam and gravel; for the till ascending to higher levels than the gravel, and often forming mounds which nearly block up the drainage of certain glens and straths; for its constituting, with a capping of stratified matter, narrow ridges, which frequently surround lake-swamps and peat-mosses; and for the total absence of organic remains in the till. Since, however, Professor Agassiz's extension to Scotland of the glacial theory, and its attendant phænomena, Mr. Lyell has re-examined a considerable portion of Forfarshire, and having become convinced that glaciers existed for a long time in the Grampians, and extended into the low country, many of his previous difficulties have been removed. (Lyell 1841, pp. 337–338)

Specific examples referred to in the published summary include: detailed accounts of the South Esk and Prosen Valleys and features that he interpreted as lateral, medial and terminal moraines, the abundant evidence of glacial erosion, transport and moraines in the upper Clova Valley, including the Loch Brandy and Loch Whorral sites already referred to above; and abundant evidence of glacial striation and transported erratics across the region, which could, he now argued, be explained only in terms of glacial transport, rather than marine submergence. The extension of glacial deposits and other features into the lowland areas of Forfarshire, down to and including the coastal areas at Lunan Bay, was also described and discussed in detail. For example, Lyell returned to the subject of his first major paper, on the series of lowland 'marl-lochs' in the Tay Valley (Lyell 1826), this time interpreting these as the result of disrupted drainage due to a combination of local morainic features and the generally very uneven surface of the glaciated topography. He also described and interpreted a remarkable high ridge of till and gravel between the lower parts of the Glen Prosen and Carity Burn valleys, just north of Kinnordy, as a median moraine formed between two glaciers descending from the Highlands into the Tay Valley. The final part of his study focused particularly on the widespread dispersal of erratics in the region, including the highest part of the Sidlaw Hills. Lyell's turnabout and his rejection of the submergence theory he had been advocating so strongly only a matter of months before could hardly have been more explicit, as the published abstract shows:

> Mr Lyell objects to a general submergence of that part of Scotland, since the till and erratic blocks were conveyed to their present positions; as the stratified gravel is too partial and at too low a level to support such a theory; and he would rather account for the existence of the stratified deposits, by assuming that barriers of ice produced extensive lakes, the waters of which threw down ridges of stratified material on tops of the moraines. (Lyell 1841, p. 343)

He concluded by arguing that there were significant differences between the present-day Swiss glaciers and the extensive glaciation of Scotland that he was describing, and argued that 'it is to South Georgia, Kergulen's Land and Sandwich Land that we must look for the nearest approach to that state of things which must have existed in Scotland during the glacial epoch' (Lyell 1841, p. 344).

Geological controversy

In the absence of the full texts of both Agassiz's and Lyell's papers, and with the uncertainty about how complete or accurate is the surviving draft of Buckland's at Oxford, it is fortunate that we have one remarkable insight into one of the three sessions at the Geological Society. The Society's (salaried) Sub-Curator and Assistant Secretary, Samuel P. Woodward, appears to have recognized the very exceptional significance of the events, above all perhaps the totally unexpected and indeed sensational conversion of Lyell to the distinctly non-uniformitarian, neo-catastrophist even, glacial theory. Therefore, contrary to the Society's strict rule that there should be no reporting or recording whatsoever of any of the often vigorous debates which traditionally followed the presentation of papers at its meetings, Woodward (presumably secretly) made detailed notes on the debate that followed the Buckland and Lyell contributions of 18 November 1840. These were preserved in his private papers and in 1883 (by which time all the main protagonists and antagonists were dead) these were published verbatim by his son, Horace B. Woodward (1883).

This account shows that Murchison opened the attack with an appeal for the glacial theory to be denounced by the 'mathematicians and physical geographers present' and immediately descended to sarcasm, asking whether scratches and polishing on London streets had now to be attributed to glaciation: 'the day will come when we shall apply it to all. Highgate Hill will be regarded as the seat of a glacier, and Hyde Park and Belgrave Square will be the scene of its influence' (Woodward 1883, p. 226).

Though Woodward's account is not completely explicit on the point, Lyell had evidently argued that the glacial theory had to be accepted because no alternative theory or argument could explain the abundant evidence presented. However, Lyell's methodology and argument provoked a severe rebuke from William Whewell, who had strongly supported and made extensive use of Lyell's work in his monumental study of the philosophy of the inductive sciences only three years earlier (Whewell 1837). Referring to Lyell's arguments, Whewell insisted that 'it does seem to me that the way in which Mr. Lyell has treated it is not the most fair and legitimate. He says, "If we do not allow the action of glaciers, how shall we account for these appearances?" This is not the way in which we should be called upon to receive a theory' (Woodward 1883, p. 228).

Eventually, Buckland 'resigned' the (traditionally impartial) President's chair to George B. Greenough so that he could enter the fray from the ordinary debating benches, stressing the importance of his own former position as a 'sturdy' opponent of Agassiz: 'and having set out from Neuchâtel with the determination of confounding and ridiculing the Professor. But he went and saw all these things, and returned converted' (Woodward 1883, p. 229).

Woodward concluded his notes on the final exchanges of the long evening, in which Buckland

> ... referred to Professor Agassiz's book and condemned the tone in which Mr. Murchison had spoken of the 'beautiful' terms employed by the Professor to designate the glacial phenomena. That highly expressive phrase 'roches moutonnés ,'[sic] which he had done so well to revive, and that other 'beautiful designation' the glacier remanié! remanié! remanié! continued the Doctor most impressively, amidst the cheers of the delighted assembly, who were by this time elevated by the hopes of soon getting some tea (it was a quarter to twelve P.M.), [the meeting having started as usual at 6.00 p.m.] and excited by the critical acumen and antiquarian allusions and philological lore poured forth by the learned Doctor, who, after a lengthened and fearful exposition of the doctrines and discipline of the glacial theory, concluded – not as we expected,

by lowering his voice to a well-bred whisper, 'Now to,' etc., – but with a look and tone of triumph he pronounced upon his opponents who dared to question the orthodoxy of the scratches, and grooves, and polished surfaces of the glacial mountains (when they should come to be d——d) the pains of *eternal itch*, without the privilege of scratching!' (Woodward 1883, p. 229)

Lyell's change of mind

Lyell appears to have been just as robust a proponent of the glacial theory as Agassiz and Buckland both in presenting his own paper to the Geological Society and in the acrimonious debate on the three papers. On returning to Switzerland in December 1840, Agassiz wrote in triumphant terms to Alexander von Humbolt: 'J'ai acummulé tant de preuves que personne en Angleterre ne doubte maintenant que les glaciers n'y aient existé' [I have gathered so many proofs that nobody in England still doubts that glaciers have existed there] (Herries Davies 1969, p. 286).

However, this was far from the case, and Lyell in particular appears to have been very shaken by the vehemence of the criticisms from virtually every one of his closest allies. Whewell's philosophical objections were particularly damaging, and William Conybeare was equally forthright on the same point, arguing that the glacial theory was 'a glorious example of hasty unphilosophical & entirely insufficient induction' (Herries Davies 1969, p. 288). Though he submitted the (now lost) full text of his paper formally to the Geological Society for publication, Lyell, plainly shaken by the reaction, appears to have abandoned the glacial theory almost as quickly as he had adopted it few months earlier. Herries Davies (1969, p. 291) noted that he had reverted to the glacial submergence theory and dismissed the idea of a Scottish land glaciation almost totally in the 1841 second edition of the *Elements of Geology*, whose preface is dated 10 July 1841. In fact, Lyell's apostasy must have been at least two months earlier, since his application to the Geological Society requesting permission to withdraw his paper was approved and recorded in the Society's Council Minutes of 5 May 1840 (Geological Society of London Archives, CM1/5).

The Minutes further record that the referee's report on Buckland's 1840 paper was received by the Council under its new President, Murchison, on 17 November 1841, but contrary to the Society's normal practice and rules the required secret ballot to decide whether to publish the paper was not held. At the following Council meeting, on 1 December 1841, Agassiz's papers of June and November 1840 were similarly 'referred' for a second time without any explanation being recorded (Geological Society of London Archives, CM1/5). By this time a further paper by Buckland, on evidence for glaciation in north Wales (Buckland 1842) had also been read to the Society and submitted for full publication, evidently adding to the embarrassment over what to do with these major papers, one of them read from the presidential chair by Buckland, and the other by Agassiz, one of the Society's most highly regarded Foreign Members. The Council clearly was unwilling to publish anything supporting the glacial theory, and yet was reluctant to take the final step of rejecting them. The deadlock lasted to beyond the end of the winter session (which had been marked by a vitriolic and extended attack on the glacial theory and its protagonists by Murchison in his presidential Anniversary Address (Murchison 1842).

Buckland finally broke the deadlock with a terse note to Lonsdale, the Secretary to the Society dated 28 June 1842, which survives in the Society's Letter Books. This read, 'I beg to apply to withdraw my papers read some time since on Glaciers in Scotland and Glaciers in N. Wales' (Geological Society of London Archives, LR7/193). The Council under Murchison quickly granted the necessary 'permission', no doubt with a sigh of relief. The fate of Agassiz's two papers was never formally resolved, in that they were never balloted on and either officially accepted or rejected by the Council. These four pioneering papers of the glacial theory are further examples of the Geological Society's failure to support and publish important research, leading in some cases to the loss of key historical documents, due to what today at least would be regarded as a policy of censorship of unorthodox or otherwise unacceptable views or findings, a matter raised by Professor W. A. S. Sarjeant in the discussion of Thackray's paper during the Lyell Bicentenary Conference.

On 20 July 1841, less than three months after his withdrawal of the Geological Society paper and a month after he completed the second edition of the *Elements of Geology*, Lyell left Britain for the first of his four extended visits to North America. Important new insights into the extent of his retreat from the glacial theory have emerged from Dott's major study of Lyell in America (1996 and this volume). Particularly informative and relevant is the evidence Dott has uncovered about the syllabus and content of the series of 12 celebrity 'Lowell Lectures' that Lyell gave between 19 October and 27 November 1841, which were so popular that, despite using a public theatre with a seating capacity of 2000, Lyell had to present each lecture twice. Dott has reconstructed the programme and outline contents from surviving fragmentary information, including local press reports and

Lyell's rough notes in the Edinburgh University Library. This material leaves little doubt that in Lectures 11 and 12 Lyell had reverted to attributing the 'Boulder Formation' to the 'transporting power of floating glacier ice' (Dott 1997, pp. 106–107). This interpretation is explicit in the outlines of later public lecture series in both North America and Britain, for example an 1852 lecture series in Boston, also investigated by Dott (1997, p. 123). Lyell's notes for these include a diagram showing deep submergence of the Berkshires in Massachusetts and of adjacent areas of New York State. The well known trains of distinctive erratics (Dott 1997, pp. 123–124) were attributed to 'masses of floating ice carrying fragments of rock', while there are various reports of Lyell spending many hours on his successive transatlantic crossings on deck looking out for icebergs loaded with boulders and rock debris that would prove his theory (see Dott this volume).

In 1858, during one of his many visits to Switzerland, Lyell finally conceded that it was after all necessary to accept the 'great extension to the ancient Alpine glaciers' to account for the dispersal of the abundant large erratic blocks so frequently found at great distance from their source outcrops and often at high altitude, across the Jura. However, he still continued to insist that in Great Britain, Scandinavia and the United States glacial submergence and transport by floating ice was still the explanation (Davies 1969, pp. 291–292). In 1863, Lyell published his highly successful review of the growing evidence for the great age of humanity, and of other aspects of the Pleistocene period, under the title *The Antiquity of Man from Geological Evidences,* (Lyell 1863; Cohen this volume). This offered a wide range of evidence and conflicting views on the 'Ice Age', though it is difficult if not impossible to ascertain what Lyell's own view was on most of the controversial points. The *Natural History Review's* anonymous, though well informed, contemporary reviewer claimed with justification (specifically in relation to Darwinism, though the complaint is equally relevant to other parts of the book), 'We are, however, unable to discover that Sir Charles anywhere expresses his own opinion' (Anon. 1863, p. 213).

Lyell's final words on the subject were in 1873 in the fourth edition of *Antiquity of Man* (Lyell 1873). One of many examples discussed was the transportation of erratics from the Scottish Highlands across the Sidlaw Hills, a central part of his 1840 evidence for a land glaciation. In 1873, Lyell once again offered the possibility that the low-lying areas of 'Strathmore was filled up with land-ice' which could have extended to the summit of the Sidlaws (around 1500 feet at the highest points). He

even included here a footnote reference to his 1840 paper on the glaciation of Forfarshire (which of course argued forcibly that there was no rational alternative to the glacial hypothesis to explain this). However, in the very next paragraph Lyell completely contradicted himself:

> Although I am willing, therefore, to concede that the glaciation of the Scottish mountains, at elevations exceeding 2,000 feet, may be explained by land ice, it seems difficult not to embrace the conclusion that a subsidence took place not merely of 500 or 600 feet ... but to a much greater amount, as shown by the present position of erratics and some patches of stratified drift. (Lyell 1873, p. 289)

The 1873 edition of *Antiquity of Man* also included a full-page map (as fig. 42) illustrating his view that most if not all of the characteristic glacial period deposits and transported erratics in the British Isles could be attributed to the submergence of the land to depth of not less than 2000 feet in Scotland, and to a minimum of 1300 feet in England and Wales north of a line from London to Gloucester, while arguing that the land south of the Thames 'alone remained above water' (Lyell 1873, p. 325). However, he offered no serious discussion of how such extraordinary simultaneous differences in sea level took place, particularly the postulated massive (and very recent) 1300 foot differential movement (presumably from faulting) across the few miles' width of the Thames Valley.

By this time Lyell was in reality almost totally isolated in his continued rejection of the evidence that he himself had presented so effectively and forcefully to the Geological Society in November 1840. Of the leading British geologists of his generation, only Murchison and Lyell went to their graves rejecting most of the evidence of a recent large-scale terrestrial glaciation of much of the British Isles and of vast areas of northern Europe and North America. Perhaps the greatest irony of all was that in his efforts to defend the 'doctrine of uniformity' in its purest form, Lyell, the founder and most effective advocate of this principle, turned to an unsubstantiated and ultimately insupportable theory of rapid differential changes in sea level, which was ultimately far more 'catastrophist' than the glacial hypothesis for which Lyell had been, if only very briefly in late 1840, one of the most powerful advocates.

The author would like to acknowledge the pioneering work of Gordon Herries Davies and the late George W. White, particularly in unravelling the events of 1840. He is indebted to both, and to Leonard G. Wilson and David Q. Bowen, for much helpful advice and discussion over many years. He is also indebted to Lord Lyell for allowing him to use the Lyell notebooks in the Kinnordy House archives and to Leonard Wilson for supplying copies of relevant parts of these from the microfilm copies at the University of Minnesota, at Lord Lyell's request. Thanks are also due to the Council and the Honorary Archivist, John Thackray, of the Geological Society of London, for access to and permission to quote from the Society's Council Minutes and Letter Books.

References

AGASSIZ, E. C. 1886. *Louis Agassiz. His Life and Correspondence*. 2 vols. Houghton, Mifflin & Company, Boston.

AGASSIZ, L. 1837. Discours prononcé à l'ouverture des séances de la Société Hélvétique des Sciences Naturelles à Neuchâtel le 24 juillet 1837. *Actes de la Société Hélvétique des Sciences Naturelles,* **22**, 369–394.

—— 1838. Upon glaciers, moraines and erratic blocks; being the address delivered at the opening of the Helvetic Natural History Society, Neuchâtel, on the 24th of July 1837, by its President, M. L. Agassiz. *Edinburgh New Philosophical Journal,* **24**, 364–383.

—— 1841a. On the polished and striated surfaces of the rocks which form the beds of glaciers in the Alps. *Proceedings of the Geological Society of London,* **3**(71), 321–322.

—— 1841b. On glaciers and boulders in Switzerland. *Reports and Transactions of the British Association for the Advancement of Science (for 1840),* 113–114.

—— 1841c. On glaciers, and the evidence of their once having existed in Scotland, Ireland and England. *Proceedings of the Geological Society of London,* **3**(72), 327–332.

—— 1842. The glacial theory and its recent progress. *Edinburgh New Philosophical Journal,* **33**, 217–283.

ANON. 1863. The Antiquity of Man from Geological Evidences. *[Review]. Natural History Reviews,* April, 211–219.

BOYLAN, P. J. 1978. The role of William Buckland (1784–1856) in the recognition of glaciation in the British Isles. *In*: INHIGEO *VIII Symposium. Zusammenfassung – Abstract – Résumé.* INHIGEO, Münster, 33.

—— 1981. The role of William Buckland (1784–1856) in the recognition of glaciation in the British Isles. *In*: NEALE, J. W. & FLENLEY, J. *The Quaternary in Britain.* Pergamon Press, Oxford, 1–8.

—— 1984. *William Buckland, 1784–1856: Scientific Institutions, Vertebrate Palaeontology, and Quaternary Geology.* PhD thesis, University of Leicester, Leicester.

BUCKLAND, W. 1831. On the occurrence of the remains of elephants, and other quadrupeds, in cliffs of frozen mud, in Eschscholtz Bay, within Beering's [*sic*] Strait, and in other distant parts of the Arctic seas. *In*: BEECHEY, F. W. *Narrative of a Voyage to the Pacific and Beering's Strait Performed in His Majesty's Ship Blossom, ... in the Years 1825, 1826, 1827, 1828. Part I.* Colburn & Bentley, London, 593–612.

—— 1840. Anniversary Address of the President, February 1840. *Proceedings of the Geological Society of London,* **3**(68), 210–267.

—— 1841a. On the evidences of glaciers in Scotland and the north of England. *Proceedings of the Geological Society of London,* **3**(72), 333–337, 345–348.

—— 1842. On diluvio-glacial phænomena in Snowdonia and adjacent parts of North Wales. *Proceedings of the Geological Society of London,* **3**, 579–584.

COHEN, C. 1998. Charles Lyell and the evidences of the antiquity of man. *This volume.*

DOTT, R. H. 1996. Lyell in America – his lectures, fieldwork, and mutual influences, 1841–1853. *Earth Sciences History,* **15**, 101–140.

—— 1998. Charles Lyell's debt to North America: his lectures and travels from 1841 to 1853. *This volume.*

ESMARK, J. 1826. Remarks tending to explain the geological theory of the Earth. *Edinburgh New Philosophical Journal,* **2**, 107–121.

EYLES, N. & MCCABE, A. M. 1989. The Late Devensian (22,00 B.P.) Irish Sea basin: the sedimentary record of a collapsed sheet margin. *Quaternary Science Reviews,* **8**, 304–351.

GORDON, E. O. 1894. *The Life and Correspondence of William Buckland, D.D., F.R.S.* Murray, London.

HAMBREY, M. J., HUDDART, D., BENNETT, M. R. & GLASSER, N. F. 1997. Genesis of "hummocky moraine" by thrusting in glacier ice: evidence from Svalbard and Britain. *Journal of the Geological Society,* **154**, 623–632.

HERRIES DAVIES, G. L. 1968. The tour of the British Isles made by Louis Agassiz in 1840. *Annals of Science,* **24**, 131–146.

—— 1969. *The Earth in Decay. A History of British Geomorphology 1578–1878.* Macdonald, London.

HOPPE, G. 1952. Hummocky moraine regions, with special reference to the interior of Norrbotten. *Geographiska Annaler,* **34**, 193–212.

IMBRIE, J. & IMBRIE, K. P. 1979. *Ice Ages: solving the mystery.* Macmillan, London.

KUHN, T. 1960. *The Structure of Scientific Revolutions.* Chicago University Press, Chicago.

LYELL, C. 1826. On a recent formation of freshwater limestone in Forfarshire ... *Transactions of the Geological Society of London* ser. 2, **2**, 73–96.

—— 1830. *Principles of Geology.* Vol. 1. (Facsimile reprint 1990. University of Chicago Press, Chicago.

—— 1840. On the Boulder Formation or drift and associated freshwater mud cliffs of eastern Norfolk. *London & Edinburgh Philosophical Magazine,* **16**, 345–380; *Proceedings of the Geological Society of London,* **3**(67), 171–179.

—— 1841. On the geological evidences of the former existence of glaciers in Forfarshire. *Proceedings of the Geological Society of London,* **3**(72), 337–345.

—— 1863. *The Antiquity of Man from Geological Evidences.* Murray, London.

—— 1873. *The Antiquity of Man from Geological Evidences.* 4th edn. Murray, London

MORRELL, J. 1976. London institutions and Lyell's career: 1820–1841. *British Journal for the History of Science,* **9**, 132–146.

MURCHISON, R. I. 1842. Anniversary Address of the

President. *Proceedings of the Geological Society of London,* **3**, 637–687.

RUDWICK, M. J. S. 1963. The foundations of the Geological Society of London: its scheme for co-operative research and its struggle for independence. *British Journal for the History of Science*, **1**, 325–355.

SHARP, R. P. 1949. Studies of superglacial debris on valley glaciers. *Journal of Science*, **67A**, 213–220.

SISSONS, J. B. 1967. *The Evolution of Scotland's Scenery.* Oliver & Boyd, Edinburgh.

—— 1974. A late glacial icefield in the central Grampians. *Transactions of the Institute of British Geographers*, **62**, 95–114.

THACKRAY, J. C. 1998. Charles Lyell and the Geological Society. *This volume.*

WHITE, G. W. 1970. Announcement of glaciation in Scotland. *Journal of Glaciology*, **9**, 143–145.

WHEWELL, W. 1837. *A History of the Inductive Sciences from the Earliest to the Present Times.* 3 vols. J. W. Parker, London.

WOODWARD, H. B. 1883. Dr. Buckland and the glacial theory. *Midland Naturalist*, **6**, 225–229. Reprinted in WOODWARD, H. B. 1907. *The History of the Geological Society of London.* Geological Society, London.

Charles Lyell and climatic change: speculation and certainty

JAMES RODGER FLEMING

Science, Technology and Society Program, Colby College, Waterville,
ME, 04901, USA

Abstract: In the first edition of the *Principles of Geology*, Charles Lyell announced his theory of the geographical determination of climate and speculated on possible climatic changes during the geological and historical past. In light of the subsequent discovery of ice ages, the proliferation of theories of climatic change, and the great climate debates of his time, Lyell's theory remained remarkably stable. This paper examines Lyell's appropriation, modification and rejection of the views of his contemporaries. It provides perspectives on elite and popular ideas of climate and climatic change from the late eighteenth century to 1875, examines Lyell's position on climatic change in geological and historical times, and explores in some detail the mutual influences of Lyell and James Croll, the proponent of an astronomical theory of ice ages.

I have often told (and been told) humorous stories about climate and human affairs. This one, from 1865, comes from a letter to Lyell from John Carrick Moore, FRS, field geologist and long-time member of the Geological Society. Moore writes:

> I fear you have not time to read the *Reader*, I must call your attention to the last number, in which there is a true story of a Physician who warned a Unitarian preacher that he would make no proselytes in Northern Virginia because the people all had fair complexions and therefore were Calvinists. If he wished to preach against the eternity of punishment, he should go to the Hill Country. A map of the world is evidently much wanted to show the influence of Climate on Creeds, with contour lines – [Buddhism] below the 50 foot level, Calvinists near the Snow line, and Papists principally on the Volcanic tufts... (J. C. Moore to C. Lyell, 5 March 1865 in Lyell papers)

So you see, climate is not only a complicated issue, it is also a cultural one; even more so for climatic change.

Apprehending climatic change

In pursuing historical research on climate change, I have had to ask several crucial questions. How do people (scientists included) gain awareness and understanding of phenomena that cover the entire globe, and that are constantly changing on time-scales ranging from geological eras to centuries, decades, years and seasons? How was this accomplished by individuals immersed in and surrounded by the phenomena? How were privileged positions created and defined? The answers are varied and worthy of extended reflection. In the absence of means to observe the climate system in its entirety, (as an astronomer might view a star or planet) or to experiment on it directly (as a chemist might view a reaction), how did scientific understanding of it emerge?

One approach, popular in the eighteenth century, was through appeals to authority – references to historical literature, first impressions of explorers or the memory of the elderly. This was the rhetorical strategy of Enlightenment writers who wanted to support a particular theory of cultural development or decline. I will say more about this shortly.

Another way of approaching the issue was to collect massive amounts of meteorological data over large areas and extended time periods in the hope of deducing climatic patterns and changes. Individual observers in particular locales dutifully tended to their journals, and networks of cooperative observers gradually extended the frontiers of meteorology. Although many of the basic meteorological instruments were invented in the seventeenth century, they were not standardized or widely distributed until well into the mid-nineteenth century.

During Lyell's lifetime, meteorology emerged as an organized, if not yet fully disciplined, observational science. Observations were tabulated, charted, mapped and analysed to provide representative climatic inscriptions. This process profoundly changed climate discourse and established the foundations of the science of climatology (Fleming 1990). National weather services were established in Europe, Russia and the United States in the third

FLEMING, J. R. 1998. Charles Lyell and climatic change: speculation and certainty. *In*: BLUNDELL, D. J. & SCOTT, A. C. (eds) *Lyell: the Past is the Key to the Present*. Geological Society, London, Special Publications, **143**, 161–169.

quarter of the century, and by 1872, within Lyell's lifetime, regular meetings were being held of the directors of national weather services (Fleming 1997).

A third approach to privileged knowledge was to establish from first principles what the climate ought to be and how it ought to change. Joseph Fourier, John Tyndall and James Croll, to name but a few, engaged in such speculative and theoretical practices. These approaches – based on mathematical, physical and astronomical principles – tended to be most satisfying to those scientists working within a particular disciplinary perspective; most only grudgingly admitted other possible secondary causes of climate change. Lyell, of course, had his own favourite causal mechanism which was solidly grounded in geological field evidence.

In the twentieth century, climatic phenomena have been rendered three dimensional by the development of upper-air observations, extended into the indefinite past by palaeoclimatic techniques and, finally, globalized in the era of satellite remote sensing. Many climate scientists today are working on links between remote sensing and more sophisticated computer models. They are hoping, through advances in technology, to provide new privileged positions. For most scientists the goal is better understanding of climate; for some it is also prediction and, ultimately, control. I might add that an additional strategy for claiming privileged knowledge is the consensus method, for example as currently practised by the Intergovernmental Panel on Climate Change (IPCC 1995).

Perceptions of climatic change in the eighteenth century

Climate – from the Greek term *klima*, meaning slope or inclination – was originally thought to depend only on the height of the Sun above the horizon, a function of the latitude. A second tradition, traceable to Aristotle, linked the quality of the air (and thus the climate) to the vapours and exhalations of a country. The Hippocratic tradition further linked climate to health and national character. Enlightenment ideas linking climate change and culture were grounded in the work of the diplomat, historian and critic Abbé Jean-Baptiste Du Bos, perpetual secretary of the French Academy, who argued that the rise and fall of creative genius was not due primarily to 'les causes morales' (education, cultivation, governance), but was largely attributable to changes in 'les causes physiques' (the nature of the air, land, soil and especially, climate). These ideas influenced

Montesquieu's ideas on climate and governance, David Hume's ideas on recent climate change in the Americas, and of course, generations of colonial settlers and revolutionary patriots (Fleming 1998).

As late as 1779 – in other words in the prime of Hutton's life – the *Encyclopédie* of Diderot and D'Alembert still defined 'climat' in the ancient way, geographically, as a 'portion or zone of the surface of the Earth, enclosed within two circles parallel to the equator', within which the longest day of the year on the northern and southern boundaries differs by some quantity of time, for example one half hour. The *Encyclopédie* provided a medical definition of climate as well, understood primarily through the effects of climate on the health and well-being of the inhabitants of various climes. It also mentioned Montesquieu's position on the influence of climate on people's mores, character and forms of governance (Diderot & D'Alembert 1751–1765).

With no established science of climatology, authors such as Du Bos, Montesquieu and Hume appealed directly to cultural sensibilities and prejudices, the authority of their positions residing in their considerable literary skills and the lack of other evidence to prove them wrong. Collectively, they generated a powerful vision of the climates of Europe and America, shaping the course of empire and the arts; the concerted efforts of innumerable individuals in turn shaping the climate itself. By the end of the century, physiocrats had come to the following general conclusions on climate change, culture and cultivation:

1. Cultures are determined or at least strongly shaped by climate.
2. The climate of Europe had moderated since ancient times.
3. These changes were caused by the gradual clearing of the forests and by cultivation.
4. The American climate was undergoing rapid and dramatic changes caused by settlement.
5. The amelioration of the American climate would make it more fit for European-type civilization and less suitable for the primitive native cultures.

This was the dominant popular understanding of climate at the dawn of the nineteenth century (Fleming 1998).

Lyell's position

In the first edition of his *Principles of Geology* (1830–1832), Charles Lyell announced his theory of the geographical determination of climate; a theory that influenced generations to follow, including G. F. Wright (1889), M. Ramsay (1909–1910), and C. E. P. Brooks (1926). He systematically rejected catastrophic agents of climatic change, arguing patiently, systematically and forcefully, as Martin Rudwick has recently reminded us, that 'modern causes', acting at their present intensities, were 'entirely adequate' to explain the evidence of the past (Rudwick 1990). Lyell also maintained that geology should remain independent of cosmogony, just as history had been divorced from myths of human origins (Bailey 1962). For Lyell, the geographical arrangements of oceans and continents, currents and winds were sufficient to explain the immense variety of climatic zones being revealed by meteorologists, such as Heinrich Wilhelm Dove, and scientific travellers, such as Alexander von Humboldt. As Lyell perceptively noted, the ocean tempered the climate, 'moderating alike an excess of heat or cold', while elevated land, extending into the colder regions of the atmosphere, 'becomes a great reservoir of ice and snow, arrests, condenses, and congeals vapour, and communicates its cold to the adjoining country'. Lyell made additional perceptive comments on the role of particular large-scale features such as the African continent – 'an immense furnace' that distributes its heat to Asia and Europe – and ocean currents such as the Gulf Stream – which 'maintains an open sea free from ice in the meridian of East Greenland and Spitzbergen' (Lyell 1830–1832).

Lyell used his climate theory to demonstrate that the past history of the Earth was 'one uninterrupted succession of physical events, governed by the laws now in operation' (Wilson 1972). Such a position assumes that geologists *know* all the laws currently in operation – a precarious assumption in 1830 or even today. Nevertheless, if it is accepted, Lyell's position has important implications. Over immense geological time, in this view, gradual processes shaped the distribution of land and sea, which in turn determined the climates of the world. The geographical distribution of species, which depends greatly on the climate and geographical conditions, was thus shaped by natural laws. In Lyell's terminology, 'transportations of climate' contributed to 'local extermination of species', while other species better suited to the new conditions eventually took their places (Bailey 1962).

Lyell introduced a substantial amount of evidence indicating that the climate of the northern hemisphere was 'formerly hotter'. He included proofs from analogy derived from extinct quadrupeds; and direct proofs from the organic remains of the Sicilian and Italian strata, from fossil remains in Tertiary and Secondary rocks and from the plants of the coal formation. He argued that the climate of Siberia and other Arctic regions had been formerly temperate, but had become subjected to 'extremely severe winters' due to changes in landforms. These changes were theorized to account for animal migration and evolutionary changes as animals adapted to different climates. Lyell concluded:

> . . . the remains both of the animal and vegetable kingdom preserved in strata of different ages, indicate that there has been a great diminution of temperature throughout the northern hemisphere, in the latitudes now occupied by Europe, Asia, and America. The change has extended to the Arctic circle, as well as to the temperate zone. The heat and humidity of the air, and the uniformity of climate, appear to have been most remarkable when the oldest strata hitherto discovered were formed. The approximation to a climate similar to that now enjoyed in these latitudes, does not commence till the era of the formations termed tertiary, and while the different tertiary rocks were deposited in succession, the temperature seems to have been still farther lowered, and to have continued to diminish gradually, even after the appearance of a great portion of existing species upon the earth. (Lyell 1830–1832, p. 103).

Lyell rejected, however, the notion of a secularly cooling Earth. Six years earlier, in 1824, Joseph Fourier had determined that the internal heat of the Earth had decreased no more than 3/100 of a degree during the course of recorded history (Fourier 1824). Instead, Lyell fixed his thoughts on gradual processes occurring steadily and repeatedly at the Earth's surface – 'on the connection at present between climate and the distribution of land and sea; and if we then consider what influence former fluctuations in the physical geography of the earth must have had on superficial temperature, we may perhaps approximate to a true theory' (Lyell 1830–1832, p. 105).

For Lyell, the driving forces of climatic change were due to continuous changes in the distribution of land and sea:

> When land is massed in equatorial and tropical latitudes polar climates are mild. The land, heated to an excess under the equatorial sun, gives rise to warm currents of air that sweep north. On the other hand, land massed around the poles produces the reverse effect. There is no land at the equator to soak up heat and no warm winds coming into polar regions.

Lyell challenged his readers to imagine the Himalaya Mountains, 'with the whole of Hindostan', sinking down and being replaced by the Indian Ocean, while an equal extent of mountainous lands rose up, extending from North Greenland to the Orkney Islands. He pointed out that under such altered circumstances 'it seems difficult to exaggerate the amount to which the climate of the northern hemisphere would now be cooled down'. Lyell's imagined refrigeration, however, did not stop there. Icebergs would find their way into southern waters, their melting creating vapour, fogs and clouds that would reduce solar insolation by half, causing the Earth to cool further, wrapping large portions of the northern hemisphere in a 'winding sheet of continental ice' – a phrase of ominous significance for the organic world. When in the course of geological time, conditions had reversed and continents again dominated the equatorial regions, snow would be a rarity, the Earth's crust would be heated to considerable depths, and springs and surface waters would run hotter, even in the winter (Lyell 1830–1832).

In pondering the vicissitudes of climate, Lyell made the following four assumptions, quoted here from the eleventh (1872) edition of the *Principles*. I cite the last edition published in his lifetime because Lyell's principles remained basically unchanged from the first edition.

I shall assume, 1st, that the proportion of dry land to sea continues always the same. 2ndly, That the column of the land rising above the level of the sea is a constant quality; and not only that its mean, but that its extreme height, is liable only to trifling variations. 3rdly, That on the whole, and in spite of local changes, both the mean and extreme depth of the sea are invariable; and 4thly, That the grouping together of the land in continents is a necessary part of the economy of nature. (Lyell 1872, p. 264).

F. F. Cunningham found it astonishing that 'Lyell could consider it consistent with his "uniformity" that in recent times there had been a large rise of sea level and an even more recent withdrawal of the sea of similar dimensions' (Cunningham 1990). What Cunningham found unusual – what he called a 'catastrophic fluctuation' in sea level – Lyell would surely have explained, in accordance with his four principles, by a rearrangement of continents and oceans, and a gradual yet dramatic *local* (but not mean global) increase in the depth of the sea.

As described earlier, it was widely held that humans might have altered the climate of the Old and New Worlds by clearing the forests and cultivating the fields. Lyell was dismissive of such notions, in particular that climate had changed much *for any reason* in historical times; he considered the time period 'insufficient to affect the leading features of the physical geography of the globe'. Lyell acknowledged popular perceptions of the variability of the seasons, but cited recent analyses of long series of meteorological observations which indicated the relative constancy of the mean temperatures of particular locations.

He admitted, however, that in certain locations 'the labours of man have, by the drainage of lakes and marshes, and the felling of extensive forests', caused minor changes in the climate system (Lyell 1830–1832). Waxing speculative (and realizing he was doing so), Lyell explored the possibility of future climatic influences caused by the progressive development of human power, 'or perhaps by some other new relations, which may hereafter spring up between the moral and material worlds' (Lyell 1853). He did not speculate, however, on what these relations were. Nor did he venture an opinion, for example, on the recent rise of industrial power.

Undoubtedly, many other aspects of Lyell's climate arguments constituted gross speculation. For example, as Patrick Boylan reminds us (this volume), in 1840 Lyell briefly joined forces with Louis Agassiz and William Buckland in decidedly non-uniformitarian speculations on glaciation. Another example comes from Lyell's well known maps 'showing the position of land and sea which *might* produce the extremes of heat and cold in the climates of the globe' (Fig. 1).

These maps appear in all eleven editions of the *Principles* published in Lyell's lifetime. The maps depict, recognizably, the seven current continents all bunched up near the equator to represent 'extreme of heat', and then shifted to polar regions to represent 'extreme of cold'. Of course, there was no discussion of a possible mechanism to cause such 'continental drift.' Lyell added a note in the ninth edition saying, 'These maps are intended to show that continents and islands having the same shape and relative dimensions as those now existing, might be placed so as to occupy either the equatorial or polar regions' (Lyell 1853, p. 111). While this exercise resembles nothing more than a child's map game, the result is spookily familiar in the contemporary era of plate tectonics.

By 1853 Lyell had examined and rejected the notion that changing sunspot abundances, as reported by Schawbe and Sabine, had any influence on climate (Lyell 1853). In 1861 John Tyndall began to popularize the results of his experiments on the absorption of radiant heat by gases. He noted that changes in the amount of any of the radiatively active constituents of the atmosphere – water vapour, carbon dioxide, ozone and hydrocarbons – could have produced 'all the mutations of climate which the researches of geologists reveal. . . they constitute true causes, the *extent* alone of the operation remaining doubtful' (Tyndall 1861). Neither Tyndall nor anyone else pursued this hypothesis, however, until the turn of the century (Fleming 1998). Lyell's biggest challenge came in 1864, when James Croll introduced his astronomical theory of the glacial epochs. As J. C. Moore wrote to Lyell in March 1865, 'Who would have thought fifty years ago, after astrology had gone out of fashion, that the stars were to enlighten us upon what is going on here.' (Lyell papers).

Lyell, Croll and the glacial epoch

James Croll (1821–1890), proponent of an astronomical theory of ice ages, was a self-educated Scotsman who was employed, after 1867, by the Scottish Geological Survey. The outlines of his life are well documented in his touching auto-biography, a chronicle of poverty, physical suffering and neglect (Irons 1896). In 1864, Croll published a paper in the *Philosophical Magazine*

'On the physical cause of the change of climate during geological epochs'. In this paper Croll introduced revolutions in the Earth's orbital elements as likely periodic and extraterrestrial mechanisms for initiating multiple glacial epochs.

Inspired by the *Révolutions de la mer* of Joseph Adhèmar (1842), and employing the calculations of Leverrier and Lagrange for the maximum eccentricity of the Earth's orbit, Croll proposed that this 'eccentricity was sufficiently great to account for every extreme of climatic change evidenced by geology' (Irons 1896). Croll's theory of ice ages took into account both the precession of the equinoxes and variations in the shape of the Earth's

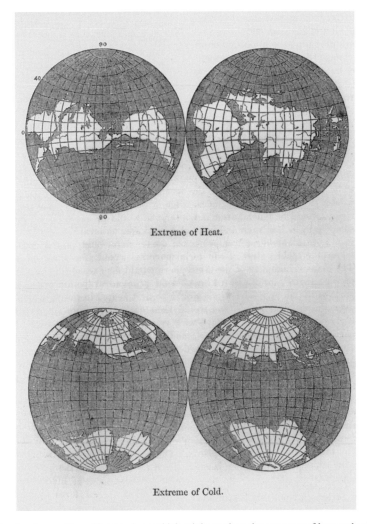

Extreme of Heat.

Extreme of Cold.

Fig. 1. Maps showing the position of land and sea which might produce the extremes of heat and cold in the climates of the globe. *Top*: Extreme heat occurs when land masses are concentrated near the equator. *Bottom*: Extreme cold occurs when land masses occupy polar regions. (From Lyell 1853, p. 111.)

orbit. It predicted that one hemisphere or the other would experience an ice age whenever two conditions occur simultaneously: 'a markedly elongate orbit, and a winter solstice that occurs far from the sun' (Imbrie & Imbrie 1979). Croll rejected two astronomical notions of climate change: that the Earth had passed through hotter and colder regions of space and that the Earth's axis had shifted. He assumed only the well established variations in orbital excentricity and the obliquity of the ecliptic. This provided a mechanism for multiple glacial epochs and alternating cold and warm periods in each hemisphere. In other words, when the northern hemisphere was in the grips of an ice age, the southern hemisphere would be in an interglacial. As the Earth's orbital elements varied, this situation would eventually be reversed. This potentially serious challenge to the geographical theory caused an uproar among Lyell and his associates.

In 1864, as he was preparing the important tenth edition of his *Principles* (1866–1868), Lyell sought expert advice on how to deal with the new contender. He asked Sir John Herschel's opinion on the reliability of Croll's 'facts and reasons', adding:

> Of their applicability to Geology I may perhaps form an independent opinion. ... I feel more than ever convinced that changes in the position of land & sea have been the principle cause of past variations in climate, but astronomical causes must of course have had their influence & the question is to what extent have they operated?

Lyell also perceptively noted what was to become a fatal flaw in Croll's theory – that, according to the geological record, the glacial periods of the southern hemisphere coincided with those of the northern, which would not be the case if the eccentricity of the orbit were the controlling factor (Lyell to Herschel, 31 January 1865 in Herschel papers). Herschel replied that astronomical causes could provide huge temperature fluctuations, 'quite enough to account for any amount of glacier and coal fields' (Herschel to Lyell, 6 February 1865, copy in Herschel papers).

> Suppose a distribution of land favorable to cold, suppose an extreme e[xcentricity], and suppose the aphelion to coincide with the winter first in one hemisphere and then in the other, and any amount of glacier you can want is at your disposal... (Herschel to Lyell, 15 February 1865, copy in Herschel papers)

Lyell was very serious about this issue and responded with a 22 page letter to Herschel explaining why geographical causes had to predominate over astronomical ones. He *knew* the enormous influence on climate of varying configur-

ations of land and sea from direct accounts and observations; the effects of varying eccentricity had yet to be proven (Lyell to Herschel, 11 February 1865, in Herschel papers). Lyell conducted a similar correspondence with the Astronomer Royal, Sir George Biddell Airy, concerning the 'ancient state of the Earth's orbit' (Airy to Lyell, 27 March 1865 in Lyell papers).

In response to Lyell's queries, J. C. Moore responded:

> The more I think of it, the more I feel puzzled to understand how Astronomic causes can give us the conditions required for glaciation. Mr. C[roll] talks of cold periods, but a winter of −17° F followed by a summer of +119° F is not what I should call a cold but an extreme climate. ... [There is nothing on the Globe which approaches such a state] ... I cannot believe in these monstrous results, and I think, as I suppose you do, that a vast extension of land about the S. Pole is at the bottom. (Moore to Lyell, 20 April 1865 in Lyell papers).

By 1866 Lyell, on the advice of Herschel and Airy, had tentatively accepted Croll's theory as a true, but minor cause of climatic change. He wrote to Darwin:

> ... the whole globe must at times have been superficially cooler. Still, during extreme excentricity the sun would make great efforts to compensate in perihelion for the chill of a long winter in aphelion in one hemisphere, and a cool summer in the other.

Lyell also incorporated into his explanation aspects of Tyndall's work on radiative transfer, noting that plants requiring heat and moisture could be saved from extinction during an ice age 'by the heat of the earth's surface, which was stored up in perihelion, being prevented from radiating off freely into space by a blanket of aqueous vapour caused by the melting of ice and snow'. Here he was grasping at straws, aware of new theoretical problems, but taking from them only the aspects that reinforced his own preconceptions. Lyell's letter to Darwin concluded:

> But though I am inclined to profit by Croll's maximum excentricity for the glacial period, I consider it quite subordinate to geographical causes or the relative position of land and sea and abnormal excess of land in polar regions. (Lyell to Darwin, 1 March 1866, in Darwin & Seward 1903)

By this time Darwin had adopted Croll's conclusion that 'whenever the northern hemisphere passes through a cold period the temperature of the southern hemisphere is actually raised'. He gently teased Lyell, pointing out that he had generally been a 'good and docile pupil', but he could not believe 'in change of land and water being more than a subsidiary agent' of the glacial period (Darwin to Lyell, 8 March 1866 in Darwin & Seward 1903). Darwin also agreed with Croll that the advocates of the iceberg theory (Lyell) had formed 'too extravagant notions regarding the potency of floating ice as a striating agent', and that the 'scored rocks throughout the more level parts of the United States result from true glacier action' (Darwin to Croll, 24 November 1868, in Darwin & Seward 1903). Of course Darwin's old nemesis, Agassiz, was still actively pursuing his defeat with the argument that extensive glaciation, in the equatorial Amazon valley and over the entire continent of North America, would have prevented the descent of any terrestrial life form from the Tertiary period (Darwin to Lyell, 8 September 1866, in Darwin & Seward 1903).

In September 1866, Croll received Lyell's proof sheets for the climate chapters of the tenth edition of the *Principles* (Croll to Lyell, 24 September 1866 in Lyell papers). After seeing Lyell's summary of his theory, which was fair but non-committal, Croll wrote back immediately that he had altered his position considerably in his latest manuscript. Croll now argued that the glacial epoch could not possibly have been caused *directly* by any change in the eccentricity of the Earth's orbit, but by the combined physical effects of 'certain agencies which were brought into operation by means of the change' (Croll to Lyell, 28 September 1866 in Lyell papers). For example, an early edition of Herschel's *Astronomy* pointed out that the amount of direct heat received by the Earth over the course of a year is independent of eccentricity. However, Croll pointed out in letters to both Lyell and Herschel that the *climate* would not be so independent because of the latent heat effects of snow cover (Croll to Lyell, 23 April 1866 in Lyell papers). This is just one example of the ways Croll's rough calculations and constant modifications of his theory to incorporate geographical feedback served to keep Lyell from dismissing the astronomical theory altogether. Croll received the first volume of the tenth edition of the *Principles* in November 1866 (Croll to Lyell, 30 November 1866 in Lyell papers). He thanked Lyell for the handsome gift and for the 'highly complimentary way' in which his astronomical theory had been treated. Lyell had agreed with Croll on many points. Although they still had deep disagreements, Croll attributed their differences to basic incompatibilities in the approaches of physics and geology (Croll to Lyell, 12 December 1866 in Lyell papers).

Compared with previous editions, the tenth and subsequent editions devoted more than twice as much space to climatic change. While the ninth edition had had about 58 pages on climate and its vicissitudes, the tenth edition had 130 pages, including a new chapter on astronomical influences with a 37 page section on Croll:

> Mr. Croll's suggestion as to the probable effects of a large excentricity in producing glacial epochs is fully discussed, and the question is entertained whether geological dates may be obtained, by reference to the combined effect of astronomical and geographical causes. (Lyell 1866–1868, vol. 1, viii).

Lyell had complimentary things to say about Adhèmar's *Révolutions de la mer*, even though most contemporaries considered it to be extremely speculative, not to mention catastrophic in its view of dramatic oceanic flooding. According to Lyell, Adhèmar 'called attention to a *vera causa* hitherto neglected' (the precession of the equinoxes) and reopened the question of historical climate changes, for example in understanding the advance of the Swiss glaciers since the thirteenth century. Lyell added that Croll's primary mechanism – changes in the eccentricity of the Earth's orbit – was also a *vera causa* and could result in a 20 per cent reduction or augmentation of the entire heat the Earth received from the Sun.

> Upon this difference of heat Mr. Croll has founded a theory which attempts to account for former changes of climate by the tendency which a maximum excentricity would have to exaggerate the cold in that hemisphere in which winter occurred in aphelion. (Lyell 1872, p. 277)

Lyell concluded his review of Croll's theory by pointing out that precession of the equinoxes would cause the alternate glaciation of only that hemisphere in which winter occurred at aphelion. As a consequence, Croll had supposed that a vast ice cap on one side of the Earth 'would so derange the earth's centre of gravity as to draw the ocean towards that pole, and cause the submergence of part of the land'. The depth of the submergence he supposed was on the order of 500 feet. Lyell pointed out to Croll that sea level would be *lowered* during a glacial epoch, since an enourmous amount of ocean water would now be deposited as snow on the ice cap (Croll to Lyell, 6 January 1866 in Lyell papers). Croll stuck to his theory, which was in fact derived from Adhèmar, but corrected his calculations to show that sea-level rise (in the northern hemisphere only) would be 500 feet *minus* the amount subtracted to build the continental glaciers (Croll to Lyell, 16 January 1866 in Lyell papers).

Lyell faulted Croll for not paying sufficient attention to abnormal geographical conditions. He pointed out that astronomical causes alone could not account for the storing up of ice when deep oceans prevailed at both poles. He further noted that during extreme eccentricity, the minor axis of the ellipse would be shortened, causing the amount of heat received from the Sun to exceed its present value, working against the formation of a glacial period. He rested his case by reasserting the power of geographical causes over astronomical causes: 'I consider the former changes of climate and the quantity of ice and snow now stored up in polar latitudes to have been governed chiefly by geographical conditions' (Lyell 1872, p. 284).

Conclusion

During Lyell's lifetime many of the major mechanisms of climatic change were proposed, if not yet fully explored: changes in solar output, changes in the Earth's orbital geometry, changes in terrestrial geography, and changes in atmospheric transparency and composition. New climate theories were introduced and new work was done on heat budgets, spectroscopy and the rising carbon dioxide content of the atmosphere. Although older theories of human agency were disproved, new ideas about industrial impacts were beginning to circulate. Through such tempestuous theoretical waters, Lyell kept a steady course, providing his readers with reasoned arguments why they should keep faith in actualist – that is mid-nineteenth century actualist – geological processes. Lyell's ideas on climatic change can be understood only in the context of the times; but we can understand the times themselves better by studying Lyell's negotiated responses to new theories of climatic change.

If Lyell eschewed the deplorable speculation of the geological system builders who had preceded him, he also engaged, quite systematically, in *disciplined* speculation of his own on climate matters. If, as he pointed out, we are prejudiced by our limited conception of geological time and our surficial habitats, so too are we *immersed in the climate itself*, heavily dependent on imagination and the assistance of others if we are ever to envision a system as vast as the Earth's climate and to apprehend its secular changes. Lyell followed the speculations of others, up to a point, but he tempered his judgements with solid evidence gathered from the record of the rocks. If being thoroughly Lyellian meant being scientific in such matters, then it also meant being cautious and judgemental. If he was speculative on some issues, he was very very certain about many others. These were not bad attributes to possess on a topic as nebulous as climatic change.

References

ADHÈMAR, J. A. 1842. *Révolutions de la mer, deluges periodiques*. Carilian-Goeury & Dalmont, Paris.

BAILEY, E. 1962. *Charles Lyell*. Nelson, London.

BROOKS, C. E. P. 1926. *Climate through the Ages: A Study of the Climatic Factors and Their Variations*. R. V. Coleman, New York.

CROLL, J. 1864. On the physical cause of the change of climate during geological epochs. *Philosophical Magazine*, **28**, 121–137.

CUNNINGHAM, F. F. 1990. *James David Forbes: Pioneer Scottish Geologist*. Scottish Academic Press, Edinburgh.

DARWIN, F. & SEWARD, A. C. (eds) 1903. *More letters of Charles Darwin*. 2 vols. Murray, London.

DIDEROT, D. & D'ALEMBERT, J. (eds) 1751–1765. *Encyclopdédie, ou Dictionnaire raisonné des sciences, des arts et des métiers*. Briasson, Paris, vol. 8, 280–286.

FLEMING, J. R. 1990. *Meteorology in America, 1800–1870*. Johns Hopkins University Press, Baltimore.

—— 1997. Meteorological observing systems before 1870 in England, France, Germany, Russia, and the USA: a review and comparison. *World Meteorological Organization Bulletin* **46**, 249–258.

—— 1998. *Historical Perspectives on Climate Change*. Oxford University Press, Oxford & New York.

FOURIER, J. 1824. Remarques générales sur les températures du globe terrestre et des espaces planétaires. *Annales de Chimie et de Physique,* ser. 2, **27**, 136–167.

HERSCHEL, J. F. W. Papers. Royal Society Library, London.

IMBRIE, J. & IMBRIE, K. P. 1979. *Ice Ages: Solving the Mystery*. Enslow, Short Hills, NJ.

IPCC 1995. Intergovernmental Panel on Climate Change, Geneva, Switzerland. http://www.ipcc.ch/.

IRONS, J. C. 1896. *Autobiographical Sketch of James Croll with Memoir of His Life and Work*. E. Stanford, London.

LYELL, C. 1830–1832. *Principles of Geology*. 1st edn. Murray, London.

—— 1853. *Principles of Geology*. 9th edn. Murray, London.

—— 1866–1868. *Principles of Geology*. 10th edn. Murray, London.

—— 1872. *Principles of Geology*. 11th edn. Murray, London.

—— Papers. University of Edinburgh Library, Edinburgh.

RAMSAY, M. 1909–1910. Orogenesis und klima. *Ofversigt af Finska Vetenskaps-Societetens Förhandlingar,* **53**, 1-48.

RUDWICK, M. 1990. Introduction. *In:* LYELL, C. *Principles of Geology*. University of Chicago Press, Chicago, vii–lviiii.

TYNDALL, J. 1861. On the absorption and radiation of heat by gases and vapours, and on the physical connexion of radiation, absorption, and conduction. *Philosophical Magazine,* ser. 4, **22**, 169–194, 273–285.

WILSON, L. G. 1972. *Charles Lyell. The Years to 1841:*

The Revolution in Geology. Yale University Press, New Haven.

WRIGHT, G. F. 1889. *The Ice Age in North America and its* *Bearings upon the Antiquity of Man, with an appendix on The Probable Cause of Glaciation by Warren Upham.* D. Appleton, New York.

Catastrophism and uniformitarianism: logical roots and current relevance in geology

VICTOR R. BAKER

Department of Hydrology and Water Resources, The University of Arizona,
Tucson, AZ 85721-0011, USA

Abstract. Catastrophism in the Earth sciences is rooted in the view that Earth signifies its causative processes via landforms, structures and rock. Processes of types, rates and magnitudes not presently in evidence may well be signified this way. Uniformitarianism, in contrast, is a regulative stipulation motivated by the presumed necessity that science achieves logical validity in what can be said (hypothesized) about the Earth. Regulative principles, including simplicity, actualism and gradualism, are imposed *a priori* to insure valid inductive reasoning. This distinction lies at the heart of the catastrophist versus uniformitarian debates in the early nineteenth century and it continues to influence portions of the current scientific program. Uniformitarianism, as introduced by Charles Lyell in 1830, is specifically tied to an early nineteenth century view of inductive inference. Catastrophism involves a completely different form of inference in which hypotheses are generated retroductively. This latter form of logical inference remains relevant to modern science, while the outmoded notions of induction that warranted the doctrine of uniformitarianism were long ago shown to be overly restrictive in scientific practice. The latter should be relegated solely to historical interest in the progress of ideas.

On 4 July 1997, the Mars Pathfinder spacecraft made a highly successful landing in the Ares Vallis region of Mars. The spectacular images transmitted back to eager scientists at the Jet Propulsion Laboratory, California, were immediately subjected to scrutiny. Within days, rather firm genetic hypotheses were presented to the crowds of news reporters eagerly waiting to share the excitement of science-in-the-making with millions of non-scientists worldwide.

The interpretation of the Pathfinder landing scene (Fig. 1) was completely geological. Large boulders, both angular and subrounded, were seen to be organized into distinct sedimentary patterns. The smooth slopes of distant hills on the horizon were also interpreted. In neither case was the causative process inferred to be one presently operative on the surface of that planet. Indeed, the processes now acting on the extremely cold, dry and nearly airless planet had done little to erase the bold imprint of processes last active hundreds of million years ago. The shaped hills and boulder trains were clearly analogous to similar features in the Channelled Scabland region of Earth. The origin of landforms in that region (Baker 1981) is the same origin proposed for the Martian landscape seen on television by more than a hundred million people: catastrophic flooding.

Only a few months before the Pathfinder landing, a meteorite from Mars, collected from Antarctic ice, was interpreted as showing various indicators of possible relic biogenic activity (McKay *et al.* 1996). That this meteorite could be from Mars is itself a profound legacy of catastrophic processes at human-centred scales of time and space, but a fact completely consistent with the physics of impact mechanics (Melosh 1984, 1985) and confirmed by geochemical measurement (McSween 1994). Indeed, the various organic geochemical and petrographic interpretations of biogenesis, processes for which we can observe ample extant modern operations, are far more controversial in interpretation (McSween 1997) than the immense meteor impact catastrophe that launched pieces of Mars to Earth.

The ease with which these invocations of catastrophism flowed for explaining Martian features stands in sharp contrast to debates over origins of Earth's valleys, the genesis of clastic sediments and the nature of fluid-shaped landforms that contributed to the intellectual origins of geology in the eighteenth and nineteenth centuries (Davies 1969). Some of the rhetoric in these debates led, in the earliest nineteenth century, to the mistaken conceptual association of cataclysmic flooding hypotheses with religious dogma, when the real dogma lay in the arbitrary stipulation of attributes for laws, processes and rates actually operative in nature (Gould 1987). By substituting theoretical standards of nature for religious ones,

BAKER, V. R. 1998. Catastrophism and uniformitarianism: logical roots and current relevance in geology. *In*: BLUNDELL, D. J. & SCOTT, A. C. (eds) *Lyell: the Past is the Key to the Present*. Geological Society, London, Special Publications, **143**, 171–182.

Fig. 1. Mars Pathfinder landing site imaged in July 1997 from the lander spacecraft. This mosaic of images shows boulders in trains and quasi-imbricate arrangement. The hills on the horizon have rounded slopes and scarps that were shaped by large-scale fluid flow. Ancient catastrophic flooding produced both the landforms and the sedimentary patterns.

well-meaning geological reformers blinded subsequent workers with the importance of rare, great floods in Earth history. It was not until the Channelled Scabland debates of the 1920s that cataclysmic flooding re-emerged for consideration as an important geomorphological process. In those debates J Harlen Bretz, the flood advocate, insisted upon drawing attention to the flood evidence at field sites in eastern Washington State, regardless of the numerous theoretical arguments to the contrary (Baker 1978). The controversy was such that Bretz's outrageous hypothesis of cataclysmic flood origins for scabland terrains of eastern Washington did not achieve general acceptance until the 1960s and 1970s. By then it was realized that a whole suite of landscape features was present that could only be explained by the high-energy physics of immense flood flow (Baker 1973). The diagnostic landforms included great streamlined hills, multiple channelways deeply scoured into rock, inner channels headed by great rock cataracts

and immense bars of flood-transported bouldery gravel (Baker 1981).

Nearly coincident with the attainment of geological respectability by Bretz's hypothesis, the Mariner 9 spacecraft produced images of Mars displaying ancient channelways of immense size, containing nearly the same landform assemblage as the Channelled Scabland (Baker 1982). It was the geological understanding of the ancient Martian floods (Baker & Milton 1974; Baker 1982) that led to such ease of interpretation for the Pathfinder landing site.

Giant glacial outburst floods are now recognized as characteristic of the terminal phases for the immense ice sheets that covered much of North America and Eurasia in the last ice age. Late-glacial age flooding in the Altay Mountains of central Asia has recently been documented to rival or exceed that of the Channelled Scabland (Baker *et al.* 1993). Landforms formerly attributed to glacial action alone are now recognized as the products of close association between flooding and glacial ice (Shaw 1994).

It is well to remember that all these revelations were achieved not by theoretical elegance in explaining the Earth, but by overcoming restrictions posed by existing theories. In their attempts to enshrine fundamental principles for their science, nineteenth century advocates of a 'more scientific' geology confused simplicity of logical expression with intrinsic qualities of nature. The resulting doctrine, named 'uniformitarianism', asserted that the relatively low-intensity, frequently occurring processes in evidence today must be the class of processes generally operating in the past. This dogma sometimes proved an even greater impediment to understanding the past than the religious motivation that it purported to replace. The mistaken need to stipulate attributes for laws, processes and rates actually operative in nature persists in science even to this day. Theories, which are increasingly emphasized in proportion to computational power, serve not to tell geomorphologists about Earth. Rather, it is the signs of Earth itself, the dirt, land and rocks, that are interpreted by geologists, employing, of course, all manner of theoretical and mechanical devices to aid in that interpretation.

Galileo Galilei once observed that the book of the universe is written in the symbolic language of mathematics. To read the book of the universe one must learn the language. However, Earth and Mars are not the entire universe, nor is symbolic logic the only language worthy of scientific inquiry. The language of indices, which are signs directly representing causative processes, comprises the critical text for geological reading. The indices for cataclysmic flooding processes are the characteristic landforms and deposits emplaced by that flooding (Fig. 1). This language must be learned from nature herself. Unlike the symbolic language of mathematics, it is not found in textbooks. It is the role of the geologist to understand nature's text, not to impose upon the original document a theoretical 'overstanding,' no matter how elegant or logical.

Lyell's logic

The doctrine of uniformitarianism in geology was persuasively advocated by Charles Lyell (1797–1875) in his influential book *Principles of Geology: Being an Attempt to Explain the Former Changes of the Earth's Surface by Reference to Causes Now in Operation* (Lyell 1830–1833). Lyell had entered Oxford University in 1816 and there received his first exposure to geology. He became very interested in a debate then raging between adherents of two alternative theories for the geological operation of the Earth. The first of these theories had been espoused by James Hutton (1726–1797), a financially successful medical doctor and chemical industrialist. Hutton devoted much of his later life to managing his farms and to various writings, including his *Theory of the Earth*. The latter denied a role of catastrophic forces by instead invoking the action of existing processes as sufficient, acting over long time scales, to shape the surface of the planet.

The second of the great theories prevailing in the early nineteenth century was that of the German mineralogist, Abraham Gottlob Werner (1749–1817). Werner believed that rocks were laid down in a primordial ocean, which convulsed and subsided both intermittently and catastrophically. Variations in the intensity and nature of processes explained the various succession of strata, precipitation of crystalline rocks, and the like. The theory was tied strictly to observed consequences of the presumed processes, which were thought merely to be much more intense variants of processes that could be observed today.

The Hutton and Werner controversy centred on substantive issues of alternative 'systems' for the operations of our planet through geological time. These were both theoretical constructs, so it was natural that concern would turn towards issues of methodology in how to construct scientific theories. The issue of methodology may have been particularly attractive to Lyell, who went on from his Oxford BA in Classics to the practice of Law. He was particularly active as an advocate before the bar during 1825–1827, precisely during the period when he was also writing the methodological polemic that comprises volume I of his *Principles of Geology*. In 1827 Lyell abandoned his law career to devote himself full time to his geological interests.

The lawyer in Lyell must have been particularly attracted to James Hutton's advocacy of order in the system espoused in his opus *Theory of the Earth with Proofs and Illustrations* (Hutton 1795):

Chaos and confusion are not to be introduced into the order of Nature, because certain things appear to our partial views as being in some disorder. Nor are we to proceed in feigning causes when those seem insufficient which occur in our experience.

It is clear, even from his somewhat tortured writing style, that Hutton greatly admired the work of Sir Isaac Newton (Fig. 2). Newton devised an ideal mechanical theory for the solar system in which the planets eternally cycled the Sun in timeless perfection. Hutton (1788, p.304) envisioned a revolutionary and cyclic system of an Earth history, 'without vestige of a beginning or prospect of an end'. Hutton's theory became widely

Fig. 2. Sir Isaac Newton statue in the chapel of Trinity College, Cambridge University, where he was a fellow and professor.

in which an attempt was made to dispense entirely with all hypothetical causes, and to explain the former changes of the Earth's crust by reference exclusively to natural agents.'

While he shared with Hutton, via Playfair, the goal of establishing a system for the Earth, Charles Lyell had the additional vision of a specific method whereby this would be achieved. On the eve of publication of volume I of his *Principles*, Lyell wrote in his letter of 15 January 1829 to Roderick Murchison (K. Lyell 1881, vol. 2, p. 234):

> My work ... will endeavour to establish the principle of reasoning in the science; and all my geology will come in as illustration of my views on those principles, and as evidence strengthening the system necessarily arising out of the admission of such principles ... (the principles being) that no causes whatever have ... ever acted, but those now acting; and that they never acted with different degrees of energy from that which they now exert ...

In establishing a system one must always be concerned with one's theoretical bias. How one controls this bias is essential to the quality of the theorizing. Lyell expressed his theoretical bias in a letter of 7 March 1837 to William Whewell (K. Lyell 1881, vol. 2, p. 6–7):

> ... I was taught by Buckland the catastrophic or paroxysmal theory, but before I wrote my first volume, I had come round, after considerable observation and reading, to the belief that a bias towards the opposite system was more philosophical ...

Lyell firmly believed that the hypothesizing of paroxysms or catastrophes involved a presumption that 'ordinary forces and time could never explain geological phenomena' (K. Lyell 1881, vol. 2, p. 3). To avoid this purported bias it is necessary to stipulate the true nature of those forces. Lyell also set forth this principle in his letter of 7 March 1837 to Whewell (K. Lyell 1881, vol. 2, p. 3):

> The reiteration of minor convulsions and changes is, I contend, a *vera causa*, a force and mode of operation which we know to be true. The former intensity of the same or other terrestrial forces may be true; I never denied its possibility; but it is conjectural. I complained that in attempting to explain geological phenomena ... there had always been a disposition to reason *a priori* on the extraordinary violence and suddenness of changes ... instead of attempting strenuously to frame theories in accordance with the ordinary operations of nature ...

The notion of *vera causae*, or 'true causes,' is highly appropriate to a science (Newtonian

known to Charles Lyell and many others, not in its original form, but in a more physical version described by John Playfair (1748–1819). Trained in both mathematics and physics, Playfair treated Hutton's theory as Newtonian science, not as the philosophy and theology implied by Hutton's original (Dean 1992). This led to several very significant deviations. Whereas Hutton (1795) considered his *Theory* to be a discovered working of nature, Playfair (1802) cast it as a human construct of knowledge, describing physical processes, and capable of improvement. Playfair further presumed scientific principles in the theory *a priori*, in the manner of geometric axioms (Dean 1992).

Lyell (1830), almost certainly following the interpretation of Playfair (1802), ascribed to Hutton the goal of according fixed principles to geology in the manner that Newton did for astronomy (Dean 1992). Lyell (1830, p. 61) writes that Hutton's *Theory of the Earth*, '... was the first in which geology was declared to be in no way concerned about questions as to the origin of things; the first physics) that eschews the Aristotelian search for knowledge of causes in a Platonic quest for essential principles. Newton himself established the principle as one of the 'Rules of Reasoning' in his highly acclaimed *Principia*: 'We are to admit no more causes of natural things than such as are both true and sufficient to explain their appearances'. For geology in the early nineteenth century this principle assumed a slightly modified form (Laudan 1987): 'No causes should be invoked in our geological reasoning unless they have real existence (i.e. we have directly observed them) and that they be adequate to produce the purported effect'.

Clearly Lyell viewed the task of the geologist 'to explain geological phenomena'. This is accomplished via principles or logical propositions foremost of which is the following *vera causa* (K. Lyell 1881, vol. 2, p. 5): '... the adequacy of known causes as parts of one continuous progression to produce mechanical effects resembling in kind and magnitude those which we have to account for ...'.

Uniformity and induction

The American geologist Joseph LeConte (1877, 1895) recognized that the source of Lyell's methodology was physics, not geology. LeConte (1895, p. 315) wrote:

> The basis of modern geology ... was undoubtedly laid down by Lyell in the idea that the study of 'causes now in operation' producing structure under our eyes is the only sound basis of reasoning ... According to this view, things have gone on from the beginning at a uniform rate, much as they are going on now ... (this) view was conceived in the spirit of the physicist ... and may be called physical rather than geological.

Rachel Laudan (1987) argues persuasively that Lyell composed his *Principles* at a time when five methodologies were competing in the science of geology. These methods, characterized in Table 1, include hypothesis, eliminative induction, enumerative induction, analogy, and *vera causa*. In common with his friend, physicist and philosopher of science John Herschel (1792–1871), Lyell viewed the method of hypothesis with particular scorn. Of the advocates of the Wernerian 'system' of Earth history and the catastrophist hypotheses attendant thereto, Lyell (1830, p. 224) writes:

> The popular reception of these ... sophisms ... has hither to thrown stumbling-blocks in the way of those geologists who desire to pursue the science according to the rules of inductive philosophy ... if authors may thus dogmatize, with impunity, on subjects capable of being determined with

Table 1. *The methods of geology, c. 1830 (Laudan 1987)*

Method	Characteristics
Hypothesis	Presumption of a causative state of affairs from observation of its consequent phenomena
Eliminative induction	Observational data are used to refute all conceivable rival explanations save one
Enumerative induction	Simple generalizations are derived from the data (facts)
Analogy	Combining hypothesis and induction
Vera causa	Causes existing and sufficient to produce the effect

considerable degree of precision, can we be surprised that they who reason on the more obscure phenomena of remote ages, should wander in a maze of error and inconsistency.

The way to avoid 'error and inconsistency' in geology was through strict adherence to logic. In science, according to Herschel among others of Lyell's contemporaries, this was possible only via Newton's *vera causa* (Laudan 1987) combined with enumerative induction, perhaps allowing for some analogical reasoning. Lyell (1830, p. 165) explicitly states this position:

> ... the value of all geological evidence, and the interest derived from the investigation of the Earth's history, must depend entirely on the degree of confidence which we feel in regard to the permanency of the laws of nature. Their immutable constancy alone can enable us to reason from analogy, by the strict rules of induction, respecting the events of former ages ... to arrive at the knowledge of general principles in the economy of our terrestrial system.

In embracing induction, and particularly its enumerative variant, as the method of geology Lyell was also embracing an empiricist philosophical tradition, beginning with John Locke and Bishop Berkeley, extending to Hutton's Edinburgh contemporary David Hume, and Lyell's own contemporary John Stuart Mill. These philosophers held, in contrast to continental idealists like Descartes and Kant, that all inquiry starts with experience, which is then treated axiomatically. One must have a logical basis to reason from what is observed (sense perceptions or 'facts') to what is unobserved. But this reasoning, loosely termed 'induction', has a fundamental difficulty that was made explicit by David Hume. Hume reasoned that it was impossible ever to justify a theory or law in science by experiment or observation. Just because one sees the sun rise each day does not, by itself, require that it will rise the next day. This is the famous logical problem of induction, which Sir Karl Popper (1959, p. 54) summarized in terms of three seemingly incompatible principles:

> (a) Hume's discovery ... that it is impossible to justify a law by observation or experiment, since it 'transcends experience'; (b) the fact that science proposes and uses laws 'everywhere and all the time' ... To this we add (c) the principle of empiricism which asserts that in science, only observation and experiment may decide upon the acceptance or rejection of scientific statements, including laws and theories ...

Popper (1959, 1969) claimed to solve the problem of induction through his famous principle of falsification (Lindh 1993). In the early nineteenth century a common resolution was achieved through the invoking of a 'uniformity'. Explanation of why this was necessary now requires some logical discourse.

Lyell's problem of geological induction is, in essence, that of the ascertaining of an objective probability concerning the nature of all geological causes (processes), i.e. the sampling of a genus (all geological causes) and observing how many of that genus fall in a certain species, and thence concluding the probability that, in that genus, any given individual will belong to that species. Note that an objective probability is simply a ratio, within a certain course of experience, between (a) the number of individuals in a species and (b) the number of individuals in a genus over that species. This is what is meant by induction. (The foregoing wording derives from unpublished writings by the great logician Charles S. Peirce in 1900.)

A uniformity may be most succinctly defined as a high objective probability of an objective probability (Peirce 1902, p. 727). Here an objective probability is defined as a ratio, with a certain course of experience, between (a) the number of individuals in a species and (b) the number of individuals in a genus over that species. In Lyell's geological problem the genus is all geological causes (or processes), and the species of his interest is the observed set of attributes characteristic of those causes (or processes), observed generally because those causes are presently in operation. Lyell's goal was to achieve the most valid possible induction, which is simply the ascertaining of the highest possible probability concerning the nature of all geological causes (the genus).

Given that the above concept of uniformity was prevalent in Lyell's day (e.g. Mill 1843), it is not difficult to see the reasoning that led Lyell to be so positive in his invocation of uniformity as a necessity for geological reasoning. The argument had all the force of the following implied syllogism:

Major premise: Enumerative induction is the only valid reasoning process in geology
Minor premise: Uniformity (*vera causa*) provides the only basis for valid enumerative induction
Conclusion: An *a priori* doctrine of uniformity (uniformitarianism) is necessary for geology to be a valid science

Do geologists reason inductively?

In embracing the empiricist philosophies and inductive methodologies of Herschel, Mill, and others, Lyell rejected methodologies advocated by other contemporary scholars. Historical accounts have sometimes ignored these methodological issues, since many of the alternative methodologies were held by those espousing catastrophist hypotheses. Ironically, it took nearly 150 years for many of Lyell's critics to be recognized as 'actualistic catastrophists' (Hooykaas 1970). Sedgwick (1831), Conybeare (1830) and Whewell (1832) were all highly critical of Lyell's *a priori* specification of the nature of causes and indeed of his whole view of the nature of geological reasoning.

Of Lyell's critics William Whewell (1794–1866) was the most important (Fig. 3). Whewell believed that one discovered the logic of science through study of its actual practice, and he devoted extensive historical research to this end (Whewell 1837, 1840). His research led him to the partly idealist position that the intuition of 'truths' in science is progressive and evolving. However, unlike other idealists like Descartes, Whewell held that any axioms or principles were products of scientific inquiry, not starting points for such inquiry. These ideas were vehemently criticized by John Stuart Mill (1843), who argued on a detached logical basis that the real work of science lay in the establishment of knowledge by inductive proof. This position was subsequently embraced by Bertrand Russell and other founders of logical empiricism and related strands of modern analytical philosophy. Whewell's ideas were declared defeatist because he made science dependent upon ingenuity and luck. As Wettersten and Agassi (1991, p. 345) observe: '... because philosophers ignored him and scientists did not write histories of philosophy, he was forgotten'.

Of course, it was Whewell (1832) who, in his review of Lyell's principles, coined the terms

Fig. 3. William Whewell statue in the chapel of Trinity College, Cambridge University, where he was Master of the College.

'catastrophism' and 'uniformitarianism.' Whewell considered geology to be a 'palaetiological science', concerned with '... the study of a past condition, from which the present is derived by causes acting in time'. Therefore, it was inappropriate to specify, via *vera causa*, the nature of those causes *a priori*. Whewell (1837, vol. 2, p. 593) writes, in a vein similar to other catastrophists:

In truth, we know causes only by their effects; and in order to learn the nature of the causes which modify the Earth, we must study them through all ages of their action, and not select arbitrarily the period in which we live as the standard for all other epochs ...

Whewell (1837, vol. 2, p. 592) goes much further than any of his contemporaries, however, in a complete rejection of Lyell's inductive logic:

(Lyell's) '*earnest and patient endeavor to reconcile* the former indication of change', with *any* restricted class of causes, — a habit which he enjoins, — is not, we may suggest, the temper in which science ought to be pursued. The effects

must themselves teach us the nature and intensity of the causes which have operated ...

Comparison of this statement to the logical description of induction will show that Whewell is denying that the kind of induction advocated by Lyell, and borrowed from physics, is appropriate to geology as a 'palaetiological science'. Note also that the point here concerns 'methodological uniformitarianism', a doctrine which many contemporary scholars hold to be valid in modern geology.

To understand what Whewell was aiming at we will have to return to a notion of 'hypotheses' that was summarily dismissed in the physics-based philosophy of inductions and uniformity. Following Charles Peirce, a scientific hypothesis may be considered to be the starting point of a question. A phenomenon is observed to have something peculiar about it. One then infers *if* a certain state of affairs existed, then that phenomenon would in all probability occur. Hypothesis is the presumption of this state of affairs.

Charles Peirce developed a logic of hypothesis and credited William Whewell for anticipating this logic (Peirce 1898). Whewell (1840) recognized that science was concerned with the forming of antithetical couplings between (1) the objective facts of nature and (2) new concepts suggested to scientific minds. Whewell considered this process to be a colligation ('binding together') of existing facts that are unconnected in themselves but become connected through mental concepts. In his treatises on logic, Whewell (1858, 1860) referred to this process as 'induction', but Charles Peirce distinguished this form of synthetic inference from the 'induction' of which Hume (and later Popper) had spoken. Peirce accorded it the various names 'hypothesis', 'abduction', 'retroduction', and 'presumption'.

If we take Whewell's methodology to be a part of abductive or retroductive geology, we can now contrast it with the inductive methodology of Lyell and others (Table 2). The details of these approaches have been more fully described elsewhere (Von Engelhardt & Zimmerman 1988; Baker 1996a,b), but here the distinction will be succinctly drawn in terms of understanding versus 'overstanding' nature. The notion of 'overstanding' may be grasped by considering the answer of the physicist Niels Bohr to a question concerning the reality of nature revealed by his theories. Bohr replied as follows (Petersen 1985, p. 305): 'It is wrong to think that the task of physics is to find out how nature is. Physics concerns what we can say about nature'. This is overstanding; the explanations of science are judged by logical validity. Indeed verification of theories against nature is logically precluded (Popper 1959, 1969).

Table 2. *Comparison of scientific reasoning styles for some early nineteenth century geologists*

	Uniformitarian overstanding (Lyell, Herschel, Playfair)	Catastrophist understanding (Whewell, Sedgwick)
Observe	Effects of geological causes	Effects of geological causes
Assume	Axiomatic aspects of causes (*vera causa*, uniformity)	Axiomatic principles (laws of physics)
Discover	Principles of geology (logically valid geological explanation)	Causes actually operative in nature
Goal	To be logically valid (true) in what we can say about nature	To find out what nature says to us

Geological understanding has to do with what nature says to us. This is interpreted through a process that begins with retroduction or abduction. The distinction of abduction from induction, so important in geology, has only been made clear by Charles Peirce, who has shared Whewell's neglect by most philosophers of science, despite his clear relevance of geology (Von Engelhardt & Zimmerman 1988; Baker 1996*a*,*b*). Consider Peirce's distinction of these two models of reasoning (in Burks 1958, pp. 136–137):

> Nothing has so much contributed to present chaotic or erroneous ideas of the logic of science as failure to distinguish the essentially different characters of different elements of scientific reasoning; and one of the worst of these confusions, as well as one of the commonest, consists in regarding abduction and induction taken together (often mixed also with deduction) as a simple argument. Abduction and induction have, to be sure, this common feature, that both lead to the acceptance of a hypothesis because observed facts are such as would necessarily or probably result as consequences of that hypothesis. But for all that, they are the opposite poles of reason ... The method of either is the very reverse of the other's. Abduction makes its start from the facts, without, at the outset, having any particular theory in view, though it is motivated by the feeling that a theory is needed to explain the surprising facts. Induction makes its start from a hypothesis which seems to recommend itself, without at the outset having any particular facts in view, though it feels the need of facts to support the theory. Abduction seeks a theory. Induction seeks for facts.

The structure of Lyell's *Principles* is clearly inductive. It starts with a *vera causa*, and it develops the facts that support this initial proposition. The peculiar backwards organization of the *Principles* follows this task. The book begins with the youngest geological periods and develops facts further down the geological column from Tertiary to Secondary to Primary. The progress is from the most secure facts to the least. This is the order of physics-based overstanding, rather than that of geology-based understanding.

How do geologists reason?

What guides are required in the reasoning process of science? What determines the value of a theory or hypothesis in geology? For Charles Lyell and his intellectual successors this value must be established by some principle of reasoning. Lyell sought to rid geology of error and inconsistency, to allow it precision in explanation according to strict rules of logic, indeed to put geology on the same strong logical grounds as the sciences of controlled experimentation. Physics was the exemplar science for Lyell, and it remains so today for nearly all philosophers of science. Indeed the heritage of Lyell's appropriation of physics reasoning has proven far more durable than the specifics of his uniformitarian doctrine. The current physics-based philosophical fashion of hypothetico–deductive scientific method, developed by Popper (1959, 1969), is strongly advocated for geological research through appeals to logic and the probability of truth (Cowan *et al.* 1997).

A widely-held methodological principle is that of simplicity. For example, the highly respected analytical philosopher Nelson Goodman (1967, p. 93) writes, 'The Principle of Uniformity dissolves into a principle of simplicity that is not peculiar to geology but pervades all science and even daily life'. Indeed, the mathematician John Playfair praised the geological theory of his friend James Hutton for its simplicity (Playfair 1802, p. 136), much as a physicist would praise a theory in that science. Because simplicity is a principle held in great esteem by analytical philosophers (Russell 1929) and by physicists (Bridgman 1961), one commonly encounters arguments that cite the principle of simplicity in support of geological explanations (Newell 1967) or claims that simpler scientific explanations are somehow 'better' than more complex ones. Such claims have received a more skeptical reception from other geologists (Anderson 1963).

Uniformitarian simplicity, or scientific parsimony, might be expressed as follows: no extra, fanciful or unknown causes should be invoked if known causes (those presently in operation and/or observed) will do the job. The substantive consequences of this principle may be as innocuous as the claim that the same laws of mechanics apply on Mars as on Earth. The successful landing of the Pathfinder spacecraft attests to the practical value of this claim. However, the success of this very limited view does not warrant the extension of the principle to other claims, such as the following (Clifton 1988, p. 4):

> In every case, responsible scientific procedure dictates that we accept the most probable, generally simplest explanation for any phenomenon in the geologic record ... Parsimony demands that we attribute phenomena in the sedimentary record to the most probable explanation, and convulsive geologic events are, by nature, improbable.

Note how the above quote conflates notions of induction (probable inference), simplicity, uniformity and factual observation (the improbability of catastrophic processes) into a proscriptive methodological statement. Is this really how geologists wish to interpret the Earth? The catastrophist philosophers cited above would hold that in a natural science, in tune with nature, the only valid demands on our explanations are those made by nature. If nature contradicts our philosophy, parsimony included, it is nature that our explanations should follow, no matter how elegant and simple the philosophy.

Louis Agassiz (1859) recognized a principle of 'naturalness' in geological reasoning and applied it to the problem of classifying the divisions of animals. Clearly this is a more complex problem of induction (sampling genus and assigning to species) than that perplexing Charles Lyell, but it is logically equivalent. However, whereas Lyell appealed to a uniformity (an asserted probability to the induction), Agassiz proposes that any order in the divisions must be natural, not artificial. Although he ascribes that order to 'the Divine Intelligence', such deification is not necessary for the operation of his warrant for induction. He writes (Agassiz 1859, p. 9):

> Is this order the result of the exertions of human skill and ingenuity; or is it inherent in the objects themselves, so that the intelligent student of Natural History is led unconsciously, by the study of the animal kingdom itself, to these conclusions ...? To me it appears indisputable, that this order and arrangement of our studies are based upon the natural, primitive relations of animal life ... The human mind is in tune with nature, and much that appears as a result of the working of our intelligence is only the natural expression of that preestablished harmony.

Note that Agassiz's scientific reasoning allows just as much order and precision to induction as does Lyell's uniformity. The difference is in from whence that order and precision will derive. Lyell, following Newton and the various interpreters of his philosophy of physics, believes that the warrant for induction lies in the precision of a logic that is objectively detached from the objects represented in its symbols. But a superprinciple is required to regulate the relationship of that logic to the natural world mirrored in its explanations. Agassiz's 'human mind in tune with nature' needs no such principle. Nature is the source of any order that we discover, and it is impossible to detach our logic from the connection of its symbols to that order. Rather than suppressing that connection in a quest for 'knowledge of general principles', Agassiz would have the scientist learn the lesson that nature has to teach.

The American polymath Charles S. Peirce probably devoted the most intense philosophical effort to understanding Agassiz's approach to induction. Peirce actually studied classification with Agassiz and may have influenced some important writings on philosophy of geology (Baker 1996a). In his 1898 lectures on *Reasoning and the Logic of Things* Peirce describes the issue as follows (Peirce 1898, pp. 176–177):

> The only end of science, as such, is to learn the lesson that the universe has to teach it. In Induction it simply surrenders itself to the force of facts. But it finds, at once, — I am partly inverting the historical order in order to state the process in its logical order, — it finds I say that this is not enough. It is driven in desperation to call upon its inward sympathy with nature, its instinct for aid, just as we find Galileo at the dawn of modern science making his appeal to *il lume naturale*. But insofar as it does this, the solid ground of fact fails it. It feels from that moment that its position is only provisional. It must then find confirmations or else shift its footing. Even if it does find confirmations, they are only partial. It still is not standing upon the bedrock of fact. It is walking upon a bog, and can

only say, this ground seems to hold for the present. Here I will stay till it begins to give way. Moreover, in all its progress science vaguely feels that it is only learning a lesson. The value of *Facts to it*, lies only in this, that they belong to Nature; and Nature is something great, and beautiful, and sacred, and eternal, and real, — the object of its worship and it aspiration.

The spirit of hypothesizing in geology, captured so well by Peirce, is far more relevant to the practice of the discipline (Baker 1996*b*) than is Charles Lyell's induction and uniformity. This spirit has been captured in the writing of Gilbert (1886), Chamberlin (1890) and Davis (1926). It was also expressed by many of Lyell's catastrophist contemporaries. Fortunately, discoveries like those on Mars continue to reveal the inadequacies of various uniformitarian dogmas, both substantive and methodologic, as applied to geology.

Conclusions

It has been the thesis of this essay that the 'new catastrophism' is rooted in a very old idea, one held by many of the old catastrophists: geology is about what Earth has to say to us. It is true that many of the early nineteenth century catastrophists interpreted their task as one of translating the thoughts of God, but the principles of logical inference, i.e. the methodological components, of their science are quite independent of any particular notion of God. Despite neglect by nearly all the modern philosophers of so-called 'science', this view of inference has just as much power to ensure scientific progress as does the notion of methodological uniformitarianism, including doctrines of simplicity and actualism.

As a general observation, working scientists are prone to many misconceptions as to the relationship of philosophy to science. Perhaps the most pernicious of these is the advocacy of foundational principles to explain the success of science, to function as a framework for correct action, or to justify the results of science. Such presumptions of foundational principles must be held on faith; they always involve antithetical formulations; and they always have substantive or 'strong' forms as well as methodological or 'weak' forms. The substantive forms make ontological claims about how the Earth actually behaves, while the methodological forms provide guidance for reasoning about the Earth. The early debates on uniformitarianism and catastrophism by Lyell and his contemporaries perpetuated notions of substantive or ontological elements that are misplaced in modern Earth science.

Because catastrophism, strictly speaking,

contrasts only with the substantive doctrine of gradualism, it is not surprising to see modern scientists embracing its position. There is nothing contradictory in adhering to uniformity of law plus uniformity of process (actualism), while also preferring catastrophist to gradualist explanations of geological phenomena. The term 'catastrophic' only applies to the intensity and duration of a particular geological process. It does not necessarily have anything to do with whether or not such a process is manifest today (actualism) or even with the well-known methodological claim that simpler explanations are to be preferred to more complex ones. This latter claim, sometimes called the 'principle of scientific parsimony', is not to mean that some sort of high-order preference must be accorded simpler explanations, presumably gradualistic, in contrast to more complex explanations, presumably catastrophic. Rather, it is simplicity in terms of the more natural explanation that has proven to be the most productive methodological guide to scientific reasoning. Current trends that invoke catastrophic hypotheses for geological phenomena are best explained as a naturalistic turn to reasoning bolstered by pragmatic approaches that deny the older foundational concerns (Baker 1996*b*).

Geology is a realistic science, not an actualistic one. A science that would limit itself to using the present as the arbitrator of what counts as natural evidence condemns itself to being actualistically unrealistic. The realism in geology derives not so much through inductive experimental contiguity as through coherence and consistency of observation with hypothesis. The latter, which William Whewell termed the 'colligation of facts', occurs in the complexity whereby nature is studied 'as is', rather than in the artificially defined simplified 'systems' so as to be amenable to controlled experimentation. To the extent that its methodology need not mimic that of mathematical physics, geology does not require notions of uniformity for its successful pursuit.

My research on catastrophic flooding has been supported over the years by grants from the National Aeronautics and Space Administration and the National Science Foundation. This essay is AUMIN contribution number 10.

References

AGASSIZ, L. 1859. *An Essay on Classification*. Longman, Brown, Green, Longmans, & Roberts, and Trübner & Co., London.

ANDERSON, C. A. 1963. Simplicity in structural geology. *In*: ALBRITTON, C. C. (ed.) *The Fabric of Geology*. Freeman, Cooper & Co., San Francisco, USA, 175–183.

BAKER, V. R. 1978. The Spokane Flood controversy and the Martian outflow channels. *Science*, **202**, 1249–1256.

—— 1973. *Paleohydrology and Sedimentology of Lake Missoula Flooding in Eastern Washington*. Geological Society of America Special Paper 144.

—— (ed.) 1981. *Catastrophic Flooding: The Origin of the Channeled Scabland*. Dowden, Hutchinson & Ross, Stroudsburg, Pennsylvania, USA.

—— 1982. *The Channels of Mars*. University of Texas Press, Austin, Texas, USA.

—— 1996a. The pragmatic roots of American Quaternary geology and geomorphology. *Geomorphology*, **16**, 197–215.

—— 1996b. Hypotheses and geomorphological reasoning. *In*: RHOADS, B. L. & THORN, C. E. (eds) *The Scientific Nature of Geomorphology*. Wiley, NY, USA, 57–85.

—— & MILTON, D. J. 1974. Erosion by catastrophic floods on Mars and Earth. *Icarus*, **23**, 27–41.

——, BENITO, G. & RUDOY, A. N. 1993. Paleohydrology of late Pleistocene superflooding, Altay Mountains, Siberia. *Science*, **259**, 348–350.

BRIDGMAN, P. W. 1961. *The Way Things Are*. Viking, NY, USA.

BURKS, A. W. (ed.) 1958. *The Collected Papers of Charles Sanders Peirce*, vol. 7. Harvard University Press, Cambridge, USA.

CHAMBERLIN, T. C. 1890. The method of multiple working hypotheses. *Science*, **15**, 92–96.

CLIFTON, E. 1988. *Sedimentological relevance of Convulsive Events*. Geological Society of America Special Paper 229, 1–5.

CONYBEARE, W. D. 1830. An examination of those phaenomena of geology, which seem to bear most directly on theoretical speculations. *Philosophical Magazine and Annals of Philosophy, new series* **8**, 359–362.

COWAN, D. S., BRANDON, M. T. & GARVER, J. I. 1997. Geologic tests of hypotheses for large coastwise displacements – a critique illustrated by the Baja British Columbia controversy. *American Journal of Science*, **297**, 117–173.

DAVIES, G. L. 1969. *The Earth in Decay: A History of British Geomorphology 1578–1878*. MacDonald, London, UK.

DAVIS, W. M. 1926. The value of outrageous geological hypotheses. *Science*, **63**, 463–468.

DEAN, D. R. 1992. *James Hutton and the History of Geology*. Cornell University Press, Ithaca, NY, USA.

GILBERT, G. K. 1886. The inculcation of scientific method by example. *American Journal of Science*, **31**, 284–299.

GOODMAN, N. 1967. Uniformity and simplicity. *In*: ALBRITTON, C. C. JR (ed.) *Uniformity and Simplicity*. Geological Society of America Special Paper 89, 93–99.

GOULD, S. J. 1987. *Time's Arrow, Time's Cycle: Myth and Metaphor in the Discovery of Geological Time*. Harvard University Press, Cambridge, USA.

HOOYKAAS, R. 1970. Catastrophism in geology, its scientific character in relation to actualism and uniformitarianism. *Koninklijke Nederlandse Akademie van Loefenschappen, afd. Letterkunde, Med. (n.r.)*, **33**(7), 271–316.

HUTTON, J. 1788. Theory of the Earth; or an investigation of the laws observable in the composition, dissolution, and restoration of land upon the globe. *Transactions of the Royal Society of Edinburgh*, **1**, 209–304.

—— 1795. *Theory of the Earth with Proofs and Illustrations*. Creech, Edinburgh, UK.

LAUDAN, R. 1987. *From Geology to Mineralogy*. University of Chicago Press, Chicago, USA.

LECONTE, J. 1877. On critical periods in the history of the Earth and their relation to evolution. *American Journal of Science and the Arts*, (third series), **14**, 99–114.

—— 1895. Critical periods in the history of the Earth. *Bulletin of the Department of Geology, University of California*, **1**, 313–336.

LINDH, A. G. 1993. Did Popper solve Hume's problem? *Nature*, **366**, 105–106.

LYELL, C. 1830–1833. *Principles of Geology*. 3 vols. Murray, London, UK.

LYELL, K. M. 1881. *Life, Letters and Journals of Sir Charles Lyell, Bart*. Murray, London, UK.

MCKAY, D. S. ET AL. 1996. Search for past life on Mars: possible relic biogenic activity in Martian meteorite ALH84001. *Science*, **273**, 924–930.

MCSWEEN, H. Y. JR 1994. What we have learned about Mars from SNC meteorites. *Meteoritics*, **29**, 757–779.

—— 1997. Evidence for life in a Martian meteorite? *GSA Today*, **7**(2), 1–7.

MELOSH, H. J. 1984. Impact ejection, spallation and the origin of meteorites. *Icarus*, **59**, 234–260.

—— 1985. Ejection of rock fragments from planetary bodies. *Geology*, **13**, 144–147.

MILL, J. S. 1843. *A System of Logic: Ratiocinative and Inductive*. Longmans and Green, London.

NEWELL, N. D. 1967. Revolutions in the history of life. *In*: ALBRITTON, C. C. JR (ed.) *Uniformity and Simplicity*. Geological Society of America Special Paper 89, 63–91.

PEIRCE, C. S. 1898. *Reasoning and the Logic of Things*. The Cambridge Conference Lectures of 1898 edited by K. L. KETNER and published in 1992 by Harvard University Press, Cambridge, USA.

—— 1902. Uniformity. *In*: BALDWIN, J. M. (ed.) *Dictionary of Philosophy and Psychology*, vol. 2. Macmillan, NY, USA, 727–731.

PETERSEN, A. 1985. The philosophy of Niels Bohr. *In*: FRENCH, A. P. & KENNEDY, P. J. (eds) *Niels Bohr: A Centenary Volume*. Harvard University Press, Cambridge, USA.

PLAYFAIR, J. 1802. *Illustrations of the Huttonian Theory of the Earth*. Cadell and Davis, London, UK.

POPPER, K. 1959. *The Logic of Scientific Discovery*. Basic Books, NY, USA.

—— 1969. *Conjectures and Refutations*. Routledge and Kegan Paul, London, UK.

RUSSELL, B. 1929. *Our Knowledge of the External World*. W. W. Norton, NY, USA.

SEDGWICK, A. 1831. Address to the Geological Society. *Proceedings of the Geological Society of London*, **1**, 281–316.

SHAW, J. 1994. A qualitative view of sub-ice-sheet landscape evolution. *Progress in Physical Geography*, **18**, 159–184.

VON ENGELHARDT, W. & ZIMMERMAN, J. 1988. *Theory of Earth Science*. Cambridge University Press, Cambridge, UK.

WETTERSTEN, J. & AGASSI, J. 1991. Whewell's problematic heritage. *In*: FISCH, M. & SCHAFFER, S. (eds) *William Whewell: A Composite Portrait*. Oxford University Press, Oxford.

WHEWELL, W. 1832. Review of Lyell, 1830–3, vol. ii. *Quarterly Review*, **47**, 103–132.

—— 1837. *History of the Inductive Sciences*. Cass, London, UK.

—— 1840. *The Philosophy of the Inductive Sciences, Founded Upon Their History*. Cass, London, UK.

—— 1858. *Novum Organon Renovatum*. J. W. Parker and Son, London, UK.

—— 1860. *On the Philosophy of Discovery*. J. W. Parker and Son, London, UK.

From William Smith to William Whitaker: the development of British hydrogeology in the nineteenth century

JOHN MATHER

Lyell Professor, Department of Geology, Royal Holloway, University of London,
Egham, Surrey TW20 0EX, UK

Abstract: Some of the earliest applications of the principles of geology to the solution of hydrologic problems were made by William Smith, who used his knowledge of strata succession to locate groundwater resources to feed the summit levels of canals and supply individual houses and towns. The industrial revolution led to a huge demand for water resources to supply new towns and cities. Nottingham, Liverpool, Sunderland and parts of London all relied on groundwater. By the middle of the nineteenth century James Clutterbuck had already recognized that groundwater was a finite resource and that if abstraction was more rapid than replenishment by rain, water levels would decline and quality would be affected by saline intrusion. In 1851 Prestwich produced the first British geological map that included groundwater information. Before 1870 the Geological Survey had shown little concern for groundwater, perhaps because its Director, Murchison, had little interest in the economic applications of geology. After his retirement in 1871 there was an explosion of activity. Lucas introduced the term 'hydrogeology' in 1874 and produced the first real hydrogeological map in 1877 after leaving the Survey to work as a consultant water engineer. De Rance was for 20 years the secretary of a British Association Committee set up in 1874 to inquire into the underground circulation of water and in 1882 produced a 600 page volume on the water supply of England and Wales. William Whitaker, sometimes described as the 'father of English hydrogeology', was a collecter of well records and his work led to the inclusion of page after page of well records in survey publications in the southeast of England and in 1899 to the first water supply memoir – *The Water Supply of Sussex from Underground Sources*. From Smith to Whitaker, knowledge of groundwater grew throughout the nineteenth century, providing the basis for the sophisticated models of today.

The part that groundwater has played in the historical development of the United Kingdom is obvious from any survey of place names. The incorporation of such words as 'well', 'spring', 'bourne' and 'spa' into the names of farms, villages and larger settlements is ample evidence of the importance of groundwater. However, until relatively recent times our forebears had rather primitive ideas about its origin and groundwater phenomena were often surrounded by mystique. Thus many early wells were associated with saintly patrons and Robins (1946) suggests that the wise men of early Christianity found the aura of sanctity the best means of ensuring that a perfectly good water supply was treated with the care and respect necessary to preserve it from pollution and abuse.

The intermittent streams peculiar to the Chalk aquifer in England were long the subject of fear and superstition. Known as 'bournes' (Hertfordshire and Surrey), 'nailbournes' (Kent), 'levants' (Sussex), 'winterbournes' (Dorset and Hampshire), or 'gypsies' (Yorkshire), they are the result of the infiltration of winter rainfall causing source springs

to break out higher up valleys as the water table rises. Because of the low specific yield of the Chalk, after very wet winters it is possible for rivers to break out some distance up the valley from their normal source and, for example, the source of the River Ver in Hertfordshire has varied by at least 5 miles (Tomkins 1969). In medieval times all bourne flows were viewed with suspicion and were regarded as a token of death or pestilence or as an omen of disaster. The earliest reference goes back to 1473 (Latham 1904) and refers to the bourne flow in the upper Ver valley as the Womere or Woewater (Fig. 1). Flows of this woe-water (or brook of woe) were thought to presage a calamity of some sort. As the Revd J. Childrey pointed out in 1661 (Latham 1904),

That the sudden eruption of springs in places where they use not always to run should be a sign of death is no wonder. For these unusual eruptions are caused by extreme gluts of rain, or lasting wet weather, and never happen but in wet years in which years wheat and most other grain

MATHER, J. 1998. From William Smith to William Whitaker: the development of British hydrogeology in the nineteenth century. *In*: BLUNDELL, D. J. & SCOTT, A. C. (eds) *Lyell: the Past is the Key to the Present.* Geological Society, London, Special Publications, **143**, 183–196.

183

Fig. 1. The now dry bed of the River Ver at Markyate where the bourne flow was described as the Womere or Woe-water in chronicles of 1473 (Latham 1904).

thrive not well and therefore death succeeds the following year.

It is interesting to note that Childrey recognized that the spring flows were the result of heavy rain some 13 years before Perrault first demonstrated experimentally that rainfall was more than adequate to account for the flow of rivers and springs (Perrault 1674).

The significance of groundwater in the development of settlements and towns is well illustrated by the growth of London to the beginning of the nineteenth century. The first building took place along the water-bearing alluvial gravels of the Thames flood plain, following the outcrop of the gravels and terminating when the London Clay came to the surface (Prestwich 1872). Water was obtained both from shallow wells and from springs which issued where the gravels had been cut down by shallow valleys to the London Clay. A good deal of water was also drawn from the Thames and from tributaries which flowed through the settlement to the river. Pollution of the tributary streams and the shallow wells gradually spread and many of these sources were abandoned.

In the early part of the thirteenth century water started to be supplied to conduits or public fountains from springs outside the populated area. The first springs to be used were those at Tyburn

(Tybourne) where water was conveyed to the first or Great Conduit where water was supplied 'for the poor to drink and the rich to dress their meat' (Stow 1603). This was followed by a number of other conduits, one in Aldgate receiving its water from springs at the base of the Bagshot Sands in Hampstead, and by the end of the sixteenth century London had at least 16 conduits (Robins 1946).

The sixteenth century marked the beginning of a new era in water supply with the commencement of large-scale schemes. London was no exception and in 1582, Peter Morice (Morrys or Morris), a Dutchman, established a pump, worked by water wheels, to bring water from the Thames to the city. However, from the viewpoint of the hydrogeologist the pioneer undertaking was the construction of the New River to bring water in an open trench from springs near Ware in Hertfordshire. The springs were the Chalk springs of Chadwell and Amwell in the valley of the River Lea and in its original winding course the New River was more than 40 miles long (Fig. 2). The river was an open channel, 10 feet wide with an average depth of 4 feet and the elevation of the springs allowed water to flow by gravity, following the contour line, to a circular pond, known as the New River Head at Islington to the north of the city. The excavation of the channel was completed in April 1613 and the New River Company remained in existence until the formation

Fig. 2. The source of the New River at Chadwell (from an old print dated 1810).

of the Metropolitan Water Board in 1904 (Robins 1946).

Over the next two centuries further waterworks were developed mostly taking their supplies from the Thames. The establishment of these works made it possible to distribute water at any point within the metropolis and thus removed the restriction in the growth of London to the area underlain by gravel. Development at once spread over the area underlain by London Clay (De Rance 1882). According to Woodward (1922) the earliest known deep well near London was sunk in 1725 and obtained water from the sands of the Tertiary Woolwich and Reading Beds. However, exploitation of deeper groundwaters did not make much headway for many years and deep wells did not become commonplace until early in the nineteenth century. This is probably because of the presence of running sands and/or lenses of well cemented sands and conglomerates (puddingstones) in the Tertiary strata underlying the London Clay (Barrow & Wills 1913). Thus the availability of groundwater was both a reason for growth and restricted expansion during the early history of the development of London.

It is the objective of the present paper to review how the science of hydrogeology developed in the UK during the nineteenth century. Over this period the impetus was provided by the needs of an expanding population and only to a lesser extent by scientific curiosity. The review commences with the work of William Smith at the beginning of the century and ends with the publication, by the Geological Survey, of the first memoir on underground water supply, written by William Whitaker and Clement Reid (1899). The paper is divided into three sections: the first ends with the death of William Smith in 1839 and the second with the death of Murchison, the second director of the Geological Survey in 1871. Each of these dates

marks the beginning of significant new developments in British hydrogeology. The historical research owes much to the bibliographies appended to each of the 28 water supply memoirs published between 1899 and 1938 and to chronological lists of references compiled by Whitaker in 1888 and 1895.

William Smith and the period to 1839

According to Biswas (1970) one of the earliest applications of the principles of geology to the solution of hydrological problems was made by the Englishman William Smith. His early work was as a surveyor undertaking surveys for both canals and colliery workings. During his work he noted, in excavations, the various soils and the character of the rocks from which they were derived. By 1793 he had grasped the principle of rock succession, as his notes of this period show (Robson 1986). In 1799 Smith had a disagreement with his employers and became an independent consulting engineer. The subsequent failure of a stone quarry on his land near Bath compelled him to concentrate on consulting work to pay off his debts and he applied his geological knowledge to many different problems, including the location of groundwater resources. Smith's interest in applying geology to water supply is clear from his original table of strata in the vicinity of Bath produced in 1799. On this table he indicates those formations which give rise to springs (Sheppard 1920). He is also known to have been active in restoring the dwindling flow of the hot springs at Bath (Kellaway 1991).

Sheppard (1920) records a reference to Smith's hydrogeological work which is given in a letter from John Farey to Sir Joseph Banks dated 24 February 1808. In 1802 a Buckinghamshire cleric had sunk a well at his parsonage to a depth of more than 100 feet but had found only clay. Smith, on consulting his geological map, assured the cleric that if he persevered through the dry clay he would strike water in a limestone which cropped out about 8 miles to the northwest. Limestone was reached at 235 feet but yielded little water. An auger hole was drilled to a second limestone which produced a plentiful supply of water which filled the well almost to the surface of the ground. The advice that he gave the Canal Company in their efforts to provide groundwater to feed the summit level of the Wilts and Berks Canal near Swindon is recorded by Phillips (1844). Although the yield of the borehole drilled was not sufficient to supply the down lockage the advice he gave demonstrated that Smith knew a considerable amount about hydrogeology, including the occurrence of water table and confined conditions and the significance of hydraulic head.

Some years later, in a paper to the Yorkshire Philosophical Society, Smith (1827) discussed a method for supplementing the water supply of Scarborough using water from a borehole, drilled several years previously for draining the land, which was found to overflow yielding a small volume of water. An open channel was subsequently cut up to the borehole and later deepened, increasing the discharge significantly. He suspected that the water came from a confined aquifer and suggested the damming of this spring for summer use as the previous supply was more than sufficient for the town in winter.

In contrast, Charles Lyell was the son of a Scottish landowner and did not need to earn his own living. He had much freedom to travel and, stimulated as an Oxford undergraduate by the lectures of Buckland, he applied himself in his travels to an intensive study of the rocks and structure of much of western Europe (George 1976). Lyell's 'Principles of Geology' (Lyell 1830–1833) contains almost nothing on groundwater. Chapter 12 of Volume 1 of the Principles is concerned with springs, but only in their role as a means of transferring material in solution from depth and not in their role as a source of potable water. However, Lyell clearly recognized that springs 'are in general, ascribable to the percolation of rainwater through porous rocks, which meeting at last with argillaceous strata, is thrown out to the surface'. Although most of his examples were taken from his travels in Europe, the chalybeate waters of Tunbridge Wells were cited as an example of ferruginous springs in which the iron was present as carbonate. He recognized that some of the sulphate in these groundwaters was derived from the decomposition of pyrite.

Smith and Lyell were not the only geologists writing about groundwater in the early years of the nineteenth century. Many of the published accounts are merely descriptions of wells or boreholes together with the strata intersected (e.g. Yeats 1826; Donkin 1836) and contribute little to the development of hydrogeological thinking. However, the controversies which were to dominate discussion later in the century were already beginning to surface.

Although it was recognised that some potential recharge was lost by evaporation it was considered by some that infiltration to permeable formations such as the Chalk of southern England was so rapid that hardly any loss occurred and almost all rainfall was available for subsequent extraction. Thus there was felt to be a vast resource available which only needed to be pumped out. However, some engineers already recognized that the resource was limited and that surface and groundwaters were inextricably linked. For example Seaward (1836) states:

That the bowels of the earth contain springs of water in abundance, there can be no doubt ... but we know full well that those same springs, if they have sufficient natural force, must find their way to the surface of the earth somewhere, without any boring, and then form rivers and flowing brooks. Why then delve a great depth at an infinite expense, to procure that which we can generally obtain so readily and economically on the surface of the earth?

He recognized that new wells often reduced the yield of adjacent wells and that underground springs 'do not furnish that inexhaustible supply of water which some persons imagine.'

Population growth and the period to 1871

Economic and technical developments, together with a rapid increase in population (from about 10 million in 1800 to 19 million in 1837), brought unprecedented changes to Britain in the first few decades of the nineteenth century. East and west of the Pennines, the presence of coal and water resulted in opportunities for employment which attracted people from other parts of the country and from Ireland. Housing was provided rapidly and cheaply and new working-class districts grew up close to mills and factories. Water supplies and sewage disposal became totally inadequate and, when cholera first reached England in 1831, these overcrowded settlements were ideal breeding grounds. London, although not directly affected in the same way by the industrial revolution, was also growing rapidly. Reports began to reveal the relationship between water supply and ill health but it was not until 1854 that Dr John Snow, working in Westminster, demonstrated beyond doubt that cholera was spread by drinking well water contaminated with sewage effluent derived from local cesspools. As pointed out by Price (1996), Snow carried out one of the earliest recorded investigations into a case of groundwater pollution.

In order to improve the quality of drinking water in towns and cities sources of supply had to be established. In some of the northern and midland cities it was possible to develop upland catchment areas. However, in other areas, particularly in the southeast of England, this was impossible and other sources of supply had to be obtained. The Report of the Royal Commission on Water Supply (1869 a, b) shows that at this time a number of large conurbations were supplied by groundwater. Nottingham, parts of Liverpool, and Birkenhead relied on wells in the Permo-Triassic Sandstones, Sunderland and South Shields on the Permian Magnesian Limestone and Croydon and parts of

London on the Chalk. London was at that time supplied by eight separate companies, only one of which, the Kent Company, derived its supplies from Chalk wells. However, government offices and several public establishments around Westminster, together with the fountains in Trafalgar Square, were supplied by wells, sunk to a depth of about 400 feet in the Chalk in 1844 (Abel & Rowney 1849; Amos 1860).

The increase in interest in groundwater resulted in considerable advances in our understanding of groundwater movement and the problems encountered in its exploitation. Some of the earliest systematic observations of groundwater levels were made by a Hertfordshire cleric, James Clutterbuck, and were reported in three papers to the Institution of Civil Engineers (Clutterbuck 1842, 1843, 1850). Using accurate measurements and sections drawn to scale he demonstrated that, in the London Basin, what he termed the 'chalk water level' was described by a line drawn from the highest level at which water accumulates in the chalk to the lowest discharge point or vent. He recognized that the only apparent vent for these waters was mean tide level in the Thames below London. He defined the term 'chalk water level' as 'the height, at any point, or continuous series of points, to which the water rises in the chalk, or to which it will rise from the chalk, in perforations through the London and plastic clays above the chalk'. The practical conclusions of his work are reviewed in the summing up of the discussion of his 1850 paper as follows:

– the natural drainage and replenishment of the chalk stratum might be accounted for by observing the alternation of level in various localities and at different seasons

– any large quantity of water abstracted from the chalk stratum, at any given point, causes a depression of level around the point of such abstraction

– in the outcrop districts, any such abstraction of water would interfere with, and diminish the supply of the streams, by which the drainage of the district was regulated

– the depression of the level under London, by pumping from boreholes, had proved, that the demand already exceeded the supply, and that any attempt to draw a large additional quantity for public use, would be attended with disastrous consequences.

He also recognized that, as the Thames was the natural discharge point for the chalk groundwaters under London, as water levels were lowered 'the natural outfall is converted into a source of supply, and the drainage is reversed' (Clutterbuck 1850).

Clutterbuck's ideas were received with considerable scepticism amongst both water engineers and chemists. His original work seems to have been initiated in response to a scheme proposed by Robert Stephenson in 1840 to supply northwest London from a well at Bushey Meads near Watford, which was rejected. Other engineers felt that the chalk could supply the large volume of water which London required and initial results from the Trafalgar Square wells were used to support this view (Homersham 1855). Over the period from December 1847 to December 1858, after yielding daily a considerable and increasing quantity of water, the water level in the wells remained the same and the water was slightly less saline than formerly (Amos 1860). Homersham, in discussion of Clutterbuck (1863), went so far as to state that 'he had no confidence in any statements, with respect to the permanent lowering of the surface of the water in wells sunk in the chalk' and that Clutterbuck 'was not justified in his views.' It was also pointed out by Clark with respect to the infiltration of sea water into the wells under London that 'if they took chalk-water and sea-water, and mixed them together, it was physically impossible to make out of the two, a water corresponding to the chalk-water under London; the theory, therefore, was simply a mistake' (Braithwate 1855).

However, there was also support for Clutterbuck's views. Prestwich (1851) recognized that if abstraction was more rapid than replenishment by rain, and this continued from year to year, the result must be an exhaustion of the groundwater reservoir, a gradual fall in the water level and a constant decrease in the supply. He recognized that by 1850 this was happening in the Chalk beneath London, with a fall in water level of 40–60 feet within the previous 30 years. He considered that this could not be 'made a valid ground of objection' against the use of groundwater generally and concluded: 'Let the demand upon any series of strata be carefully regulated not to exceed the mean annual supply by rain, and then the yield will not fluctuate.' He recognized that supplies from the Chalk, though of considerable value, were limited and insufficient to meet the wants of a city the size of London. However, he felt that problems identified with the Chalk were not applicable to the underlying Lower Greensand and that these sands were extensive enough to contribute a very important proportion of the quantity of water required. This was based on the mistaken view that the Lower Greensand was continuous in the Thames Basin. Napier (1851) also proposed the Lower Greensand, to the south in Surrey, as a source of water supply. His report to the General Board of Health devotes much attention to the merits of the soft water of the Greensand compared to the hard water of the Chalk. Clutterbuck's views were also supported by the Royal Commission on Water Supply which included Prestwich as a member. The Royal Commission was presented with evidence which suggested that immense quantities of water could be obtained from the Chalk to supply London. However, they recognized that the supply must obviously be limited by the amount of rainfall. Moreover, they stated that as water within the aquifer ultimately finds its way by springs into streams at lower elevations any water abstracted from wells will most probably be at the expense of those streams. (Royal Commission on Water Supply 1869a, b).

The debate about the supply of London from the chalk continued throughout the rest of the nineteenth century (see for example, Harrison 1891; Hopkinson 1891) and many of the discussions were extremely acrimonious. The strong feelings of many geologists in their opposition to the schemes of civil engineers are well summarized by John Evans in his Presidential Address to the Geological Society (Evans 1876):

> It will, I think, come within the province of the geologist to point out not only where spring water of good quality is to be obtained, but also what will be the effect of its abstraction upon the districts where it now exists in sufficient abundance to overflow into the streams. It will be for him to show what will be the effect of producing a void below the level at which the drainage of the country naturally escapes; how what are now fertile and even irrigated meadows will be converted into arid wastes; how watercress beds, now of fabulous value, will be brought to the resemblance of newly metalled turnpike roads; how in such a district all existing wells, many of them already some hundreds of feet in depth, will be dried, the mill-streams disappear, and even the canals and navigable rivers become liable to sink and be lost in their beds.

Evans would have been horrified to see that some of his worst fears have since been realized (Fig. 3).

The work involved with London's water supply produced other advances in hydrogeology. Prestwich's small book (Prestwich 1851) contains a map and sections illustrating the relative positions and areas of the principal water-bearing strata around London. In the key to the map, individual deposits are marked according to their permeability (Fig. 4). This map marks the first British geological map to show hydrogeological information. Playfair, in discussion of Clutterbuck (1850), describes analyses of some chalk wells in and around London. From the analyses he was able to show that when outcrop water, which contained carbon-

Fig. 3. The original source of the River Hiz in Hertfordshire showing the now dry spring pool. A willow, which once overhung the pool, now stands forlornly on a slope that once formed its bank.

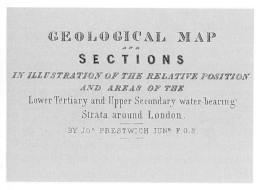

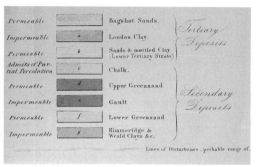

Fig. 4. Title (top) and key (bottom) to the map showing the water-bearing strata of the country around London (Prestwich 1851).

ate of lime in solution, filtered throught the chalk, and came into contact with sodium silicate minerals 'its carbonic acid seized the alkali and formed carbonate of soda'. What he was describing was cation exchange whereby calcium in the groundwater was replaced by sodium from the clay minerals within the chalk, although it was to be another 60 years before this was recognized (Thresh 1912).

It was not only in south-east England that groundwater was abstracted. The Permo-Triassic Sandstones in Central England (Hull 1865) and Merseyside (Cunningham 1847; Stephenson 1850; Roberts 1869) were used for water supply, as was the Magnesian Limestone in the North East. During development of the latter by the Sunderland and South Shields Water Company between 1846 and 1868 potential interference between pumping wells was recognized and an empirical rule was developed that no new station should be within 2 miles of any other (Binnie 1981).

Cunningham (1849) was perhaps the first to recognize the effect urbanization was having on recharge to the sandstone aquifer. He remarks,

The formation of macadamized roads, paved streets, and other impervious media through which no water can penetrate into the strata, but is immediately carried off by sewers and other artificial channels into the sea, must of course diminish, in a corresponding ratio, the quantity of water that would otherwise sink down into their mass.

Stephenson (1850) prepared a report for the Water Committee of Liverpool Town Council on the town water supply. This report uses the terms 'permeability' and 'porosity' in a modern context and provides a correct qualitative description of a cone of depression around a pumping well:

The effect of pumping from a well under such conditions will be to drain the adjacent rock, producing a comparative dryness on all sides, in such a manner as would be represented by an inverted cone; the bottom of the well being the apex of that cone, the sloping sides would represent the inclined surface of the water, flowing towards the well in all directions: and, as the pumping is continued, the sides of the cone will become more and more obtuse, or, in other words, more nearly horizontal, until an inclination is established where the friction of the water, in moving through the pores and fissures of the rock, is in equilibrium with the gravity upon the plane.

Hull (1865) described wells deriving water from the Permo-Triassic sandstones in Liverpool, Manchester, Birmingham and Nottingham. He recognized that faults were not impervious and recommended the line of the fault as the best site for a well as it was certain to draw water from a long distance. He also recognized the excellent quality of the groundwaters and eulogized about the sandstone as a wonderful natural filter:

Receiving as it does on its surface water from various sources, and charged with the impurities of various kinds, it imbibes a portion, allows it to percolate downwards in a slow and gradual descent, every instant extracting some noxious particle, till the liquid is freed from every substrate injurous to human life, and is returned to us limpid as the waters of a brook which gurgles along the rugged bed of a Highland glen.

Post 1871: the work of the Geological Survey and the period to 1900

The Geological Survey may be said to have started as a one-man Department of the Ordnance Survey in 1835 (Wilson 1985). The man was Henry De la Beche and over the next 20 years the organization developed and expanded with De la Beche as its

Director. Early reports emphasized the applied aspects of geology and for example the first report on the geology of Cornwall, Devon and West Somerset contains a 163 page chapter on the economic geology of the region (De la Beche 1839) and includes information on the water problems experienced in the mines. However, even under De la Beche, the Survey made little contribution to the general question of water supply. In the discussion of his paper, Clutterbuck (1850) suggested

> that the geological survey now being carried on by Government, in a remote district of North Wales, where no urgent need existed for early geological information, and where no new works of paramount importance were in progress, or in contemplation, should be at once transferred to the metropolitan districts, with a view to throw light on the real structure, mechanical and chemical, of the deep water-bearing strata, relative to which such conflicting opinions had been advanced.

The second Director, Murchison, who took over in 1855, was the epitome of the amateur geologist – a man of position and means who had travelled extensively and done significant geological research in England and Wales (Wilson 1985). According to Flett (1937), in taking over the work of De la Beche, he had one disqualification. He had little interest in the economic applications of geology which for De la Beche had been of prime importance.

The only survey geologist with an interest in water supply during this period seems to have been Edward Hull who was on the English staff from 1850, prior to Murchison's appointment as Director. According to Whitaker (1888) his survey memoir on the *Geology of the Country around Bolton-le-Moors, Lancashire* is the first to contain well sections and he was also publishing on the Permo-Triassic Sandstones as a source of water supply (Hull 1865).

Whether or not it was Murchison who was holding back the development of groundwater studies within the Survey is a topic for debate but there was certainly an increase in activity after his death. Whitaker's memoir on the geology of the London Basin, published in 1872, a year after Murchison's death, contained 141 pages of well sections (Whitaker 1872) and thereafter all memoirs on southeast England geology had sections on water supply. Wilson (1985) suggests that the application of the science of geology to the study of underground water supplies was 'a field pioneered by the Geological Survey more than a century ago'. However, this is a view which is difficult to sustain and in fact the Survey geologists were rather late on the scene.

When it came, the Survey contribution resulted principally from the work of three men: Joseph Lucas, Charles de Rance and William Whitaker. Although the latter has been given much of the credit and has been designated the 'father of English hydrogeology' it was Lucas who was the innovative member of the trio. Lucas and de Rance were appointed to the Survey as a consequence of the large increase in geological staff obtained by Murchison in 1867–1868. For nine years Lucas mapped the Carboniferous rocks of the West Riding and later an area in northeast Yorkshire. In the West Riding he was much impressed with the large volume of water yielded by the Lower Carboniferous sandstones and, according to his obiturist (Anon. 1926), this probably led to his taking up the study of water supply when he left the Survey in 1876.

His first paper was on the use of horizontal wells in the Chalk and Lower Greensand of Surrey to supply London and was published while he was still on the staff of the Survey (Lucas 1874). These horizontal wells, or more correctly galleries, were to be driven along the strike at the base of the water-bearing formations. The idea was far-fetched and by 1879 Lucas had ceased to advocate such wells (in discussion of Lucas 1879); however, the paper is important for other reasons. In it he first used the term 'hydro-geological survey' and drew the first map which showed 'contours of the upper surface of water in the Chalk'. This map was of an area to the south and east of Croydon in Surrey and the contours were drawn at 10 foot intervals. The paper contained a lot of his own gauging information and must have involved a considerable amount of work. It is difficult to understand how anyone could have published privately such a large volume while supposedly mapping in Yorkshire. Unfortunately he was not able to devote much attention to the editing and two pages of errata correcting 27 separate errors are included.

Lucas resigned from the Geological Survey in 1876 to pursue the profession of water engineer, concentrating his work in the Thames Valley. His work was 'marked by a distinct originality of thought which at once brought him to public notice' and he published extensively over the next five years (Anon. 1926). He read a paper on the Chalk in November 1876, the extensive research for which must have been carried out whilst he was still with the Survey (Lucas 1877a). Indeed he records in this paper that his observations extended over four years, ranging over about 200 square miles on which almost every accessible well had been measured (Fig. 5). Papers on the artesian system of the Thames (Lucas 1877b) the hydrogeology of Middlesex and part of Hertfordshire (Lucas 1878a) and the Lower Greensands of Surrey

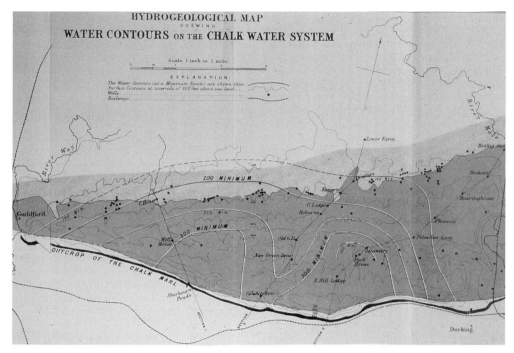

Fig. 5. Contours on the Chalk water table between Guildford and Dorking, Surrey (Lucas 1877a).

and Hampshire (Lucas 1880) followed. The information contained in these papers was abstracted and, together with other data, was used to prepare the first real hydrogeological maps (Gray 1968) in which colour ornament was used to illustrate various features of the hydrogeology of parts of the central Thames Basin (Lucas 1877c, 1878b). The maps were accompanied by an explanatory leaflet (Lucas 1878c).

He defined the new subject of hydrogeology, stating, 'hydrogeology takes up the history of rainwater from the time that it touches the soil, and follows it through the various rocks which it subsequently percolates' (Lucas 1877d). Lucas made a major contribution to two conferences on National Water Supply organized by the Society of Arts in 1878 and 1879 and was awarded a silver medal for his essay in response to a request for the best suggestions, founded upon evidence already published, for dividing England and Wales into watershed districts, for the supply of pure water to the towns and villages in each district (Lucas 1879). It might be imagined that this wealth of published research would lead to even greater achievements. However, after 1881, when some details of his consulting work were published (Lucas, 1881), he largely disappeared from sight and, his obiturist records, that 'although he continued to practice as a water engineer for some years, he ceased, after this,

to make any further publications of general scientific interest' (Anon. 1926).

His colleague Charles de Rance also worked in the north of England and in 1874 was appointed secretary of a Committee appointed by the British Association for the Advancement of Science at its forty-fourth meeting in Belfast for the purpose of investigating the circulation of underground waters in the New Red Sandstone and Permian formations of England and the quantity and character of the water supplied to various towns and districts from these formations. Edward Hull, by now Director in Ireland, was the chairman of this committee and an initial sum of £10 was placed at their disposal to conduct their investigations. The remit of the Committee was later extended to cover all the aquifers in England. A circular was drawn up asking for information and after committee approval nearly 1000 copies distributed. Individual committee members were responsible for different areas of the country and Committee reports principally consist of well records sent in by correspondents. However, in some of the reports, e.g. those for 1877 (British Association 1878) and 1878 (British Association 1879), other hydrogeological information is provided and the latter report contains an appendix on the filtration of sea water through the Triassic Sandstone.

The Committee continued in existence for 20

years and de Rance was secretary for the whole of this time. According to the nineteenth and 20th reports (British Association 1894a and 1894b) as secretary de Rance was to produce a digest of the previous reports giving details grouped in geological formations and counties, to be issued as a separate publication. However, this seems never to have appeared and although this may be because of the 'pressure of official and other duties' as pleaded by de Rance in 1893 (British Association 1894a) it may also be connected with the reason for which he was sacked by the Geological Survey in 1898 – for inefficiency and addiction to drink (Wilson 1985). The twenty-first and final report (British Association 1895) contains a number of general results but perhaps the most valuable suggestion is that well records collected by local societies should be sent annually to the Geological Survey so that they could be collated centrally.

De Rance's main legacy is the first comprehensive overview of the hydrogeology of England and Wales (De Rance 1882). According to the preface of this book, which is 623 pages in length, it is based on his contribution to the National Water Supply Congress convened in 1878 by the Council of the Society of Arts together with lectures on water supply which he gave to the Wigan Mining School in 1876. However, it seems much more likely that the basis is the returns which he received as Secretary of the British Association Committee. The volume describes the character and quantity of the water supplied to every town and urban sanitary authority and by describing the area of the principal geological formations, with the amount of rainfall in each river basin, it provides the data necessary for estimating the water available. Thus the work is the first attempt in the UK to estimate groundwater resources. It contains a hydrogeological map in which the rocks of England and Wales are divided into four categories: impermeable, partially porous, supra-pervious and permeable (Fig. 6). The supra-pervious areas were those where pervious rocks were overlain by clay. The text also contains many of his own observations, for example in the Folkestone district he comments that fissures in the Chalk are rarer and joints less open at depth and he doubted that fissures would occur which might conduct the sea into the workings of the proposed Channel Tunnel for which in 1882 preliminary headings were being constructed (Slater & Barnett 1957).

De Rance also contributed to the National Water Supply Congresses of 1878 and 1879 and to a third Congress held in 1884. His paper to the 1884 Congress contains a covert advertisement for his book when he writes,

the amount of information that has been accumulated is very large, but, investigated by Royal Commissions, inquired into by scientific societies, it is spread over a wide range of literature. It is difficult for any one individual to focus the stores of information already available, still more for him to follow up the numerous lines of investigation these inquiries suggest. (de Rance 1884)

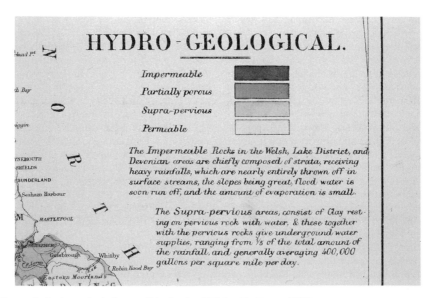

Fig. 6. Key to the hydrogeological map of England and Wales (De Rance 1882).

Fig. 7. William Whitaker in his later years.

He could have quite easily added, – 'But read my book: I have done it all for you!'

Whitaker was the senior colleague of Lucas and de Rance and had been appointed to the Geological Survey in 1857 at the age of 21 (Fig. 7). In contrast to the innovative contribution of Lucas, that of Whitaker might be described as worthy. According to his obituaries (Anon. 1925*a* and 1925*b*) he was well liked and 'made many friends but never an enemy' with a 'genial helpful attitude to his fellow-workers'. He had a passion for collecting records of sections of wells and temporary exposures and it is said that in later life he would willingly accept the presidency of any local society which was prepared to reproduce a presidential address consisting largely of well sections (Bailey 1952). Certainly he afforded considerable assistance to amateur scientific societies and was twice President of the Geologists Association.

He must have started collecting well records early in his career and, although the Geological Survey did not publish them until after Murchison's death, Bailey (1952) records that Whitaker had already in 1866 prevailed upon the Medical Officer of the Privy Council to allow him to produce an appendix of this character in a note on the surface geology of London. His revised and extended edition of the 1872 memoir had a separate volume of 352 pages devoted to lists of well sections,

investigation boreholes and temporary exposures (Whitaker 1889). These serried ranks of well records described by Bailey (1952) as being 'as dull as they are useful' came to dominate the geological memoirs such that in 1899 the first Water Supply Memoir, on Sussex, was produced. Whitaker had by then retired but was still entrusted with the work. This first memoir (Whitaker & Reid 1899) contains mostly well records with little supporting text, in contrast to later publications such as the Water Supply of Kent (Whitaker 1908) where 63 pages are devoted to such matters as shafts and galleries, geology, rainfall, springs, swallow holes and intermittent streams.

Whitaker was also involved, as a consultant, in an early case of groundwater pollution (Whitaker 1886). A brewery well in Brentford west of London was found to be polluted from another well 297 feet away which had been turned into the drainage for the privy belonging to a printing works. The case, Ballard v. Tomlinson, was finally settled in the Court of Appeal, which found that no owner had the right to pollute a source of water supply common to his own and other wells.

Although the science of hydrogeology had advanced rapidly, many old ideas still prevailed in parts of the country. Thus in Liverpool some individuals still found it difficult to accept that rainfall was sufficient to supply the wells. For example 'Mr Robert Bostock, an excellent practical geologist, of Birkenhead, believes that sea water is decomposed by filtration through the rock, and that the water of the sea is the main source of supply' (British Association 1878). However, it was now generally accepted that groundwater was not an infinite resource and that heavy abstraction could lead to reduced water levels and saline intrusion. By the end of the century evidence for a reduction in water levels beneath London was irrefutable.

Conclusions

This paper has followed the development of hydrogeology from the essentially practically-orientated work carried out by William Smith through to the first water supply memoir authored by William Whitaker. Smith provided advice in his role as a consultant but has left no lasting impact on British hydrogeology.

The first individual to make systematic observations of groundwater and apply them to develop theoretical concepts was James Clutterbuck. His three papers, published between 1842 and 1850, laid the foundations for the later work of Joseph Lucas. Although his name is largely unknown to modern hydrogeologists he was the first to plot hydrogeological sections and recognized that it was possible to estimate recharge from fluctuations in water levels. He realized that groundwater abstraction would affect the flow of Chalk streams and predicted that overabstraction beneath London would result in saline intrusion.

His contemporary, Joseph Prestwich, is much better known and eventually became Professor of Geology at Oxford University. His map, published in 1851, is probably the first to include hydrogeological information. Arguments between engineers and scientists became quite acrimonious over the amount of water which the Chalk would yield. However, as pointed out by Bramwell (in discussion of Lucas 1877a),

> although ... there seemed to be among the philosophers a great discord as to what could and what could not be done in the Chalk, the fact was, that while scientific men, had been making the discord, engineers ... had been quietly getting the water and supplying the populations. That was a sufficient answer.

The result of such a policy was reduced yields and saline intrusion, exactly as predicted by Clutterbuck.

Before 1871 the Geological Survey was not at the forefront of developments in hydrogeology, although Edward Hull was involved with water supplies derived from the Permo-Triassic Sandstones. However, the death of Murchison, the Director from 1855 to 1871, saw a blossoming of activity. Joseph Lucas first used the term 'hydrogeological survey' and defined the new science of hydrogeology. He drew the first hydrogeological maps to show water table contours on the Chalk and over a period of six years from 1874 to 1880 produced a wealth of innovative research, only to disappear from sight shortly after taking up practice as a water engineer. His colleague Charles de Rance prepared the first comprehensive overview of the groundwater resources of England and Wales, including the first hydrogeological map dividing the country's rocks on the basis of their permeability.

William Whitaker had a passion for collecting records of wells and temporary exposures and produced the serried ranks of well records which adorned most of the geological memoirs covering southeastern England after 1871. He was the author in 1899, along with Clement Reid, of the first Water Supply Memoir of the Geological Survey covering the county of Sussex.

Individuals such as Clutterbuck, Prestwich, Lucas and Whitaker laid the foundations of the subject of hydrogeology in the nineteenth century. They developed the concepts and the scientific principles on which the sophisticated models of today are based. It is hoped that this paper will bring their names and their contributions to the attention of modern geologists.

References

ABEL, F. A. & ROWNEY, T. H. 1849. Analysis of the water of the artesian wells, Trafalgar Square. *Quarterly Journal of the Chemical Society*, **1**, 97–103.

AMOS, C. E. 1860. On the government waterworks in Trafalgar Square. *Proceedings of the Institution of Civil Engineers*, **19**, 21–52.

ANON. 1925a. William Whitaker (obituary). *Geological Magazine*, **62**, 240.

—— 1925b. William Whitaker (obituary). *Quarterly Journal of the Geological Society*, **81**,61–62.

—— 1926. Mr. Joseph Lucas (obituary). *Nature*, **117**, 730.

BAILEY, E. B. 1952. *Geological Survey of Great Britain*. Thomas Murby & Co., London.

BARROW, G. & WILLS, L. J. 1913. *Records of London Wells*. Memoir of the Geological Survey, HMSO, London.

BINNIE, G. M. 1981. *Early Victorian Water Engineers*. Thomas Telford, London.

BISWAS, A. K. 1970. *History of Hydrology*. North Holland Publishing Co., Amsterdam.

BRAITHWAITE, F. 1855. On the infiltration of salt-water into the springs of wells under London and Liverpool. *Proceedings of the Institution of Civil Engineers*, **24**, 507–523.

BRITISH ASSOCIATION 1878. 3rd report of the Committee for investigating the circulation of the underground waters in the New Red Sandstone and Permian formations of England and the quantity and character of the waters supplied to various towns and districts from these formations. *In*: *Report 47th Meeting, Plymouth, August 1877*, John Murray, London, 56–81.

—— 1879. 4th report of the Committee for investigating the circulation of the underground waters in the Jurassic, New Red Sandstone and Permian formations of England and the quantity and character of the waters supplied to various towns and districts from these formations; with appendix by Mr. Roberts on the filtration of water through Triassic Sst. *In*: *Report of 48th Meeting Dublin, August 1878*, John Murray, London, 382–419.

—— 1894a. The circulation of underground waters. nineteenth Report of the Committee. *In*: *Report of the 63rd Meeting, Nottingham, September 1893*, 463–464.

—— 1894b. The circulation of underground waters. 20th Report of the Committee. *In*: *Report of the 64th Meeting, Oxford, August 1894*, John Murray, London, 283–302.

—— 1895. The circulation of underground waters. 21st Report of the Committee, including as an appendix, 2nd list of works by W. Whitaker. *In*: *Report of the 65th Meeting, Ipswich, September 1895*, John Murray, London, 393–402.

CLUTTERBUCK, J. C. 1842. Observations on the periodical drainage and replenishment of the subterranean reservoir in the Chalk Basin of London.

Proceedings of the Institution of Civil Engineers, **2**, 155–165.

—— 1843. Observations on the periodical drainage and replenishment of the subterraneous reservoir in the Chalk Basin of London – continuation of the paper read at the Institution, May 31st 1842. *Proceedings of the Institution of Civil Engineers*, **3**, 156–165.

—— 1850. On the periodical alternations, and progressive permanent depression, of the Chalk water level under London. *Proceedings of the Institution of Civil Engineers*, **9**, 151–180.

—— 1863. The perennial and flood waters of the Upper Thames. *Proceedings of the Institution of Civil Engineers*, **22**, 336–370.

CUNNINGHAM, MR 1847. On the geological conformation of the neighbourhood of Liverpool, as respects the water supply. *Proceedings of the Literary and Philosophical Society of Liverpool*, **3**, 58–74.

DE LA BECHE, H. T. 1839. *Report on the Geology of Cornwall, Devon and West Somerset*. Longman, London.

DE RANCE, C. E. 1882. *The Water Supply of England and Wales*. Stanford, London.

—— 1884. On a possible increase of underground water supply. *Journal of the Society of Arts*, **33**, 851–854.

DONKIN, J. 1836. Some accounts of borings for water in London and its vicinity. *Transactions of the Institution of Civil Engineers*, **1**, 155–156.

EVANS, J. 1876. The anniversary address of the President. *Quarterly Journal of the Geological Society*, **32**, Proceedings, 53–121.

FLETT, J. S. 1937. *The First Hundred Years of the Geological Survey of Great Britain*. HMSO, London.

GEORGE, T. N. 1976. Charles Lyell: the present is the key to the past. *Philosophical Journal*, **13**, 3–24.

GRAY, D. A. 1968. Hydrogeological maps. *Proceedings of the Geological Society of London Session 1966–67*, no. 1644, 288–290.

HARRISON, J. H. 1891. On the subterranean water in the chalk formation of the Upper Thames, and its relation to the supply of London. *Proceedings of the Institution of Civil Engineers*, **105**, 2–25.

HOMERSHAM, S. C. 1855. The chalk strata considered as a source for the supply of water to the metropolis. *Journal of the Society of Arts*, **3**, 168–182.

HOPKINSON, J. 1891. Water and water supply with special reference to the supply of London from the chalk of Hertfordshire. *Transactions of the Hertfordshire Natural History Society*, **6**, 129–161.

HULL, E. 1865. On the New Red Sandstone as a source of water supply for the central towns of England. *Quarterly Journal of Science*, **2**, 418–429.

KELLAWAY, G. A. 1991. The work of William Smith at Bath (1799–1813). *In*: KELLAWAY, G. A. (ed.) *Hot Springs of Bath, Investigations of the Thermal Waters of the Avon Valley*. Bath City Council, Bath, 25–55.

LATHAM, B. 1904. Croydon bourne flows. *Proceedings and Transactions of the Croydon Natural History Society* (Special Supplement).

LUCAS, J. 1874. *Horizontal Wells, a New Application of Geological Principles to Effect the Solution of the Problem of Supplying London with Pure Water*. Stanford, London.

—— 1877a. The Chalk water system. *Proceedings of the Institution of Civil Engineers*, **47**, 70–167.

—— 1877b. The artesian system of the Thames Basin. *Journal of the Society of Arts*, **25**, 597–619.

—— 1877c. *Hydrogeological Survey. Sheet 1 (South London)*. Stanford, London.

—— 1877d. Hydrogeology: one of the developments of modern practical geology. *Transactions of the Institution of Surveyors*, **9**, 153–184.

—— 1878a. The hydrogeology of Middlesex and part of Hertfordshire, showing the original position of the artesian plane and its present position over the metropolitan area of depression, as lowered by pumping. *Transactions of the Institution of Surveyors*, **10**, 279–316.

—— 1878b. Hydrogeological survey. *Sheet 2 (North London)*. Stanford, London.

—— 1878c. *Hydrogeological Survey, Explanation Accompanying Sheet 1 Second Edition and Sheet 2*. Stanford, London.

—— 1879. Suggestions for dividing England into watershed districts. *Journal of the Society of Arts*, **27**, 715–727.

—— 1880. The hydrogeology of the Lower Greensands of Surrey and Hampshire. *Proceedings of the Institution of Civil Engineers*, **61**, 200–227.

—— 1881. Rural water supply; with especial reference to the objects of the Public Health Water Act, 1878. *Transactions of the Institution of Surveyors*, **13**, 143–178.

LYELL, C. 1830–1833. *Principles of Geology*. 3 vols. Murray, London.

NAPIER, W. 1851. *Suggestions for the Supply of the Metropolis from the Soft Water Springs of the Surrey Sands, Addressed to the General Board of Health*. Smith, Elder & Co., London.

PERRAULT, P. 1674. *De l'origine des fontaines*. (*On the Origin of Springs*, trans. A. La Roque 1967. Hafner Publishing Co., New York).

PHILLIPS, J. 1844. *Memoirs of William Smith, LL.D., Author of the 'Map of the Strata of England and Wales.'* Murray, London.

PRESTWICH, J. 1851. *A Geological Inquiry Respecting the Water-Bearing Strata of the Country around London, with Reference Especially to the Water-Supply of the Metropolis and Including some Remarks on Springs*. John Van Voorst, London.

—— 1872. The anniversary address of the President. *Quarterly Journal of the Geological Society*, **28**, 29–90.

PRICE, M. 1996. *Introducing Groundwater*. 2nd edn. Chapman & Hall, London.

ROBERTS, I. 1869. On the wells and water of Liverpool. *Proceedings of the Liverpool Geological Society*, **10**, 84–97.

ROBINS, F. W. 1946. *The Story of Water Supply*. Oxford University Press, London.

ROBSON, D. A. 1986. *Pioneers of Geology*. Special Publication, Natural History Society of Northumbria, Hancock Museum, Newcastle.

ROYAL COMMISSION ON WATER SUPPLY 1869a. *Report of the Commissioners.* HMSO, London.

—— 1869b. *Appendix to the minutes of Evidence Together with Maps and Plans and an Index.* HMSO, London.

SEAWARD, J. 1836. On procuring supplies of water for cities and towns, by boring. *Transactions of the Institution of Civil Engineers*, **1**, 145–150.

SHEPPARD, T. 1920. *William Smith: His Maps and Memoirs.* Brown & Sons, Hull.

SLATER, H. & BARNETT, C. 1957. *The Channel Tunnel.* Allan Wingate, London.

SMITH, W. 1827. On retaining water in rocks for summer use. *Philosophical Magazine*, **1**, 415–417.

STEPHENSON, R. 1850. *Report of Robert Stephenson, Civil Engineer, on the Supply of Water to the Town of Liverpool*, Liverpool Town Council, Liverpool, UK.

STOW, J. 1603. *A Survey of London, Written in the Year 1598.* 2nd edn. 1994 edition ed. H. Morley. Alan Sutton Publishing, Stroud.

THRESH, J. C. 1912. The alkaline waters of the London Basin. *The Chemical News*, **105**, 25–27 and 37–44.

TOMKINS, M. 1969. Will 1969 be the year of the woe-water? *Hertfordshire Countryside*, **24**, 42–44.

WHITAKER, W. 1872. *The Geology of the London Basin. Part 1 – The Chalk and the Eocene Beds of the Southern and Western tracts.* Memoir of the Geological Survey, vol. 4. Longmans, London.

——. 1886. On a recent legal decision of importance in connection with water supply from wells. *Geological Magazine*, **3**, 111–114.

—— 1888. Chronological list of works referring to underground water, England and Wales, Appendix in 13th Report of the British Association Committee appointed for the purpose of investigating the circulation of underground waters in the permeable formations of England and Wales and the quantity and character of water supplied to various towns and districts from these formations. *In*: *Report of the 57th meeting of the British Association, Manchester, August/September 1887*, John Murray, London, 384–414.

—— 1889. *The Geology of London and Part of the Thames Valley.* Memoir of the Geological Survey, vol. 2, appendices. HMSO, London.

—— 1895. Second chronological list of works referring to underground water, England and Wales. (In British Association 1895, 394–402.)

—— 1908. *The Water Supply of Kent with Records of Sinkings and Borings.* Memoir of the Geological Survey. HMSO, London.

—— & REID, C. 1899. *The Water Supply of Sussex from Underground Sources.* Memoir of the Geological Survey. HMSO, London.

WILSON, H. E. 1985. *Down to Earth: One Hundred and Fifty Years of the British Geological Survey.* Scottish Academic Press, Edinburgh.

WOODWARD, H. B. 1922. *The Geology of the London District.* 2nd edn, revised by C. E. N. Bromehead & C. P. Chatwin. Memoir of the Geological Survey. HMSO, London.

YEATS, T. 1826. Section of a well sunk at Streatham Common, in the county of Surrey. In a letter addressed to - Brown Esq. secretary to the Westminster Fire-Office; and by him communicated to the Geological Society. *Transactions of the Geological Society*, ser. 2, **2**, 135–136.

Part 3.   The legacy of Lyell

Lyell would have approved of plate tectonics. Indeed, by implication, he almost anticipated it. Recognizing the connection between climate and the global distribution of land and sea, as Fleming points out (pp. 164–165 of this volume), Lyell explained climatic change in the geological past in terms of changes in the geographical distribution of land and sea – continental drift in disguise. Furthermore, since the paradigm of plate tectonics is founded upon an understanding of processes active at present, whose rates can be measured, then applying them to the past, it is entirely in accord with Lyell's principle that geological 'phenomena should be explained only in terms of causal agencies that are observably effective, both in kind and degree' (Rudwick p. 4 of this volume). In a new synthesis of the Lower Palaeozoic tectonic evolution of the northern Appalachians and British Caledonides, Cees van Staal and his colleagues draw upon the modern analogue of the southeast Asia region. Intriguingly, they are able to turn the comparisons around to predict that the modern tectonic activity of southeast Asia will eventually end up in structures looking much like those of the Appalachian–Caledonian orogenic belt. This carefully argued work, paying close attention to field observations, is in the true tradition of Lyell and is firmly set within his principles.

The legacy of Lyell is equally evident in Andrew Scott's paper in which he has literally followed in Lyell's footsteps across to North America. Scott discusses on-going research into the occurrence of reptiles preserved inside the upright trunks of Upper Carboniferous trees at Joggins, Nova Scotia, first discovered by Lyell and Dawson in 1852. Scott examines Lyell's ideas on the formation of coal and brings them up to date using modern analogues from southeast Asia. However, he sounds a note of caution in the use of modern analogues since it has now been established that Carboniferous coal-forming plants had life habits and growth mechanisms that were radically different from those of modern peats. The recognition by Lyell of charcoal in coal deposits resulted only many years later in an understanding of its origin from ancient forest fires. Scott explains how current research on ancient charcoal is providing new insights into climatic change, atmospheric composition, and processes of erosion and sedimentation.

Lyell visited Nova Scotia twice during his visits to North America. The influence of Lyell on Nova Scotian geology and the use that Lyell made of his fieldwork in his publications is examined by John Calder. Calder provides us with a succinct account of the Carboniferous evolution of Nova Scotia and places Lyell's observations and conclusions in context. Clearly Nova Scotia provides an important link between Europe and North America. Lyell was impressed by the geology of Nova Scotia and his enthusiasm was reciprocated by the local geologists. Particularly important was the influence that Lyell had on Sir William Dawson who was to become one of Nova Scotia's most important and influential geologists. Lyell used many of his discoveries in Nova Scotia to illustrate subsequent editions of his volumes including his discovery of tetrapods within sandstone-filled lycophyte trunks at Joggins.

Unravelling the stratigraphic record was always a fundamental objective of Lyell. His desire to understand rates of sedimentation, rises and fall of land and sea level changes as seen both in ancient rocks and as active processes in the modern world is evident from his field observations as recorded in his diaries, travel, and geological books and papers. Our current understanding of the stratigraphic record has been revolutionized in the past twenty years with the advent of sequence stratigraphy as shown by Chris Wilson. The paper by Wilson looks at how the sequence stratigraphic model evolved and its utility. Wilson argues that the application of sequence stratigraphy has provided us with a revolutionary new way of interpreting the stratigraphical record. However, Wilson adds caution in that a new global stratigraphy, allowing global correlations based upon eustatic signals, is not yet a reality as their recognition can not yet be detected unequivocally in the record. Clearly again Lyell takes us back to looking at the rock record but as he also maintained, we must also consider the processes involved that were responsible for what we observe.

This point is amplified by Chris Talbot in a review of salt tectonics in the Zagros mountains of Iran. Talbot notes the parallels between the controversy in Lyell's day about whether ice could flow uphill and the more recent controversy about whether salt can flow across the surface of the land or, indeed, the sea bed. Talbot takes the argument a stage further in demonstrating how field measurements of the rates of salt flow and observations of the domal shapes of extrusions are used to quantify the dynamics of the process. He hints at the possible analogy between salt extrusion and the extrusion and gravity spreading of metamorphic core complexes. This richly illustrated, authoritative account of salt tectonics in a classic area is yet further evidence of Lyell's influence on the approach to modern research.

Lyell was much concerned with the most tangible expressions of current geological activity in the form of volcanic eruptions and earthquakes. He spent a major part of his life in the field studying volcanoes in Italy and elsewhere, as Wilson describes on pp. 24–31 of this volume, and he was well aware of their destructive power. Mount Etna was of particular significance, so it is an essential part of Lyell's legacy that volcanic activity on Etna should be closely monitored, both to give greater quantitative understanding of the process of eruption and to give predictive power to mitigate the risks from future eruptions. Hazel Rymer is a member of an international, multidisciplinary team of scientists who have set up a huge array of instrumentation to monitor every waking moment of this volcanic giant. Their paper gives a historical account of volcano monitoring and describes the modern techniques being utilized on Etna, which are advancing all the time. This is especially so with satellite measurements, including observations of ground deformation using GPS and SAR, which they confidently expect to revolutionize monitoring methods within the next decade.

Bruce Bolt brings a wealth of experience to review the advancement of seismology since Lyell's day. Arguably the most significant advance was the construction of the Milne-Shaw seismograph and its use in a global network of observatories set up at the end of the nineteenth century. This duly advanced to the Worldwide Standardized Seismograph Network in the 1950s which has now evolved into the Global Digital Network of seismograph stations. Bolt traces the history of seismology from the valuable descriptive accounts of earthquakes in Lyell's time to the present day analysis of seismic waveforms which gives detailed information about both the nature of the earthquake itself and the nature of the Earth's interior along the pathways of the seismic waves. Seismology has become one of the most powerfol tools for exploring and imaging the Earth's interior, well beyond anything that Lyell could have dreamed of, and continues to be the main means of monitoring earthquakes, in efforts to mitigate their risks. Earthquake prediction, as Bolt explains, remains elusive despite huge research efforts in recent years. The current view is that it is likely to remain so because of the frictional nature of the earthquake mechanism and the heterogeneity of the Earth's crust. A better strategy for risk mitigation is to concentrate on reducing the vulnerability of the population at risk by means of planning, education and the construction or upgrading of buildings and other structures to appropriate levels of earthquake resistance.

Lyell was intensely interested in the antiquity of Man, as Cohen (pp. 83–93) and other authors in Part 1 of this book have made clear. Lyell was also intensely interested in the natural environment since this is the laboratory in which to record and measure geological phenomena active at present which are the basis for understanding the geological past. Lyell was well aware of the power of nature over Man and of human dependence on the Earth for its resources, but there was little regard in Lyell's day for the need to protect the Earth from human activities. Nowadays it is a major issue and Sir John Knill completes this book with a telling essay on Man and the modern environment to make the point that, more than ever, we need to apply Lyell's principles. It is now imperative to understand and quantify the mechanisms of global climate change in the geological past, especially over the last 500 000 years, in order to predict future climate change including the anthropogenic contribution. Equally, geological processes on all scales must be better understood in order to build strategies for the environmental sustainability of the Earth. There is thus a critical role for geologists in modern society. The old textbook shorthand for Lyell's principles that 'the present is the key to the past' has to be turned around into the title of this book: the past is the key to the present – and, indeed, to the future.

Derek J. Blundell
Andrew C. Scott

The Cambrian–Silurian tectonic evolution of the northern Appalachians and British Caledonides: history of a complex, west and southwest Pacific-type segment of Iapetus

C. R. VAN STAAL[1], J. F. DEWEY[2] , C. MAC NIOCAILL[2] & W. S. MCKERROW[2]

[1] Geological Survey of Canada, 601 Booth Street, Ottawa, Ontario, K1A OE8 Canada

[2] Department of Earth Sciences, Oxford University, Parks Road, Oxford OX1 3PR, UK

Abstract: This paper presents new ideas on the Early Palaeozoic geography and tectonic history of the Iapetus Ocean involved in the formation of the northern Appalachian–British Caledonide Orogen. Based on an extensive compilation of data along the length of the orogen, particularly using well-preserved relationships in Newfoundland as a template, we show that this orogen may have experienced a very complicated tectonic evolution that resembles parts of the present west and southwest Pacific Ocean in its tectonic complexities. Closure of the west and southwest Pacific Ocean by forward modelling of the oblique collision between Australia and Asia shows that transpressional flattening and non-coaxial strain during terminal collision may impose a deceptively simple linearity and zonation to the resultant orogen and, hence, may produce a linear orogen like the Appalachian–Caledonian Belt. Oceanic elements may preserve along-strike coherency for up to several thousands of kilometres, but excision and strike-slip duplication, as a result of oblique convergence and terminal collisional processes, is expected to obscure elucidation of the intricacies of their accretion and collisional processes. Applying these lessons to the northern Appalachian–Caledonian belt, we rely principally on critical relationships preserved in different parts of the orogen to constrain tectonic models of kinematically-related rock assemblages.

The rift–drift transition, and opening of the Iapetus Ocean took place between c. 590–550 Ma. Opening of Iapetus was temporally and spatially related to final closure of the Brazilide Ocean and amalgamation of Gondwanaland. During the Early Ordovician, the Laurentian margin experienced obduction of young, supra-subduction-zone oceanic lithosphere along the length of the northern Appalachian–British Caledonian Belt. Remnants of this lithosphere are best preserved in western Newfoundland and are referred to as the Baie Verte Oceanic Tract. Convergence between Laurentia and the Baie Verte Oceanic Tract was probably dextrally oblique. Slab break-off and a subsequent subduction polarity reversal produced a continental magmatic arc, the Notre Dame Arc, on the edge of the composite Laurentian margin. The Notre Dame Arc was mainly active during the late Tremadoc–Caradoc interval and was flanked by a southeast- or south-facing accretionary complex, the Annieopsquotch Accretionary Tract. Southerly drift of Laurentia to intermediate latitudes of c. 20–25°S was associated with the compressive (Andean) nature of the arc and the accompanying backthrusting of the already-accreted Baie Verte Oceanic Tract further onto the Laurentian foreland. Equivalents of the Notre Dame Arc and its forearc elements in the British Isles have been preserved as independent slices in the Midland Valley and possibly the Northern Belt of the Southern Uplands.

During the late Tremadoc (c. 485 Ma), the passive margin on the eastern side of Iapetus also experienced obduction of primitive oceanic arc lithosphere. This arc is referred to as the Penobscot Arc. The eastern passive margin was built upon a Gondwanan fragment (Ganderia) that rifted off Amazonia during the Early Ordovician and probably travelled together with the Avalonian terranes as one microcontinent. The departure of Ganderia and Avalonia from Gondwana opened the Rheic Ocean. Equivalents of the Penobscot Arc may be preserved in New Brunswick and Maine, Leinster in eastern Ireland, and Anglesey in Wales. An arc-polarity reversal along the Ganderian margin after the soft Penobscot collision produced a new arc: the west-facing Popelogan–Victoria Arc, which probably formed a continuous arc system with the Bronson Hill Arc in New England. The Popelogan–Victoria Arc transgressed from a continental to an oceanic substrate from southern to northeastern Newfoundland. Rapid roll back rifted the Popelogan–Victoria Arc away from Ganderia during the late Arenig (c. 473 Ma) and opened a wide back-arc basin; the Tetagouche–Exploits back-arc basin. The Popelogan–Victoria Arc was accreted sinistrally oblique to the Notre Dame Arc and, by implication, Laurentia during the Late Ordovician. After accretion, the northwestward-dipping subduction zone stepped eastwards into the Tetagouche–Exploits back-arc basin. Equivalents of the Popelogan–Victoria Arc in the British Isles may be preserved as small remnants in the Longford Down Inlier in Ireland. The Longford

VAN STAAL, C. R., DEWEY, J. F., MAC NIOCAILL, C. & MCKERROW, W. S. 1998. The Cambrian–Silurian tectonic evolution of the northern Appalachians and British Caledonides: history of a complex, west and southwest Pacific-type segment of Iapetus. In: BLUNDELL, D. J. & SCOTT, A. C. (eds) Lyell: the Past is the Key to the Present. Geological Society, London, Special Publications, **143**, 199–242.

Down Arc is not preserved in Scotland, although its presence has been inferred there on the tenuous basis of arc detritus. The suture between the Notre Dame Arc and the Popelogan–Victoria–Longford Down Arc system is the Red Indian Line in the Northern Appalachians, but in the British Isles the position is not clear. The fault-bounded Grangegeeth Arc terrane in eastern Ireland, immediately to the south of the Longford Down inlier, may be a displaced piece of the Popelogan–Victoria–Longford Down Arc system. Diachronous closure of the Tetagouche–Exploits basin during the Ashgill to the Wenlock finally caused the collision between Ganderia/Avalonia and Laurentia, whereas the Lake District Arc is related to an earlier closure of the Tornquist Sea between Baltica and Avalonia. After arrival of Avalonia at the Laurentian margin, continuous, dextral oblique convergence between Gondwana and Laurentia was accommodated by another northwest-dipping subduction zone, this time in the Rheic Ocean. The Acadian orogeny in both North America and the British Isles occurred in the Early to Mid-Devonian and is probably related to the collision of Gondwana and/or peri-Gondwanan elements (Meguma, Armorica etc.) with the northern continents.

Many Palaeozoic and older orogens, such as the Urals and the Appalachian–Caledonide belt (inset of Figs 1–3), although narrowing and widening and possessing modest, very open oroclines, are remarkably linear or very gently arcuate over thousands of kilometres. Within these orogens, it has been normal practice to draw zones and subzones characterized by particular tectonic elements and styles in semi-linear fashion along great strike-lengths (e.g. Dewey 1969, H. Williams 1978); i.e. we may have tended to 'impose' an along-strike correlative unity and continuity of elements and zones. This is in striking contrast to many Mesozoic orogens such as the Tethysides with their great complexity of internal loops, oroclines and syntaxes. It is, also, in profound contrast to modern convergent plate boundary systems such as the west and southwest Pacific (Fig. 4) where very great geometric and kinematic complexity is accompanied by rapid temporal and spatial tectonic change. From the typical array of tectonic indicators of plate boundary zones such as rifted margins, ophiolites, blueschists and calc-alkaline volcanism, we are, nevertheless, confident that these older, now intracontinental, orogens were formed by plate tectonic processes (Bird & Dewey 1970; Hamilton 1970). Hence, we must ask the question: is substantial along-strike continuity a reality or is it a consequence of a linearity imposed by terminal continental collision following an earlier long and complicated continental margin and oceanic tectonic history?

This problem is particularly pertinent to the Appalachian–Caledonide Orogen, because faunal provinciality, palaeomagnetic data and structural studies indicate that it consists of along-strike segments that experienced different tectonic histories. This is largely the result of the contemporaneous convergence of several continents and microcontinents, Baltica, Avalonia and Carolinia, with Laurentia. The various oceanic elements preserved in the Northern Appalachian–British Caledonide segment (Figs 1–3) have few, if any, comparable equivalents in the southern Appalachians. Some related oceanic element(s) may be involved in the formation of the British and Scandinavian Caledonides, but the collisonal histories of these two segments are markedly different. Because several tectonic events took place temporally so closely in each of the segments, these differences have often not been fully appreciated and there has been a tendency to relate these tectonic events to one unifying kinematic process. For example, evidence of tectonic loading and orogenesis of the Laurentian margin during the Early to Mid-Ordovician (Taconic Orogeny) can be found from Spitsbergen in the Arctic through the British Caledonides (Fig. 3) and Northern Appalachians (Figs 1 and 2) into the Southern Appalachians in the United States, with the notable exception of East Greenland. The subsequent kinematic correlation along the length of the Appalachian–Caledonide Orogen led to a recent controversy whether the Taconic Orogeny of the northern Appalachians resulted from collision between Laurentia and the South American margin of Gondwana (Dalziel *et al.* 1996) or from an arc-continent collision. The former model is incompatible with the geological relationships preserved in the northern Appalachians and the British Caledonides because it is difficult to see how the Gondwanan margin could have bypassed the independent systems of oceanic arcs in the Iapetus Ocean that were converging and/or colliding with the Laurentian margin during the Ordovician (e.g. Mac Niocaill *et al.* 1997). Moreover, the faunal distinctions between Laurentia and Gondwana are not consistent with any close convergence before the Late Ordovician (Cocks & Fortey 1990)

In this paper, we present new ideas on the tectonic evolution of the northern Appalachian–British Caledonide Belt and their implications for the history of Iapetus. The rapidly-growing geological database for this segment of the

Appalachian–Caledonide Orogen, particularly the large number of recent, high quality, U–Pb age dates and isotope tracer and provenance studies combined with detailed field studies, has indicated that its tectonic history was much more complex than previously thought and resembles parts of the modern west and southwest Pacific Ocean. Despite identification of the internal complexities, we will argue that the northern Appalachian–British Caledonide belt largely represents a tectonic/kinematic entity, i.e. all parts of this segment of the orogen experienced more or less the same major tectonic–kinematic events, although some tectonic elements appear to be confined to restricted parts of the orogen. For example, the Mid- to Late Ordovician arc volcanics in the English Lake District, Wales and south-east Ireland have no obvious equivalents further west in the northern Appalachians.

Tectonic reconstructions are hindered by structural excision, particularly during large-scale orogen-parallel strike-slip motion. For example, some Irish and Scottish accretionary terranes may have been moved for large distances along the eastern margin of Laurentia. Elsewhere, post-Ordovician cover sequences may mask important Early Palaeozoic structures. These problems can be overcome only through an overview of the whole orogen, so that segments with missing or additional components can be recognized. We use the constraints imposed on different parts of the orogen to model the overall tectonic processes responsible for some regions where local evidence is lacking or poorly preserved. The focus of this paper is on tectonic processes active during the Cambrian and Ordovician, and their implications for understanding the destruction of the Iapetus Ocean. We will reaffirm earlier notions that the Ordovician orogenesis in the northern Applachians and British Caledonides (Taconic and Grampian orogenies, respectively) cannot be related exclusively to interaction of one single volcanic/magmatic arc with the Laurentian margin during the Ordovician. Palaeomagnetic and geological data (H. Williams *et al.* 1988; Colman-Sadd *et al.* 1992*a*; van der Pluijm *et al.* 1995; Mac Niocaill *et al.* 1997), orogenesis along the Avalonian margin (van Staal 1987, 1994; Colman-Sadd *et al.* 1992*b*) and significant inconsistencies between timing of obduction of ophiolites, Mid-Ordovician Taconic deformation and metamorphism, and age of classic Taconic arc magmatism (e.g. Tucker & Robinson 1990; Karabinos *et al.* 1993; Pinet & Tremblay 1995) make such a model untenable.

In this paper, we follow Fortey *et al.* (1995) in the coincidence of the base of the Caradoc series with the base of the *N. gracilis* zone (*c.* 458 Ma; Tucker & McKerrow 1995 and references therein);

the Llandeilo series is reduced to a stage (Llandeilian) of the Llanvirn series. The expanded Llanvirn now ranges from 470 to 458 Ma. The Tremadoc (the base of the Ordovician) starts at *c.* 495 Ma and ends at *c.* 480 Ma (Landing *et al.* 1997). The Tremadoc and Arenig together define the Early Ordovician (495–470 Ma). The composition and tectonic setting of volcanic rocks is indicated by their commonly-used abbreviations. MORB stands for mid ocean ridge basalt; the prefix N, T or E refers to normal, transitional or enriched, respectively. OFB and OIB refer to ocean floor and ocean island basalts. IAT and CAB refer to island arc tholeiite and calc-alkaline basalt.

The west and southwest Pacific and a future Australia–Asia sinistral oblique collision

We have constructed an extremely simplified version (Fig. 5a) of the tectonic map of the west and southwest Pacific (Fig. 4) and allowed the present relative motions (De Mets *et al.* 1990) among the Asian, Pacific and Indian/Australian Plates and smaller plates to run to 'completion', that is the collision of Australia with the Asia continent along a broad zone of sinistral oblique convergence (Fig. 5b). The model assumes a 55 million year constancy of relative plate motions. We recognize, of course, that the Pacific has been a long-lived ocean since the late Proterozoic and may not fully close in the short-lived way that Iapetus opened and closed in *c.* 160 Ma. This exercize was carried out mainly to test whether the superficial simple zonation and linearity of the Appalachian–Caledonide Orogen could be a result of terminal collisional processes. In making this prognostic reconstruction, we have had to make a series of 'iterative tectonic decisions', especially in the siting of transform faults and how volcanic arcs are transported and rotated. Others might have made different tectonic decisions but the final collided result would be unlikely to be fundamentally different from the one shown in Fig. 5b in its general form of sliced linear component terranes that are exotic with respect to the cratons and to each other. We have run the model forward four times, each time making different iterative decisions; the results differ slightly in details but not in essentials. There are four basic conclusions to be drawn from this exercise. First, plate boundary zones of immense complexity may be swept up and transpressionally flattened, sheared and rotated into semi-linear orogenic zones resulting in a pseudo-simplicity that conceals an earlier complicated history and geometry. Secondly, contemporaneous volcanic arcs with quite different original palaeogeographic position and orientation may give an illusion of along-strike continuity. Thirdly, fragments of a once continuous tectonic element, like the Ontong-Java/Caroline-Truk Plateau may be scattered throughout thousands of kilometres of the final collisional system. Fourthly, the interval from the present to the terminal Australia–Asia collision is only *c.* 45 Ma, about the same interval as between the first interaction of oceanic elements with the Laurentian and Avalonian margins in the Early Ordovician and Late Silurian terminal closure of Iapetus. The western Pacific would take only

c. 100 Ma to close at present rates of relative plate motion. Given the final collisional result (Fig. 5b), we are faced with an intractable facet of inverse science, i.e. we can forward model with 'impunity' but backward modelling is a near impossibility given the multiple pathways that may lead to a final result. Perhaps the greatest real continuity exists along the edges of an orogenic belt where the results of collision between the old rifted margins and early oceanic elements (e.g. arcs) have been preserved. The further one goes into the interior of an orogen with progressively more oceanic

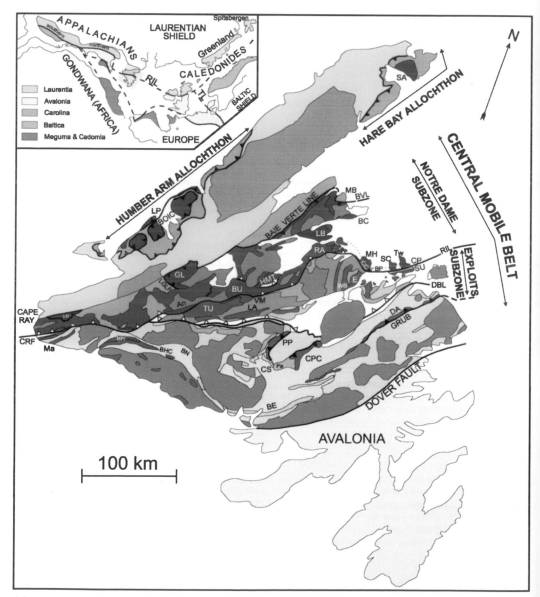

(a)

elements, the more one faces the destructive complexity of plate tectonics. In orogenic interiors, two particular problems raise further difficulties, namely significant 'terraning' and bulk rotation of whole arcs and/or other tectonic elements.

Quite apart from the flattening and smearing imposed by oblique terminal collision, the earlier tectonic collages are beset by a number of substantial complexities and subtleties that are inherent in plate kinematics but which are unlikely to be preserved in the final collisional product. When we analyse old orogens, we try to apply the broad tenets and geological corollaries of plate tectonics but the difficulties are many. First, very rapid spatial changes in tectonics may occur along a single plate boundary depending upon the way in which it crosses rotational latitudes. Secondly, the enormous variety of triple junctions (McKenzie & Morgan 1969) and their evolution can yield fast-changing tectonic environments that allow the rapid superposition of quite different strain fields and consequent polyphase deformation and rapid igneous, metamorphic and stratigraphic changes (Dewey 1975). Third, in a mosaic consisting of three or more plates, slip vectors across plate boundary zones must change progressively so that rates and directions of plate motion change and may eventually cause a change in plate boundary type (Dewey 1975). One especially important example is the nucleation of intraoceanic arcs on

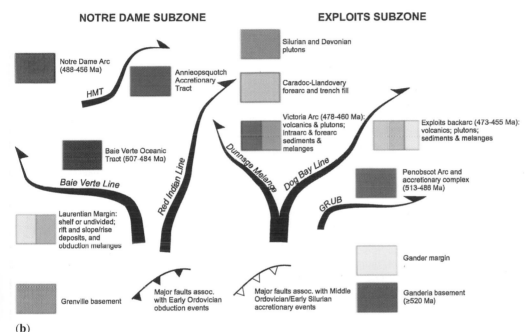

(b)

Fig. 1. Simplified geological map (**a**) of Newfoundland after Colman-Sadd *et al.* (1990) with the various tectonic elements discussed in this paper. The legend (**b**) represents a summary of the structural relationships between the various tectonic elements. This legend also applies to Figs 2 and 3, where the Newfoundland-defined elements are replaced by their equivalents in New England and the British Isles, respectively. The white areas in the Central Mobile Belt represent Silurian or younger volcanic or sedimentary rocks. An = Annieopsquotch ophiolite; BC = Betts Cove Ophiolite Complex; BE = Baie d'Espoir Group; BH = Baggs Hill Granite; BHC = Blue Hills of Couteau Ophiolite Complex; BOIC = Bay of Island Ophiolite Complex; BN = Bay du Nord Group; BP = Burlington Pluton; BU = Buchans Group; CC = Cotrell's Cove; CM = Carmanville Melange; CP = Chanceport Group; CPC = Coy Pond Complex; CS = Cold Spring Pond Formation; CRF = Cape Ray Fault; DA = Davidsville Group; DBL = Dog Bay Line; E = Exploits Group; GL = Grand Lake Complex; GRUB = Gander River Ultrabasic Belt; HMT = Hungry Mountain Thrust; LA = Lake Ambrose Volcanic Belt; LB = Lushs Bight Group; LGLF = Little Grand Lake Fault; LP = Little Port Complex; LR = Long Range Ophiolite Complex; Ma=Margaree Complex; MB = Mings Bight; MH = Moreton's Harbour Group; Pa = Partridgeberry Hill Granite; PP = Pipestone Pond Ophiolite Complex; RA = Robert's Arm; RIL = Red Indian Line; SA = St Anthony Ophiolite Complex; SC = Sleepy Cove Group; SU = Summerford Group & Dunnage Mélange; TU = Tulks Hill Volcanic Belt; Tw = Twillingate Trondhjemite; VM = Victoria Mine Volcanic Belt; WB = Wild Bight Group. Inset shows the Appalachian-Caledonian Belt with the approximate distribution of the various (micro)continents involved in this orogen. TL is Tornquist Line and represent the suture between Avalonia and Baltica.

oceanic transform/fracture zones (Casey & Dewey 1984). Fourth, the relative motion of oceanic arcs may be wholly unrelated to motion among the larger bounding plates. For example, in the western Mediterranean, (Dewey *et al.* 1989) arcs roll and spread radially, with substantial longitudinal extension, driven by subduction roll-back (Dewey 1980) such that they invade other oceanic tracts leaving young back arc basins in their wake to die only when they collide with other arcs and rifted margins. Thus, such arcs tend to fill and obliterate older oceans and generate young rear-arc oceanic lithosphere and develop complex looping and oroclinal shapes. Fifth, major plate reorganization may result from large-scale continental collision whereby new plate boundaries are formed and major changes occur in the direction and rates of relative plate motion. Sixth, we point to the great geometric and kinematic complexity of present west and southwest Pacific tectonics (Fig. 4) that are hardly decipherable today let alone in their likely condition 45 Ma hence. Our principal conclusion is that the inherent complexities of plate tectonics and the strains imposed by terminal collision make difficult the unique elucidation of the tectonic history of orogenic belts, such as the Appalachian/Caledonian Orogen, that have resulted from the progressive closure of large oceans.

This is not an exclusive subscription to pessimism for understanding pre-Mesozoic tectonics. The original rifted margins of a closed oceanic tract may well preserve an along-strike continuity relating to the timing and geometry of rifting; oceanic elements that collide early with these rifted margins may provide continuity for up to several thousand kilometres. Rather, it is a warning that we must not expect to unravel, uniquely, the tectonic history of an orogen, especially its pre-terminal

collisional history. We do our best, but solutions are exceedingly unlikely to be unique. However, in spite of these *caveats* in attempting pre-Mesozoic palaeotectonic syntheses, areas such as the west and southwest Pacific (Fig. 4) give us deep insights into the kinematics and histories of older orogenic belts and into the wide variety of tectonic environments and their kinematic relationships that permit portions of those older belts to be explained rationally in terms of plate boundary zone tectonics.

We now, briefly, outline ten features of the tectonics of the southwest Pacific that, we believe, may be useful as analogues in explaining various features of, and providing a backdrop to, the Early Palaeozoic evolution of the northern Appalachians and the British Caledonides. First, passive, rifted margins, that form the edges of orogens and are destroyed commonly by collision with arcs, may be of several types and ages in relation to the progressive closure history of oceans. The northern edge of Australia is a Jurassic rifted margin in collision with the Timor/Banda/New Guinea arc complex, beginning in the Miocene, diachronous from east to west, and still progressing westwards. This might be termed a grand-scale collision, leading to subduction polarity flip (Flores/Wetar Thrust and New Guinea/North Solomons Trench) and the gross destruction of a rifted margin on a massive scale, similar to the Early Ordovician destruction of the Laurentian margin (Dewey & Shackleton 1984; Dewey & Ryan 1990). In contrast, the Tasman Sea and the Coral Sea opened to form the eastern rifted margin of Australia during the Late Cretaceous to early Tertiary; its destruction during the next 45 Ma will probably be by collision with the Lord Howe Rise and Vanuatu arc, respectively (Fig. 5b). In even greater contrast are the rifted margins of the

---

Fig. 2. Simplified geological map of maritime Canada and New England with the continuation of the Newfoundland tectonic elements displayed in Fig. 1. Geology largely based on tectonic lithofacies map of H. Williams (1978) and plate 2 of Rankin *et al.* (1990) with tectonic additions by the authors. An = Annidale Belt; AW = Ascot Weedon Volcanic Belt; At = Attean Pluton; B = Brookville Terrane; BHA= Bronson Hill Arc; BMC = Boil Mountain Ophiolite Complex; BdO = Bras d'Or Terrane; BRM = Belledune River Melange; C = Caledonia Terrane; CL = Chain Lakes Massif; CM= Caucomgomoc Melange; F = Fournier Group; FT = Fredericton Trough; M = Mira Terrane; HM = Hurricane Mountain Mélange; KC = Kingston Complex; Mu = Munsungun basalts; Ma = Massabesic Gneiss; N = Nashoba Terrane; P = Pelham Dome; Po = Popelogan Inlier; RHB = Rowe-Hawley Belt; Ro = Rockabema Diorite; S = Shelburn Falls Dome; SQOB=Southern Quebec Ophiolite Belt;T = Tetagouche Group; U= Upsalquitch Gabbro; W = Winterville Basalts; WL = Weeksburo Lunksoos Inlier. (1) Continuation of Baie Verte Line; (2) Approximate trace of Red Indian Line (the second suture immediately to the south of the Winterville and Munsungun Basalts has been added to emphasize our interpretation that these basalts are remnants of seamounts. Analogous to the Summerford Group in Newfoundland these basalts are thought to have accreted to the Popelogan Arc before accretion to Laurentia); (3) Inferred continuation of Dog Bay Line and/or associated major structures (this suture is poorly defined because the fault system associated with collision between Ganderia and Laurentia in Maritime Canada and New England, during the Early Silurian, probably involved more than one collisional/ accretionary event due to the piecemeal arrival of small continental blocks with Ganderian basement; hence other important sutures probably occur southeast of line 3). Inset same as Fig. 1.

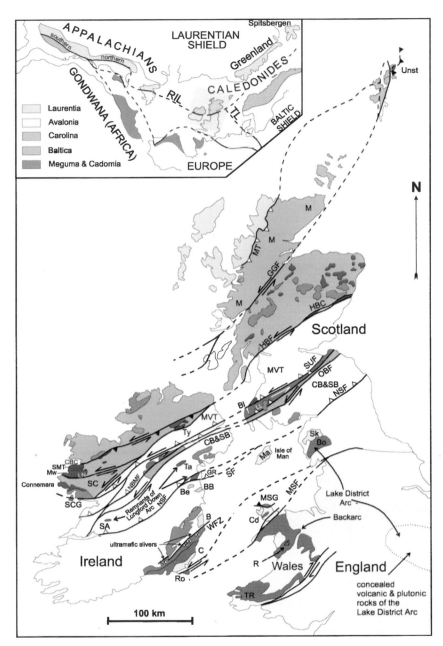

Fig. 3. Simplified geological map of the British Isles with the continuation of the Newfoundland tectonic elements displayed in Fig. 1. Map patterns are given by the legend in Fig. 1b. B = Bray Group; BB = Dalbriggan, Be = Bellewstown; BL = Ballantrae Ophiolite; Bo = Borrowdale; C = Cahore Group; CB = Central Belts; CBC = Clew Bay Complex; Cd = Coedana Complex; CT = Cullenstown Formation; GGF = Great Glen Fault; GR = Grangegeeth Terrane; HBC = Highland Border Complex; HBF = Highland Boundary Fault; LL = Leadhills Line; M = Moine; Ma = Manx Group; MSF = Menai Strait Fault; MSG = Monian Super Group; MT = Moine Thrust; MVT = Midland Valley Terrane; MW = Mweelrea Ignimbrites; NBMF = Northern Belt Median Fault; NSF = Navan-Silvermines Fault; OBF = Orlock Bridge Fault; R = Rhobell Volcanic Complex; RDG = Ribband and Dungannon Groups; Ro = Rosslare Complex; SA = Slieve Aughty; SB = Southern Belt; SC = South Connemara Terrane; SF= Slane Fault; SCG = South Connemara Group; Sk = Skiddaw Group; SMT = South Mayo Trough; SUF = Southern Uplands Fault; Ta = Tattinlieve; Tr = Treffgarne; Ty = Tyrone Ophiolite Complex; WFZ = Wicklow Fault Zone. Inset same as Fig. 1.

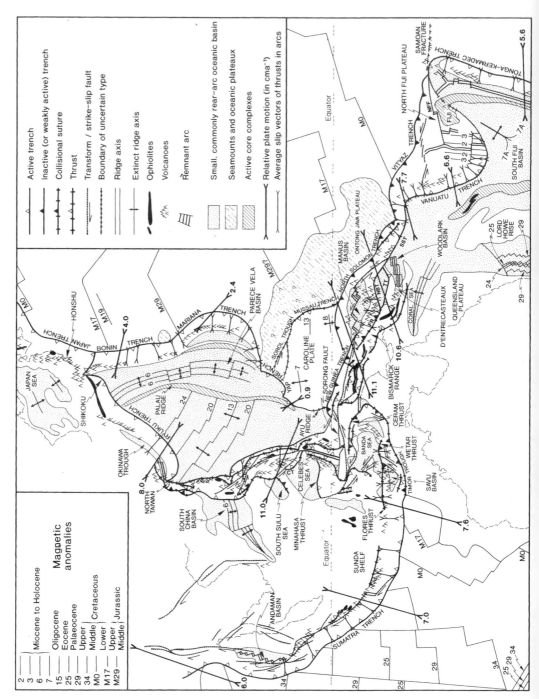

Fig. 4. Simplified tectonic map of the southwest Pacific. FR = Finisterre Ranges, NBT = New Britain Trench, NFF = North Fiji Fracture Zone, SST = South Solomon Trench & TT = Tobriand Trench. Data, principally, from the Geological World Atlas (1976), Hall (1996, 1997), McCaffrey (1996) and the Tectonic Map of the World (1985).

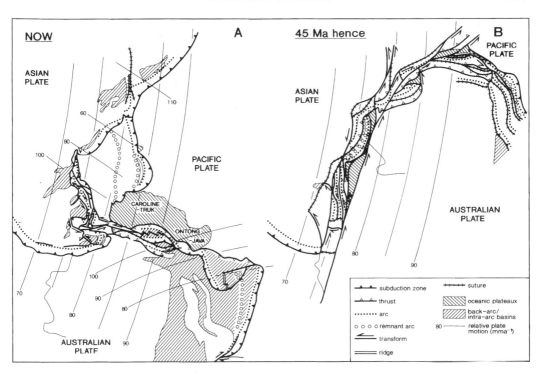

Fig. 5. Interpretative tectonic map of the future orogen produced by the Australia-Asia collision. (**a**) illustrates the present day configuration. (**b**) illustrates the plate configurations projected 45 Ma into the future.

small back-arc basins of the Okinawa Trough, the Japan Sea, the South Sulu Sea and the Celebes Sea, young rifted margins behind arcs whose ages span the closing ages of the big oceans.

Secondly, the collisions between rifted margins and arcs that lead eventually to subduction polarity flips commonly generate short-lived Barrovian metamorphic events in the substantial nappe sheets formed beneath major obducted ophiolite complexes and accretionary prisms. There is a strong resemblance between the New Guinea ophiolite obduction and the related metamorphism of the Bismarck Range, with the Grampian Barrovian metamorphism of the Scottish Highlands and the obduction of the Shetlands Ophiolite (Dewey & Shackleton 1984). An important deduction from these early metamorphic complexes, adjacent to early rifted margins, is that they formed and were also unroofed very rapidly, if diachronously, over great distances along the orogen.

Thirdly, cognate arcs can vary greatly in length, polarity and mode of termination. Typically, they have lengths of more than 1000 km and may terminate in triple junctions (Shikoku, Honshu, Izu–Bonin arcs), by slip vector changes in rate and direction (south end of Yap/Palau Arc), or by

truncation by a strike-slip (transform) fault (Sorong Fault terminates the double Celebes Arc). Arc polarity may change along strike length across transform or transpressional zones as in the Philippines. Also substantial lengths of arcs may have no magmatism for long periods of time (e.g. non-volcanic segments of the Andes; Dewey & Lamb 1992) and individual arc volcanic centres may be separated by hundreds of kilometres as in the Izu–Bonin–Marianas Arc. Thus, the absence of arc volcanic or plutonic rocks in a particular part of a fossil arc system does not necessarily imply the absence or cessation of subduction.

Fourth, remnant arcs may become isolated from the active arc by rapidly-opening back-arc oceanic tracts but still showing magmatic ages very close to those of the active arc. When remnant arcs and related, only slightly younger, active arcs are, subsequently, packed together in collisional systems, they may be regarded, too easily and incorrectly, as separate, unrelated arcs with their own subduction systems.

Fifth, face-to-face arc collision is occurring in two areas of the west and southwest Pacific. In the Molucca Sea, the east-facing northeast Sulawesi arc is colliding with the west-facing Halmahera arc with an intervening, growing collision zone. Collision-induced uplift raised the tiny islands of Mayu and Tifore above sea-level. Across the Solomon Sea, the New Britain and Trobriand Trenches, with opposed polarities, join to the east to form the south-facing San Cristobal Trench and, to the west, to form the southward-vergent collision zone between the Finistere Range and the main mobile core of the Bismarck Range. The Trobriand Trench and associated fore-arc is being subducted diachronously beneath the Finistere–South Solomon arc; the double subduction zone beneath eastern New Guinea can be seen in the seismicity (Pegler *et al.* 1995).

Sixth, a single arc, for example the Banda Arc, may pass through an 180° orocline, the same arc facing in opposite directions. If subsequently, the joining loop is obscured or tectonically removed, the opposing-facing segments could be regarded as two independent arcs.

Seventh, arc-rifted margin collision, followed by arc polarity reversals, may be diachronous as, for example, between northern Australia and the Banda arc allowing propagation overlap that generates a zone with synchronous double subduction. Collision, as in New Guinea during the Early Miocene, is commonly accompanied or predated by ophiolite obduction, that pushed and carried a subjacent nappe pile across the old rifted margin to form a short-lived Barrovian metamorphic nappe complex, the Bismarck Range. The subduction polarity reversal allows the superposition of a post-obduction magmatic arc to form on the deformed nappe complex. However, the line of subduction polarity reversal (i.e. the new trench) is likely to substantially overlap with the colliding arc-continental margin zone as in Taiwan. This might allow calc-alkaline mafic plutons to inject a deforming nappe pile, a relationship seen in the Caledonides of Western Ireland in Connemara. In Connemara, substantial calc-alkaline mafic plutons were injected into shear zones within a northward (towards and onto Laurentia) vergent Barrovian nappe pile (Wellings 1998), driven by the collision of a north-facing oceanic arc with the Laurentian margin and the northward obduction of its ophiolitic fore-arc and accreting prism (Dewey & Shackleton 1984). Clearly it is difficult to relate, directly and simply, the calc-alkaline plutonism to southward subduction during this event because the deforming nappe pile was part of the subducting continental footwall. There may, however, be a better solution (Fig. 6) than propagating and overlapping subduction zones to explain such kinematic calc-alkaline magmatism. Figure 6 illustrates, schematically, map (a), sequential cross-sectional (b), and block diagram (c) views of a diachronous arc-continental margin collision zone

with six lines of cross-section. Prior to collision (1) the oceanic arc (Lough Nafooey Arc) with its fore-arc, supra-subduction-zone ophiolite and accretionary prism is about to collide with the Laurentian rifted margin (Dalradian sediments with a Grenville basement) at *c.* 475 Ma in the Mid-Arenig. Progressively, from stages 2 to 5 in Fig. 6, the arc and ophiolite assemblage and subjacent Dalradian nappe pile are driven across the Laurentian margin. The collisional tightening of the suture zone leads to subduction polarity flip (stages 2 to 6), retrocharriage as seen in nappe root rotation in the Clifden steep belt, and southward thrusting, as along the Mannin Thrust (stage 5). However, for northward subduction to be established, the older southward-subducting slab must break off to allow the asthenospheric gap through which the north-ward-subducting slab can be inserted (stages 3 to 6). Just as subduction zones are likely to overlap, so the break-off 'tear' in the southward subducting slab is likely to propagate beneath the active collision zone (3 to 4). Thus, for a short, diachronous, syncollisional, period, asthenospheric mantle wells through the break-off tear beneath the deforming nappe-pile (4) and allows the syn-kinematic intrusion of calc-alkaline basaltic magma.

Eighth, the partitioning of oblique plate convergence into orthogonal thrust- and plate-boundary-zone-parallel strike-slip components (Fitch 1972; Dewey 1980; Molnar 1992) may result in intra-arc ophiolite generation and obduction. For example, in New Guinea, southward ophiolite obduction in the Miocene was followed by a sub-duction polarity flip with the subsequent choking of the North Solomon Trench by the Ontong-Java Plateau. This led to partitioning of the oblique ENE-convergence into orthogonal and arc-parallel slip components; the arc-parallel strike-slip component of the sinistral convergence was partly accommodated by the opening of the oceanic, pull-apart Manus Basin. Thus, a new potential 'ophiolite' is being created in the Manus Basin within a post-collisional arc complex. Other small ocean basins behind (South China Basin, South Sulu Sea, Celebes Sea, Woodlark Basin and Coral Sea) or within (Parece Vela Basin, North Fiji Plateau) arcs yield another potential source for 'young' ophiolites that may be obducted. The critical criterion appears to be that obduction must occur shortly after generation to allow the detachment of a thin (<15 km) slice of oceanic crust and mantle above the 900°C isotherm (Dewey & Shackleton 1984). Oceanic lithosphere older than *c.* 20 Ma is likely to be subducted rather than obducted and, therefore, most of the back and intra-arc basins mentioned above, with the exception of the Manus Basin, are unlikely to form future

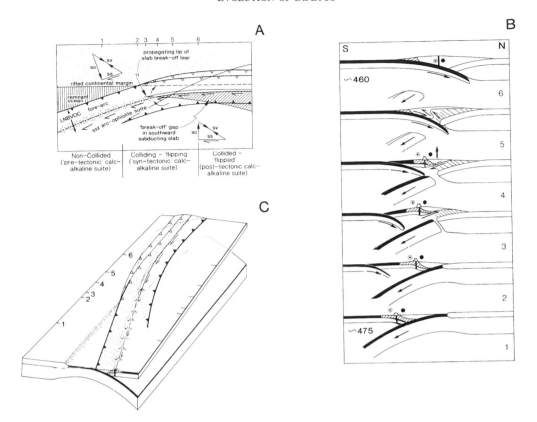

Fig. 6. Schematic tectonic model of diachronous, dextral collision between Laurentian margin and Baie Verte oceanic tract and the arc-polarity reversal. Note that subsequent strike-slip dissection of this collision/arc complex may juxtapose segments where the reversal in subduction had taken place at different times. (**a**) Map view. SV = slip vector of relative plate motion; SO = component of slip vector orthogonal to plate boundary zone; SS = component of slip vector parallel with plate boundary zone. Lines of section 1–6 are those depicted in (**b**) and (**c**). (**b**) Sectional view illustrating the evolution of any one section through time. (**c**) Block diagram view illustrating that sectional views also vary spatially; i.e. all sections are not at the same stage of their temporal evolution (**b**) at the same time.

ophiolite complexes. Ninth, an important feature of orthogonal and strike-parallel partitioning of oblique plate convergence is that orogen or arc-normal thrusting is not a guide to relative plate motion, for which the partitioned arc-parallel strike-slip motion is a critical component.

Tenth, massive relief changes are related with rapid lateral changes in tectonic environments. For example, the extension and uplift of the D'Entrecasteaux metamorphic core complex results from the westward propagation of the Woodlark Basin spreading centre. The resulting geomorphic gradients yield rapid lateral facies changes and slopes that drive instability and mass flow. When such already complicated geology is intensely strained during terminal collision, it seems unlikely that unequivocal reconstructions could be made.

Rifting of Rodinia, the opening of Iapetus and Gondwanan elements in the northern Appalachians/British Caledonides

Most recent palaeogeographic reconstructions generally assume that the proto-Andean margin of Amazonia formed the conjugate margin to Laurentia during Late Neoproterozoic times (c. 750 Ma); a period when Laurentia formed the core of the Rodinia supercontinent (e.g. Hoffman 1991). The breakup of Rodinia supposedly took place in several stages (Dalziel 1992), which is consistent with the longevity of rift-related magmatism and formation of several large Neoproterozoic rift basins in Laurentia (e.g. Moine and Dalradian). Relevant to this paper, however, is the final break-up, which includes the separation of Laurentia from Amazonia at some stage between

590 and 550 Ma. This separation together with the departure of Baltica led to the opening of the Iapetus Ocean (Fig. 7).

The opening of the Iapetus Ocean and the origin of the Laurentian margin are best constrained by the onset of thermal subsidence-driven sedimentation of the continental margins (Bond *et al.* 1984). Calibration of the rift–drift transition (e.g. Williams & Hiscott 1987) against the new Cambrian time scale (Bowring *et al.* 1993; Tucker & McKerrow 1995) would indicate that this took place between 590 and 550 Ma in the northern Appalachians, which is consistent with the youngest known U–Pb zircon ages of synrift magmatism along the Laurentian margin (554 +4/–2 Ma, Tibbit Hill volcanic rocks in Quebec, Kumarapeli *et al.* 1989; 555 +3/-5 Ma Lady Slipper pluton, Cawood *et al.* 1996; 550 Ma+3/–2 Ma Skinner Cove volcanics in Newfoundland, McCausland *et al.* 1997). The 590 Ma U–Pb zircon age of the Ben Vuirich granite in Scotland (Rogers *et al.* 1989) and the deformation it cuts have been reinterpreted as extension-related magmatism and deformation, respectively (Ryan & Dewey 1991; Soper 1994). When this model is combined with the clear stratigraphic and structural continuity between the southern Highland Group of the Dalradian Supergroup and the Cambrian Leny Limestone (Harris 1969) we conclude that the rift–drift transition in Scotland also occurred during the very latest pre-Cambrian. Hence, the final breakout of Laurentia took place in a relatively short period (<40 Ma). Given these age constraints on the opening of the Iapetus Ocean, we suggest that the *c.* 576 Ma and 540 Ma ages of, respectively, an eclogite block in a mélange of the Ballantrae Complex (Hamilton *et al.* 1984) and garnet amphibolite in the Highland Border Complex represent cooling ages of deep lithospheric rocks exhumed during low-angle extensional faulting in a manner analogous to that observed along the Galicia margin (Boillot *et al.* 1988), rather than representing a Late Neoproterozoic/earliest Cambrian, syn-rifting obduction event (Dempster & Bluck 1991). All pre-Cambrian tectono-metamorphic events in the Scottish and Irish Caledonides that postdate the Grenville orogenic events are probably related to intracontinental extension (Soper & Harris 1997). The first post-Grenville contractional event to affect the British Caledonides was in the Early Ordovician (Dewey & Shackleton 1984). However, separation of contractional from extensional structures is complicated because older structures are commonly reactivated during younger events. For example, we believe that the present Moine Thrust roughly coincides with the original position of the Grenville front in the British Isles, separating a foreland with Lewisian basement from a basement involved in Grenville orogenesis as witnessed by Grenville-age eclogites in the Glenelg inlier (Sanders *et al.* 1984). Accordingly, the Grenville front became an extensional fault during formation of the Moine basins and was reactivated as a major structural front during Caledonian contractional events.

The voluminous, pre-600 Ma part of the Dalradian Supergroup may not extend into Newfoundland. Alternatively, it occurs only as a small cryptic unit within the Fleur de Lys Supergroup (cf. Hibbard 1988) or, more likely, much of the Fleur de Lys Supergroup is the stratigraphic equivalent of the Late Neoproterozoic–Cambrian southern Highland Group. Thus the Moine/Dalradian extensional basins narrowed significantly somewhere between the British and Canadian segments of the Laurentian margin and, hence, the presence or absence of Dalradian rocks in a displaced piece of the Laurentian margin (such as Connemara) constitutes a tectonic tracer (cf. Hutton 1987). The diminution in space and time, of the long-lived, wide and diffuse extensional basin complex of the Moine and Dalradian in the British Caledonides (referred to herein as the Dalradian margin) to the short-lived, very narrow and sharply-defined rifted margin of the northern Appalachians with its very abrupt facies change from Cambro–Ordovician platform carbonate to continental rise strata immediately above Grenville crystalline basement (referred to herein as the Grenville margin) from Newfoundland to New York appears to have been of substantial tectonic importance in

Fig. 7. (**a**) Late Neoproterozoic paleogeography as proposed in this paper. Laurentia and Baltica are positioned after the compilation of palaeomagnetic data of Torsvik *et al.* (1996); Siberia after the compilation of Smethurst *et al.* (1998); Gondwana is positioned after the data compilation of Meert & Van der Voo (1997). No palaeomagnetic constraints are available for the position of Amazonia and our positioning here is based on geological grounds (see text). Av = Avalonia; Ga = Gander. (**b**) Middle to Late Cambrian paleogeography. Pc is the Precordilleran Laurentian fragment incorporated into South American part of Gondwana during the Late Ordovician (see Dalziel *et al.* 1994, 1996 and references therein). Note that opening of Iapetus is related to the closure of the Brazilide Ocean. (**c–f**) Ordovician to Silurian palaeogeographic/tectonic evolution of the Iapetus Ocean. Laurentia is positioned throughout on the basis of compilation of palaeomagnetic data of Mac Niocaill & Smethurst (1994), Siberia after the palaeomagnetic compilation of Smethurst *et al.* (1998), Avalonia, Baltica and Gondwana after Torsvik *et al.* (1996). Note that the connection between the various subduction zones is poorly known and speculative. ARM: Armorica.

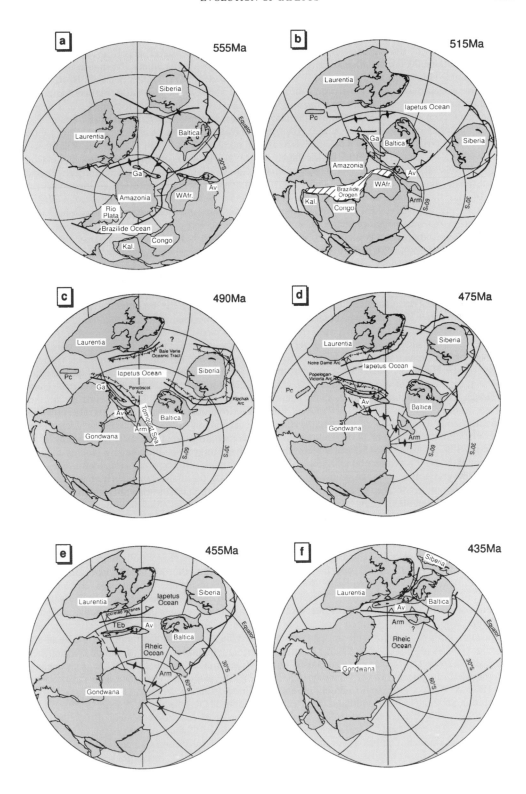

controlling the structural style of the Laurentian margin during Ordovician orogenesis, a point further discussed in the section on the tectonic model for the Baie Verte Oceanic Tract and Notre Dame Arc (see below).

The time constraints on the rift–drift transition of the Laurentian margin and, hence, the opening of the Iapetus Ocean have important implications for palaeogeographic reconstructions. A key issue is whether Amazonia assembled with the African cratons before or after the opening of the Iapetus Ocean. If this assembly took place before the opening of the Iapetus Ocean, it implies the existence of another supercontinent in the late Neoproterozoic (Dalziel 1992). However, such a configuration is inconsistent with the present palaeomagnetic database because it predicts palaeolatitudes of 75–85°S for the western margin of Amazonia if Amazonia formed part of the Gondwanan assembly at 550 Ma, whereas Laurentia is situated at intermediate latitudes (Fig. 7a), i.e. they cannot be adjacent at this time. These restraints are lifted if Amazonia accreted to the African cratons after 550 Ma, an option that we prefer (Fig. 7b) and which was also recently proposed by Hoffman (1996), who argues for a closure of the Brazilide Ocean between 550 and 545 Ma at the latitude of the Kalahari Craton (Fig. 7a). Hence, our interpretation implies the existence of two major landmasses at the end of the Neoproterozoic (Fig. 7a), Gondwana without Amazonia and the remnants of Rodinia: namely Laurentia–Baltica–Amazonia–Siberia which separated from each other between 600 and 550 Ma (see also Pelechaty 1996).

The development of the eastern Gondwanan margin of the Iapetus Ocean in the northern Appalachians and British Caledonides is poorly understood because there is no direct sedimentary link between the Avalon and the immediately adjacent Gander Zones of Williams (1978). However, isotope (Nd, Pb and O) and U–Pb zircon studies indicate that the Cambrian to Early Ordovician, continentally-derived, quartz-rich clastic sediments of the Gander Zone in the northern Appalachians formed part of a Gondwanan passive margin with an Amazonian basement (van Staal *et al.* 1996*a*). Such a basement is probably exposed in the Brookville–Bras d'Or–Hermitage Flexure terranes of the Canadian Avalon Zone (Barr & White 1996), some tectonic inliers in the Exploits Subzone in Newfoundland (S. O'Brien *et al.* 1996), New Brunswick and New England (Figs 1 and 2); e.g. Late Neoproterozoic/Early Cambrian Upsalquitch Gabbro (van Staal *et al.* 1996*a*) and Massabesic Gneiss (Aleinikoff *et al.* 1995). Combined, this basement and its Gander cover are referred to as Ganderia to distinguish it

from the late Neoproterozoic Avalonian arc terranes and their Palaeozoic cover (Avalonia) that make up the eastern part of the Avalon Zone in the northern Appalachians. These terranes (Avalon Peninsula, Mira, Caledonia and Boston) experienced, at least in part, a different geological history during the Late Neoproterozoic and Cambrian (Barr & White 1996). Ganderia basement has been correlated with the Rosslare and Coedana Complexes (Fig. 3) in Ireland and Wales (van Staal *et al.* 1996*a*). An Amazonian provenance for Ganderia is consistent with the high Ordovician palaeomagnetic latitudes (Liss *et al.* 1993), Early Ordovician cold-water, peri-Gondwanan faunas (S. Williams *et al.* 1995) and the chemical composition of its Lower Ordovician black shales (Fyffe & Pickerill 1993). Ganderia contains a late Neoproterozoic to Early Cambrian (*c.* 600–530 Ma) continental arc and associated metamorphic rocks constructed on Mesoproterozoic and older crust typical of Amazonia that, overall, are markedly different from the dominantly juvenile Avalonian arc rocks (generally older than 600 Ma). The time of juxtaposition of Ganderia against Avalonia is poorly constrained. It may have occurred at any time between the Late Neoproterozoic (forming the continuous Avalonian superterrane or arc of Gibbons 1990; Horak *et al.* 1996) and the Devonian (Holdsworth 1994). The presence of the Acado–Baltic fauna in all terranes of the Canadian Avalon Zone and an Arenig overstep sequence between the Monian Terrane and Welsh Avalonia in the British Isles suggest a linkage by at least the Late Cambrian, a linkage that remained in place during most of the Palaeozoic such that Ganderia and Avalonia behaved as one microcontinent during the Ordovician and Silurian, albeit with significant internal modification by strike-slip motion. Hence, for the sake of simplicity, we refer to this composite microcontinent as Avalon when discussing Palaeozoic events shared by both elements.

Closure of the Brazilide Ocean after 550 Ma also provides a simple mechanism to explain the young ages of the Late Neoproterozoic–Early Cambrian magmatic arc on Ganderia and its amalgamation with Avalonia. The arc and accompanying low-pressure metamorphism simply formed as a result of subduction of the Brazilide Ocean beneath Amazonia and were shut-off when the ocean was finally closed in the Early Cambrian rather than a termination as a result of transform activity as proposed by Murphy & Nance (1989). Therefore, opening of Iapetus is related to the closure of the Brazilide Ocean, a hypothesis recently also proposed by Grunow *et al.* (1996). We speculate that the Brazilide convergence system extended to northern Africa and may have been responsible for the Early Cambrian magmatic ages (Egal *et al.*

1996) in Armorica, Iberia and Bohemia (Cadomia, inset of Figs 1–3). These terranes may have formed a promontory extending north from North Africa or a microcontinent peripheral to it (Van der Voo 1988).

The timing of the Early to Mid-Ordovician separation of Avalonia and Ganderia from West Africa and Amazonia, respectively, is not well constrained by palaeontology (Cocks & Fortey 1982) or by palaeomagnetism (Trench *et al.* 1992). Cladistic analyses have shown that, until the Arenig, shelf benthic faunas of Avalonia clustered with those from Gondwana (Fortey & Cocks 1992), but, by the Late Llanvirn, separation was wide enough (>1000 km) to reduce faunal interchange with Gondwana and allow interchange with Laurentia and Baltica. Palaeomagnetic (Trench *et al.* 1992) and Sm/Nd sediment provenance (Thorogood 1990) studies suggest that separation had taken place earlier in the Arenig. The latter study shows that there was a drastic change from a large continental provenance to a juvenile source in the Arenig to Ashgill Avalonian sediments. Subsidence analysis (Prigmore *et al.* 1997) suggests that it may correspond with a Late Cambrian to early Tremadoc or to an Arenig/Llanvirn interval. The latter interval is favoured because the basal Arenig Stiperstones Quartzite in Shropshire, a coarse shallow water sandstone, is probably correlative with the Armorican Quartzite of Brittany (Noblet & Lefort 1990). An Arenig age of, probably rift-related, magmatism in west Avalonia represented by a within-plate rhyolite in the Caledonian Highlands of southern New Brunswick (U–Pb zircon age of 479 ± 8 Ma, Barr *et al.* 1994) is consistent with this. Anorogenic bimodal magmatism in Ganderian terranes in southern Newfoundland (Hermitage Flexure) and Nova Scotia (Bras d'Or terrane) yielded Late Cambrian to Tremadoc U–Pb zircon ages between 499 and 493 Ma (Dunning & O'Brien 1989; Dunning *et al.* 1990) and may be related also to the onset of rifting.

Tectonic elements and evolution of Iapetus

Most information on the pre-Silurian tectonic history of the British Caledonide–northern Appalachian Orogen is preserved in the Dunnage Zone of Newfoundland (Williams 1978), where the post-Ordovician cover sequences are least extensive and a large number of high quality U–Pb age dates have been assembled as parts of detailed mapping projects during the last ten years. Because Newfoundland (Fig. 1) also forms the link between the northern Appalachians and the British Caledonides, its geology imposes important constraints upon tectonic modelling of the whole northern Appalachian–British Caledonide Belt.

The Dunnage Zone of Newfoundland has been subdivided into two major subzones by H. Williams *et al.* (1988), mainly on the basis of their marked Lower to Middle Ordovician stratigraphic and faunal differences, namely the Notre Dame Subzone with the Arenig Toquima–Table Head (Laurentian), low-latitude fauna to the northwest and the Exploits Subzone with an Arenig high-latitude, peri-Gondwanan fauna to the southeast (e.g. see also S. H. Williams *et al.* 1992, 1995). The latter includes elements of the 'Celtic fauna' of Neuman (1984). The Celtic elements may never have formed part of one coherent ecological group because they lack internal cohesion. Of the 41 genera in the 'Celtic fauna' of Neuman (1984), 34 are endemic and occur in only one or two localities, while the remaining seven genera include some that occur also in Baltica and/or Laurentia. We agree, however, that most elements of the 'Celtic fauna' have no affinities with the near-equatorial, Laurentian Toquima–Table Head fauna (Harper *et al.* 1996) and in general are indicative of high(er) latitude, cold water conditions.

The tectonic boundary between the Notre Dame and Exploits Subzones is the Red Indian Line (Fig. 1), which is regarded commonly as the main Iapetus suture zone in the Canadian Appalachians. The notion of one main suture zone in a complex ocean like Iapetus, which contained several different basins and arcs (van Staal 1994) is, of course, unrealistic and further highlighted by the subdivision of the two major Dunnage subzones into smaller subdivisions (H. Williams 1995). Nevertheless, the recognition of the Red Indian Line as a fundamental tectonic boundary is very important in understanding the tectonic evolution of the Canadian Appalachians. The Red Indian Line had a complex and protracted history of movements ranging from thrusting to strike slip (Nelson 1981; Thurlow *et al.* 1992; Lafrance & Williams 1993; Lin *et al.* 1994) and should, therefore, not be regarded as a single, continuous fault but, rather, as a complex movement zone that was reactivated continuously during most of the lifespan of the northern Appalachian Orogen.

Marked contrasts in the Pb-isotope contents of the synvolcanic massive sulphide deposits of both subzones (Swinden & Thorpe 1984) and the palaeomagnetic data in general (van der Pluijm *et al.* 1995) support this subdivision of the Dunnage Zone. The main lithological difference between the two major subzones is the presence of a Mid-Ordovician black shale and conformably overlying Late Ordovician/Early Silurian upward-coarsening turbidites in the Exploits Subzone and their absence in the Notre Dame Subzone, which is characterized by a sub-Silurian unconformity. However, locally in the tectonic boundary zone between these two subzones this stratigraphic contrast may not be always diagnostic (Thurlow *et al.* 1992; Dec & Swinden 1994).

On the basis of continuity of the specific geological characteristics, some elements of each of these subzones can be traced into Maritime Canada, New England (H. Williams 1995; van Staal 1994) and the British Isles (Colman Sadd *et al.* 1992*a*; Winchester & van Staal 1995), suggesting some degree of continuity of kinematically-related belts along the length of the northern Appalachian–British Caledonide segment.

Notre Dame Subzone

The Notre Dame Subzone as defined by H. Williams *et al.* (1988) is separated from the Laurentian margin (Humber Zone) to the west by the Baie Verte Line and from the Exploits Subzone to the east by the Red Indian Line (Fig. 1). However, the easterly-derived oceanic allochthons such as the Bay of Islands Ophiolite Complex and associated ophiolitic mélanges emplaced on the Laurentian margin rocks (Humber Zone) are also considered part of this subzone, i.e. they represent large structural outliers (Fig. 1). This is consistent with seismic reflection data that indicates that the whole of the Notre Dame Subzone is allochthonous and structurally underlain by the Grenville margin (Keen *et al.* 1986).

The ophiolitic rocks of the Notre Dame Subzone have been subdivided into a western and eastern belt (Fig. 1) on the basis of age and characteristics of the ophiolites (Colman-Sadd *et al.* 1992*a*). The western belt contains mainly Upper Cambrian to Tremadoc (505–489 Ma) supra-subduction-zone ophiolites (e.g. St Anthony, Lushs Bight, Betts Cove, Grand Lake) and juvenile ensimatic volcanic–plutonic complexes, which commonly contain consanguineous, locally sheeted mafic dyke suites (Twillingate-Sleepy Cove-Moreton's Harbour Group, Little Port) (Dallmeyer 1977; Coish *et al.* 1982; Dunning & Krogh 1985; Swinden *et al.* 1997; Elliott *et al.* 1991; Kean *et al.* 1995; Cawood *et al.* 1996); characteristic are various combinations of boninite, IAT, trondhjemite or plagiogranite and some MORB, OIB and alkalic basalt, which are collectively referred to herein as the Baie Verte Oceanic Tract. Overall, these characteristics suggest formation in an extensional regime. The supra-subduction-zone assemblages are overlain by Arenig to Llanvirn calc-alkaline volcanic rocks and are intruded by Lower to Upper Ordovician (488–456 Ma) magmatic arc plutons, such as the Cape Ray and Hungry Mountain plutons, which range in composition from tonalite to granite. The plutons contain xenocrystic zircons and have moderate to high ε_{Nd} values; hence they ascended through Laurentian crust (Whalen *et al.* 1987, 1997*a*,*b*; Dunning & Krogh, 1991; Dubé *et al.* 1996). These post-Tremadoc arc plutons and associated volcanic rocks are referred to as the Notre Dame Arc. The Baie Verte Oceanic Tract was progressively emplaced westward or northwestward onto the Grenville margin during the Tremadoc to Llanvirn. Obduction onto the margin was diachronous and started in the Tremadoc opposite promontories on the Laurentian margin. Obduction was synchronous with the generation of some supra-subduction-zone ophiolites (e.g. *c.* 484 ± 5 Ma Bay of Islands Complex; Jenner *et al.* 1991; Cawood & Suhr 1992). Hence, the Baie Verte Oceanic Tract contains pre- and syn-obduction ophiolites. Furthermore, most of the calc-alkaline volcanic and plutonic rocks postdate obduction, confirmed by a stitching relationship between Tremadoc arc plutons, and obducted ophiolites and their underlying mélanges (Hall *et al.* 1994; van Staal *et al.* 1996b; Whalen *et al.* 1997*a*,*b*). An implication of ophiolite obduction and interaction with an irregular margin is that obduction will be early and most pronounced opposite promontories, which may explain the presence of numerous windows of metamorphosed continental margin rocks (Fleur de Lys Supergroup) south of the Little Grand Lake fault (LGLF in Fig. 1) opposite the St Lawrence Promontory. Such windows are less common or absent in the northern part of the Baie Verte Oceanic Tract, although the isolated Ming's Bight Group at the eastern end of Baie Verte may also be a tectonic window rather than representing a close flexure (cf. Hibbard 1982).

The narrower, eastern ophiolite belt (e.g. Annieopsquotch ophiolite) is composed mainly of younger Arenig, fault-bounded, MORB-like ophiolitic fragments, comprising various combinations of pillow basalt, gabbros, sheeted dykes and plagiogranite (*c.* 481–478 Ma, Dunning *et al.* 1987). Mantle harzburgites have not been preserved anywhere in this belt, suggesting that these oceanic rocks represent scraped-off rather than obducted ophiolites. The uniform MORB character of the Annieopsquotch Ophiolite (Dunning 1987) indicates that this piece of oceanic crust (layers 1, 2 and 3) is potentially the only near-complete ophiolitic fragment in the Newfoundland Appalachians that does not have a supra-subduction-zone character. The eastern ophiolite belt is, spatially, closely associated with coeval or slightly younger upper Tremadoc to upper Arenig (484–473 Ma) calc-alkaline volcanic and arc tholeiitic rocks of the

Roberts Arm, Cottrell's Cove, Chanceport and Buchans Groups (Dec *et al.* 1997). The original relationships between the ophiolitic rocks and volcanic arc rocks are poorly defined. This results, at least in part, from a poor understanding of the structural relationships between the ophiolitic bodies and associated volcanic rocks. However, fault-bounded slivers of MORB pillow basalts and cherts are, locally, structurally interleaved with island arc tholeiites (Dec & Swinden 1994) whereas, in the vicinity of the Buchans Mine and Red Indian Lake, seismic and structural studies have shown that the ophiolitic Skidder Basalt (Pickett, 1987) structurally underlies the Late Arenig calc-alkaline volcanic rocks of the Buchans Group in a complex, Mid-Ordovician, east-directed thrust wedge (Thurlow *et al.* 1992). It is uncertain whether the ophiolitic bodies form part of the basement to the Roberts Arm belt arc volcanic rocks or the relationships are dominantly structural. Negative ε_{Nd} values for most of the calc-alkaline volcanic rocks indicate involvement of continental crust in their petrogenesis (Swinden *et al.* 1997). Therefore, we suspect a dominantly structural relationship between most of the ophiolitic fragments and the bulk of the arc volcanic rocks. The Red Indian Line deformation zone in central Newfoundland is marked by a section several kilometres thick of structural mélange and deformed Mid-Ordovician siltsone, limestone and basalt (Harbour Round Formation) between the overlying, intensely-imbricated Buchans Group and Skidder Basalt and the structurally underlying Victoria mine sequence of the Exploits Subzone. The overall character of this structural assemblage suggests an accretionary complex. The Harbour Round Formation was grouped into the Notre Dame Subzone by Thurlow *et al.* (1992), although these rocks are atypical of this subzone. O'Brien (1991) described the Red Indian Line in Notre Dame Bay as a 2–3 km wide zone of mélange (Sops Head-Boones Point Complex; Nelson 1981) containing variably-sized exotic blocks including basaltic rafts with N-MORB compositions (Dec & Swinden 1994). The mélange is imbricated with parts of the arc volcanic rocks of the Buchans–Roberts Arm belt and metamorphic assemblages with actinolite and pumpellyite occur in rocks of the Chanceport Group close to the Lukes Arm Fault (S. Armstrong pers. comm.) These types of relationship suggest that the eastern ophiolite belt and associated arc rocks were incorporated into an east-facing subduction/accretionary complex, here referred to as the Annieopsquotch Accretionary Tract. High-level thrusting, largely accommodated by mélange formation, represents the earliest history of the Red Indian Line deformation zone but these structures have been complexly

overprinted during a subsequent polyphase folding and faulting history in part related to dextral transpression (e.g. Lafrance & Williams 1992). A major unconformity, representing uplift and exhumation, separates the Mid-Ordovician and older rocks from a generally thin cover of a discontinuous and probably diachronous sequence of Late Ordovician to Late Silurian (*c.* 453–422 Ma) dominantly terrestrial to shallow marine calc-alkaline to alkaline volcanic and sedimentary rocks (Whalen *et al.* 1987; Chandler *et al.* 1987; Dubé *et al.* 1996) throughout the Notre Dame Subzone.

Sparse low-latitude, Laurentian, faunas occur in sediments of both the Baie Verte Oceanic Tract and the Annieopsquotch Accretionary Tract (S. H. Williams *et al.* 1995 and references therein). The low-latitude faunas are consistent with the near-equatorial palaeomagnetic latitudes (*c.* 11°S, Johnson *et al.* 1991) obtained for the Upper Cambrian or Tremadoc Moreton's Harbour Group (Dec *et al.* 1997), which is regarded generally to form part of or to overlie the Upper Cambrian Sleepy Cove–Twillingate trondhjemite complex (Williams & Payne 1975) of the Baie Verte Oceanic Tract. This palaeolatitude overlaps within error with the Early Ordovician latitude of the Appalachian margin of Laurentia (*c.* 18°S) and suggests that the Baie Verte Oceanic Tract formed very close to the Laurentian margin. On the other hand, the late Arenig volcanic rocks of the Roberts Arm and Chanceport groups yielded mainly intermediate latitudes (*c.* 30°S, van der Voo *et al.* 1991). Whether this difference is real and represents significant Arenig latitudinal separation (15–20°) between the Baie Verte Oceanic Tract and Annieopsquotch Accretionary Tract, a southerly drift of Laurentia during the Arenig to Llanvirn or a combination of both is unclear. Dec and Swinden (1994) claim that the oldest parts of the Cottrell's Cove and Chanceport groups of the Annieopsquotch Accretionary Tract locally conformably overlie the Moreton's Harbour Group, which suggests that the Chanceport fault is not a terrane boundary (cf. Lafrance & Williams 1992). This suggests a southerly drift of Laurentia to intermediate latitudes (20–25°S) rather than a major separation in the Arenig to Llanvirn. Such a drift is permitted, within analytical error, by the present palaeomagnetic dataset of Laurentia (Mac Niocaill & Smethurst 1994; Torsvik *et al.* 1996) (Fig. 8). One should also take into account that the Laurentian margin was expanding southwards by the addition of the accreted Baie Verte Oceanic Tract. Other supporting arguments in favour of southerly drift rather than latitudinal separation of both ophiolite belts are (i) low-latitude faunas in both the Baie Verte Oceanic- and Annieopsquotch Accretionary Tract suggest development near the

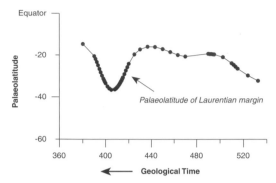

Fig. 8. Predicted palaeolatitude of the Newfoundland margin of Laurentia (based on a present-day reference locality at 49°N, 57°W) versus time. The predicted palaeolatitudes are derived from the palaeomagnetic compilation of Mac Niocaill & Smethurst (1994). Errors on the predicted palaeolatitudes are of the order of 10°.

Laurentian margin, and (ii) oceanic elements of both tracts have been intruded by Tremadoc to Llanvirn magmatic arc plutons and/or include coeval calc-alkaline volcanic rocks that formed on Laurentian crust (Dunning *et al.* 1987; Swinden *et al.* 1997; Whalen *et al.* 1997*a,b*; Dec *et al.* 1997). These relationships suggest that by late Tremadoc/ early Arenig time all calc-alkaline rocks formed part of one coherent magmatic arc (Notre Dame Arc) built on the Laurentian margin.

Correlatives of the Baie Verte Oceanic Tract outside Newfoundland

The Notre Dame Subzone can be traced into Quebec and New England (e.g. Williams & St Julien 1982). Age and lithological equivalents of the Baie Verte Oceanic Tract in southern Quebec are represented mainly by the Southern Quebec Ophiolite Belt (Fig. 2). This belt comprises supra-subduction-zone ophiolites that include a significant amount of boninitic rocks (Oshin & Crocket, 1986; Laurent & Hebert 1989). IAT and boninites, preserved as structural lenses in mélange in the Ascot structural complex (Tremblay & St Julien 1990; Tremblay 1992), are also considered equivalents of the Baie Verte Oceanic Tract. In general, the complex magmatic assemblages preserved in the oceanic complexes closely resemble those in Newfoundland. A Late Cambrian U–Pb zircon age of 504 ± 3 Ma (David *et al.* 1993) for the Mt Orford Ophiolite overlaps with the oldest ages from Newfoundland. As in Newfoundland, obduction probably started in the Tremadoc (Pinet & Tremblay 1995; Whitehead *et al.* 1996). Unlike most parts of the Baie Verte Oceanic Tract in

Newfoundland, most oceanic fragments of the southern Quebec Ophiolite Belt are structurally intercalated with mélange (St Daniel Mélange) and disconformably overlain by Llanvirn–Caradoc chromite-bearing flysch of the Magog Group. In this respect, the southern Quebec Ophiolite Belt resembles the Bay of Islands Ophiolite Complex of the Humber Arm Allochthon, with the syn-obduction, Middle-Ordovician Crabb Brook Group (Casey & Kidd 1981) representing the Newfoundland equivalent of the Magog Group. A plagiogranite of the ophiolitic Thetford Mines Complex yielded an age 479 ± 3 Ma (Dunning & Pedersen 1988), which makes the latter the youngest known ophiolite in the southern Quebec Ophiolite Belt, provided the plagiogranite is consanguineous with the remainder of the ophiolite complex. Hence, the ophiolite may be older. Considering the position of the Thetford Mines within the southern Quebec Ophiolite Belt and the overlap within error between the age of the plagiogranite and a gabbro of the Bay of Islands Ophiolite Complex (484±5 Ma), we interpret the Thetford Mines Complex as a syn-obduction ophiolite. The volcanic–plutonic domains of the Middle Ordovician Ascot Complex (460–455 Ma, David *et al.* 1993), which lack mélanges (Tremblay & St Julien 1990), are the Quebec analogues of the calc-alkaline Middle Ordovician volcanic/magmatic arc rocks of the Notre Dame Arc. Nd-isotope data also indicate development on Laurentian crust (Tremblay *et al.* 1994). The southern Quebec Ophiolite Belt continues into the Rowe–Hawley Belt in New England (Stanley & Ratcliffe 1985) with its relative abundance of boninite and primitive arc volcanic rocks (Kim & Jacobi 1996). Intrusive arc plutons (488–462 Ma, Karabinos & Tucker 1992; Williamson & Karabinos 1993; Karabinos *et al.* 1993) suggest that the oceanic elements are also Lower Ordovician or older in age, consistent with Late Cambrian to Tremadoc $^{40}Ar/^{39}Ar$ ages (505–490 Ma) of blueschists developed in ophiolitic rocks (Laird *et al.* 1993). A characteristic feature of the boundary zone between the oceanic element of the Notre Dame Subzone and rocks of the Humber Zone from Newfoundland to Massachussets is a belt of continental margin-derived albite schists tectonically interleaved with ultramafic pods (e.g. Rowe and Bennett schists, Fleur de Lys Supergroup) in New England, Quebec and Newfoundland (Stanley & Ratcliffe 1985; Hibbard 1988).

Continuation of the Baie Verte Oceanic Tract into the British Caledonides (Fig. 3) is also probable (Dewey & Shackleton 1984; Colman-Sadd *et al.* 1992*a*). The oceanic structural fragments of the Highland Border Complex, Midland Valley (Ballantrae) and Northwestern

Terrane (South Mayo and Tyrone) are associated with sedimentary rocks containing Laurentian faunas (Cocks & Fortey 1982; S. H. Williams *et al.* 1995) and are characterized by low palaeomagnetic latitudes (Torsvik *et al.* 1990). Of these three terranes, the southern Midland Valley most closely resembles the geology of the Notre Dame Subzone of Newfoundland. The supra-subduction-zone Ballantrae Ophiolite Complex has a composite nature like those in the Baie Verte Oceanic Tract in Newfoundland. Incorporation of volcanic rocks of diverse tectonic settings may provide the explanation for the wide variety of ages, which range almost continuously from the Late Cambrian to late Tremadoc (505–484 Ma; Bluck *et al.* 1980; Hamilton *et al.* 1984). Furthermore, the Ballantrae Ophiolite Complex, like the Bay of Islands Ophiolite Complex, was obducted onto the margin during the Arenig. Obduction was followed immediately by intrusion of Arenig diabases with island arc affinity (Holub *et al.* 1984). In general, the Arenig to Caradoc history of the Midland Valley is dominated by formation of a south-facing volcanic/magmatic arc (Midland Valley arc; Bluck 1984), which was unroofed during the Llanvirn–Caradoc. The Midland Valley arc terrane also appears to have a Grenville rather than a Dalradian basement (Aftalion *et al.* 1984; Parnell & Swainbank 1984; Upton *et al.* 1984), suggesting that this terrane represents a far-travelled piece of the Grenville margin of Laurentia, i.e. the Midland Valley was carried by post-Ordovician, margin-parallel, sinistral strike slip faults into its present position to the south of rocks underlain by Dalradian basement (Dewey & Shackleton 1984; Hutton 1987; Ryan *et al.* 1995), consistent with provenance studies that do not show a sedimentary linkage with Dalradian rocks until the Devonian (Bluck 1984). Early Ordovician northward-directed ophiolite obduction also characterized the Dalradian part of the Laurentian margin. Shetland Ophiolite plagiogranite yielded a high-quality U–Pb zircon age of 492 ± 3 Ma (Spray & Dunning 1991), whereas the underlying hornblende schist gave a slightly older $^{40}Ar/^{39}Ar$ plateau age of 503 ± 6 Ma (Flinn *et al.* 1991). This suggests that the plagiogranite either formed off-axis during the earliest stages of obduction in an oceanic realm or the older ages of the hornblende schist represent a remnant of an oceanic transform that was later tectonically juxtaposed with a supra-subduction-zone ophiolite.

In Tyrone, obduction took place before 471 Ma, the age of a cross-cutting tonalite pluton (Hutton *et al.* 1985), whereas sediment provenance studies suggest that ophiolite obduction onto the Dalradian margin was well advanced by late Arenig times in South Mayo (Dewey & Ryan 1990). Continuity of

Early Ordovician structures from the Dalradian margin into ophiolitic mélanges of the Clew Bay Complex supports this conclusion (Harris 1995). South Mayo is unique, compared with its equivalents in the northern Appalachians, in preserving a continuous section from a primitive north-facing oceanic, Tremadoc–Arenig arc to a well developed fore-arc basin/accretionary complex (Dewey & Ryan 1990; Clift & Ryan 1994), characterized by sporadic blueschists and other high-pressure assemblages (Gray & Yardley 1979). Another critical key to the geology of western Ireland is the recognition of an Arenig to late Llanvirn (*c.* 480–460 Ma, Cliff *et al.* 1996) continental arc in the Dalradian margin, syntectonic with D2-D3 Grampian deformation and low-P metamorphism (Yardley & Senior 1982; Leake 1989). Recent detailed structural and U–Pb studies of the Cashel–Currywongaun calc-alkaline gabbros in Connemara showed they also intruded syn-D2 at *c.* 468–460 Ma (Wellings 1998; Friedrich *et al.* 1997), which is similar to the *c.* 468 Ma zircon age of the probably correlative Aberdeenshire gabbro suite in Scotland (Rogers *et al.* 1994). The 471 Ma ophiolite-cutting tonalite pluton in Tyrone probably also forms part of this continental arc. The age and geological characteristics of the Dalradian-margin arc and the Midland Valley arc bear a strong resemblance to the Notre Dame Arc developed on the Grenville margin in Newfoundland after initial ophiolite obduction in the Tremadoc and early Arenig. Furthermore, in all three areas (Newfoundland–Ireland–Scotland), unroofing of the magmatic arc and associated metamorphic rocks started in the Mid- to Late Ordovician, reflecting a once continuous tectonic setting and kinematic history.

Tectonic model for the Baie Verte Oceanic Tract and the Notre Dame Arc

The Late Cambrian–Tremadoc (508–486 Ma) oceanic fragments of the Baie Verte Oceanic Tract and correlatives throughout the northern Appalachians and British Caledonides appear to be characterized mainly by an assemblage of supra-subduction-zone rocks which commonly include IAT, boninite, MORB and minor bodies of juvenile silicic igneous rocks. Any tectonic model for the Baie Verte Oceanic Tract has to explain the following: (1) it appears to have formed close to the Laurentian margin in an extensional setting; (2) obduction took place shortly after formation; some ophiolites formed while obduction was in process at nearby promontories; (3) the oldest oceanic fragments locally contain mafic xenoliths with strong pre-entrainment foliations (e.g. Williams &

Payne 1975) and were deformed during trans-current shear prior to and/or during obduction (e.g. Elliott *et al.* 1991); (4) coeval arc, fore-arc or back-arc volcanic-sedimentary assemblages are rare or absent; (5) stitching Early Ordovician arc plutons ascended through 'Grenville' crust; and (6) high pressure assemblages (e.g. blueschists) occur in several places in rocks structurally beneath the obducted ophiolitic rocks (New England, Laird *et al.* 1993; Quebec, Trzcienski 1976; Newfoundland, Jamieson 1977; Ireland, Gray & Yardley 1979; Harris 1995).

These criteria do not favour recent tectonic models that place the Baie Verte Oceanic Tract in an arc–back-arc setting above a westward-directed subduction zone underneath Laurentia, with the back-arc basin closing in Early to Mid-Ordovician times (e.g. Swinden *et al.* 1997; van der Pluijm *et al.* 1995). Particularly, formation of syn-obduction ophiolites, blueschists and eclogites are hard to reconcile with closure of a very young back-arc basin. Furthermore, boninite occurs dominantly in the fore-arc areas of modern oceanic subduction systems and is thought to be indicative of subduction zone initiation (Casey & Dewey 1984; Stern & Bloomer 1992). Some have suggested that formation of the Baie Verte Oceanic Tract boninites took place during intra-arc rifting (e.g. Coish *et al.* 1982). However, where boninite occurs in a modern back-arc position such as in parts of the Lau basin, it is either regarded as products of fore-arc lithosphere remelted during propagation of 'back-arc' spreading centres into the fore-arc, e.g. where initial rifting of the Tonga arc occurred trenchward of the arc volcanic front, or by remelting of pre-existing refractory back-arc mantle (Clift 1995). The Tonga Arc–Lau marginal basin system is, in some ways, a pertinent example, because the magmatic history of the Tonga Arc–Lau basin, varying from alkaline volcanism at the still active remnant arc to MORB or MORB transitional to IAT in the rift zone (Clift 1995) shows similarities with the magmatic assemblages preserved in the Baie Verte Oceanic Tract, although no good equivalent of the pre-rifting Tonga arc is present. Hence, a rifted arc setting is not a satisfying analogue for the Baie Verte Oceanic Tract.

We concur that the original model of Karson & Dewey (1978), subduction initiation in a dextral oceanic transform fault or fracture zone close to the Laurentian margin at *c.* 508 Ma, is the most appropriate to explain the formation of the Baie Verte Oceanic Tract when comparing these rocks with recent supra-subduction-zone equivalents (e.g. Stern & Bloomer 1992; Bloomer *et al.* 1995). The syn-magmatic shear zones in the Little Port Complex (505+3/-2 Ma; Jenner *et al.* 1991) and amphibolite xenoliths with a pre-entrainment

foliation in the 507 Ma Twillingate trondhjemite (Williams & Payne 1975; Elliott *et al.* 1991) represent the oldest known ages in the Baie Verte Oceanic Tract and correlatives in the northern Appalachian–Caledonian Orogen and may represent remnants of such oceanic transform faults. The transform fault–subduction intiation model could also explain the anomalous Late Cambrian ages (503 Ma) of hornblende schists in the Unst Ophiolite in Shetland (Flinn *et al.* 1991) that predate the Tremadoc plagiogranite. A dextral obliquity of convergence, after the transform fault became a subduction zone, is consistent with the structural data of Elliott *et al.* (1991), Goodwin & Williams (1990) in other parts of Newfoundland, and Harris (1995) in western Ireland. However, we deviate from the Karson & Dewey (1978) model, where it concerns generation of the relatively young (*c.* 484 Ma), syn-obduction ophiolite of the Bay of Islands Ophiolite Complex in an active ridge segment of the transform fault. McCaig & Rex (1995) presented evidence that deformation and cooling of the Little Port Complex in the western Lewis Hills had taken place before, not during, generation of the intruding ultramafic-mafic rocks of the Bay of Islands Ophiolite Complex as is required by the Karson & Dewey (1978) model. Instead, we propose that the Bay of Islands Ophiolite Complex formed in a dextral transtensional basin (rhombochasm) in the over-riding plate above a re-entrant after collision had started. The former existence of such a re-entrant may still be outlined by the tight bend in the Baie Verte Line near Grand Lake (marked by the trace of the Grand Lake Complex) where it changes from a dominantly northeast trend into an east–west strike marked by the Little Grand Lake Fault (LGLF in Fig. 1). A possible modern analogue of ophiolite generation within a colliding or post-collisional arc complex is the Manus Basin north of New Guinea (Fig. 4). The northwards translation of the Bay of Islands Ophiolite Complex during its subsequent Arenig emplacement onto the Laurentian margin (Suhr & Cawood 1993) may explain the conflicting dextral and sinistral kinematic indicators observed by McCaig & Rex (1995) in the Little Port Complex of the western Lewis Hills, because it acted then as a sinistral lateral ramp. The more MORB-like character of the Bay of Islands Ophiolite Complex and the Arenig Snooks Arm Group in Newfoundland, compared with the more typical supra-subduction-zone geochemistry of most older basalts of the Baie Verte Oceanic Tract (Jenner & Fryer 1980; Jenner *et al.* 1991) is consistent with this model.

One of the characteristics of modern analogues of supra-subduction-zone magmatism is the absence of a well-defined volcanic front; instead, based on studies in the Izu–Bonin–Mariana system, subduction initiation created a wide area (>200 km) of extension and widely-distributed zone of fore-arc magmatism (infant arc magmatism of Stern & Bloomer 1992) in the upper oceanic plate. Such juvenile supra-subduction-zone magmatism may have lasted until *c.* 489 Ma in Newfoundland. Equivalent supra-subduction-zone magmatism may have lasted longer in the British Isles although geo-chronological control does not allow much diachroneity. With the possible exception of the amphibolite xenoliths in the Twillingate trond-hjemite, evidence for older, pre-subduction, oceanic crust on which the forearc supra-sub-duction-zone magmatism was built has not been detected yet in the Baie Verte Oceanic Tract; how-ever, neither has much clear evidence been found for pre-subduction oceanic crust in the Bonin–Mariana or Tonga forearcs. This suggests that most existing oceanic crust was, largely, replaced or displaced during supra-subduction-zone forearc magmatism (Bloomer *et al.* 1995). Obduction of the Baie Verte Oceanic Tract onto promontories in the Laurentian margin must have started in the early Tremadoc between 500–488 Ma (Dallmeyer 1977; Cawood & Suhr 1990; Dewey & Ryan 1990; Pinet & Tremblay 1995; Swinden *et al.* 1997) soon after subduction was initiated, presumably while fore-arc extension, due to rapid hinge retreat of the sinking slab, was still active and before stable lithospheric subduction was established (see Stern & Bloomer 1992); hence supra-subduction-zone fore-arc magmatism may have had a lifespan of up to *c.* 20 Ma in the Baie Verte Oceanic Tract, which is not unlike the period of supra-subduction-zone magmatism in the Izu–Bonin–Mariana fore-arc (10–15 Ma; Bloomer *et al.* 1995). The polarity direction of subduction associated with the Baie Verte Oceanic Tract therefore must have been towards the east or southeast in present coordinates.

The oldest known Notre Dame Arc pluton that cuts obducted ophiolites of the Baie Verte Oceanic Tract in Newfoundland is the 488 ± 3 Ma (Dubé *et al.* 1996) Cape Ray Granodiorite (Hall *et al.* 1994). This pluton, which ascended through Grenville margin crust (Whalen *et al.* 1997*b*) is no exception as similar, stitching continental arc plutons of Tremadoc to Arenig age have been identified throughout the Notre Dame Subzone of Newfoundland (Whalen *et al.* 1997*a*), while Swinden *et al.* (1997) even describe crustally contaminated high-Mg andesite dikes with $^{40}Ar/^{39}Ar$ ages between 501 and 490 Ma that cut deformed rocks of the Lushs Bight ophiolite. The ages of the bulk of the Notre Dame Arc plutons fall in the same range as the oldest arc plutons in New England (*c.* 488–484 Ma; Karabinos & Tucker

1992; Williamson & Karabinos 1993). Such plutons and associated volcanic rocks occur from the British Isles to New England and signal a change in polarity of subduction along the Grenville and Dalradian margins in the Early Ordovician (Fig. 9). An alternative model, that the Notre Dame Arc formed on a little-travelled, rifted-off sliver of the Grenville margin is thought to be unlikely. Such a model requires that the auto-chthonous, well-developed passive margin in Western Newfoundland and New England formed partly in a back-arc environment, for which there is no evidence, and does not provide a satisfactory explanation for the abrupt change from supra-subduction-zone oceanic to continental arc mag-matism in the late Tremadoc to Arenig.

The arc-polarity reversal along the Laurentian margin must have been very rapid and diachronous because westward-directed obduction and ophiolite formation in parts of the Baie Verte Oceanic Tract postdate or overlap with generation of the oldest members of the Notre Dame Arc elsewhere. This is particularly apparent in western Ireland where the polarity reversal recorded by the synorogenic arc magmatism and related metamorphism in Connemara (480–465 Ma; Cliff *et al.* 1996) over-laps, in part, with lower Arenig oceanic arc mag-matism and fore-arc emplacement onto the Dalradian margin in the adjacent South Mayo Trough (see Figs 3 and 6). The first possible elements of continental arc volcanism in the South Mayo trough are the Llanvirn Mweelrea ignim-brites (Figs 3 and 6); hence the juxtaposition of Connemara with the South Mayo Trough took place as a result of post-Llanvirn but pre-upper Llandovery transcurrent movement (Dewey & Ryan 1990). The first possible sedimentary link between Connemara and the South Mayo Trough is southerly-derived high-grade metamorphic detritus in the Late Ordovician Derryveeny conglomerate (Dewey & Ryan 1990). It is logical to link this detritus to strike-slip docking of the Connemara terrane because this terrane experienced rapid uplift and erosion after 460 Ma (Elias *et al.* 1988).

After the polarity reversal was complete along the length of the orogen in the Arenig, obduction onto the margin continued into the Mid-Ordovician. Mutual cross-cutting relationships between Ordovician arc plutons, folds and amphibolite-facies shear fabrics in the Fleur de Lys rocks of SW-Newfoundland also corroborate that the Notre Dame Arc plutons were roughly syn-tectonic with the Mid-Ordovician Taconic Orogeny (van Staal *et al.* 1996*b*). Therefore, the Taconic west-directed thrusting mainly represents back-thrusting behind an east or south-facing, compres-sive Notre Dame arc (Bird & Dewey 1970) (Figs 6 and 9).

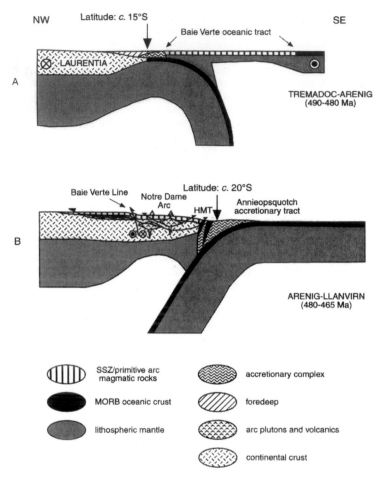

Fig. 9. Early to Middle Ordovician tectonic evolution of the Grenville margin. Note that west-directed transport and associated deformation during middle Arenig to Caradoc in (**b**) is due to backthrusting. Evidence for Lower to Middle Ordovician, east-directed thrusting in the Notre Dame Subzone near the western boundary with the Annieopsquotch accretionary tract is locally well preserved (e.g. Hungry Mountain Thrust, see Thurlow *et al.* 1992 and references therein). Also note that the syntectonic Notre Dame Arc plutons are both deformed by these structures and also cut them (Hall *et al.* 1994). The general principles of this model are also applicable to the Dalradian margin although the deformation history generally is more complex.

In the northern Appalachians, the Early to Mid-Ordovician westward translation of the Baie Verte Oceanic Tract led to a rather simple, narrow, dominantly westward-vergent orogenic belt beneath the ophiolite and the transport of the latter together with subjacent continental rise rocks onto a foreland basin successor of the carbonate platform. In contrast, a more complex structural history is preserved in the Dalradian margin of the British Isles. Although a suture is recognized in western Ireland (Clew Bay Zone) and a high-level ophiolite nappe and/or accretionary prism was probably emplaced above the Dalradian (Dewey & Shackleton 1984), there is no evidence as to

whether it did or did not travel across the Moine/Dalradian complex onto the Durness carbonate platform northwest of the Moine Thrust (Fig. 3) or any successor foreland basin above it. The structural and metamorphic style and sequence of the Moine and Dalradian are generally more complex and intense than the Fleur de Lys rocks and its equivalents in Quebec and New England, although there are also parallels (e.g. Tremblay & Pinet 1994). Structural polarity and bulk vergence are variable both on the Dalradian margin of the Caledonides and Grenville margin of the Appalachians but the rapid lateral changes from flat belts to intensely-deformed steep belts characteristic of the central Highlands of Scotland, are rare or absent in the northern Appalachians. Diachroneity and changes in style of structures within one type of margin are another complication. For example, most of the Early Grampian deformation (D1–D3) in Connemara, western Ireland (Fig. 3), was characterized by northwards-directed tectonic transport, the main southward retrocharriage taking place during the Mid- to Late Ordovician (D4). On the other hand, tectonic transport was both to the north and south during the Early to Mid-Ordovician (D2) in the Dalradian of the Grampian Highlands (Krabbendam *et al.* 1997). These contrasts suggest a complex response to the transpressional strains (D2–D4) superimposed on the initial obduction and collision structures (D1) during and following the polarity reversal along the Dalradian margin. The marked ductile behaviour and thickening of the Dalradian contrasted with the Grenville margin is perhaps due to a combination of a higher degree of underthrusting and convergence with a more extended part of the Laurentian crust with its thick sedimentary cover of the Moine/ Dalradian. Syntectonic magmatic arc activity continued until the Caradoc both in the northern Appalachians and British Caledonides, after which the Notre Dame–Midland Valley arc experienced uplift, exhumation and a temporary magmatic shutoff. This coincides with accretion of the Bronson Hill–Popelogan–Victoria–Longford Down island arc system with the Notre Dame Arc approximately along the present day trace of the Red Indian Line (van Staal 1994; see below).

Correlatives of the Annieopsquotch Accretionary Tract in New England and the British Caledonides and their tectonic setting

Equivalents of the Annieopsquotch Accretionary Tract are less well defined in New England. A recent U–Pb zircon age of 477 ± 1 Ma for a tonalite in the Boil Mountain Ophiolite Complex in Maine (Kusky *et al.* 1997) suggest that the latter and onstrike soapstone belt, immediately west of the Bronson Hill anticlinorium in the Connecticut Valley in New Hampshire and Massachusetts (Fig. 2; Lyons *et al.* 1982) are their probable equivalents in New England. If the tonalite is consanguineous with the rest of the Boil Mountain Ophiolite Complex, the latter cannot represent the root zone of the older southern Quebec ophiolites as proposed by Pinet & Tremblay (1995). As in Newfoundland, we interpret this New England equivalent of the Annieopsquotch Accretionary Tract to mark approximately the trace of the Red Indian Line. Robinson & Hall (1980) inferred this line to represent their Taconic suture, implying that the Bronson Hill Arc forms part of the Exploits Subzone, a conclusion entirely consistent with the presence of Caradoc to Ashgill black shales and volcanics. The Boil Mountain Ophiolite Complex is in tectonic contact to the west with the Chain Lakes Massif (Boudette 1982). Seismically, the Chain Lakes Massif forms basement to the Mid-Ordovician Ascot Complex (Spencer *et al.* 1986) and is intruded by Middle Ordovician, probably consanguineous, arc-like plutons (e.g. Attean pluton: 463 Ma, Holtzman *et al.* 1996). The Chain Lakes Massif contains typical Laurentian Grenville detrital zircons and experienced Taconic high-grade contact metamorphism (U–Pb monazite of 468 ± 2 Ma) from the arc plutons; hence, it probably represents an exposed window of metamorphosed rocks of the Laurentian margin (Dunning & Cousineau 1990; Trzcienski *et al.* 1992). These characteristics strongly resemble the relationships observed in structural inliers of Fleur de Lys clastics in Southwest Newfoundland (Fig. 1), which experienced amphibolite and locally even granulite facies, Taconic metamorphism coeval with intrusion of the syntectonic Notre Dame Arc plutons (Currie *et al.* 1992; Hall *et al.* 1994; van Staal *et al.* 1996*b*). Thus, the bulk of the juvenile oceanic elements of the Baie Verte Oceanic Tract and equivalents in Quebec and New England were obducted onto the Laurentian margin before formation of the Early to Mid-Ordovician Notre Dame Arc plutons and related volcanics along the whole length of the northern Appalachians, i.e. the Tremadoc and older elements of the Baie Verte Oceanic Tract represent a rootless oceanic klippe first obducted before the classic Mid-Ordovician Taconic Orogeny. Absence of the Attean contact metamorphism in the Boil Mountain Ophiolite Complex and tonalite in the Chain Lakes Massif suggest that tectonic juxtaposition of the ophiolite complex with the massif, took place after 468 Ma (Trzcienski *et al.* 1992; Kusky *et al.* 1997).

Equivalents of the Annieopsquotch Accretionary Tract in the British Caledonides are not well defined and their existence is uncertain. Colman Sadd *et al.* (1992*a*) and S. H. Williams *et al.* (1992), correlated all of the Southern Uplands in Scotland with the Exploits Subzone in Newfoundland on the basis of the presence of Caradoc black shale, strictly adhering to the original definition of H. Williams *et al.* (1988). However, we suggest that, although there may be a general spatial correlation along a zone that contains both the Exploits Sub-Zone and the Southern Uplands, and both contain Caradoc black shale, the Southern Uplands has a substantial southward-growing accretionary prism that has no direct correlative in Newfoundland. The dominantly Arenig–Llanvirn age rocks of the

Annieopsquotch Accretionary Tract, which lies immediately west of the Red Indian Line in the Notre Dame Subzone, may have British and Irish equivalents in the northern Belt of the Southern Uplands in Scotland (Leggett *et al.* 1982; Armstrong *et al.* 1996) and Ireland (Morris 1987). The northern Belt continues across Ireland to the South Connemara Group on the west coast (D. Williams *et al.* 1988; Harper & Parkes 1989), which exhibits a series of slices involving trench-fill sediments and 'clipped-off' seamounts (Dewey & Ryan, in prep.). Palaeocurrent patterns and provenance studies have shown that the rocks of the northern Belt had markedly variable source areas, although the bulk comes either from the northeast or northwest. Important provenance tracers are early Caradoc (*N. gracilis*) conglomerates in tracts 1 and 2 in the Scottish Southern Uplands, which were derived from the northwest. These conglomerates are considered to represent trench deposits formed when the uppermost part of oceanic crust represented by Early to Mid-Ordovician ocean floor basalts (MORB to OIB, possibly seamounts, Lambert *et al.* 1981; Phillips *et al.* 1995) and oceanic cherts entered a trench and were scraped-off and incorporated into an accretionary complex resembling the Shimanto Belt in eastern Japan (Taira *et al.* 1982). Similar conglomerates, with Dalradian metamorphic detritus, overlie the Arenig–Llanvirn South Connemara Group pillow basalts, which have MORB-like compositions (Ryan *et al.* 1983; Winchester & van Staal 1995), cherts and turbidites. The Rb/Sr ages of the clasts in the northern and Central Belts of Scotland suggest a provenance that encompassed the Grenville margin and an Early to Mid-Ordovician magmatic arc (Elders 1987). Miller & O'Nions (1984) suggested, on the basis of Sm–Nd studies, that the Dalradian margin was not, in general, a source area for the detritus but recent work by G. Oliver (pers. comm.) on garnet chemistry suggests a possible derivation from the Dalradian. The provenance of hornblende-biotite granite clasts of *c.* 600 ± 40 Ma in the Corsewall conglomerate are problematic, but tonalites and granodiorites of the Lady Slipper Pluton of the Grenville margin of Newfoundland have yielded a U–Pb zircon age of *c.* 555 Ma (Cawood *et al.* 1996). Although these rocks fall just outside the error range of the Rb/Sr ages, such rocks or undated, perhaps slightly older equivalents elsewhere may have provided detritus to conglomerates during the Caradoc. Rare Earth Element studies (Williams *et al.* 1996) and detrital micas (Kelley & Bluck 1989) in Caradoc sandstones of the Kirkcolm Formation of tract 2, which also has a northerly provenance, confirm a source area mainly comprising an Early to Mid-Ordovician magmatic arc and associated metamorphic country rocks,

while Nd-isotope studies also indicate a significant juvenile component consisting of ophiolitic debris (Stone & Evans 1995) in the rocks of tract 1. Collectively, the most obvious source of most rocks in tract 1 and 2 is provided by the obducted ophiolites of the Baie Verte Oceanic Tract, unroofed Dalradian metamorphic and equivalent tracts, and the Notre Dame Arc and its Grenville margin basement such as is exposed in Newfoundland and probably also exists beneath the Upper Palaeozoic cover of the Midland Valley, consistent also with the presence of allochthonous Laurentian faunas (Owen & Clarkson 1992) in tract 2. A comparison of the tract 1 clasts with those in coeval conglomerates of the Midland Valley suggest that these had different source areas. Hence, Elders (1987) and McKerrow & Elders (1989) suggested that the northern and Central Belts were part of an allochthonous terrane originally situated adjacent to the Grenville margin of Newfoundland rather than the Midland Valley, and moved into its present position by sinistral strike-slip translation. If this is correct, the Midland Valley and parts of the Southern Uplands were displaced independently by sinistral translation along the Laurentian margin.

In summary, the remnants of a once-continuous Shimanto-type accretionary complex (Annieopsquotch Accretionary Tract; Figs 9 and 10) occur along the south or south-eastern edge of the Notre Dame–Midland Valley magmatic arc in the northern Appalachians and British Caledonides. The sedimentary component of the Annieopsquotch Accretionary Tract in Newfoundland is significantly less than in the Southern Uplands. However, overprinting by post-Middle Ordovician folding and faulting was much more severe in the northern Appalachians (e.g. P. F. Williams *et al.* 1988; Lafrance & Williams 1992; O'Brien 1991, 1993). Hence, its original architecture may have been largely destroyed and removed by erosion in the Newfoundland Appalachians; alternatively, and perhaps more realistically, parts of it may have been translated along the active Laurentian margin (Elders 1987), making the Newfoundland Appalachians an area of strike-slip excision and the British Isles one of strike-slip duplication (Dewey & Shackleton 1984).

Tectonic setting of the Exploits Subzone of Newfoundland and New Brunswick

The Exploits Subzone (Figs 1 and 2) is composite and contains elements that are unrelated tectonically. These elements have in common that they developed at high or intermediate latitudes and contain Gondwanan or peri-Gondwanan faunas in the Early Ordovician. During the Mid- to Late Ordovician, fossils in sediments interlayered with

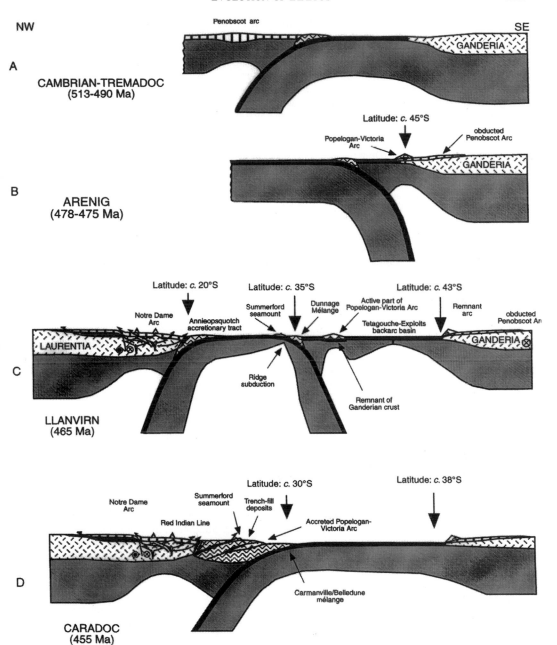

Fig. 10. Early Ordovician to Late Ordovician tectonic evolution of the Gander margin and its convergence with the Laurentian margin. See also Fig. 9.

volcanic rocks of the Exploits Subzone in the northern Appalachians began to acquire Laurentian or mixed affinities (Scoto-Appalachian), or contain low-latitude, deep-water faunas (e.g. *Foliomena*), signalling arrival of Exploits elements at similar latitudes to the Laurentian margin (Neuman 1984,

1994; Cocks & Fortey 1990; S. Williams *et al.* 1992, 1995; Colman-Sadd *et al.* 1992*a*; van Staal 1994).

The oldest known Exploits subduction-related, ensimatic rocks occur in the Middle Cambrian Lake Ambrose Volcanic Belt (513 ± 2 Ma, Dunning *et al.*

1991) of the Victoria Lake Group east of the Red Indian Line in Newfoundland (Fig. 1). The Victoria Lake Group is a composite volcanic–sedimentary complex with an apparently conformable cover of Mid-Ordovician limestone, shale and chert, and Late Ordovician to Silurian greywackes of the Badger Belt (Williams et al. 1993). The volcanic component also includes Upper Cambrian / lower Tremadoc (498 +6 / –4 Ma ensimatic Tulks Hill Belt, Evans et al. 1990) and Llandeilo (462 +4 / –2 Ma Victoria River rhyolite, Dunning et al. 1987) calc-alkaline felsic volcanic rocks. Therefore, subduction of Iapetus oceanic crust had started by at least the Mid-Cambrian at high latitudes and continued for more than 50 Ma. Formation of ensimatic and ensialic arc complexes at high latitudes thus overlaps for a considerable period with formation of the Baie Verte Oceanic Tract and Notre Dame Arc at low latitudes; hence, Iapetus must have contained several kinematically-independent subduction zones. It is significant that, towards the east, the Victoria Lake Group is tectonically juxtaposed with two late Neoproterozoic quartz-monzonite plutons (c. 563 Ma). The plutons experienced an earlier Cambrian metamorphism (c. 545 Ma) and were amalgamated with the Victoria Lake Group by at least the Early Silurian (Evans et al. 1990). The late Neoproterozoic ages, nature of plutonism and associated metamorphism are typical of Ganderian basement (S. J. O'Brien et al. 1996; van Staal et al. 1996a). These relationships suggest pre-Silurian interaction between elements of the Victoria Lake Group and Ganderia. Inherited pre-Cambrian zircons in the Victoria River rhyolite (Dunning et al. 1987) and Pb-isotope studies (Swinden & Thorpe 1984) suggest that amalgamation took place during the Early Ordovician Penobscot obduction (see below) because the older Cambrian and Tremadocian volcanic rocks of the Victoria Lake Group appear to be ensimatic with no indication of interaction with continental crust in their petrogenesis (Dunning et al. 1991).

Temporal correlatives of the Early and Mid-Ordovician parts of the Victoria Lake Group occur in the Wild Bight and Exploits Groups along strike (Fig. 1) towards the north (B. H. O'Brien et al. 1997; MacLachlan & Dunning in press). Detailed stratigraphic and petrological studies have revealed the formation of a south- or southeast-facing Tremadoc, juvenile volcanic–plutonic arc complex (c. 490–486 Ma) with an ophiolitic base (B. H. O'Brien 1992). This volcanic–plutonic arc complex experienced rifting and was exhumed during the Arenig. After a marked decrease or absence in volcanism during the early and middle Arenig (volcanic shut-off), volcanism flared up again in the late Arenig (c. 472–470 Ma), recording formation

of a north- or northwest-facing arc–back-arc complex (B. H. O'Brien et al. 1997; MacLachlan & Dunning in press), implying a reversal in arc polarity. Mafic calc-alkaline arc magmatism continued at least until c. 464 Ma, coeval with felsic calc-alkaline arc volcanism in the Victoria Lake Group (c. 462 Ma).

Formation of the Tremadoc arc volcanic rocks in the Victoria Lake, Exploits and Wild Bight Groups partly overlaps formation of Tremadoc ophiolites (e.g. 494 ± 2 Ma Pipestone Pond Ophiolite, Dunning & Krogh 1985). These ophiolites have been interpreted as having formed during rifting of an ensimatic arc (Jenner & Swinden 1993), remnants of which are probably represented by the Lake Ambrose and Tulks Hill Belts in the Victoria Lake Group and temporal equivalents in the Wild Bight and Exploits groups. Our interpretation is that all Tremadoc and older ensimatic magmatism in the Exploits Subzone represent part of a single late Mid-Cambrian to late Tremadoc south-east-facing, ensimatic arc–back-arc complex (Penobscot Arc of van Staal, 1994; Fig. 10). The late Arenig–Llanvirn (473–460 Ma), northwest-facing arc rocks in the Victoria Lake, Wild Bight and Exploits groups are referred to herein as the Victoria Arc (see below). Penobscot age (493 ± 2 Ma) arc–back-arc elements occur as a fault-bounded belt (Annidale Belt) in the Gander Zone of southern New Brunswick and adjacent Maine (Fig. 6), immediately north of the boundary between the Gander and Avalon zones (McLeod et al. 1992, 1994). Extension of the Penobscot Arc complex from the Red Indian Line to the line of the Gander River Ultrabasic Belt in eastern Newfoundland (Fig. 1) and into the Annidale Belt of southeastern New Brunswick and Maine (Fig. 2) requires that the oceanic complex was obducted for at least several hundred kilometres over the Gander margin. Subsequent deformation, erosion and Mid-to Late Ordovician cover sequences have largely destroyed or obscured this configuration, although mantles of Penobscot ophiolitic rocks and mélanges around large structural inliers of Gander Zone rocks in the Exploits Subzone of central Newfoundland leave little doubt that this process has happened (Fig. 1). Despite its considerable tectonic transport, obduction of the Penobscot arc was a relatively shallow 'soft' event, mainly characterized by mélanges (Williams & Piasecki 1990; van Staal 1994) and minor shear zones and folding but not by any significant regional metamorphism. Unconformities, stitching plutons and tightly-dated movement zones constrain this event to the late Tremadoc/early Arenig (485–478 Ma) in the northern Appalachians (Neuman 1967; Williams & Piasecki 1990; van Staal & Williams 1991; Colman-Sadd et al. 1992b; Dunning et al. 1993;

Tucker *et al.* 1994). Originally, the Penobscot Orogeny was defined in east-central Maine near the border with New Brunswick by an unconformity that is also present in central and northern New Brunswick. The unconformity separates Cambrian, *Oldhamia*-bearing sandstones of the Gander Zone and mélange (Boone *et al.* 1989) from overlying middle Arenig–Llanvirn volcanic and sedimentary rocks of the Exploits Subzone. The latter typically contain high latitude, peri-Gondwanan shelly faunas (Neuman 1984). The Mid-Ordovician rocks (including their peri-Gondwanan fossils) of the Exploits Subzone were also deposited unconformably above the Penobscot ophiolites and the Gander sandstones in Newfoundland and thus form an important overstep sequence (van Staal & Williams 1991; Colman-Sadd *et al.* 1992*b*).

After the Penobscot collision, arc magmatism, represented by ensialic arc plutonic–volcanic complexes, flared up again during the Arenig (479–474 Ma) in central and northern New Brunswick and adjacent Maine. Rifting and migration of this arc towards the northwest formed the Arenig–Llanvirn Popelogan Arc–Tetagouche back-arc complex (*c.* 473–455 Ma) in New Brunswick (van Staal *et al.* 1991), demanding that subduction was towards the southeast and that subduction polarity along the Gander margin had reversed after the Penobscot collision (van Staal 1994). Equivalents of the Arenig part of the Popelogan Arc in southwestern Newfoundland are the large Baggs Hill granite pluton (478 Ma, Tucker *et al.* 1994), which cuts the Penobscot structures and obducted ophiolites, and the nearly coeval tonalites, diorites and granodiorites of the Margaree Complex (*c.* 474 Ma) near Port aux Basques (Van Staal *et al.* 1994; Valverde-Vaquero *et al.* 1998). The Baggs Hill Pluton was partially exhumed before the Mid-Ordovician (van Staal *et al.* 1996*c*), probably as a result of arc rifting. The late Arenig–Llanvirn arc and rifted arc rocks of the Victoria Lake, Exploits and Wild Bight Groups (Victoria Arc) in central and northern Newfoundland (Fig. 1) are the obvious kinematic equivalents of the Mid-Ordovician part of the New Brunswick Popelogan Arc. The Mid-Ordovician felsic volcanic rocks (and their contained syn-genetic massive sulphides) of the Bay du Nord and Baie d'Espoir Groups in southern and southeastern Newfoundland (Fig. 1) are lithologically, temporally and isotopically similar to coeval rocks in the Tetagouche Group (Swinden & Thorpe 1984; Dunning *et al.* 1990; Colman Sadd *et al.* 1992*b*). Together with the equivalent volcanic rocks in the Tetagouche Group in New Brunswick they are considered to represent magmatism generated during the early phases of ensialic back-arc basin formation (referred to as the Exploits back-arc basin in Newfoundland). Both

sets of felsic volcanic rocks in New Brunswick and Newfoundland were deposited on Ganderia basement (Colman Sadd *et al.* 1992*b*; van Staal *et al.* 1996*a* and references therein). In New Brunswick, the Tetagouche back-arc basin opened up sufficiently to produce oceanic crust (470–460 Ma MORB-like Fournier Ophiolite Complex (van Staal *et al.* 1991; Sullivan *et al.* 1990; Sullivan & van Staal 1996). Although no Mid-Ordovician ophiolite has been recognized as such in Exploits back-arc basin in Newfoundland, supposedly Mid-Ordovician ocean floor basalts (E- or T-MORB) occur together with silicic volcanic rocks in the Cold Spring Pond and North Steady Pond Formations of Central Newfoundland (Fig. 1; Swinden 1988). Their geological and geochemical characteristics are reminiscent of the fragments of transitional and oceanic crust preserved in the nappes that form the boundary between the ensialic and ensimatic elements of the Tetagouche backarc basin in New Brunswick (van Staal *et al.* 1991).

Accretion of the Popelogan–Victoria Arc with the Annieopsquotch Accretionary Tract of the Notre Dame Subzone and, by implication, Laurentia is reasonably well constrained in the Canadian Appalachians. First, there is the shut-off and locally also uplift of the Popelogan Arc in the Caradoc (van Staal *et al.* 1991) and, second, there is a sedimentary linkage represented by arc and ophiolitic detritus and resedimented fossils derived from the Notre Dame Arc in the late Caradoc–Ashgill rocks of the adjacent Exploits Subzone (Nelson 1981; Arnott 1983; Colman-Sadd *et al.* 1992*a*; O'Brien *et al.* 1997). The Upper Ordovician, generally north-younging, upward-coarsening greywackes and conglomerates of the Badger Group that overlie the rocks of the Victoria Arc probably also date accretion. Their deposition appears to be diachronous from west to east and likely to have been structurally controlled because there are consistent stratigraphic differences between each thrust sheet (Arnott 1983; Arnott *et al.* 1985; P. F. Williams *et al.* 1988). We interpret these rocks as clastic trench-fill wedges (McKerrow & Cocks 1976) deposited when the Victoria Arc together with an accreted seamount (Summerford Group: E-MORB – limestone association, Wasowski & Jacobi 1985) entered the trench in front of the Annieopsquotch Accretionary Tract. We propose that the Summerford seamount was already accreted to the Victoria Arc prior to collision with the Annieopsquotch Accretionary Tract when it entered the west-facing subduction complex that must have fringed this arc immediately to the west (Fig. 10). Remnants of this accretionary complex are represented by the late Arenig–early Llanvirn Dunnage Mélange, which contains blocks of the Summerford seamount and the Early Ordovician Exploits Group (Wasowski & Jacobi 1985). The presence of the Cambrian trilobite *Baliella* in a limestone block in the Dunnage Mélange is consistent with the Dunnage having formed at high latitudes near the Gondwanan rather than the Laurentian side of the Iapetus Ocean (Dean 1985). The Dunnage accretionary complex was probably small, like the accretionary complex off the Mariana Arc, because the supply of clastic material to intraoceanic trenches is generally low. The Dunnage Mélange was intruded late syn- to post-kinematically by the early Llanvirn (*c.* 467 Ma) Coaker Porphyry (Elliott *et al.* 1991). The Coaker Porphyry has been interpreted as an S-type granite, although it contains many ultramafic xenoliths (Hibbard & Williams 1979; Lorentz 1984; Currie 1995). This porphyry probably formed by melting of sediments in the accretionary wedge, either during subduction of a spreading ridge (Kidd *et al.* 1977) shortly after accretion of the Summerford seamount to the Victoria Arc or as a result of trenchward migration of this arc. Other possible remnants of accreted seamounts or ridge subduction along the approximate trace of the Red Indian Line are the Caradoc Winterville and Munsungun within-plate alkalic and tholeiitic basalts in northern Maine near the New Brunswick border (Winchester & van Staal 1994; Fig. 2), which are associated with Caradoc black shales typical of the Exploits Subzone. Their low palaeomagnetic latitudes (10–20°S, Potts *et al.* 1993, 1995) indicates that they must have formed close to the Laurentian margin in the Iapetus Ocean; hence, formation in the Tetagouche back-arc basin as earlier postulated by Winchester & van Staal (1994) is probably incorrect. A Late Ordovician collision between the Popelogan–Victoria Arc and the Notre Dame Arc also explains the breakdown in faunal provinciality in the late Llanvirn–Caradoc (replacement of peri-Gondwanan faunas by Scoto-Appalachian faunas in Exploits Subzone rocks) and poor preservation of the Exploits accretionary complex because most of the latter was probably subducted beneath the Notre Dame Arc. Modern analogues of face-to-face arc collisions occur in the Molucca Sea (Fig. 4) and immediately east of New Guinea where the Trobriand Trench and associated fore-arc are being subducted diachronously beneath the Finistere–south Solomon arc (Pegler *et al.* 1995).

When the Notre Dame and Victoria Arcs had collided in the Late Ordovician, the active Laurentian margin (Notre Dame Arc) was situated at intermediate latitudes (*c.* 20–30°S), separated from the Gander margin by the Tetagouche–Exploits back-arc basin. The width of this marginal basin is constrained solely by palaeomagnetic data

obtained from the Tetagouche Group by Liss *et al.* (1993) who determined an average palaeolatitude of 52° + 21/–16°S, mainly from Llanvirn-age (*c.* 470–466 Ma) rocks. We consider that their data probably derives from the passive margin side of the back-arc basin. Considering its large error, and that Avalon was consistently moving northwards such that the Avalon peninsula of eastern Newfoundland acquired a palaeolatitude of 32° ± 10°S (Hodych & Buchan 1994) at 441 Ma (Ashgill, Greenough *et al.* 1993), suggests that the calculated average of Liss *et al.* (1993) probably over-estimates the palaeolatitude of Ganderia, consistent with a Caradoc palaeolatitude of *c.* 43°S for Avalon in the British Isles (Trench *et al.* 1992; Channell & McCabe 1992). The Tetagouche–Exploits back-arc basin thus achieved a width of at least 1100 km by the Caradoc (*c.* 10° of latitudinal separation) but not more than 2000 km. An upper value of 2000 km is constrained by the occurrence of low-latitude, midcontinent, conodont faunas in late Caradoc to Ashgill limestones (Nowlan *et al.* 1997) deposited on Avalon Zone rocks (Ganderia) in southern New Brunswick (Fig. 2). A Tetagouche–Exploits basin wider than 2000 km would also imply unrealistic convergence rates of 20 cm/a or higher. The large width of the Tetagouche–Exploits Basin (Iapetus II of van der Pluijm & van Staal 1988) was achieved probably by slab rollback (hinge retreat; Dewey 1980), dragging the Popelogan–Victoria Arc across Iapetus to the Laurentian margin in a manner analogous to the movement of the Cimmerian microcontinent across Palaeo-Tethys, opening up Neo-Tethys in its wake (Sengör 1979).

Closure of the Tetagouche–Exploits Basin took place by west or northwestward-directed sub-duction (Fig. 10) during the Ashgill and Lower Silurian (van Staal 1994; Currie 1995); basically the northwest-dipping Benioff zone beneath the Notre Dame Arc/Annieopsquotch Accretionary Tract, jumped eastward behind the accreted Popelogan–Victoria Arc. Hence, southeast-directed thrusting continued in the accreted Victoria–Exploits Arc in Newfoundland during the Early Silurian (e.g. Lafrance & Williams 1992) and propagated southeastward when progressively more of the Tetagouche–Exploits Basin was subducted (van Staal 1994; Currie 1995). Continuation of sub-duction is also supported by the flare-up of arc magmatic activity (now dominantly subaerial) in the Notre Dame Subzone of Newfoundland and northern Quebec during the Ashgill and early Llandovery (453–430 Ma, Whalen 1989; David & Gariepy 1990; Dunning *et al.* 1990; Dubé *et al.* 1996). The suture zone of the Tetagouche–Exploits Basin is marked by Ashgill blueschists (van Staal *et al.* 1990) and Late Ordovician to Early Silurian ophiolitic mélanges in New Brunswick and

Newfoundland (e.g. Duder Complex–Carmanville Mélanges in northeast Newfoundland and Belledune River Mélange in northern new Brunswick, Williams *et al.* 1993; van Staal 1994; Currie, 1995; Lee & Williams 1995).

The Exploits Subzone in New England and the British Isles

The Mid-Ordovician Popelogan Arc can be traced into adjacent Maine (Winchester & van Staal 1994) and probably is equivalent to the Mid- to Late Ordovician Bronson Hill magmatic arc further along strike in New Hampshire and Massachusetts (e.g. Leo 1985; Tucker & Robinson 1990). Late Proterozoic basement (613 Ma) and detrital zircon suites in associated quartzites (Robinson & Tucker 1996) are similar to those found in Ganderian basement in New Brunswick and Newfoundland (see van Staal *et al.* 1996*a* and references therein). Another link between Ganderian basement and arc plutons is preserved possibly in the Weeksburo–Lunksoos inlier in central Maine (Fig. 2). The Upper Ordovician Rockabema Diorite (Neuman 1967) probably also forms part of the Popelogan Arc. It stitches Cambrian (*Oldhamia*) sandstones of the Grand Pitch Formation and Early Caradoc tholeiitic basalts. The basalts acquired a Late Ordovician palaeolatitude of 20°S (Wellensiek *et al.* 1990), which is consistent with their accretion to and/or forming part of the Popelogan Arc, because the arc had docked or nearly so by this stage with the active Laurentian margin.

The Bronson Hill Arc was accompanied by a back-arc basin, the closure of which telescoped the Partridge and Ammonoosuc volcanic rocks in Massachusetts with the magmatic arc (Hollocher 1993). Rifting of the Bronson Hill magmatic arc and formation of oceanic-like back-arc crust must have started by at least 467 ± 3 Ma in New Hampshire (Fitz & Moench 1996), whereas volcanic rocks in Massachusetts are in part as young as 453–449 Ma (Tucker & Robinson 1990). Combined, these data suggest a tectonic setting very similar to that represented by the Popelogan Arc–Tetagouche back-arc system in New Brunswick and Maine, although arc–back-arc activity appears to have continued slightly longer in southern New England.

The Boil Mountain Ophiolite and the on-strike soapstone belt, immediately west of the Bronson Hill Anticlinorium in the Connecticut Valley in New Hampshire and Massachusetts (Lyons *et al.* 1982), are inferred to mark the approximate trace of the Red Indian Line in New England. Equivalents of the Boones Point and Exploits subduction complexes *per se* have not been identified but could be buried beneath the Connecticut Valley or obscured/destroyed during Acadian tectonism. The best record of the Ordovician accretionary events is preserved in Maine, although the precise affiliation of the numerous subduction-related mélanges is not well understood. The Hurricane Mountain Mélange in western Maine supposedly has been thrusted above the Boil Mountain Ophiolite Complex (Boone *et al.* 1989). It contains mafic clasts with low temperature–high pressure assemblages. A sodic–calcic amphibole in one of the entrained mafic bodies yielded an $^{40}Ar/^{39}Ar$ age of 485 ± 4 Ma, which predates the 477 Ma tonalite of the Boil Mountain Ophiolite Complex (Kusky *et al.* 1997). Juxtaposition of the Hurricane Mountain Mélange and the Boil Mountain Ophiolite Complex probably took place after the Early Ordovician because tonalite bodies or clasts do not occur in the mélange and high pressure–low temperature assemblages do not occur in the ophiolite. The Hurricane Mountain Mélange is unconformably overlain by Mid-Ordovician volcanic and sedimentary rocks. The latter have been interpreted to form part of the Popelogan–Tetagouche arc/back-arc system (Boone *et al.* 1989; Winchester & van Staal 1994). The Hurricane Mountain Mélange probably represents a remnant of an accretionary complex formed during the Penobscot Arc collision with Ganderia. An undated Ordovician subduction-related mélange (Fig. 2, Caucomgomoc Mélange), exposed as a basement window in the Connecticut Valley synclinorium, is associated with east-directed thrusting and has been correlated with the Hurricane Mountain Mélange (Pollock 1993). However, it lies much further north and not on strike with the Hurricane Mélange, nor does it have the same unconformable cover. This mélange may be an equivalent of the mélanges situated along the Red Indian Line in Newfoundland (e.g. Boones Point Complex) and, if correct, may mark the trace of the Red Indian Line through Maine (Fig. 2).

A possible continuation of the Popelogan–Victoria Arc into the British Isles has important implications for understanding the tectonic setting of the Southern Uplands (e.g. McKerrow 1987). Some Caradoc turbiditic sandstones of tract 2 in the northern Belt show a sudden influx of fresh andesitic detritus. Palaeocurrent directions in these sediments are variable (Leggett *et al.* 1982) but at least some of the volcanic detritus appears to have been derived from the south. It has been argued that the variability in palaeocurrent directions may have been the result of meandering currents on the trench floor (e.g. Leggett 1987) or the presence of a southerly island arc offshore (e.g. Hutton & Murphy 1987; Stone *et al.* 1987; Morris 1987). However, in the Southern Uplands, while there is evidence of the accretion of seamounts (e.g. Bail

Hill, Wrae), there is no exposed evidence of a major arc. Furthermore the smooth, diachronous, progressively younger southwards facies change from oceanic shales to trench turbidites seems to preclude such an arc in the Southern Uplands. Remnants of an arc may, however, be preserved in the Longford Down inlier of Ireland (Fig. 3). John Morris of the Irish Geological Survey has mapped calc-alkaline basalts and andesites in close proximity to Caradoc (*N. gracilis*) black shales in Slieve Aughty and Tattinlieve (Winchester & van Staal 1995), south of the Leadhills Line–Northern Belt Median Fault (LL-NBMF in Fig. 3). Implicit in the island arc models of Stone *et al.* (1987) and Morris (1987) is that the northern Belt sediments were deposited in a back-arc basin, contrary to the evidence for their formation in a trench. If, on the other hand, the southerly-derived andesite detritus reflects an approaching island arc, similar to the rapidly northwards-drifting Popelogan–Victoria arc in the northern Appalachians (Fig. 10), the sudden influx of andesite detritus may mark the unroofing of this arc as a result of its accretion to the active Laurentian margin during the Caradoc. Such a model removes the necessity of invoking a back-arc basin to explain the development of the northern Belt.

The Caradoc Grangegeeth Arc-related basalts and andesites (Winchester & van Staal 1995) in eastern central Ireland (Fig. 3), a small terrane bounded on all sides by splays of the Navan–Silvermines Fault (the classical trace of the Iapetus Suture in the British Isles), may represent a displaced part of the Popelogan–Victoria or Longford Down Arc; that is, it may have been situated originally to the north of the central and southern belts. We propose this model, because the Grangegeeth Arc Terrane, like the Popelogan–Victoria Arc in the Appalachians, shows a change in biogeographical affinity during the Ordovician. It contains high-latitude Atlantic province graptolites in the Arenig but mixed Laurentian–Baltic brachiopods in the Caradoc (referred to as Scoto-Appalachian by Harper & Parkes 1989 and Owen & Clarkson 1992). These faunas are different from the Anglo–Welsh faunas of the Lake District and the Laurentian faunas of the Southern Uplands during the Caradoc, although the affinities are much stronger with the Laurentian than the Anglo–Welsh faunas. The adjacent Bellewstown Terrane contains strikingly different within-plate mafic and felsic volcanic rocks (Winchester & van Staal 1995) and is characterized by Anglo–Welsh faunas during the Caradoc (cf. Harper & Rast 1964). The presence of a Llanvirn shelly fauna in the Bellewstown Terrane (Harper *et al.* 1990) similar to those in the Exploits Subzone of Newfoundland suggests that this terrane probably formed near the

outboard edge of the Gander margin. The oceanic basin in which the Ashgill–Lower Silurian rocks of the central and southern belts were deposited, i.e. the basin south of the accreted but poorly preserved Longford Down Arc, is the logical continuation of the Tetagouche–Exploits back-arc basin (Iapetus II) in the British Caledonides. Hence, the closest analogue to the Silurian Navan–Silvermines Suture of the British Isles in Newfoundland is the Early Silurian Dog Bay Line, just east of the Reach Fault where Arnott *et al.* (1985) located their Iapetus suture in Newfoundland. Continuation of north- or northwestward-directed subduction during the Ashgill and Early Silurian is much more widely accepted in the British Isles (e.g. Leggett *et al.* 1982) than in the northern Appalachians, despite preservation of high pressure–low temperature metamorphic rocks (e.g. blueschists) and subduction-related mélanges of Ashgill to Early Silurian age in New Brunswick (van Staal *et al.* 1990) and Newfoundland (e.g. Williams *et al.* 1993). Not surprisingly, similarly to the northern Appalachians, calc-alkaline, dominantly subaerial volcanism also continued into at least the Wenlock on the active Laurentian margin of the British Caledonides (McKerrow & Campbell 1960).

Underthrusting of Avalonian crust beneath the Southern Uplands accretionary prism began in the Wenlock in England (Kneller *et al.* 1993; King 1994) but may have started earlier in Ireland (Hutton & Murphy 1987) suggesting a west to east diachroneity. Williams *et al.* (1993) also presented evidence for a Silurian closure of the last vestiges of Iapetus along the Dog Bay Line in Newfoundland. Van Staal (1994), on the other hand, argued that the first arrival of highly extended Ganderian crust in the Brunswick subduction complex in the Ashgill marked the onset of collision between Ganderia and Laurentia in New Brunswick. Although this is consistent with a west-to-east diachroneity, the Brunswick subduction complex did not breach sea level before the late Llandovery. Considering that the earliest-accreted piece of Ganderian crust was situated on the arc side of the Tetagouche Basin (Rogers & van Staal 1997), the latter event rather than the first arrival of this highly-extended continental crust or a thick sediment pile on oceanic crust is, therefore, better regarded as the real start of collision. Moreover, the extended crust that entered the subduction zone during the Ashgill comprised a relatively large amount of E-MORB pillow lavas interlayered with felsic dacite (van Staal *et al.* 1991; Rogers & van Staal 1997) and probably had properties comparable with transitional crust. Analogous to the British Isles, an extensive foreland basin formed on the Ganderian margin during the Early to Late Silurian in the northern Appalachians (e.g.

Fredericton Trough, Fyffe 1995; van Staal & de Roo 1995), the sediments of which are lithologically similar to those in the Windemere Group in England. Like the Windemere Group, the Fredericton Trough was inverted and strongly deformed before the end of the Silurian (West *et al.* 1992; Fyffe 1995).

The Cambrian–Early Ordovician Penobscot Arc and its accretionary complex probably also continued into the British Isles (van Staal *et al.* 1996*a*). In Leinster (southeast Ireland, Fig. 3) ophiolitic, serpentinized ultramafic–mafic slivers and pods are tectonically juxtaposed with the Ribband and Duncannon Groups along the Wicklow Fault Zone (Gallagher 1989; Gallagher *et al.* 1994; Max *et al.* 1990; Winchester & van Staal 1995), while serpentinites, gabbros and basalts occur in the New Harbour Group and Gwna mélanges of the Monian Supergroup together with quartzite clasts of the underlying Cambrian South Stack Group (Gibbons & Ball 1991; Gibbons *et al.* 1994). The basalts have, at least in part, an island arc affinity (Thorpe *et al.* 1984). Lithostratigraphic correlations and sedimentary linkages suggest that the ophiolitic lenses and the highly tectonised Ribband Group structurally overlie a Cambrian substrate mainly consisting of turbiditic quartzose sandstones and shales, containing *Oldhamia*, of the Cahore, Cullenstown and Bray Groups (Max *et al.* 1990). On the basis of lithostratigraphic correlations with the above units in Ireland, Tietzsch-Tyler & Phillips (1989) also interpreted the South Stack Group as Cambrian, consistent with the possible presence of *Skolithos* burrows (see Gibbons *et al.* 1994 for review). The lithological characteristics of the latter rock units invite correlation with the Cambrian–Lower Ordovician Gander Zone sandstones and shales of the northern Appalachians, while the mafic–ultramafic rocks and mélanges are temporal and tectonic equivalents of the obducted late Tremadoc/early Arenig Penobscot Arc-accretionary complex. The thrust slices of the New Harbour Group and Gwna mélange above the South Stack Group quartzites on Anglesey and ultramafic rocks along the Wicklow fault zone are thus the British analogues of the Gander River Ultrabasic Belt line in Newfoundland. The unexposed basement to the British Gander Zone equivalents is linked commonly with the Rosslare and Coedana complexes of the Monian terrane (Gibbons *et al.* 1994). The latter complexes show similarities to exposed Ganderian basement in the northern Appalachians (van Staal *et al.* 1996*a*) and, similarly with the northern Appalachians, basement and structural cover are linked by an Arenig overstep sequence containing the 'Celtic' fauna. Hence, we consider these Lower Palaeozoic clastic rocks and the Skiddaw and Manx

groups in northern England to represent equivalents of the Ganderian margin in the British Isles. Common to all Lower Ordovician units of the Ganderian and Avalonian margin in the British Isles is a pre-Llanvirn deformation event represented by ductile folding, mélanges and/or olistostromes (Kokelaar 1986; Bennet *et al.* 1989; Max *et al.* 1990; Cooper *et al.* 1995) and late Tremadoc/early Arenig (*c.* 480 Ma) mylonites in the Rosslare Gneisses (Max & Roddick 1989). The locally-intense, pre-Llanvirn recumbent folding and olisto-stromes of the Skiddaw Group are generally interpreted to have formed in response to significant slope instability. We are not convinced that all the strong folding took place when the rocks were still unlithified but we suggest that this deformation is related to loading and obduction of the margin by the ensimatic Penobscot Arc-accretionary complex, although remnants of the latter are not exposed or preserved in northern England. However, the former existence of the Penobscot Arc is consistent with the juvenile arc detritus, represented by among others relatively high contents of Cr and Ni, in the upper Arenig sediments of the Skiddaw Group (Cooper *et al.* 1995). If early Ordovician southeastward ophiolite obduction took place along the southeast margin of Iapetus, the ultramafic rocks along the Wicklow Fault Zone must be part of an ophiolite nappe rooted to the west of the present Leinster Palaeozoic inlier, because the Bray Series lies to the west of the fault zone yet belongs to the Gander Zone.

Deformation of the Avalonian margin in mainland Wales occurred somewhat earlier than in northern England and Ireland. Folding in the Tremadoc was followed rapidly by the formation of a late Tremadoc/early Arenig ensialic arc (Kokelaar 1986), which may be correlated with the early Arenig part of the Popelogan/Victoria Arc in the northern Appalachians prior to its rifting from Ganderia. This requires some diachroneity of the Penobscot collision and arc polarity reversal, which is permissible considering the strike-slip motion along the Avalonian and Ganderian terranes in the British Isles (Kokelaar 1988; Gibbons 1990; Horak *et al.* 1996).

The late Llanvirn–Caradoc Lake District Arc in northern England (Kokelaar 1988) has no direct temporal equivalent in a similar tectonic setting in the northern Appalachians. Correlatives of the Lake District volcanic rocks occur in southeastern Ireland, Wales and as concealed bodies (Fig. 3) in eastern England (Noble *et al.* 1993). However, the Llanvirn–Caradoc volcanic rocks in both Wales and southeastern Ireland were formed in a supra-subduction extensional setting, probably an ensialic backarc basin (Kokelaar *et al.* 1984; McConnell *et al.* 1991; Winchester & van Staal 1995) rather than in an arc *senso stricto*. Kokelaar (1988) argued that the Lake District–Leinster Terrane was unlikely to have been adjacent to Wales in the Ordovician; instead, he suggested that these two terranes were juxtaposed by sinistral strike-slip motion. The alternative, that the Lake District Arc and Wales backarc basin were in their present relative positions with respect to one another, would require that the late Tremadoc arc of Wales had migrated northwards for more than 150 km and had intruded its own fore-arc–accretionary complex. There is no evidence for such an arc migration in England and Wales, nor does the Skiddaw Group resemble a fore-arc–accretionary complex. An alternative solution is that the Lake District Arc does not have a southwesterly strike continuing into Ireland but is related to south- or southwesterly-directed subduction of the Tornquist Sea beneath Avalonia (McKerrow *et al.* 1991; Noble *et al.* 1993). This model is attractive because the Late Ordovician cessation of the Lake District Arc coincides with amalgamation of benthic faunas between Baltica and Avalonia (Cocks & Fortey 1982), both suggesting a Late Ordovician start of collision between Baltica and Avalonia. This putative collision was probably 'soft' because there is no obvious supply of coarse clastics from Baltica during this time in Avalonia (Soper & Woodcock 1990). Collision probably continued into the late Llandovery when all physical barriers (seaways) for benthic ostracods were removed (Berdan 1990). The latter is consistent with Upper Ordovician to Lower Silurian calc-alkaline volcanic rocks in the Brabant massif of Belgium (André *et al.* 1986) and SW-dipping seismic reflectors in the subsurface across the Caledonian deformation front in Denmark and northern Germany (Tanner & Meissner 1996).

Synthesis of the tectonic evolution of Iapetus

The continental reconstructions of the southern hemisphere between 555 and 435 Ma shown in Fig. 7 summarize the opening and closing history of the Iapetus and neighbouring oceans. These reconstructions are based on a combination of palaeomagnetic, faunal and geological data, most of which has been discussed above. Opening of the Iapetus Ocean took place in the latest Neoproterozoic (590–550 Ma) and continued to widen throughout the Cambrian at *c.* 7.5 cm/a across the Amazonian–Laurentian sector. While Amazonia, Avalonia and Baltica moved to the south and Siberia moved east with respect to Laurentia, the latter continued migrating further northwards towards the equator (Mac Niocaill & Smethurst 1994). The absence of any significant differential

movement between Amazonia, Avalonia and Baltica during the Cambrian suggests that these were more or less fellow travellers. Spreading rates in the Tornquist Sea were therefore probably low or zero soon after the rift–drift transition. Baltica, however, was far enough from Avalonia and by implication, Amazonia, during the Early to Mid-Cambrian to have significantly more diverse trilobite faunas than Avalonia (McKerrow *et al.* 1992). On the other hand, Siberia moved independently and far enough east to allow faunal exchange (McKerrow *et al.* 1992) with the Cadomian elements of southern Europe (Bohemia, Amorica, Iberia). Because the south pole was situated in northwest Africa during most of the Cambrian (Meert & Van der Voo 1997), we infer that the majority of Morocco (north of the South Atlas Fault) containing warm water Early Cambrian archaeocyathans was, at the time, at a lower latitude than adjacent Mauritania; hence Morocco was situated further eastward (present coordinates) along the Gondwanan margin and may have been connected with Cadomia (Figs 1 and 7).

Intra-oceanic subduction east of Siberia appears to have been active during most of the Cambrian and formed the complex Kipchak Arc, an arm of which, the Tuva–Mongol Arc, may have extended into the Iapetus Ocean west of Siberia (Sengör & Natal'in 1996). Intra-Iapetus subduction on the Gondwanan side must have started by at least 513 Ma to account for the initiation of the Penobscot Arc. Soon after, subduction started along the Laurentian margin forming the Baie Verte Oceanic Tract, which may have linked up with, or perhaps formed by southward propagation of, the Tuva–Mongol Arc (Fig. 7b, c). Obduction of the Baie Verte Oceanic Tract and Penobscot Arc onto, respectively, the Laurentian and the Ganderian/ Avalonian margins started nearly coevally between 490 and 485 Ma (Tremadoc). Baltica also experienced obduction at this time as evidenced by, for example, eclogite metamorphism (Andréasson & Albrecht 1995). The Tremadoc was also marked by a transgression. Combined, these events signal a major plate reorganization during the Tremadoc, which changed the dynamics of the Iapetus Ocean to one of destruction rather than one of opening. Obduction on both margins was followed by an arc polarity reversal which formed the Notre Dame and Popelogan–Victoria Arcs on, respectively, the Laurentian and Ganderian margins. The Notre Dame Arc was a compressive Andean-type continental arc resulting from southward migration of Laurentia (Fig. 9). The Popelogan–Victoria Arc was extensional and rifted-off Gondwana shortly after the arc-polarity reversal and opened the Tetagouche–Exploits back-arc basin in its wake at a rate of *c.* 5 cm/a. Initiation of subduction beneath

Avalon in the Arenig approximately coincided with its departure from Gondwana, which opened the Rheic Ocean. Hence, the rapid roll-back that caused the rifting of the Popelogan–Victoria Arc may have contributed also to the partial break-up of west Gondwana during the Lower Ordovician.

Globally, the Early Ordovician and the Mid- to Late Cretaceous were two unique periods in Phanerozoic tectonic history (Dewey 1988). They were both characterized by high sea levels, with carbonate platforms and oceans with abundant black shales, widespread supra-subduction-zone ophiolites obducted shortly after generation, and abundant blueschists and arc magmatism. Both were times of substantial plate boundary reorganization, perhaps caused by high rates of relative plate motion with widely-dispersed continents and little continental collision (Dewey 1988).

By the middle to late Caradoc, the Notre Dame and the Popelogan–Victoria Arcs had collided, implying rapid closure (c. 12.5 cm/a) of the main tract of the Iapetus Ocean during the late Arenig–early Caradoc (475–455 Ma). This is not surprising considering that Iapetus was subducting beneath both margins, i.e. both the Laurentian and Avalonian/Ganderian margins were initially active (Fig. 10). Such a tectonic scenario was originally proposed by Dewey (1969), supported by Liss et al. (1993), van der Pluijm et al. (1995) and Mac Niocaill et al. (1997) based on their palaeomagnetic data and is also supported by palaeontological evidence (e.g. Cocks & Fortey 1990).

After the Caradoc, convergence between Laurentia and Avalon continued with northwesterly-directed subduction but now involving mainly young oceanic crust of the Tetagouche–Exploits basin (Iapetus II). At the same time, subduction of Tornquist oceanic lithosphere beneath Avalon is thought to have been responsible for the Caradoc arc–back-arc volcanism in England, Wales and southeast Ireland. This volcanism ceased in the Ashgill, presumably by the onset of collision between Baltica and Avalonia.

Geological data indicate that Laurentia and Avalon collided diachronously during the Silurian. In the British Isles the progressive underthrusting of Avalonia beneath Laurentia is best reflected in the distribution of the Wenlock and Ludlow foreland basin deposits in Central Ireland and northern England (Hutton & Murphy, 1987; King, 1994), and uplift of the Midland Valley. Deformation of the foreland basin rocks during the Emsian in northern England and Wales may represent the final stage of Avalonian/Laurentian collision. The presence of deep foreland basins on the Avalonian and Ganderian margins (van Staal & de Roo 1995) may also explain the patterns of provinciality of Silurian ostracod fauna. Through-

out the Silurian Laurentian ostracods were mostly distinct from the Avalonian/Baltic ostracod province (Berdan 1990). However, some ostracod connection existed from the Llandovery, suggesting geographic proximity of these two landmasses by the Llandovery, a view supported by the interchange of fish between Baltica and Laurentia in the Silurian (Turner & Turner 1974). Persistence of narrow seaways after the start of collision such as the foreland basin between present day Timor and Australia (cf. van Staal & de Roo 1995) can explain the ostracod biogeography. Breakdown of the faunal provincialities of ostracods and fish between Baltica and Gondwana in the Emsian coincide with the formation of two short-lived brachiopod faunal provinces (Rhenish–Bohemian versus the Appalachian) following cosmopolitan brachiopod conditions during most of the Silurian. These relationships have been explained by impingement of part of Gondwana with Laurentia during the Emsian creating a narrow land connection allowing interchange of fish and ostracods but separating the Rheic Ocean into two major seas each with a distinctive brachiopod fauna, namely the Rhenish–Bohemian fauna in the east and the Appalachian fauna in the west (Scotese & McKerrow 1990). The geographical distribution of these two provinces was further influenced by the presence of a mountain range between the Gaspé–Connecticut Valley seaway and Avalon (van Staal & de Roo 1995).

In the southern Appalachians, the Amazonian margin of Gondwana may have collided with Laurentia during the Silurian (Fig. 7f), possibly following an earlier arrival of the Carolina microcontinent. Such a scenario is consistent with the evidence of distinct Ordovician and Early Silurian loading cycles in the Taconian foredeep of the central and southern Appalachians (Dorsch et al. 1994). Subsequent closure of the Rheic Ocean probably involved a significant anticlockwise rotation of Gondwana to allow convergence between northwest Africa and Laurentia; hence the Laurentian margin became the locus of significant dextral transpression during the Devonian (Dewey 1982; Dalziel et al. 1994; van Staal & de Roo, 1995). The polarity of subduction of the Rheic Ocean was probably also towards the northwest. Late Ordovician/Silurian arc-like magmatism in Avalonia (Greenough et al. 1993) and Ganderia (Doig et al. 1990; Hepburn et al. 1995) is most simply explained by such a tectonic setting because Laurentia, with its accreted terranes, represents the upper plate. Cessation of this arc magmatism during the Late Silurian coincides with the onset of collision of Meguma (lower plate), either as a microcontinent or as a promontory on the Gondwanan continent, with Avalonia during the

Devonian. This collision probably was responsible for the Acadian Orogeny.

Much of the data and ideas presented herein were developed by discussions and fieldtrips with our Appalachian/Caledonian colleagues. We thank, particularly, Sandra Barr, Garry Boone, Robin Cocks, Steve Colman-Sadd, Ken Currie, Benoit Dubé, Greg Dunning, Richard Fortey, Les Fyffe, Wes Gibbons, Lindsay Hall, Chris Hepburn, Jim Hibbard, George Jenner, Paul Karabinos, Jonathan Kim, Tim Kusky, Shoufa Lin, Michel Malo, John Morris, Bob Neuman, Brian O'Brien, Paul Ryan, Dave Scofield, Scott Swinden, Alain Tremblay, Pablo Valverde, Ben van der Pluijm, Rob Van der Voo, Joe Whalen, Hank Williams, Paul Williams, and John Winchester. We realize that not all of them will agree with our interpretations! Paul Hoffman, Steve Lucas, Jim Hibbard and Alain Tremblay are thanked for critically reading this manuscript. We would like to thank Peter Coney, Rod Gayer and Robert Hall for their thoughtful reviews. We also wish to especially thank Claire Grainger and Deborah Lemkow for their skill and immense patience in drafting several of the figures. Van Staal thanks the Department of Earth Sciences of Oxford University for hospitality during his 1996 sabbatical. Mac Niocaill gratefully acknowledges funding from an EU Marie Curie Research Fellowship. Much of this work was done through funding provided by the Canada-New Brunswick cooperation agreement on Mineral development 1990–1994: The Canada-Newfoundland cooperation agreement on Mineral development 1990–1994, 1994–1995. This is Geological Survey of Canada contribution 1997229.

References

AFTALION, M., VAN BREEMEN, O. & BOWES, D. R. 1984. Age constraints on basement of the Midland Valley of Scotland. *Transactions of the Royal Society of Edinburgh: Earth Sciences*, **75**, 53–64.

ALEINIKOFF, J. N., WALTER, M. & FANNING, C. M. 1995. U–Pb ages of zircon, monazite and sphene from rocks of the Massabesic gneiss complex and Berwick Formation, New Hampshire and Massachusetts. *Geological Society of America, Abstracts with Programs*, **27**, 26.

ANDRÉ, L., HERTOGEN, J. & DEUTSCH, S. 1986. Ordovician-Silurian magmatic provinces in Belgium and the Caledonian orogeny in Middle Europe. *Geology*, **14**, 879–882.

ANDRÉASSON, P. G. & ALBRECHT, L. 1995. Derivation of 500 Ma eclogites from the passive margin of Baltica and a note on the tectonometamorphic heterogeneity of eclogite-bearing crust. *Geological Magazine*, **132**, 729–738.

ARMSTRONG, H. A., OWEN, A. W., SCRUTTON, C. T., CLARKSON, E. N. K. & TAYLOR, C. M. 1996. Evolution of the Northern Belt, Southern Uplands: implications for the Southern Uplands Controversy. *Journal of the Geological Society, London*, **153**, 197–205.

ARNOTT, R. J. 1983. Sedimentology of Upper Ordovician-Silurian sequences on New World Island, Newfoundland: separate fault-controlled basins? *Canadian Journal of Earth Sciences*, **20**, 345–354.

——, MCKERROW, W. S. & COCKS, L. R. M. 1985. The tectonics and depositional history of the Ordovician and Silurian rocks of Notre Dame Bay, Newfoundland. *Canadian Journal of Earth Sciences*, **22**, 607–618.

BARR, S. M. & WHITE, C. E. 1996. Contrasts in Late Precambrian–Early Paleozoic Tectonothermal History between Avalon Composite Terrane sensu stricto and other possible Peri-Gondwanan Terranes in Southern New Brunswick and Cape Breton Island, Canada. *In*: NANCE, D. & THOMPSON, M. (eds) *Avalonian and related Peri-Gondwanan Terranes of the Circum-North Atlantic*. Geological Society of America Special Paper, **304**, 95–108.

——, BEVIER, M. L., WHITE, C. E. & DOIG, R. 1994. Magmatic history of the Avalon Terrane of Southern New Brunswick, Canada, Based on U–Pb (Zircon) Geochronology. *Journal of Geology*, **102**, 399–409.

BENNET, M. C., DUNNE, W. M. & TODD, S. P. 1989. Reappraisal of the Cullenstown Formation: implications for the Lower Paleozoic tectonic history of SE Ireland. *Geological Journal*, **24**, 317–329.

BERDAN, J. M. 1990. The Silurian and Early Devonian biogeography of ostracodes in North America. *In*: MCKERROW, W. S. & SCOTESE, C. R. (eds) *Palaeozoic Palaeogeography and Biogeography*, Geological Society of London Memoir, **12**, 223–231.

BIRD, J. M. & DEWEY, J. F. 1970. Lithosphere Plate-continental margin tectonics and the evolution of the Appalachian Orogen. *Geological Society of America Bulletin*, **81**, 1031–1060.

BLOOMER, S. H., TAYLOR, B., MACLEOD, C. J., STERN, R. J., FRYER, P., HAWKINS, J. W. & JOHNSON, L. 1995. Early arc volcanism and the ophiolite problem: a perspective from drilling in the Western Pacific. *In*: TAYLOR, B. & NATLAND J. (eds) *Active margins and marginal basins of the western Pacific*, Geophysical Monograph, **88**, 1–30.

BLUCK, B. J. 1984. Pre-Carboniferous history of the Midland Valley of Scotland. *Transactions of the Royal Society of Edinburgh*, **75**, 275–295.

——, HALLIDAY, A. N., AFTALION, M. & MACINTYRE, R. M. 1980. Age and origin of the Ballantrae ophiolite and its significance to the Caledonian Orogeny and Ordovician time scale. *Geology*, **9**, 331–333.

BOILLOT, G., GIRARDEAU, J. & KORNPROBST, J. 1988. Rifting of the Galicia margin: crustal thinning and emplacement of mantle rocks on the seafloor. *Proceedings of the Ocean Drilling Program, Scientific Results*, **103**, 741–756.

BOND, G. C., NICKESON, P. A. & KOMINZ, M. A. 1984. Break-up of a supercontinent between 625 Ma and 555 Ma: New evidence and implications for continental histories. *Earth and Planetary Science Letters*, **70**, 325–345.

BOONE, G., DOTY, D. T. & HEIZLER, M. T. 1989. Hurricane Mountain Formation Mélange: description and tectonic significance of a Penobscottian accretionary complex. *In*: *Studies in Maine Geology*, Maine Geological Survey, **2**, 33–83.

BOUDETTE, E. L. 1982. Ophiolite assemblage of Early Paleozoic age in central Maine. *In:* ST JULIEN, P. & BÉLAND, J. (eds) *Major Structural Zones and Faults of the Northern Appalachians*, Geological Association of Canada Special Paper, **24**, 43–66.

BOWRING, S. A., GROTZINGER, J. P., ISACHSEN, C. E., KNOLL, A. H., PELECHATY, S. M. & KOLOSOV, P. 1993. Calibrating rates of Early Cambrian evolution. *Science,* **261**, 1293–1298.

CASEY, J. F. & DEWEY, J. F. 1984. Initiation of subduction zones along transform and accreting plate boundaries, triple junction evolution, and forearc spreading centres-implications for ophiolitic geology and obduction. *In:* GASS, I. G., LIPPARD, S. J. & SHELTON, A. W. (eds) *Ophiolites and Oceanic Lithosphere*, Geological Society of London Special Paper, **13**, 131–144.

—— & KIDD, W. S. F. 1981. A parallochthonous group of sedimentary rocks unconformably overlying the Bay of Island ophiolite complex, north Arm Mountain, Newfoundland. *Canadian Journal of Earth Sciences*, **18**, 1035–1050.

CAWOOD, P. A. & SUHR, G. 1992. Generation and obduction of ophiolites: constraints from the Bay of Islands Complex, Western Newfoundland. *Tectonics*, **11**, 884–897.

——, VAN GOOL, J. A. M. & DUNNING, G. R. 1996. Geological development of eastern Humber and western Dunnage zones: Corner Brook – Glover Island region, Newfoundland. *Canadian Journal of Earth Sciences* **33**, 182–198.

CHANDLER, F. W., SULLIVAN, R. W. & CURRIE, K. L. 1987. The age of the Springdale Group, western Newfoundland, and correlative rocks-evidence for a Llandovery overlap assemblage in the Canadian Appalachians. *Transactions of the Royal Society of Edinburgh, Earth Sciences*, **78**, 41–49.

CHANNELL, J. E. T. & McCABE, C. 1992. Palaeomagnetic data from the Borrowdale Volcanic Group:volcano-tectonics and Late Ordovician palaeolatitudes. *Journal of the Geological Society, London*, **149**, 881–888.

CLIFF, R. A., YARDLEY, B. W. D. & BUSSY, F. R. 1996. U–Pb and Rb–Sr geochronology of magmatism and metamorphism in the Dalradian of Connemara, western Ireland. *Journal of the Geological Society, London*, **153**, 109–120.

CLIFT, P. & RYAN, P. D. 1994. Geochemical evolution of an Ordovician island arc. *Journal of the Geological Society, London*, **151**, 329–342.

—— 1995. Volcaniclastic sedimentation and volcanism during rifting of western Pacific backarc basins. *In:* TAYLOR, B. & NATLAND J. (eds) *Active margins and marginal basins of the western Pacific*, Geophysical Monograph, **88**, 67–96.

COCKS, L. R. M. & FORTEY, R. A. 1982. Faunal evidence for oceanic separations in the Paleozoic of Britain. *Journal of the Geological Society, London*, **139**, 467–480.

—— & —— 1990. Biogeography of Ordovician and Silurian faunas. *In:* McKERROW, W. S. & SCOTESE, C. R. (eds) *Paleozoic Paleogeography and Biogeography*, Geological Society of London Memoir, 12, 97–104.

COISH, R. A., HICKEY, R. & FREY, F. A. 1982. Rare earth element geochemistry of the Betts Cove ophiolite, Newfoundland: complexities in ophiolite formation. *Geochimica et Cosmochimica Acta*, **46**, 2117–2134.

COLMAN-SADD, S. P., HAYES, J. P. & KNIGHT, I. 1990. Geology of the Island of Newfoundland. *Newfoundland Department of Mines and Energy*, Map 90-01

——, STONE, P., SWINDEN, H. S. & BARNES, R. P. 1992*a*. Parallel geological development in the Dunnage Zone of Newfoundland and the Lower Paleozoic terranes of southern Scotland: an assesment. *Transactions of the Royal Society of Edinburgh, Earth Sciences*, **83**, 571–594.

——, DUNNING, G. R. & DEC, T. 1992*b*. Dunnage-Gander relationships and Ordovician orogeny in central Newfoundland: a sediment provenance and U/Pb study: *American Journal of Science*, **292**, 317–355.

COOPER, A. H., RUSHTON, A. W. A, MOLYNEUX, S. G., HUGHES, R. A., MOORE, R. M. & WEBB, B. C. 1995. The stratigraphy, correlation, provenance and palaeogeography of the Skiddaw Group (Ordovician) in the English Lake District. *Geological Magazine*, **132**, 185–211.

CURRIE, K. L. 1995. The northeastern end of the Dunnage Zone in Newfoundland. *Atlantic Geology*, **31**, 25–38.

——, VAN BREEMEN, O., HUNT, P. A. & VAN BERKEL, J. T. 1992. The age of high-grade gneisses south of Grand Lake, Newfoundland. *Atlantic Geology* , **28**, 153–161.

DALLMEYER, R. D. 1977. Diachronous ophiolite obduction in western Newfoundland: evidence from 40Ar/39Ar ages of the Hare Bay metamorphic aureole. *American Journal of Science*, **277**, 61–72.

DALZIEL, I. W. D. 1992. On the organization of American plates in the Neoproterozoic and the breakout of Laurentia. *GSA Today*, **2**, 237–241.

——, DALLA SALDA, L., CINGOLANI, C. & PALMER, P. 1996. The Argentine Precordillera: A Laurentian Terrane? Penrose Conference Report. *GSA Today*, **6**(2), 16–18.

——, DALLA SALDE, L. H. & GAHAGAN, L. M. 1994. Paleozoic Laurentia–Gondwana interaction and the origin of the Appalachian–Andean mountain system. *Geological Society of America Bulletin*, **106**, 243–252.

DAVID, J. & GARIEPY, C. 1990, Early Silurian orogenic andesites from the central Quebec Appalachians. *Canadian Journal of Earth Sciences*, **27**, 632–643.

——, MARQUIS, R. & TREMBLAY, A. 1993. U–Pb geochronology of the Dunnage Zone in the southwestern Quebec Appalachians. *Geological Society of America, Annual Meeting Program with Abstracts*, A-485.

DE METS, C., GORDON, R. G, ARGUS, D. F. & STEIN, S. 1990. Current plate motions. *Geophysical Journal International*, **101**, 425–478.

DEAN, W. T. 1985. Relationships of Cambro-Ordovician faunas in the Caledonide-Appalachian region, with particular reference to trilobites. *In:* GAYER, R. A. (ed.) *The Tectonic Evolution of the Caledonide–Appalachian Orogen*, Braunschweig/Wiesbaden, Fredrich Vieweg & Sohn, 17–47.

DEC, T. & SWINDEN, H. S. 1994. Lithostratigraphic model, geochemistry and sedimentology of the Cotrells Cove Group, Buchan-Roberts Arm volcanic belt, Notre Dame Subzone. *Current Research, Newfoundland Department of Mines and Energy, Geological Survey Branch Report*, **94-1**, 77–100.

——, —— & DUNNING, R. G. 1997 Lithostgratigraphy and geochemistry of the Cotrells Cove Group, Buchans-Robert Arm volcanic belt: new constraints for the paeotectonic setting of the notre dame subzone, Newfoundland Appalachians. *Canadian Journal of Earth Sciences*, **34**, 86–103.

DEMPSTER, T. J. & BLUCK, B. J. 1991. Age and tectonic significance of the Bute amphibolite, Highland Border Complex, Scotland. *Geological Magazine*, **152**, 77–80.

DEWEY, J. F. 1969. Evolution of the Appalachian–Caledonian Orogen, *Nature,* **222**, 124–129,

—— 1975. Finite plate evolution: Implications for the evolution of the rock masses at plate margins. *American Journal of Science*, **275-A**, 260–284.

—— 1980. Episodicity, sequence and style at convergent plate boundaries. *In*: STRANGWAY, D. W. (ed.) *The Continental Crust and its Mineral Deposits.* Geological Association of Canada Special Paper, **20**, 553–573.

—— 1982. Plate tectonics and the evolution of the British Isles. *Journal of the Geological Society, London,* **139**, 317–414.

—— 1988. Lithospheric stress, deformation and tectonic cycles: the disruption of Pangea and the closure of Tethys. *In*: AUDLEY-CHARLES, M. G. & HALLAM, A. (eds) *Gondwana and Tethys.* Geological Society of London Special Publications, **37**, 23–40.

—— & LAMB, S. H. 1992. Active tectonics of the Andes. *Tectonophysics*, **205**, 79–95.

—— & RYAN, P. D. 1990. The Ordovician Evolution of the South Mayo Trough, Western Ireland. *Tectonics*, **9**, 887–901.

—— & SHACKLETON, R. M. 1984. A model for the evolution of the Grampian tract in the early Caledonides and Appalachians. *Nature,* **312**, 115–121.

——, HELMAN, M. L., TURCO, E., HUTTON, D. H. W. & KNOTT, S. D. 1989. Kinematics of the western Mediterranean. *In*: COWARD, M. P., DIETRICH, D. & PARK, R. G. (eds) *Alpine Tectonics.* Geological Society of London Special Publications, **45**, 265–283.

DOIG, R., NANCE, R. D., MURPHY, J. B. & CASSEDAY, R. P. 1990. Evidence for Silurian sinistral accretion of Avalon composite terrane in Canada. *Journal of the Geological Society, London,* **147**, 927–930

DORSCH, J., BAMBACH, R. K. & DRIESE, S. G. 1994. Basin-rebound origin for the "Tuscarora unconformity" in southwestern Virginia and its bearing on the nature of the taconic Orogeny. *American Journal of Science*, **294**, 237–255.

DUBÉ, B., DUNNING, G. R., LAUZIÉRE, K. & RODDICK, J. C. 1996. New insights into the Appalachian Orogen from geology and geochronology along the Cape Ray fault zone, southwest Newfoundland. *Geological Society of America Bulletin,* **108**, 101–116.

DUNNING, G. R. 1987. Geology of the Annieopsquotch Complex, southwest Newfoundland. *Canadian Journal of Earth Sciences*, **24**, 1162–1174.

—— & COUSINEAU, P. A. 1990. U/Pb ages of single zircons from Chain Lakes Massif and a correlative unit in ophiolitic mélange in Quebec. *Geological Society of America, Abstracts with Programs, Northeastern Section*, **22**, 13.

—— & KROGH, T. E. 1985. Geochronology of ophiolites of the Newfoundland Appalachians. *Canadian Journal of Earth Sciences*, **22**, 1659–1670.

—— & —— 1991. Stratigraphic correlation of the Appalachian Ordovician using advanced U–Pb zircon geochronology techniques. *In*: BARNES, C. R. & WILLIAMS, S. H. (eds) *Advances in Ordovician Geology.* Geological Survey of Canada Paper, **90-9**, 85–92.

—— & O'BRIEN, S. J. 1989. Late Proterozoic–Early Paleozoic crust in the Hermitage flexure, Newfoundland Appalachians: U/Pb ages and tectonic significance. *Geology*, **17**, 548–551.

—— & PEDERSEN, R. B. 1988. U/Pb ages of ophiolites and arc-related plutons of the Norwegian Caledonides: Implications for the development of Iapetus. *Contributions to Mineralogy and Petrology*, **98**, 13–23.

——, BARR, S. M., RAESIDE, R. P. & JAMIESON, R. A. 1990. U–Pb zircon, titanite and monazite ages in the Bras d'Or and Aspy terranes of Cape Breton Island, Nova Scotia: implications for magmatic and metamorphic history. *Geological Society of America Bulletin*, **102**, 322–330.

——, KEAN, B. F., THURLOW, J. G. & SWINDEN H. S. 1987. Geochronology of the Buchans, Roberts Arm, and Victoria Lake Groups and Mansfield Cove Complex, Newfoundland. *Canadian Journal of Earth Sciences*, **24**, 1175–1184.

——, O'BRIEN S. J., COLMAN-SADD, S. P., BLACKWOOD, R. F., DICKSON, W. L., O'NEILL, P. P. & KROGH, T. E. 1990. Silurian orogeny in the Newfoundland Appalachians. *Journal of Geology*, **98**, 895–913.

——, ——, O'BRIEN, B. H., HOLDSWORTH, R. E. & TUCKER, R. D. 1993. Chronology of Pan-African, Penobscot and Salinic shear zones on the Gondwanan margin, Northern Appalachians. *Geological Society of America Annual Meeting, Abstracts with Programs*, 421–422.

——, SWINDEN, H. S., KEAN, B. F., EVANS, D. T. W. & JENNER, G. A. 1991. A Cambrian island arc in Iapetus; geochronology and geochemistry of the Lake Ambrose volcanic belt, Newfoundland Appalachians. *Geological Magazine,* **128**, 1–17.

EGAL, E., GUERROT, C., LE GOFF, E., THIÉBLEMONT, D. & CHANTRAINE, J. 1996. The Cadomian Orogeny revisited in northern Brittany (France). *In:* NANCE, D. & THOMPSON, M. (eds) *Avalonian and related Peri-Gondwanan Terranes of the Circum-North Atlantic.* Geological Society of America Special Paper, **304**, 281–318.

ELDERS, C. F. 1987. The provenance of granite boulders in conglomerate of the Northern and Central Belts of the Southern Uplands of Scotland. *Journal of the Geological Society, London,* **144**, 853–863.

ELIAS, E. M., MACINTYRE, R. N. & LEAKE, B. E. 1988.

The cooling history of Connemara western Ireland, from K–Ar and Rb–Sr studies. *Journal of the Geological Society, London,* **145**, 649–660.

ELLIOTT, C. G., DUNNING, G. R. & WILLIAMS, P. F. 1991. New constraints on the timing of deformation in eastern Notre Dame Bay, Newfoundland, from U/Pb zircon ages of felsic intrusions. *Geological Society of America Bulletin,* **103**, 125–135.

EVANS, D. T. W., KEAN, B. F. & DUNNING, G. R. 1990. Geological studies, Victoria Lake Group, Central Newfoundland. *Current Research, Newfoundland Department of Mines and Energy, Geological Survey Branch, Report* 90–1, 131–144.

FITCH, T. J. 1972. Plate convergence, transcurrent faults, and internal deformation adjacent to south-east Asia and the western Pacific. *Journal of Geophysical Research,* **77**, 4432–4460.

FITZ, T. J. & MOENCH, R. H. 1996. Tectonic significance of the sheeted dike complex in the Chickwolnepy intrusions in northern New Hampshire. *Geological Society of America, Annual Meeting Program with Abstracts,* **28**, 54.

FLINN, D., MILLER, J. A. & RODDOM, D. 1991. The age of the Norwick hornblendic schists of Unst and Fetlar and the obduction of the Shetland ophiolite. *Scottish Journal of Geology,* **27**, 11–19.

FORTEY, R. A. & COCKS, L. R. M. 1992. The Early Paleozoic of the north Atlantic Region as a test case for the use of fossils in continental reconstruction. *Tectonophysics,* **206**, 147–158.

——, HARPER, D. A. T., INGHAM, J. K., OWEN, A. W. & RUSHTON, A. W. A. 1995. A revision of Ordovician series and stages from the historical type area. *Geological Magazine,* **132**, 15–30.

FRIEDRICH, A. M., BOWRING, S. A. & HODGES, K. V. 1997. U–Pb geochronological constraints on the duration of arc magmatism and metamorphism from Connemara, Irish Caledonides. *Terra Nova,* **9**, abstract supplement 1, p.331.

FYFFE, L. R. 1995. Fredericton Belt. *In:* WILLIAMS, H. (ed.) *Geology of the Appalachian-Caledonian Orogen in Canada and Greenland,* Geological Survey of Canada, Geology of Canada, **6**, 351–354.

—— & PICKERILL, R. K. 1993. Geochemistry of Upper Cambrian Lower Ordovician black shale along a northeastern Appalachian transect. *Geological Society of America Bulletin,* **105**, 897–910.

GALLAGHER, V. 1989. The occurrence, textures, mineralogy and geochemistry of a chromite-bearing serpentenite. *Geological Survey of Ireland Bulletin,* **4**, 89–98.

——, O'CONNOR, P. J. & AFTALION M. 1994. Intra-Ordovician deformation in southeast Ireland: evidence from the geological setting, geochemical affinities and U–Pb zircon age of the Croghan Kinshelagh granite. *Geological Magazine,* **131**, 669–684.

GEOLOGICAL WORLD ATLAS 1976. Commision for the Geological Map of the World and UNESCO, Paris.

GIBBONS, W. & BALL, M. J. 1991. A discussion of Monian Supergroup stratigraphy in northwest Wales. *Journal of the Geological Society, London,* **148**, 5–8.

—— 1990. Transcurrent ductile shear zones and the dispersal of the Avalon superterrane. *In:* D'LEMOS, R. S., STRACHAN, R. A. & TOPLEY, C. G. (eds) *The Cadomian Orogeny.* Geological Society of London Special Publication, **51**, 407–423.

——, TIETZSCH-TYLER, D., HORAK, J. M. & MURPHY, F. C. 1994. Precambrian rocks in Anglesey, southwest Llyn and southeast Ireland. *In:* GIBBONS, W. & HARRIS, A. L. (eds) *A revised correlation of Precambrian rocks in the British Isles.* Geological Society of London, Special report, **22**, 75–84.

GOODWIN, L. B. & WILLIAMS, P. F. 1990. Strike-slip motion along the Baie Verte Lone. *Atlantic Geology,* **26**, 170.

GRAY, J. R. & YARDLEY, B. W. D. 1979. A Caledonian blueschist from the Irish Dalradian. *Nature,* **278**, 736–737.

GREENOUGH, J. D., KAMO, S. L. & KROGH, T. E. 1993. A Silurian U–Pb age for the Cape St. Mary's sills, Avalon Peninsula, Newfoundland, Canada: implications for Silurian orogenesis in the Avalon Zone. *Canadian Journal of Earth Sciences,* **30**, 1607–1612.

GRUNOW, A., HANSON, R. & WILSON, T. 1996. Were aspects of Pan-African deformation linked to Iapetus opening? *Geology,* **24**, 1063–1066.

HALL, L. M., VAN STAAL, C. R. & WILLIAMS, H. 1994. Ordovician structural evolution of SW-Newfoundland. *Geological Association of Canada / Mineralogical Association of Canada Annual Meeting Program with Abstracts, Waterloo, Ontario,* A46.

HALL, R. 1996. Reconstructing Cenozoic South-East Asia. *In:* HALL, R. & BLUNDELL, D. J. (eds) *Tectonic Evolution of Southeast Asia,* Geological Society of London Special Publications, **106**, 153–184.

—— 1997. *Cenozoic Plate Reconstructions of Southeast Asia.* Geological Society of London Special Publications, **126**, 11–23.

HAMILTON, P. J., BLUCK, B. J. & HALLIDAY, A. N. 1984. Sm–Nd ages from the Ballantrae complex, SW Scotland. *Transactions of the Royal Society of Edinburgh: Earth Sciences,* **75**, 183–187.

HAMILTON, W. 1970. The Uralides and the motion of the Russian and Siberian platforms. *Geological Society of America Bulletin,* **81**, 2553–2576.

HARPER, D. A. T. & PARKES, M. A. 1989. Palaeontological constraints on the definition and development of the Irish Caledonide terranes. *Journal of the Geological Society, London,* **146**, 413–415.

——, —— & HÖEY, A. N. 1990. Intra-Iapetus brachiopods from the Ordovician of eastern Ireland: implications for Caledonide correlation. *Canadian Journal of Earth Sciences,* **27**, 1757–1761.

HARPER, J. C. & RAST, N. 1964. The faunal succession and volcanic rocks of the Ordovician near Bellowstown, County Meath. *Proceedings of the Royal Irish Academy,* **B64**, 1–23.

HARRIS, A. L. 1969. The relationship of the Leny Limestone to the Dalradian. *Scottish Journal of Geology,* **5**, 187–190.

HARRIS, D. H. M. 1995. Caledonian transpressional terrane accretion along the Laurentian margin in Co. Mayo, Ireland. *Journal of the Geological Society, London,* **152**, 797–806.

HEPBURN, J. C., DUNNING, G. R. & HON, R. 1995. Geochronology and regional tectonic implications of Silurian deformation in the Nashoba terrane, southeastern New England, USA. *In:* HIBBARD, J., VAN STAAL, C. R. & CAWOOD, P. (eds) *Current Perspectives in the Appalachian-Caledonian Orogen.* Geological Association of Canada Special Paper, **41**, 349–366.

HIBBARD, J. & WILLIAMS, H. 1979. Regional setting of the Dunnage Melange in the Newfoundland Appalachians. *American Journal of Science,* **279**, 993–1021.

—— 1982. Significance of the Baie Verte Flexure, Newfoundland. *Geological Society of America Bulletin,* **93**, 790–797.

—— 1988. Stratigraphy of the Fleur de Lys Belt, northwest Newfoundland. *In:* WINCHESTER, J. A. (ed.) *Later Proterozoic stratigraphy of the Northern Atlantic Regions.* Blackie, London, 200–211.

HODYCH, J. P. & BUCHAN, K. L. 1994. Paleomagnetism of the Early Silurian Cape St. Mary's Sills of the Avalon Peninsula of Newfoundland, Canada. *EOS,* **75**, 128.

HOFFMAN, P. F. 1996. Evolutionary model of the Southern Brazilide Ocean. *EOS,* **77**, S87.

—— 1991. Did the breakout of Laurentia turn Gondwanaland inside-out? *Science,* **252**, 1409–1412.

HOLDSWORTH, R. E. 1994. Structural evolution of the Gander-Avalon terrane boundary: a reactivated transpression zone in the NE Newfoundland Appalachians. *Journal of the Geological Society, London,* **151**, 629–646.

HOLLOCHER, K. 1993. Geochemistry and origin of volcanics in the Ordovician Partridge Formation, Bronson Hill Anticlinorium, west-central Massachusetts. *American Journal of Science,* **293**, 671–721.

HOLTZMAN, B., TRZCIENSKI, W. & GROMET, L. P. 1996. New tectonic constraints on the Boil Mountain Complex/Chain lakes unit contact, central western Maine. *Geological Society of America Annual Meeting, Abstracts with Programs,* **28**, 65.

HOLUB, F. V., KLAPOVA, H., BLUCK, B. J. & BOWES, D. R. 1984. Petrology and geochemistry of post-obduction dykes of the Ballantrae complex, SW Scotland. *Transactions of the Royal Society of Edinburgh, Earth Sciences,* **75**, 211–224.

HORAK, J. M., DOIG, R., EVANS., J. A. & GIBBONS, W. 1996. Avalonian magmatism and terrane linkage: new isotopic data from the Precambrian of North Wales. *Journal of the Geological Society, London,* **153**, 91–100.

HUTTON, D. H. W. 1987. Strike-slip terranes and a model for the evolution of the British and Irish Caledonides. *Geological Magazine,* **124**, 405–425.

HUTTON, D. H. W., AFTALION, M. & HALLIDAY, A. N. 1985. An Ordovician ophiolite in County Tyrone, Ireland. *Nature,* **315**, 210–212.

—— & MURPHY, F. C. 1987. The Silurian of the Southern Uplands and Ireland as a successor basin to the end-Ordovician closure of Iapetus. *Journal of the Geological Society, London,* **144**, 765–772, 1987.

JAMIESON, R. A. 1977. The first metamorphic sodic amphibole identified from the Newfoundland Appalachians – its occurrence, composition and possible tectonic implications. *Nature,* **165**, 428–430.

JENNER, G. A. & FRYER, B. J. 1980. Geochemistry of the upper Snooks Arm basalts, Burlington Peninsula, Newfoundland: evidence against formation in an island arc. *Canadian Journal of Earth Sciences,* **17**, 888–900.

—— & SWINDEN, H. S. 1993. The Pipestone Pond Complex, central Newfoundland: complex magmatism in an eastern Dunnage Zone ophiolite. *Canadian Journal of Earth Sciences,* **30**, 434–448.

——, DUNNING, G. R., MALPAS, J., BROWN, M. & BRACE, T. 1991. Bay of Islands and Little Port complexes, revisited: age, geochemical and isotopic evidence confirm suprasubduction-zone origin. *Canadian Journal of Earth Sciences,* **28**, 1635–1652.

JOHNSON, R. J. E., VAN DER PLUIJM, B. A. & VAN DER VOO, R. 1991. Paleomagnetism of the Moreton's Harbour Group, Northeastern Newfoundland Appalachians: Evidence for an Early Ordovician island Arc near the Laurentian margin of Iapetus. *Journal of Geophysical Research,* **96**, 689–701.

KARABINOS, P. & TUCKER, R. D. 1992. The Shelburn Falls arc in western Massachusetts and Connecticut: the lost arc of the Taconic orogeny. *Geological Society of America Annual Meeting, Abstracts with Programs,* **24**, 288–289.

——, STOLL, H. M., HEPBURN, J. C. & TUCKER, R. D. 1993. The New England segment of the Laurentian margin: continuous tectonic activity from Early Ordovician through Early Devonian. *Geological Society of America Annual Meeting, Abstracts with Programs,* A-178.

KARSON, J. & DEWEY, J. F. 1978. Coastal complex, western Newfoundland: an Early Ordovician oceanic fracture zone. *Geological Society of America Bulletin,* **89**, 1037–1049.

KEAN, B. F., EVANS, D. T. W. & JENNER, G. A. 1995. Geology and mineralization of the Lushs Bight Group. *Newfoundland Department of Natural Resources, Geological Survey Report,* 95-2.

KEEN, C. E., KEEN, M. J., NICHOLS, B., *ET AL.* 1986. Deep seismic reflection profile across the northern Appalachians. *Geology,* **14**, 141–145.

KELLEY, S. & BLUCK, B. J. 1989. Detrital mineral ages from the Southern Uplands using Ar laser probe. *Journal of the Geological Society, London,* **146**, 401–403.

KIDD, W. S. F., DEWEY, J. F. & NELSON, K. D. 1977. Medial Ordovician Ridge subduction in central Newfoundland. *Geological Society of America Annual Meeting, Abstracts with Programs,* **9**, 283–284.

KIM, J. & JACOBI, R. D. 1996. Geochemistry and tectonic implications of Hawley Formation meta-igneous units: northwestern Massachusetts. *American Journal of Science,* **296**, 1126–1174.

KING, L. M. 1994. Subsidence analysis of eastern Avalonian sequences: implications for Iapetus closure. *Journal of the Geological Society, London,* **151**, 647–657.

KNELLER, B. C., KING, L. M. & BELL, A. M. 1993. Foreland basin development and tectonics on the northwest margin of western Avalonia. *Geological Magazine*, **130**, 691–697.

KOKELAAR, P. 1986. Petrology and geochemistry of the Rhobell Volcanic Complex: amphibole-dominated fractionation at an early Ordovician Arc Volcano in North Wales. *Journal of Petrology*, **27**, 887–914.

—— 1988. Tectonic controls of Ordovician arc and marginal basin volcanism in Wales. *Journal of the Geological Society, London*, **145**, 759–775.

——, HOWELLS, M. F., BEVINS, R. E., ROACH, R. A. & DUNKLEY, P. N. 1984. The Ordovician marginal basin of Wales. *In:* KOKELAAR, P. & HOWELLS, M. F. (eds) *Marginal Basin Geology*, Geological Society of London Special Publication, **16**, 245–269.

KRABBENDAM, M., LESLIE, A. G., CRANE, A. & GOODMAN, S. 1997. Generation of the Tay nappe, by large-scale SE-directed shearing. *Journal of the Geological Society, London*, **154**, 15–24.

KUMARAPELI, P. S., DUNNING, G. R., PINTSON, H. & SHAVER, J. 1989. Geochemistry and U–Pb zircon age of comenditic metafelsites of the Tibbit Hill Formation, Quebec Appalachians. *Canadian Journal of Earth Sciences*, **26**, 1374–1383.

KUSKY, T. M., CHOW, J. S. & BOWRING, S. A. 1997. Age and origin of the Boil Mountain Ophiolite and Chain Lakes Massif, Maine: implications for the Penobscotian Orogeny. *Canadian Journal of Earth Sciences*, **34**, 646–654.

LAFRANCE, B. & WILLIAMS, P. F. 1992. Silurian deformation in eastern Notre Dame Bay, Newfoundland: *Canadian Journal of Earth Sciences,* **29**, 1899–1914.

LAIRD, J., TRZCIENSKI, W. E. JR & BOTHNER, W. A. 1993. High-pressure, Taconian and subsequent polymetamorphism of southern Quebec and northern Vermont. Department of Geology and Geography, University of Massachusetts, Contribution, 67-2, 1–32.

LAMBERT, R. STJ., HOLLAND, J. G. & LEGGET, J. K. 1981. Petrology and tectonic setting of some Ordovician volcanic rocks from the Southern Uplands of Scotland. *Journal of the Geological Society, London*, **138**, 421–436.

LANDING, E., BOWRING, S. M., FORTEY, R. A. & DAVIDEK, K. 1997. U–Pb date from Avalonian Cape Breton Island and geochronology calibration of the Early Ordovician. *Canadian Journal of Earth Sciences*, **34**, 724–730.

LAURENT, R. & HEBERT, R. 1989. The volcanic and intrusive rocks of the Quebec Appalchians ophiolites and their island-arc setting. *Chemical Geology*, **77**, 265–286.

LEAKE, B. E. 1989. The metagabbros, orthogneisses and paragneisses of the Connemara complex, western Ireland. *Journal of the Geological Society, London*, **146**, 575–596.

LEE, C. B. & WILLIAMS, H. 1995. The Teakettle and Carmanville mélanges in the Exploits subzone of northeast Newfoundland: recycling and diapiric emplacement in an accretionary prism. *In:* HIBBARD, J., VAN STAAL, C. R. & CAWOOD, P. (eds) *Current perspectives in the Appalachian-Caledonian Orogen.* Geological Association of Canada Special Paper, **41**, 147–160.

LEGGETT, J. K. 1987. The Southern Uplands as an accretionary prism: the importance of analogues in reconstructing palaeogeography. *Journal of the Geological Society, London*, **144**, 737–752.

——, MCKERROW, W. S. & CASEY, D. M. 1982. The anatomy of a Lower Paleozoic accretionary forearc:the Southern Uplands of Scotland. *Geological Society London Special Publication*, **10**, 494–520.

LEO, G. H. 1985. Trondhjemite and metamorphosed quartz keratophyre tuff of the Ammonoosuc volcanics (Ordovician), western New Hampshire and adjacent Vermont and Massachusetts. *Geological Society of America Bulletin*, **96**, 1493–1507.

LIN, S., VAN STAAL, C. R. & DUBÉ B. 1994, Promontory-promontory collision in the Canadian Appalachians. *Geology*, **22**, 897–900.

LISS, M. J., VAN DER PLUIJM, B. A. & VAN DER VOO, R. 1993. Avalonian proximity of the Ordovician Miramichi terrane, northern New Brunswick, northern Appalachians: Palaeomagnetic evidence for rifting and back-arc basin formation at the southern margin of Iapetus. *Tectonophysics,* **227**, 17–30.

LORENTZ, B. E. 1984. Mud-magma interactions in the Dunnage melange, Newfoundland. *In:* KOKELAAR, P. & HOWELLS, M. F. (eds) *Marginal Basin Geology*. Geological Society London Special Publications, **16**, 271–277.

LYONS, J. B., BOUDETTE, E. L. & ALEINIKOFF, J. N. 1982. The Avalonian and Gander zones in Central Eastern New England. *In:* ST JULIEN, P. & BÉLAND, J. (eds) *Major Structural Zones and Faults of the Northern Appalachians*. Geological Association of Canada Special Paper, **24**, 43–66.

MAC NIOCAILL, C. & SMETHURST, M. A. 1994. Palaeozoic palaeogeography of Laurentia and its margins: a reassessment of palaeomagnetic data. *Geophysical Journal International*, **116**, 715–725.

——, VAN DER PLUIJM, B. A. & VAN DER VOO, R. 1997. Ordovician paleogeography and the evolution of the Iapetus ocean. *Geology*, **25**, 159–162.

MACLACHLAN, K. & DUNNING, G. R. 1998. U–Pb ages and tectonic setting of Middle Ordovician calc-alkaline and within plate volcanic rocks of the Wild Bight Group, northeastern Dunnage Zone, Newfoundland Appalachians: implications for the evolution of the Gondwanan margin. *Canadian Journal of Earth Sciences* (in press).

MAX, M. D. & RODDICK, J. C. 1989. Age of metamorphism in the Rosslaare Complex, S.E. Ireland. *Proceedings of the Geologists' Association*, **100**, 113–121.

——, BARBER, A. J. & MARTINEZ, J. 1990, Terrane assemblage of the Leinster Massif, SE Ireland, during the Lower Paleozoic. *Journal of the Geological Society, London*, **147**, 1035–1050.

MCCAFFREY, R. 1996. Slip partitioning at convergent plate boundaries of southeast Asia. *In:* HALL, R. & BLUNDELL, D. J. (eds) *Tectonic Evolution of*

Southeast Asia, Geological Society of London Special Publications, **106**, 3–18.

McCAIG, A. & REX, D. 1995. Argon-40/Argon-39 ages from the Lewis Hills massif, Bay of island complex, west Newfoundland. *In*: HIBBARD, J., VAN STAAL, C. R. & CAWOOD, P. (eds) *Current perspectives in the Appalachian-Caledonian Orogen.* Geological Association of Canada Special Paper, **41**, 137–146.

McCAUSLAND, P. J. A., HODYCH, J. P. & DUNNING, G. R. 1997. Evidence from western Newfoundland for the final breakup of Rodinia? U–Pb age and palaeolatitude of the Skinner Cove volcanics. *Geological Association of Canada/Mineralogical Association of Canada Annual Meeting, Program with Abstracts, Ottawa, Ontario*, A-96.

McCONNELL, B. J., STILLMAN, C. J. & HERTOGEN, J. 1991. An Ordovician basalt to peralkaline rhyolite fractionation series from Avoca, Ireland. *Journal of the Geological Society, London*, **148**, 711–718.

McKENZIE, D. P. & MORGAN, W. J. 1969. Evolution of triple junctions. *Nature,* **224**, 125–133.

McKERROW, W. S. & CAMPBELL, C. J. 1960. The stratigraphy and structure of the Lower Paleozoic rocks of north-west Galway. *Scientific Proceeding of the Royal Dublin Society*, **1**, 27–52.

—— & COCKS, L. R. M. 1976. Progressive faunal migration across the Iapetus Ocean. *Nature,* **263**, 304–306.

—— & ELDERS, C. F. 1989. Movements on the Southern Upland Fault. *Journal of the Geological Society, London*, **146**, 393–395.

—— 1987. The Southern Uplands controversy. *Journal of the Geological Society, London*, **144**, 735–736.

——, DEWEY, J. F. & SCOTESE, C. R. 1991. The Ordovician and Silurian development of the Iapetus Ocean. *In*: BASSETT, M. G., LANE, P. & EDWARDS, D. (eds) *The Murchison Symposium.* Special Paper in Palaeontology, **44**, 165–178.

——, SCOTESE, C. R. & BRASIER, M. D. 1992. Early Cambrian reconstructions. *Journal of the Geological Society, London*, **149**, 599–606.

McLEOD, M. J., RUITENBERG, A. A. & KROGH, T. E. 1992. Geology and U–Pb geochronology of the Annidale Group, southern New Brunswick: Lower Ordovician volcanic and sedimentary rocks formed near the southeastern margin of Iapetus Ocean. *Atlantic Geology*, **28**, 181–192.

——, WINCHESTER, J. A. & RUITENBERG, A. A. 1994. Geochemistry of the Annidale Group: implications for the tectonic setting of Lower Ordovician volcanism in southwestern New Brunswick. *Atlantic Geology*, **30**, 87–95.

MEERT, J. G. & VAN DER VOO, R. 1997. The assembly of Gondwana (800–550 Ma), *Journal of Geodynamics*, **23**, 223–235.

MILLER, R. G. & O'NIONS, R. K. 1984. The provenance and crustal residence age of British sediments in relationship to palaeogeographic reconstructions. *Earth and Planetary Science Letters,* **68**, 459–470.

MOLNAR, P. 1992. Bruce-Goetze strength profiles, the partitioning strike-slip and thrust faulting at zones of oblique convergence, annd the stress-heat flow paradox of the San Andreas Fault. *In: Fault*

Mechanics and Transport Properties of Rocks, Academic Press, 435–459.

MORRIS, J. H. 1987. The Northern Belt of the Longford-Down Inlier, Ireland and Southern Uplands: an Ordovician back-arc basin. *Journal of the Geological Society, London*, **144**, 773–786.

MURPHY, J. B. & NANCE, R. D. 1989. Model for the evolution of the Avalonian-Cadomian belt. *Geology,* **17**, 735–738.

NELSON, K. D. 1981. Mélange development in the Boones Point Complex, north-central Newfoundland. *Canadian Journal of Earth Sciences*s, **18**, 433–442.

NEUMAN, R. B. 1967. Bedrock geology of the Shin Pond and Stacyville Quadrangles, Penobscot County, Maine. *United States Geological Survey, Professional Paper*, **524-1**, 37pp.

—— 1984. Geology and paleobiology of islands in the Ordovician Iapetus Ocean: review and implications. *Geological Society of America Bulletin*, **94**, 1188–1201.

—— 1994. Late Ordovician (Ashgill) Foliomena Fauna brachiopods from northeastern Maine. *Journal of Paleontology*, **68**, 1218–1234.

NOBLE, S. R., TUCKER, R. D. & PHARAOH, T. C. 1993. Lower Paleozoic and Precambrian igneous rocks from eastern England and their bearing on late Ordovician closure of the Tornquist Sea: constraints from U–Pb and Nd isotopes. *Geological Magazine,* **130**, 835–846.

NOBLET, C. & LEFORT, L. P. 1990. Sedimentological evidence for a limited separation between Armorica and Gondwana during the Early Ordovician. *Geology*, **18**, 303–306.

NOWLAN, G. S., McCRACKEN, A. D. & McLEOD, M. J. 1997. Tectonic and palaeogeographic significance of Ordovician conodonts from the Letang area, Avalon Terrane, southwestern New Brunswick. *Canadian Journal of Earth Sciences*, **34**, 1521–1537.

O'BRIEN, B. H. 1991. Geological development of the Exploits and Notre Dame Subzones in the New Bay area (parts of NTS 2E/6 and 2E/11), Notre Dame Bay, Newfoundland. *Current Research, Newfoundland Department of Mines and Energy, Geological Survey Branch Report*, 91-1, 155–166.

—— 1992. Internal and external relationships of the South Lake Igneous complex, north-central Newfoundland: Ordovician and later tectonism in the Exploits Subzone? *Current Research, Newfoundland Department of Mines and Energy, Geological Survey Branch Report*, 92-1, 159–69.

—— 1993. A mapper's guide to Notre Dame Bay's folded thrust faults: evolution and regional development. *Current Research, Newfoundland Department of Mines and Energy, Geological Survey Branch Report*, 93-1, 279–291.

——, SWINDEN, H. S., DUNNING, G. R., WILLIAMS, S. H. & O'BRIEN, F. H. C. 1997. A peri-Gondwanan arc-backarc complex in Iapetus: Early–Mid Ordovician evolution of the Exploits Group, Newfoundland. *American Journal of Science*, **297**, 220–272.

O'BRIEN, S. J., O'BRIEN, B. H., DUNNING, G. R. & TUCKER, R. D. 1996. Late Neoproterozoic Avalonian and related peri-Gondwanan rocks of the

Newfoundland Appalachians. *In:* NANCE, D. & THOMPSON, M. (eds) *Avalonian and related Peri-Gondwanan Terranes of the Circum-North Atlantic.* Geological Society of America Special Paper, **304**, 9–28.

OSHIN, I. O. & CROCKET, H. 1986. The geochemistry and petrogenesis of ophiolitic volcanic rocks from Lac de l'Est, Thetford Mines Complex, Quebec, Canada. *Canadian Journal of Earth Sciences,* **23**, 202–213.

OWEN, A. W. & CLARKSON, E. N. K. 1992. Trilobites from Kilbucho and Wallace's Cast and the location of the Northern Belt of the Southern Uplands during the Late Ordovician. *Scottish Journal of Geology,* **28**, 3–17.

PARNELL, J. & SWAINBANK, I. 1984. Interpretation of Pb isotope compositions of galenas from the Midland Valley of Scotland and adjacent regions. *Transactions of the Royal Society of Edinburgh, Earth Sciences,* **75**, 85–96.

PEGLER, G., DAS, S. & WOODHOUSE, J. 1995. A seismological study of the eastern New Guinea and the western Solomon Sea regions and its tectonic implications. *Geophysical Journal International,* **122**, 961–981.

PELECHATY, S. M. 1996. Stratigraphic evidence for the Siberia-Laurentia connection and Early Cambrian rifting. *Geology,* **24**, 719–722.

PHILLIPS, E. R., BARNES, R. P., MERRIMAN, R. J. & FLOYD, J. D. 1995. The tectonic significance of Ordovician basic rocks in the Southern Uplands, Southwest Scotland. *Geological Magazine,* **132**, 549–556.

PICKETT, J. W. 1987. Geology and geochemistry of the Skidder basalt. *In:* KIRKHAM, R. V. (ed.) *Buchans Geology.* Geological Survey of Canada Special Paper, **86-24**, 195–218.

PINET, N. & TREMBLAY, A. 1995. Tectonic evolution of the Quebec-Maine Appalachians: from oceanic spreading to obduction and collision in the Northern Appalachians. *American Journal of Science,* **295**, 173–200.

POLLOCK, S. G. 1993. Terrane sutures in the Maine Appalachians, USA and adjacent areas. *Geological Journal,* **28**, 45–67.

POTTS, S. S., VAN DER PLUIJM, B. A. & VAN DER VOO, R. 1993. Paleomagnetism of the Ordovician Bluffer Pond Formation: Palaeogeographic Implications for the Munsungun Terrane of Northern Maine. *Journal of Geophysical Research,* **98**, 7987–7996.

——, —— & —— 1995. Paleomagnetism of the Pennington Mountain terrane: A near Laurentian back arc basin in the Maine Appalachians. *Journal of Geophysical Research,* **100**, 10 003–10 011.

PRIGMORE, J. K., BULLER, A. J. & WOODCOCK, N. H. 1997. Rifting during separation of eastern Avalonia from Gondwana: evidence from subsidence analysis. *Geology,* **25**, 203–206.

RANKIN, D. W., DRAKE, A. & RATCLIFFE, N. M. 1990. Plate 2: Geologic map of the U. S. Appalachians showing the Laurentian margin and the Taconic Orogen. *In:* HATCHER, R. D. JR, THOMAS, W. A. & VIELE, G. W. (eds) *The Appalachian-Ouachita Orogen in the United States.* Geological Society of

America, The Geology of North America, Boulder, Colorado, F-2.

ROBINSON, P. & HALL, L. M. 1980. Tectonic synthesis of southern New England. *In:* WONES, D. R. (ed.) *Proceedings 'The Caledonides in the U.S.A.',* IGCP Project 27: Caledonide Orogen, Virginia Polytechnic & State University Memoir, **2**, 73–82.

——. & TUCKER, R. D. 1996. The Bronson Hill magmatic arc, New England: myth and reality. *Geological Society of America, Northeastern Section, Annual meeting Program with Abstracts,* **28**, 94.

ROGERS, G., DEMPSTER, T. J., BLUCK, B. J. & TANNER, P. W. G. 1989. A high-precision U–Pb age for the Ben Vurich granite: implications for the evolution of the Scottish Dalradian Supergroup. *Journal of the Geological Society, London,* **146**, 789–798.

——, PATERSON, B. A., DEMPSTER, T. J. & REDWOOD, S. D. 1994. U–Pb geochronology of the 'newer' gabbros, NE Grampians. Symposium on Caledonian Terrane Relationships in Britain, September, 1994

ROGERS, N. & VAN STAAL, C. R. 1997. Comparing the Bathurst Mining Camp to the Japan Sea and Okinawa Trough: ancient, recent and active back-arcs. *Geological Association of Canada/ Mineralogical Association of Canada Annual Meeting, Program with Abstracts, Ottawa, Ontario,* A-127.

RYAN, P. D. & DEWEY, J. F. 1991. A geological and tectonic cross-section of the Caledonides of western Ireland. *Journal of the Geological Society, London,* **148**, 173–180.

——, MAX, M. D. & KELLY, T. J. 1983. The petrochemistry of the basic volcanic rocks of the South Connemara Group (Ordovician), Western Ireland. *Geological Magazine,* **120**, 141–152.

——, SOPER, N. J., SNYDER, D. B., ENGLAND, R. W. & HUTTON, D. H. W. 1995. The Antrim-Galway Line: a resolution of the highland Border Fault enigma of the Caledonides of Britain and Ireland. *Geological Magazine,* **132**, 171–184.

SANDERS, I. S., VAN CALSTEREN, P. W. C. & HAWKESWORTH, C. J. 1984. A Grenville Sm–Nd age for the Glenelg eclogite in northwest Scotland. *Nature,* **312**, 439–440.

SCOTESE, C. R. & MCKERROW, W. S. 1990. Revised world maps and introduction. *In:* MCKERROW, W. S. & SCOTESE, C. R. (eds) *Palaeozoic Palaeogeography and Biogeography.* Geological Society of London Memoir, **12**, 1–21.

SENGÖR, A. M. C. & NATAL'IN, B. A. 1996. Paleotectonics of Asia: fragments of a synthesis. *In:* YIN A. & HARRISON, M. (eds) *Tectonic evolution of Asia.* Cambridge Press, 486–640.

—— 1979. Mid-Mesozoic closure of permo-Triassic Tethys and its implications. *Nature,* **279**, 590–593.

SMETHURST, M. A., KHRAMOV, A. N. & TORSVIK, T. H. 1998. The Neoproterozoic and Palaeozoic palaeomagnetic data for the Siberian Platform: from Rodinia to Pangea. *Earth Science Reviews,* **43**, 1–24.

SOPER, N. J. & HARRIS, A. L. 1997. Proterozoic orogeny questioned: a view from Southern Highland fieldtrips and workshops, 1995–1996. *Scottish Journal of Geology,* **33**, 187–190.

—— & WOODCOCK, N. 1990. Silurian collision and sediment dispersal patterns in southern Britain. *Geological Magazine*, **127**, 527–542.

—— 1994. Was Scotland a Vendian RRR junction? *Journal of the Geological Society, London*, **151**, 579–582.

SPENCER, C., GREEN, A., MOREL, A. *ET AL.* 1986. The extension of Grenville basement beneath the northern Appalachians: results from the Quebec-Maine seismic reflection and refraction surveys. *Tectonics*, **8**, 677–696.

SPRAY, J. G. & DUNNING, G. R. 1991. A U/Pb age for the Shetland Islands oceanic fragment, Scottish Caledonides: evidence from anatectic plagiogranites in layer 3 shear zones. *Geological Magazine*, **128**, 667–671.

STANLEY, R. & RATCLIFFE, N. M. 1985. Tectonic synthesis of the Taconian Orogeny in western New England. *Geological Society of America Bulletin*, **96**, 1227–1250.

STERN, R. J. & BLOOMER, S. H. 1992. Subduction zone infancy: examples from the Eocene Izu-Bonin-Mariana and Jurassic California arcs. *Geological Society of America Bulletin*, **104**, 1621–1636.

STONE, P. & EVANS, J. A. 1995. Nd-isotopic study of provenance patterns across the British sector of the Iapetus Suture. *Geological Magazine*, **132**, 571–580.

——, FLOYD, J. D., BARNES, R. P. & LINTERN, B. C. 1987. A sequential back-arc and foreland basin thrust duplex model for the Southern Uplands of Scotland. *Journal of the Geological Society, London*, **144**, 753–764.

SUHR, G. & CAWOOD, P. A. 1993. Structural history of ophiolite obduction, Bay of Islands, Newfoundland. *Geological Society of America Bulletin*, **105**, 399–410.

SULLIVAN, R. W. & VAN STAAL, C. R. 1996. Preliminary chronostratigraphy of the Tetagouche and Fournier Groups in northern New Brunswick, Canada; *In*: *Radiogenic Age and Isotopic Studies*, Report 9, Geological Survey of Canada, Paper 1995-F, 43–56.

——, —— & LANGTON, J. P. 1990. U–Pb zircon ages of plagiogranite and gabbro from the ophiolitic Devereaux Formation, Fournier group, northeastern New Brunswick. *Geological Survey of Canada Paper*, 89-2, 119–122.

SWINDEN, H. S. & THORPE, R. I. 1984. Variations in style of volcanism and massive sulphide deposition in Early–Middle Ordovician island arc sequences of the New Brunswick Central Mobile Belt. *Economic Geology*, **79**, 1569–1619.

—— 1988. Geology and economic potential of the Pipestone Pond area (12A/1 NE; 12A/8 E), central Newfoundland. *Newfoundland Department of Mines, Geological Survey Branch Report*, 88-2.

——, JENNER, G. A. & SZYBINSKI, Z. A. 1997. Magmatic and tectonic evolution of the Cambrian-Ordovician Laurentian margin of Iapetus: Geochemical and isotopic constraints from the Notre Dame Subzone, Newfoundland. *In*: SINHA, K., WHALEN, J. B. & HOGAN, J. (eds) *Magmatism in the Appalachian Orogen*. Geological Society of America Memoir, **191**, 337–365.

TAIRA, A., OKADA, H., WHITAKER, J. H. McD & SMITH, A. J. 1982. The Shimanto Belt of Japan: Cretaceous to Lower Miocene active margin sedimentation. *In*: LEGGETT, J. K. (ed.) *Trench-forearc Geology: Sedimentation and Tectonics on Modern and Ancient Active Margins*. Geological Society of London Special Publications, **10**, 5–26.

TANNER, B. & MEISSNER, R. 1996. Caledonian deformation upon southwest Baltica and its tectonic implications: alternatives and consequences. *Tectonics*, **15**, 803–812.

TECTONIC MAP OF THE WORLD 1985. Exxon Production and Research Company.

THOROGOOD, E. J. 1990. Provenance of the pre-Devonian sediments of England and Wales: Sm–Nd isotopic evidence. *Journal of the Geological Society, London*, **147**, 591–594.

THORPE, R. S., BECKINSALE, R. D., PATCHETT, P. J., PIPER, J. D. A., DAVIES, G. R. & EVANS, J. A. 1984. Crustal growth and late Precambrian-early Palaeozoic plate tectonic evolution of England and Wales. *Journal of the Geological Society, London*, **141**, 521–536.

THURLOW, J. G., SPENCER, C. P., BOERNER, D. E., REED, L. E. & WRIGHT, J. A. 1992. Geological interpretation of a high resolution reflection seismic survey at the Buchans Mine, Newfoundland. *Canadian Journal of Earth Sciences*, **29**, 2022–2037.

TIETZSCH-TYLER, D. & PHILLIPS, E. 1989. Correlation of the Monian Supergroup in NW Anglesey with the Cahore Group in SE Ireland. *Journal of the Geological Society, London*, **146**, 417–418.

TORSVIK, T. H., SMETHURST, M. A., BRIDEN, J. C. & STURT, B. A. 1990. A review of Palaeozoic data from Europe and their palaeogeographical implications. *In*: McKERROW, W. S. & SCOTESE, C. R. (eds) *Palaeozoic Palaeogeography and Biogeography,.*Geological Society of London Memoir, **12**, 25–41.

——, ——, MEERT, J. G., VAN DER VOO, R., McKERROW, W. S., BRASIER, M. D., STURT, B. A. & WALDERHAUG, H. J. 1996. Continental break-up and collision in the Neoproterozoic and Palaeozoic – A tale of Baltica and Laurentia. *Earth Science Reviews*, **40**, 229–258.

TREMBLAY, A. & PINET, N. 1994. Distribution and characteristics of Taconian and Acadian deformation, southern Quebec Appalachians. *Geological Society of America Bulletin.*, **106**, 1172–1181.

—— & ST JULIEN, P. 1990. Structural style and evolution of a segment of the Dunnage Zone from the Quebec Appalachians and its tectonic implications. *Geological Society of America Bulletin*, **102**, 1218–1229.

—— 1992. Tectonic and accretionary history of Taconian oceanic rocks of the Quebec Appalachians. *American Journal of Science*, **292**, 229–252.

——, LAFLÉCHE, M. R., McNUTT, R. H. & BERGERON, M. 1994. Petrogenesis of Cambro-Ordovician subduction-related granitic magmas of the Quebec Appalachians, Canada. *Chemical Geology*, **113**, 205–220.

TRENCH, A., TORSVIK, T. H. & McKERROW, W. S. 1992. The palaeogeographic evolution of Southern Britain during early Paleozoic times: a reconciliation of palaeomagnetic and biogeographic evidence. *Tectonophysics*, **201**, 75–82.

TRZCIENSKI, W. E., JR 1976. Crossitic amphibole and its possible tectonic significance in the Richmond area, southeast Quebec. *Canadian Journal of Earth Sciences*, **13**, 711–714.

——, RODGERS, J. & GUIDOTTI, C. V. 1992. Alternative hypotheses for the Chain Lakes Massif, Maine and Quebec. *American Journal of Science*, **292**, 508–532.

TUCKER, R. D. & McKERROW, W. S. 1995. Early Paleozoic chronology: a review in light of new U-Pb zircon ages from Newfoundland and Britain. *Canadian Journal of Earth Sciences*, **32**, 368–379.

—— & ROBINSON, P. 1990. Age and setting of the Bronson Hill magmatic arc: a re-evaluation based on U–Pb zircon ages in southern New England. *Geological Society of America Bulletin*, **102**, 1404–1419.

——, O'BRIEN, S. J. & O'BRIEN, B. H. 1994. Age and implications of Early Ordovician plutonism in the type area of the Bay du Nord Group, Dunnage Zone, southern Newfoundland Appalachians. *Canadian Journal of Earth Sciences*, **31**, 351–357.

TURNER, R. D. & TURNER, S. 1974. Thelodonts from the Upper Silurian of Ringerike, Norway. *Norsk Geologisk Tidsskrift*, **54**, 183–192.

UPTON, B. G. J., ASPEN, P. & HUNTER, R. H. 1984. Xenoliths and their implications for the deep geology of the Midland Valley of Scotland and adjacent regions. *Transactions of the Royal Society of Edinburgh, Earth Sciences*, **75**, 65–70.

VALVERDE-VAQUERO, P., DUNNING, G. R. & VAN STAAL, C. R. 1998. The Margaree orthogneiss: an Arenig-early Llanvirn arc/backarc complex developed on the peri-Gondwanan margin of SW Newfoundland. *Geological Society of America Annual meeting Program with Abstracts*, **30**, 82.

VAN DER PLUIJM, B. A. 1986, Geology of eastern New World Island, Newfoundland: an accretionary terrane in the northeastern Appalachians. *Geological Society of America Bulletin*, **97**, 932–945.

——, VAN DER VOO, R. & TORSVIK, T. H. 1995. Convergence and subduction at the Ordovician margin of Laurentia. *In*: HIBBARD, J. P., VAN STAAL, C. R. & CAWOOD, P. A. (eds) *Current Perspectives in the Appalachian-Caledonian Orogen*. Geological Association of Canada Special Paper, **41**, 127–136.

—— & VAN STAAL, C. R. 1988. Characteristics and evolution of the central mobile belt, Canadian Appalachians. *Journal of Geology*, **96**, 535–547.

VAN DER VOO, R. 1988. Paleozoic paleogeography of North America, Gondwana and intervening displaced terranes: comparisons of paleomagnetism with paleoclimatological and biogeographical patterns. *Geological Society of America Bulletin*, **100**, 311–324.

——, JOHNSON, R. J. E., VAN DER PLUIJM, B. A. & KNUTSON, L. C. 1991. Palaeogeography of some vestiges of Iapetus: paleomagnetism of the Ordovician Robert's Arm, Summerford, and Chanceport Groups, Central Newfoundland.

Geological Society of America Bulletin, **103**, 1564–1575.

VAN STAAL, C. R. 1987. Tectonic setting of the Tetagouche Group in northern New Brunswick: implications for plate tectonic models of the northern Appalachians. *Canadian Journal of Earth Sciences*, **24**, 1329–1351.

—— 1994. The Brunswick subduction complex in the Canadian Appalachians: record of the Late Ordovician to Late Silurian collision between Laurentia and the Gander margin of Avalon. *Tectonics*, **13**, 946–962.

—— & DE ROO, J. A. 1995. Mid-Paleozoic tectonic evolution of the Appalachian Central Mobile Belt in Northern New Brunswick, Canada: Collision, extensional collapse and dextral transpression. *In*: HIBBARD, J., VAN STAAL, C. R. & CAWOOD, P. (eds) *Current perspectives in the Appalachian-Caledonian Orogen*. Geological Association of Canada Special Paper, **41**, 95–114.

——, SULLIVAN R. W. & WHALEN, J. B. 1996a. Provenance and tectonic history of the Gander Margin in the Caledonian/ Appalachian Orogen: implications for the origin and assembly of Avalonia. *In*: NANCE, D. & THOMPSON, M. (eds) *Avalonian and related Peri-Gondwanan Terranes of the Circum-North Atlantic*. Geological Society of America Special Paper, **304**, 347–367.

——, HALL, L., SCOFIELD, D. & VALVERDE, P. 1996b. Geology Port aux Basques, Newfoundland, scale 1: 25 000 (part of NTS 11-O/11). *Geological Survey of Canada Open File*, 3165.

——, LIN, S., HALL, L., VALVERDE, P. & GENKIN, M. 1996c. Geology Rose Blanche, Newfoundland, scale 1: 25 000 (NTS 11-O/10). *Geological Survey of Canada Open File*, 3219.

—— & WILLIAMS, H. 1991. Dunnage Zone–Gander Zone relationships in the Canadian Appalachians. *Geological Society of America Annual meeting Program with Abstracts*, **23**, 143.

——, RAVENHURST, C., WINCHESTER, J. A., RODDICK, J. C. & LANGTON, J. P. 1990. Post Taconic blueschist suture in the northern Appalachians of northern New Brunswick, Canada. *Geology*, **18**, 1073–1077.

——, WINCHESTER, J. A. & BEDARD, J. H. 1991. Geochemical variations in Ordovician volcanic rocks of the northern Miramichi Highlands and their tectonic significance. *Canadian Journal of Earth Sciences*, **28**, 1031–1049.

WASOWSKI, J. J. & JACOBI, R. D. 1985. Geochemistry and tectonic significance of the mafic volcanic blocks in the Dunnage mélange, north central Newfoundland. *Canadian Journal of Earth Sciences*, **22**, 1248–1256.

WELLENSIEK, M. R., VAN DER PLUIJM, B. A., VAN DER VOO, R. & JOHNSON, R. J. 1990. Tectonic history of the Lunksoos composite terrane in the Maine Appalachians. *Tectonics*, **9**, 719–734.

WELLINGS, S. A. 1998. Structure and metamorphism of syn-tectonic mafic intrusions from Connemara, western Ireland. *Journal of the Geological Society, London*. **155**, 25–38.

WEST, D. P., LUDMAN, A. & LUX, D. R. 1992. Silurian age for the Pocomoonshine Gabbro-Diorite, southeastern Maine and its regional tectonic implications. *American Journal of Science*, **292**, 253–273.

WHALEN, J. B. 1989. The Topsails igneous suite, western Newfoundland: an Early Silurian subduction-related magmatic suite? *Canadian Journal of Earth Sciences*, **26**, 2421–2434.

——, CURRIE, K. L. & VAN BREEMEN, O. 1987. Episodic Ordovician–Silurian plutonism in the Topsails terranne, western Newfoundland. *Transactions of the Royal Society of Edinburgh, Earth Sciences*, **78**, 17–28.

——, JENNER, G. A., LONGSTAFFE, F. J. & GARIÉPY, C. 1997*a*. Implications of granitoid geochemical and isotopic (Nd, O, Pb) data from the Cambro–Ordovician Notre Dame arc for the evolution of the Central Mobile Belt, Newfoundland Appalachians. *In*: SINHA, K., WHALEN, J. B. & HOGAN, J. (eds) *Magmatism in the Appalachian Orogen*. Geological Society of America Memoir, **191**, 367–395.

——, VAN STAAL, C. R., LONGSTAFFE, F. J., GARIEPY, C. & JENNER, G. A. 1997*b*. Insights into tectonostratigraphic zone identification in southwestern Newfoundland based on isotopic (Nd, O, Pb) and geochemical data. *Atlantic Geology*, **33**, 231–241.

WHITEHEAD, J., REYNOLDS, P. H. & SPRAY, J. G. 1996. 40Ar/39Ar age constraints on taconian and acadian events in the Quebec Appalachians. *Geology*, **24**, 359–362.

WILLIAMS, D., ARMSTRONG, H. A. & HARPER, D. A. 1988. The age of the South Connemara Group, Ireland and its relationship to the southern Uplands Zone of Scotland and Ireland. *Scottish Journal of Geology*, **24**, 279–287.

WILLIAMS, H. 1978. Tectonic-Lithofacies map of the Appalachian Orogen. *Memorial University of Newfoundland, St. John's, Newfoundland*, Map no. 1.

—— 1995. Temporal and spatial divisions. *In*: WILLIAMS, H. (ed.) *Geology of the Appalachian-Caledonian Orogen in Canada and Greenland*. Geological Survey of Canada, Geology of Canada, **6**, 21–44.

—— & HISCOTT, R. N. 1987. Definition of the Iapetus rift-drift transition in western Newfoundland. *Geology*, **15**, 1044–1047.

—— & PAYNE, J. G. 1975. The Twillingate granite and nearby volcanic groups: an island arc complex in northeast Newfoundland. *Canadian Journal of Earth Sciences*, **12**, 982–995.

—— & PIASECKI, M. A. J. 1990. The Cold Spring Melange and a possible model for Dunnage–Gander zone interaction in central Newfoundland. *Canadian Journal of Earth Sciences*, **27**, 1126–1134.

—— & ST JULIEN, P. 1982. The Baie Verte-Brompton Line: Early Paleozoic continent ocean interface in the Canadian Appalachians. *In*: ST JULIEN, P. &

BÉLAND, J. (eds) *Major Structural Zones and Faults of the Northern Appalachians*. Geological Association of Canada Special paper, **24**, 43–66.

——, COLMAN-SADD, S. P. & SWINDEN, H. S. 1988.Tectonic-stratigraphic subdivisions of central Newfoundland. *Geological Survey of Canada Paper*, **88-1B**, 91–98.

——, CURRIE, K. L. & PIASECKI, M. A. J. 1993. The Dog Bay Line: a major Silurian tectonic boundary in northeast Newfoundland. *Canadian Journal of Earth Sciences*, **30**, 2481–2494.

WILLIAMS, P. F., ELLIOTT, C. G. & LAFRANCE, B. 1988. *Structural geology and melanges of eastern notre Dame Bay, Newfoundland*. Geological Association of Canada, Guidebook Trip B2, 60pp.

WILLIAMS, S. H., BOYCE, W. D. & COLMAN-SADD, S. P. 1992. A new Lower Ordovician (Arenig) faunule from the Coy Pond complex, central Newfoundland, and a refined understanding of the closure of the Iapetus Ocean. *Canadian Journal of Earth Sciences,* **29**, 2046–2057.

——, HARPER, D. A. T., NEUMAN, R. B., BOYCE, W. D. & MAC NIOCAILL, C. 1995. Lower Paleozoic fossils from Newfoundland and their importance in understanding the history of the Iapetus Ocean. *In*: HIBBARD, J., VAN STAAL, C. R. & CAWOOD, P. (eds) *Current perspectives in the Appalachian-Caledonian Orogen*. Geological Association of Canada Special Paper, **41**, 115–126.

WILLIAMS, T. M., HENNEY, P. J., STONE, P. & LINTERN, B. C. 1996. Rare earth element geochemistry of Lower Palaeozoic turbidites in the British trans-Iapetus zone: provenance patterns and basin evolution. *Scottish Journal of Geology*, **32**, 1–8.

WILLIAMSON, B. F. & KARABINOS, P. 1993. Constraints on ages of Taconian and Acadian deformation from zircon evaporation ages of felsic plutons from western Massachusetts. *Geological Society of America, Annual meeting program with Abstracts*, **25**, 90.

WINCHESTER, J. A. & VAN STAAL, C. R. 1994. The chemistry and tectonic setting of Ordovician volcanic rocks in northern Maine and their relationships with contemporary volcanic rocks in northern New Brunswick. *American Journal of Science*, **294**, 641–662.

—— & —— 1995. Volcanic and sedimentary terrane correlation between the Dunnage and Gander zones of the Canadian Appalachians and the British Caledonides reviewed. *In*: HIBBARD, J., VAN STAAL, C. R. & CAWOOD, P. (eds) *Current perspectives in the Appalachian-Caledonian Orogen*. Geological Association of Canada Special Paper, **41**, 95–114.

YARDLEY, B. W. D. & SENIOR, A. 1982. Basic magmatism in Connemara, Ireland: Evidence for a volcanic arc? *Journal of the Geological Society, London*, **139**, 67–70.

The legacy of Charles Lyell: advances in our knowledge of coal and coal-bearing strata

ANDREW C. SCOTT

Geology Department, Royal Holloway University of London, Egham, Surrey TW20 OEX, UK

Abstract: On his travels in both Britain and North America Sir Charles Lyell paid particular attention to coalfields and their fossils. Understanding the formation of coal and associated rocks was the subject of several publications. In this contribution three aspects of this work are highlighted, all of which are the subject of ongoing modern research. Lyell was interested in modern analogues for the Carboniferous coal swamps and was amongst the first to suggest analogies with the mires of the eastern United States, such as the Dismal Swamps. A brief review of recent research on new modern analogues from Southeast Asia is presented. Lyell observed 'mineral charcoal' in some of the coals and noted its anatomical structure. Considerable advances in our understanding of the role of fires in terrestrial ecosystems and their potential as an agent of fossil preservation are addressed. One of Lyell's most important palaeontological finds was of remains of the earliest reptiles (these proved to be amphibians but Sir William Dawson found reptiles subsequently) preserved in the Upper Carboniferous tree trunks at Joggins, Nova Scotia, Canada. Whilst this occurrence of reptiles is no longer the oldest, the reasons for the remarkable tetrapod occurrences in upright sandstone-filled lycophyte trees at Joggins are currently being investigated and recent progress is presented.

Throughout his life, Sir Charles Lyell was fascinated by the study of coal and coal-bearing strata. The study of coals and coal fields was not only the subject of numerous papers but also became a major feature of his travels in North America. Indeed, coal was a strategic material at the time and was the subject of several lectures that Lyell made during his visits to North America. Some of his researches, such as the discovery with Sir William Dawson of tetrapods within upright lycophyte trunks at Joggins, Nova Scotia (Lyell 1853*a*; Lyell & Dawson 1853), remain as important now as then (Carroll 1994). Many of the problems which Lyell attempted to address are still the subject of ongoing research. While some of the problems addressed by Lyell, such as the age and structure of various coalfields, have largely been solved, others are still the subject of intense research and debate.

In this contribution three major research topics are highlighted: the origin of coal together with evidence of the fauna and flora of coal measures; the occurrence and significance of charcoal in coal seams; and modern analogues for ancient coals. In each of these three topics it is important to recognize that background data were limited. It is interesting to note, however, that Lyell understood the limits of his expertise and called upon the services of other researchers so that the descriptions of the fossils he found were undertaken by the leading palaeontologists of his day.

Historical background

The nature and origin of coal has remained a field of study for over 200 years and even today many fundamental questions concerning coal have yet to be satisfactorily resolved. Although this is indeed a very complex problem, geologists at the end of the eighteenth and beginning of the nineteenth century considered the possibility of simple solutions. One can consider the origin of coal as comprising three basic problems. First, there is its composition – whether in fact coal is formed from vegetable or mineral material. Second, if its origin is vegetable is coal formed from *in situ* or drifted plant remains? Third, there is the question of coal's diagenesis – the processes that affect the material during burial (the coalification process – the conversion of peat into coal by heat and pressure). The solution to each problem requires information from both laboratory and field observations. In the latter half of the eighteenth century many authors lacked basic experience of field work.

By 1780 a number of authors, particularly on the continent (De Buffon and De Luc in France and

SCOTT, A. C. 1998. The legacy of Charles Lyell: advances in our knowledge of coal and coal-bearing strata. *In*: BLUNDELL, D. J. & SCOTT, A. C. (eds) *Lyell: the Past is the Key to the Present.* Geological Society, London, Special Publications, **143**, 243–260.

Baron van Beroldingen in Germany) had recognized the vegetable origin of coal (Stevenson 1911; Bennett 1963). This idea of a vegetable origin was first accepted in Britain by Hutton (1795). Hutton considered that coal was formed by the deposition of vegetable material in the sea and that it was subsequently heated to form coal. Opposed to this view, however, was the Irish chemist Richard Kirwan (1799) (championing the mineral origin of coal) who was in turn vehemently attacked by James Parkinson (1804). By the time of the publication of the first edition of Lyell's *Principles of Geology* (1830–1833) the vegetable origin of coal was firmly established. In Volume 1 of the *Principles* Lyell made only four references to coal strata, in Volume 2 (1832) he made only one reference to coal formation and in Volume 3 (1833) only one reference to coal alteration.

At the time of the publication of Volume 2 of the *Principles* (1832) there was considerable support for the drift origin of coal, which would have given support to the diluvialsts. Lyell (p. 242) writes,

> Notwithstanding the vast forests intercepted by the lakes, a still greater mass of drift-wood is found where the Mackenzie reaches the sea, in a latitude where no wood grows at present except a few stunted willows. At the mouths of the river the alluvial matter has formed a barrier of islands and shoals where we may expect a great formation of coal at some distant period.

The drift versus *in situ* hypothesis of the origin of coal was of considerable interest to Lyell. His field observations on coal and coalfields played an important role in developing his ideas which were more prominently discussed with each edition of his *Principles* and later his *Elements* and *Manuals*. Following his visits to North America to study coalfields, and especially with his experiences at Joggins, Nova Scotia, Lyell expands his entries on coal to nine by the ninth edition (1853b) and the fifth edition of his *Manual of Elementary Geology* (1855b) contains 18 entries on coal and coalfields which includes detailed descriptions of *in situ* trees, the use of modern analogues and also a discussion of fire.

The origin of coal and the associated fauna and flora

By the time of the studies of Lyell, although the botanical origin of coal had been firmly established, the nature of the accumulation of the plants was widely debated. Lyell noted (1843) mention of upright fossil trees in the coal measures of Joggins in a paper published in 1829 by Richard

Brown in Haliburtons 'Nova Scotia' (Lyell 1843). The work of Brown and Logan contributed to establishing that Carboniferous coals were formed by vegetation living *in situ* (see for example Binney 1846; Brown 1846, 1848; Logan 1841; Lyell 1841). Lyell became interested in all aspects of the Carboniferous coal measures in particular, from the plants and animals preserved associated with the coals, trace fossils, sediments, to the structure of the coal basins (Lyell 1846a, b, c, d). A key to the development of his thoughts on the origin of coal-bearing sequences was the visits Lyell made to the coalfields of North America and to Nova Scotia in particular.

Lyell first visited the superb exposures of Upper Carboniferous Coal Measures along the Bay of Fundy in Nova Scotia in the summer of 1842. He writes (1845, vol. 2, p. 63),

> Just returned from an expedition of 3 days to the strait which divides Nova Scotia from New Brunswick, whether I went to see a forest of fossil coal – trees – the most wonderful phenomenon perhaps that I have seen, so upright do the trees stand, or so perpendicular to the strata, in the ever-wasting cliffs, every year a new crop being brought into view, as the violent tides of the Bay of Fundy, and the intense frost of the winters here, combine to destroy, undermine, and sweep away the old one – trees twenty-five feet high and some have been seen of forty feet, piercing the beds of sandstone and terminating downwards in the same beds, usually coal. This subterranean forest exceeds in extent and quality of timber all that has been discovered in Europe put together.

Lyell provides a detailed account of his visit to the section at Joggins (Lyell 1843). He not only notes the occurrence of several levels of trees but also some of the fauna and flora associated with the coals. Lyell (1843, p. 178) draws several conclusions from his observations of the Joggins sequence:

> 1. That the erect position of the trees, and their perpendicularly to the planes of stratification, imply that a thickness of several thousand feet of coal strata, now uniformly inclined at an angle of 24°, were deposited originally in a horizontal position [see Fig. 1].

> 2. There must have been repeated sinkings of the dry land to allow of the growth of more than ten forests of fossil trees one above the other, an inference which is borne out by the independent evidence afforded by the *Stigmaria*, found in the underclays beneath coal-seams in Nova Scotia, as first noticed in South Wales by Mr. Logan [see Fig. 1].

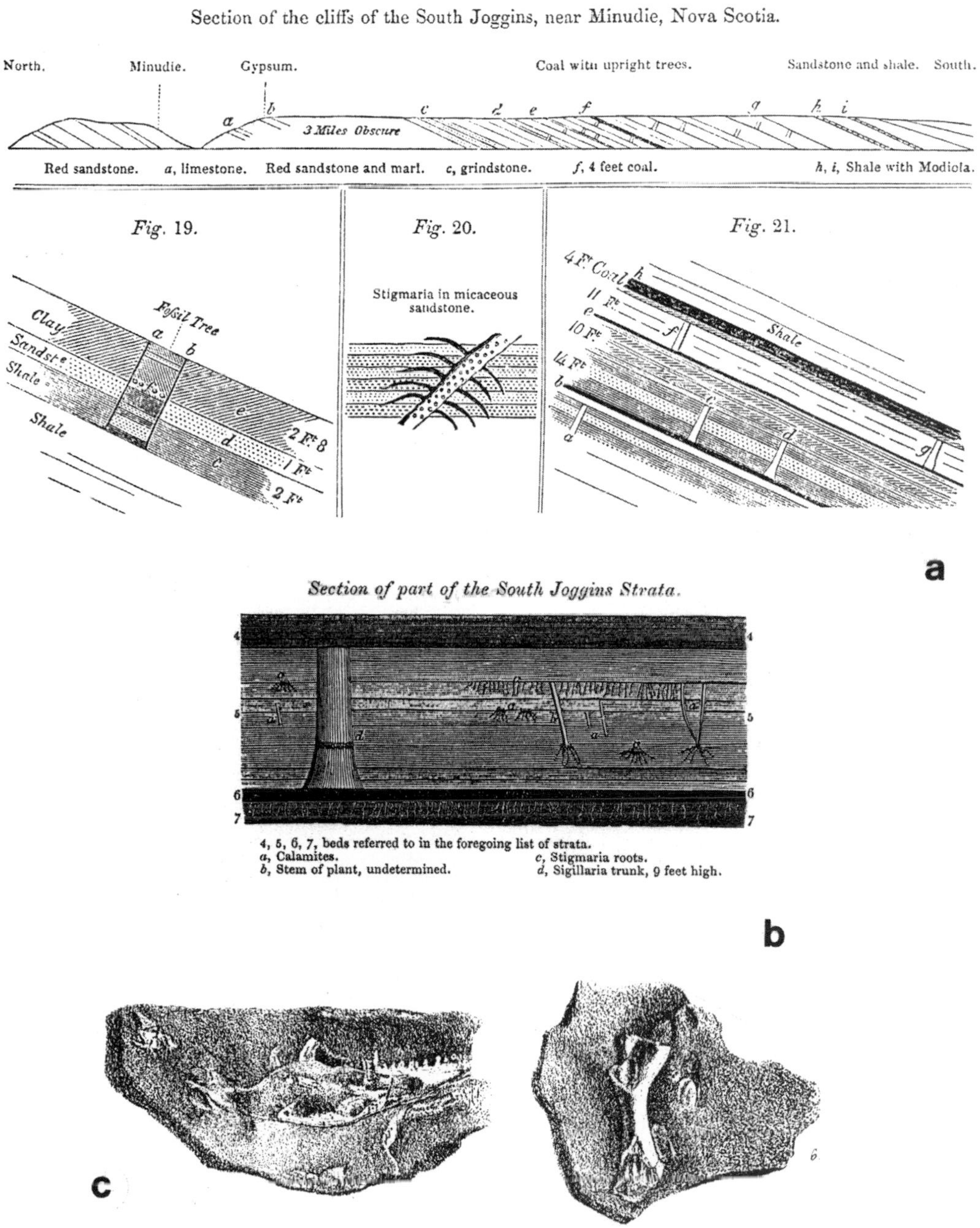
Section of the cliffs of the South Joggins, near Minudie, Nova Scotia.
North.
Minudie.
Gypsum.
Coal with upright trees.
Sandstone and shale.
South.
3 Miles Obscure
Red sandstone.
a, limestone.
Red sandstone and marl.
c, grindstone.
f, 4 feet coal.
h, i, Shale with Modiola.
Fig. 19.
Fossil Tree
Clay
Sandstone
Shale
Shale
2 Ft 8
1 Ft
2 Ft
Fig. 20.
Stigmaria in micaceous sandstone.
Fig. 21.
4 Ft Coal
11 Ft
10 Ft
14 Ft
Shale
a
Section of part of the South Joggins Strata.
4, 5, 6, 7, beds referred to in the foregoing list of strata.
a, Calamites.
b, Stem of plant, undetermined.
c, Stigmaria roots.
d, Sigillaria trunk, 9 feet high.
b
c

Fig. 1. Observations of Lyell on the Upper Carboniferous Fossil Forests of Joggins, Nova Scotia, Canada. (a) Field observations (Lyell 1845). (b) Detail of fossil trees at Coal Mine Point above Coal 15 (Lyell & Dawson 1853). (c) *Dendrerpeton* bones from inside sediment-filled lycophyte trunk (from Lyell & Dawson 1853) (now identified as microsaur tooth and a femur of *Dendrerpeton* by Carroll (1982)).

3. The correspondence in general characters of the erect trees of Nova Scotia with those found near Manchester, leads to the opinion that this tribe of plants may have been enabled by the strength of its large roots to withstand the power of waves and currents much more effectually than the *Lepidodendra* and other coal plants more rarely found in a perpendicular position.

In addition to his studies of the Joggins cliffs, Lyell also investigated the formation of Recent trace fossils made in the red muds of the Bay of Fundy (Lyell 1849*b*). He observed not only vertebrate footprints but also rain prints and recorded these in his publications (e.g. Lyell 1845, vol. 2, plate 7; 1853*b*, p. 203). He used these to compare with fossil footprints and traces which he encountered in his studies of coalfields (e.g. Lyell 1846*a*, *b*, *c*, *d*, 1849*a*) (Fig. 2(b)).

Lyell was so impressed with his visit to Joggins that he revisited the locality with Sir William Dawson in September 1852. In his letter of 12 September 1852 to Leonard Horner (Lyell 1881), he writes,

Dawson and I set to work and measured foot by foot many hundred yards of the cliffs, where the forests of erect trees and calamites most abound [see Fig. 3]. It was hard work, as the wind one day was strong, and we had to look sharp ... the rocking of living trees just ready to fall from the top of the undermined cliff should cause some of the old fossil ones to come down upon us by the run. But I never enjoyed the reading of a marvellous chapter of the big volume more ... I believe I mentioned in my last that Dawson and I found the skeleton of an animal in the middle of one of the upright trees of the Joggins Nova Scotia.

Clearly Lyell and Dawson had embarked upon a project to make more detailed observations on the Joggins sections. The discovery of the fossil animals within upright tree trunks meant that Lyell could not complete his task of measuring the Joggins section (Fig. 1). Indeed, given the time available it is unlikely that he would have done so anyway and probably concentrated on the part of the sequence yielding the greatest number of tree horizons. The publication of the Joggins section was left to Dawson alone (Dawson 1853) but the occurrence of the earliest reptile within tree stumps became the subject of a joint paper (Lyell & Dawson 1853) which has stimulated further research up to the present day. Particularly interesting is the manner of the discovery of *Dendrerpeton acadianum* (in fact an amphibian – reptiles were found by Dawson on subsequent visits, one being named *Hylonomus lyelli*) (Figs 4

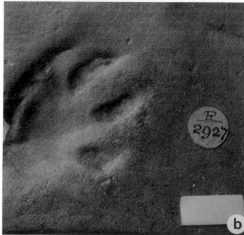

Fig. 2. Coal Measures of Joggins, Nova Scotia, Canada. (a) Section through sediment-filled lycopyte trunk, Joggins, Nova Scotia, eroded to show internal sediment stratigraphy (for sketch see Fig. 7). (b) Tetrapod footprint, (exact horizon unknown) (Natural History Museum, R2927).

Fig. 3. Upright sediment-filled trunks, Upper Carboniferous, Joggins, Nova Scotia, Canada. (a) Two successive horizons (arrowed) with upright lycophyte trunks. (b) Sandstone-filled lycophyte trunk. (c) Sandstone-filled calamite pith cast. (d) Upright sandstone-filled lycophyte trunk.

and 5) (Dawson 1859). In the original publication the discovery of the vertebrate fossils was recorded as being by chance in 1852 inside a trunk filled with sediment and found lying on the beach. The account by Dawson changed over the years so that in 1859 we read that the fossil stump containing the vertebrates was found with Lyell in 1851 (Dawson 1859, p. 268; also Dawson 1863).

In his autobiography, Dawson (Dawson 1901) notes that Lyell's poor eyesight was a great hindrance to his work and sometimes a source of personal danger. Dawson was clearly impressed with Lyell's intellect and enthusiasm. He wrote (Dawson 1901, pp. 54–55):

I remember how, after we had disinterred the bones of *Dendrerpeton* from the interior of a

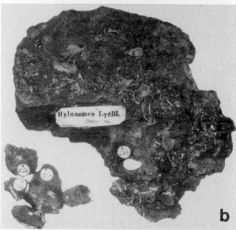

Fig. 4. Vertebrates from within Joggins trunks, Upper Carboniferous, Canada. (a) Cranial and postcranial remains of tetrapod (Natural History Museum, R299). (b) The holotype of *Hylonomus lyelli* (Natural History Museum, R4168). (c) Reconstruction of *Hylonomus lyelli* on a Canadian stamp.

large tree on the Joggins shore, his thoughts ran rapidly over all the strange circumstances of the burial of the animal; its geological age, and the possible relations to reptiles and other animals, and he enlarged enthusiastically on these points, till suddenly observing the astonishment of a man who accompanied us, he abruptly turned to me and whispered, 'the man will think us mad if I run on in this way'.

The discovery of the vertebrates persuaded Dawson to look especially for additional material. Further discoveries were made at odd intervals over the succeeding 30 years culminating in his detailed work on Coal Mine Point (Dawson 1882) where he documents details of 25 trees, 14 of which proved productive in the occurrence of vertebrates. In their original publication of 1853, Lyell and Dawson state that they had paid 'special attention to the difference of the deposits enveloping the upright trees, and those which fill the trunks themselves, forming casts within a cylinder of bark now turned to coal, the central wood of the trunk having decayed'. (p. 58) (Fig. 1). It is clear that the authors were interested not only in the stratigraphy but more so in the sedimentology and taphonomy of the plants and animals, a thoroughly modern approach. In this paper the authors offer several hypotheses to explain the occurrence of animals within the trunks. They note that in another tree the

> lower part of the cast, the sandstone contained a large quantity of vegetable fragments, as above mentioned [p. 59], principally pieces of carbonised wood [see for example Fig. 6], leaves of *Noeggerathia* or *Poacites*?, and stems of *Calamites*. With these vegetable remains we found the bones, jaws, teeth etc. before alluded to, all distinctly *within* the lower part of the cast, and scattered among the vegetable fragments contained in it, as if either washed in separate pieces, or, more probably, mixed with woody matter when the animal fell to pieces through decay [see Figs 5 and 6]. A part of the vegetable matter present must have been introduced after the tree became hollow. The creature to which the bones belonged may, therefore, either have been washed in after death, or may, when creeping on the surface, have fallen into the open pit caused by the decay of the tree, or may have crept into some crevice in the trunk before it was finally buried in the mud and sand. (pp. 62–63)

Clearly three hypotheses were being actively considered. By 1882, however, Dawson clearly favours the pit-fall theory and it is thus the one which has been favoured by most subsequent authors.

It is clear that Lyell was correct in the need to

Fig. 5. Vertebrates from sediments within lycophyte trunks, Joggins, Nova Scotia, Canada. (a) Skeletal remains of *Calligenethlon watsoni* (Natural History Museum, R442). (b) Small tetrapod limb bone with fusain fragment (Natural History Museum, R4564). (c) Tetrapod jaw fragment with fusain fragment (Natural History Museum, R434).

examine the sediments both inside and outside the trunks (Fig. 1). Re-examination of all the published data indicates that the one hypothesis that will not work is the pit-fall theory! As part of an ongoing reinvestigation of Upper Carboniferous fossil forests of Nova Scotia (Scott & Calder 1994; Calder *et al.* 1996), several of the tree-bearing levels have been reinvestigated (Fig. 3). Details of these investigations will be published elsewhere but some general points may be made. (1) Animals are frequently associated with fossil charcoal at the base of some trees (Fig. 7). This is evident from examination of the literature, museum material from the original collections and from new discoveries made by J. H. Calder and the author. The role of fire needs to be more fully explored (see below). (2) The detail of the sediment infill and surrounding sediments indicates that they would not have acted as pit falls (Figs 2 and 7). In the pit-fall theory it is supposed that the tree trunk rots in the centre and is hollow. Sediment is deposited

around the outside of the trunk until it reaches the top of the broken or rotted trunk. This 'land surface' is the one on which the animals were living. The rotted centres of the trees would then act as traps or pit-falls for any unsuspecting animals who stumbled upon them. Once the animal had fallen in it would not be able to escape and would be entombed in sediment from the next flood. For most of the trees (e.g. Fig. 7), sediment from both inside and outside the trunks can be matched and indicates that sediment was being deposited both inside and outside the trunk at the same time although the thickness of the sediment layers may vary. No soil surfaces have been found at the top of the filled trees in the surrounding sediments. Most of the tetrapods (except in a few isolated cases) occur near the base of the trunk fill (Dawson 1882 and personal observation, Fig. 6) when much of the trunk was still exposed. Several theories are possible for the occurrence of the animals but two of the most likely are that they acted as dens or

Fig. 6. Sediment-filled lycophyte trunks, Upper Carboniferous, Joggins, Nova Scotia, Canada. (a) Base of trunk with fusain and vertebrates overlain by sandstone fill. (b) Fusain fragments from base of trunk.

refuges for living animals or else the animals were washed into the hollow stumps. Clearly also of interest is the fact that so many tree horizons with trees preserved as casts up to 10 m high occur at Joggins. This is almost certainly because of the very rapid subsidence and sedimentation rates seen in the basin (Calder this volume). (3) Fossiliferous trees can be predicted, hence making it unnecessary to indiscriminately remove and break up the trees to discover fossils, which is important given the protected status of the site and the importance of

palaeoecological studies of trees *in situ* (see below).

Lyell clearly recognized that it was important to examine the sedimentology and taphonomy of the trees, a level of detail rarely undertaken today. Interest in these forests has increased over the past ten years and we are only now making progress in understanding their formation and significance (see DiMichele & De Maris 1987; Demko & Gastaldo 1992; Scott & Calder 1994; Calder *et al.* 1996).

To at least some degree Lyell could claim to be a

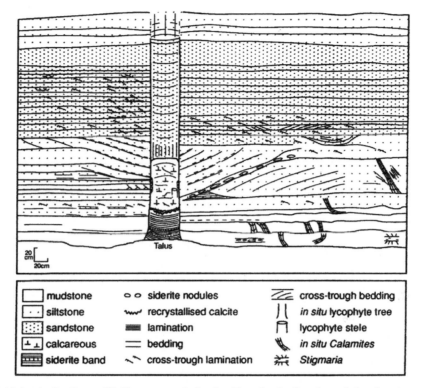

Fig. 7. Field sketch of sediment-filled lycopyte trunk, Joggins, Nova Scotia, Canada, eroded to show internal sediment stratigraphy (for photo see Fig. 2(a)).

founding father of the study of terrestrial palaeo-ecology. Clearly Lyell's discovery stimulated a wealth of research and both sedimentologists and palaeontologists should be grateful to the prominence that the discovery gave to the Joggins sequence and to the subsequent finds.

The occurrence and significance of charcoal in coal seams

Lyell (1847) was one of the first to recognise and illustrate charcoal in coal seams (Fig. 8). Fossil charcoal was also found associated with vertebrates in the base of some of the fossil trees (Fig. 6). In describing the coal of Eastern Virginia he states that some portions exactly resemble charcoal in appearance (p. 268). He notes also that similar material is observed in some Welsh coal, where the charcoal is called 'mother coal'. Much of the discussion concerned the botanical affinities of the charcoal rather than its origin from burnt vegetation but it is important to note that Lyell was aware of the significance of the anatomical structure of the material

for the identification of the plants and that such material was of palaeontological interest.

Despite the observations of Lyell on the preservation of anatomy in charcoal in coal seams, few researchers have developed this theme. Much of the debate over the subsequent 150 years has concerned the nature and origin of such mineral charcoal, called fusain by Stopes (1919). Interestingly, Lyell fails to link the occurrence of mineral charcoal in coals to the occurrence of fire. The discussion of a fire origin of charcoal did not become common until the end of the nineteenth and beginning of the twentieth century (Muck 1881; Stutzer 1929). In the volume edited by Stutzer (1929) the debate on the origin of fusain was polarised. Jeffrey (*in* Stutzer 1929) wrote, 'My opinion of the origin of fusit or fusain ... is unhesitatingly that this material is charcoal derived from the action of forest fires'. White (*in* Stutzer 1929) wrote on the other hand, 'it is quite impossible for me to conceive of fusain as exclusively the result of forest fires'. Fifty years later, many still failed to recognize the fire origin of fusain; the situation was still unresolved, with Schopf (1975) for example

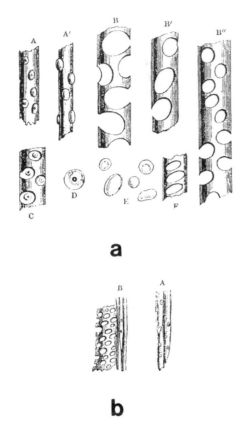

a

b

Fig. 8. Fossil charcoal (fusain) from the USA (from Lyell 1847). (a) 'Vegetable structure of mineral charcoal from Cover–hill mines, Virginia' (A–F discussed by Lyell). (b) 'Blackheath specimen (Creek mines)' (A–B discussed by Lyell).

writing, 'For some and probably for the majority of the occurrence of fusain the forest fire origin seems ruled out.'

Over the past 20 years much effort has gone into the study of fusain to understand its origin and the identity of fusain as fossil charcoal and its formation following wildfire has been firmly established (see Scott 1989a; Jones 1993; Jones et al. 1993 a, b; Scott & Jones 1991a,b, 1994; Winston 1993 and references therein).

One of the keys to understanding the nature and origin of fusain is the preservation of fine anatomical details, as observed by Lyell (1847). The advent of the scanning electron microscope has revolutionized the study of fossil charcoal. The high-fidelity three-dimensional anatomical preservation both in modern charcoals produced following wildfires (Fig. 9 (e), (f)) and in fossil fusains demonstrates a similar mode of origin (Scott 1989a) (Fig. 9 (a)–(d)). Lyell (1847) also

recognized that the anatomy of the plants may be useful for the identification of the species concerned. Even today botanical identification of fusain is not routinely undertaken. Several plants have now been described from fossil deposits preserved as fusain (e.g. Alvin 1974; Scott 1974; Friis & Skarby 1981) but many geologists still ignore this valuable data source and most palaeobotanists prefer to work on less fragmentary plant fossils.

The occurrence of fossil charcoal in coal seams provides many important clues to the ecology and environment of ancient peat deposits (Scott & Jones 1994). Peat-forming environments of the southern USA have been taken as possible analogues to ancient coal seams and recent studies indicate that fires are an integral part of many of these wetland ecosystems (e.g. Cohen 1974; Cohen et al. 1984; Rollins et al. 1993). It is, however, only in the last few years that it has recognised that studies of ancient charcoals provide data on climate, atmospheric composition and erosion and depositional events (Chaloner 1989; Cope & Chaloner 1980; Scott & Jones 1994; Nichols & Jones 1992).

In his visit to peat-forming areas of southern United States Lyell did encounter evidence of fires. He states (Lyell 1855a p. 386), 'I may mention that whenever any part of a swamp in Louisiana is dried up, during an unusually hot season, and the wood set on fire, pits are burnt into the ground many feet deep, or as far down as the fire can descend, without meeting with water, and it is then found that scarcely any residuum or earthy matter is left' (see also Lyell 1849a, vol. 2, p. 245). As already mentioned, Lyell observed fossil charcoal in the base of some of the upright lycopsid trunks but did not link these two observations. Examination of the tetrapod material collected by Lyell and Dawson from the tree-trunk fills from Joggins has shown that most are associated with fusain (fossil charcoal) (e.g. Fig. 5). Recent field observations by the author and J. Calder have indicated that several filled stumps have abundant charcoal at their base (Fig. 6). The occurrence of a large quantity of charcoal in the base of the trunks could have meant that they were hollowed out by fire with the charred plant remains accummulating at the base (Fig. 7). The trunk could then, if it remained upright, have acted as a den or refuge for animals particularly if there was a fire scarred base which created a hole through which animals might crawl. However, fire is likely to have spread through the vegetation and may have also killed the animals. Such charred debris may have been subsequently washed into the hollowed trunks during a flood event. A further possiblity is that the trees were partially rotted before the fire and the animals that were either living inside the trunk or were using it as a refuge

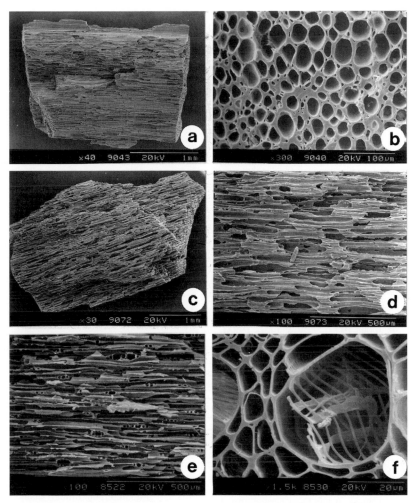

Fig. 9. Scanning electron micrographs of fossil (a)–(d) (Coal 15, Upper Carboniferous, Coal Mine Point, Joggins, Nova Scotia, Canada) and Recent (e)–(f) charcoal (Tilford, Surrey, UK). (a) Fusain fragment. (b) Detail of end walls showing homogenized cell walls. (c) Fusain fragment. (d) Detail of longitudinal section. (e) Longitudinal section of *Pinus* wood showing rays. (f) Transverse section of angiosperm wood showing homogenized cell walls and vessel end plate.

from the fire were killed by the fire. The driving of animals to their deaths by the action of fire has been suggested for other tetrapod deposits (Scott *et al.* 1994). The common fragmentation of skeletons and the mixing of tetrapod material with charcoal may suggest that this is not the case. Details of these arguments will be presented elsewhere but the observation that tetrapods occur most often in the base of trunks associated with charcoal in relatively few of the many tree horizons is important, but much more data are needed on the taphonomy of the vertebrates and the trees themselves.

For the most part, even today, the occurrence of fossil charcoal in sediments is not widely recognized. The recording of plant fossils by sedimentologists is rare in itself and the distinction between coalified plants and fossil charcoal (fusain) is rarely made. Over the past 20 years there have been several studies which have demonstrated the importance of fossil charcoal.

Palaeoatmospheres and palaeoclimates

There has been considerable recent interest in the modelling of ancient atmospheres and climates (Chaloner & McElwain 1997; Berner 1994; Berner

& Canfield 1989). It has been demonstrated that the occurrence of fossil charcoal constrains atmospheric oxygen levels over the past 400 million years to within both upper and lower limits (Cope & Chaloner 1980; Chaloner 1989), the 'fire window' of Jones & Chaloner (1991) who suggest that atmospheric oxygen has not fallen below 13 per cent and not risen above 35 per cent. It is clear that oxygen levels have risen and fallen from the Devonian to the present day. Berner and Canfield (1989) note very high (over 30per cent) oxygen levels during the Carboniferous. This, together with the occurrence of widespread lowland equatorial peat systems, may have made the Euramerican coal measures particularly fire prone (Scott & Jones 1994) resulting in Carboniferous coals having high fusain contents (Robinson 1987).

Sedimentology

Few sedimentologists working on ancient sequences have appreciated the linkage between fire and erosion/depositional cycles (but see Allen 1974; Nichols & Jones 1992; Scott & Jones 1994). Fires may result in a significant increase in erosion and hence deposition in a basin (Swanson 1981). This has been well demonstrated in Yellowstone National Park following the extensive fires of 1988 (Meyer *et al.* 1992). There have been relatively few detailed studies of Pre-Quaternary charcoal deposits (Scott & Jones 1994). Evidence of early catastrophic fires has been recorded from Lower Carboniferous nearshore marine sediments in western Ireland (Nichols & Jones 1992; Scott & Jones 1994; Falcon-Lang 1998). These fires not only resulted in significant erosion but also affected the depositional systems. It has been demonstrated that fires of different vegetation types in different climatic zones burn at different temperatures (Rundel 1981). Experimental charring of wood has shown the linkage between temperature and reflectance values of the charcoal (Scott & Jones 1991*a*). Experimental data appear consistent with data collected from charcoals from Recent fires (Scott & Jones 1994), thus offering the possibility of using details of the fossil charcoal to help understand past fire regimes.

Palaeobotany and palaeoecology

In hand specimen, fossil charcoal looks uninteresting and uninformative. Charcoal, and hence fossil charcoal, does, however, retain considerable anatomical data. Over the past 25 years, particularly with the advent of scanning electron microscopy, the anatomy of many charcoalified plants has been described (see Scott 1989 for a

Fig. 10. *Taxodium distichum*, the swamp cypress (the leaves have fallen from this deciduous conifer), in peat-forming swamp, Everglades, southern USA.

review). It is not only important to consider this in terms of botanical/anatomical data but also as yielding important ecological data. For example some Carboniferous plants are preferentially preserved as charcoal and hence may have been living in fire-prone ecosystems (Scott & Jones 1994; DiMichele & Phillips 1994).

The botanical identity of a charcoalified plant assemblage may also yield significant palaeo-ecological data when compared with associated non-charcoalified assemblages (e.g. in the Lower Carboniferous of Kingswood, Fife; Scott & Jones 1994). In the Yorkshire Jurassic, Cope (1993) has contrasted the charcoal taxonomic composition with the taxonomic composition of associated plant compression fossils noting that the differences can be explained by ecological factors and sedimentological processes acting within a delta environment subject to periodic natural wildfire.

Fossil charcoal has much to offer the geologist and perhaps we should take a lesson from Lyell when recording field observations – all information may be relevant.

Modern analogues for ancient coals

In his early work Lyell (1830) showed little concern for coal. He did, however, use the analogy of the Mackenzie River to explain his views on the drifted origin of ancient coals (p. 242). Lyell's visits to North America, especially to the wetlands of the south, together with his observations on Carboniferous coals in particular, profoundly changed his views. In his first visit to North America (Lyell 1843), Lyell describes his visit in December 1841 to the Great Dismal Swamp and makes a direct comparison with the coal measures. Lyell later noted (1855, p. 382) in comparison with upright trees of the coal measures that 'some *Coniferae* of the Coal Period grew in the same swamps as *Sigillariae,* just as now the deciduous Cypress (*Taxodium distichum*) abounds in the marshes of Louisiana even to the edge of the sea' (Fig. 10).

During his second visit to the United States in 1846 Lyell was impressed by his observations on the Mississippi. He documents his observations in some detail (Lyell 1849a, vol. 2, pp. 77–81, 190–195) and incorporates these into both the *Principles* and the *Manual.* Some of his observations relate to the problem of the absence of sediment in coal: 'It may be affirmed that generally in the "Cypress swamps" of the Mississippi no sediment mingles with the vegetable matter accumulated there from the decay of trees and semi-aquatic plants.' The problem of the nature of the Mississippi peats is a long-standing one. Some have claimed that these would make good analogues for coal (Kosters *et al.* 1987) but others (e.g. McCabe 1987) point out that the peats do in fact show considerable 'ash' on ignition and may, upon burial, only give rise to a carbonaceous shale. Combining observations on modern systems together with data from Coal Measures of Nova Scotia Lyell makes some interesting calculations on the time involved. His conclusion (Lyell 1855, pp. 286–287) is that it would take more than 2 million years to convey to the Gulf of Mexico via the Mississippi the quantity of sediment of the Nova Scotia coal measures, which is approximately the time span represented by the coal measures of Joggins (Calder, pers. comm. 1997).

The interpretation of the Carboniferous coal measures using the peats and depositional systems of the southeast United States is one that has had a long history and many authors have used this modern analogue (e.g. Dapples & Hopkins 1969; Fisk 1960; Spackman *et al.* 1966). An increase in our understanding of the nature of peat formation has demonstrated that the most appropriate analogues to Carboniferous coals may lie in the ombrogenous peats of Southeast Asia (Moore 1987; Clymo 1987). It was shown that peats form in waterlogged conditions and that the source of the water has a major influence on the nature of the resultant peat. Predominantly flow-fed systems (rheotrophic) yielded planar peats with variable ash and sulphur contents whereas rain water-fed systems (ombrotrophic) yielded raised or domed peats with low ash and sulphur content (Moore 1987). Combined with the fact that the Carboniferous coals of Euramerica (Calder & Gibling 1994) formed in the palaeotropics, tropical raised mires (peat-forming systems) are a more appropriate analogue for Carboniferous coals (Cohen *et al.* 1989; Cohen & Stack 1996; Staub & Esterle 1992; Cameron *et al.* 1989; Moore & Hubbert 1992; Esterle & Ferm 1994; Cobb & Cecil 1993; Cecil *et al.* 1993; Neuzil *et al.* 1993; Supardi *et al.* 1993; Grady *et al.* 1993; Eble & Grady 1993). Interpretation of the trophic setting of ancient peats/coals has been used as a proxy for palaeoclimate with ombrotrophic peats representing a non-seasonal regime and rheotrophic peats showing a restricted regime with a dry season (Cecil *et al.* 1985; Moore 1987; Calder 1994).

A major feature of many of these tropical analogues is that they are thick and domed with low ash and sulphur content. Detailed studies of the coals (e.g. Neuzil *et al.* 1993) has demonstrated the domed ombrogenous peats of Indonesia will result in low ash and low sulphur coal (less than 10 per cent ash and less than 1 per cent sulphur), even if marine rocks are laterally and vertically adjacent to the coal. The upsurge of interest in using domed tropical peats as analogies for Carboniferous coals in place of the planar peats of the southeast USA has led researchers to think of peats as either planar or domed, rheotrophic or ombrotrophic (Cohen & Stack 1996, Scott 1989b). It has, however, been demonstrated by several authors (both from temperate systems (e.g. Styan & Bustin 1983) and tropical systems (e.g. Cohen & Stack 1996) that many of the larger peat deposits may be partially domed and partially planar. In some cases these may record a planar–domed–planar sequence (Cohen & Stack 1996), a succession interpreted for some Carboniferous coals (Bartram 1987). The importance of the trophic status of peat has been recognised only relatively recently. The state of knowledge at the time of Lyell would not have allowed him to appreciate this. Most of the coals Lyell observed at Joggins were thin planar, rheotrophic peats deposited in a rapidly subsiding basin (Hower *et al.* 1995) where there were frequent changes in rainfall pattern as seen by the interfingering of coal zones with trees and reddened intervals. Lyell's careful observations on the coals, sediments and biota represented, however, the way forward. Teichmüller (1989) observed that the trophic status of ancestral coals is one of the most intriguing and problematic aspects of the genesis of coal. Lyellian studies of fossil forests and underlying peats are shedding light on this problem (DiMichele & Phillips 1994; Calder *et al.* 1996).

In all these studies there is the view that the present holds the key to the past and that a study of modern peats will lead to a greater understanding of Carboniferous coals. Whilst this may be true to some extent, there are several inherent problems with this approach. It has been widely recognized that coal-forming vegetation has changed radically through time (see reviews in Collinson & Scott 1987 a, b). This was a fact not obvious to Lyell. If it were simply that there were different species, genera or families of plants which had a similar growth and life habit, this might not be problematic. It is now clear, following detailed anatomical studies of Upper Carboniferous coal-forming plants using data from coal balls, that the growth strategies and life habits of the peat-forming plants were quite different from those living today. Plant construction, growth strategy, tissue types and reproductive strategy were all different (DiMichele & Phillips 1994). Even the way the resultant peat compacts is affected by types of plants forming the peat (Winston 1986). It has also been shown that the rate of plant growth may be radically different between Carboniferous and Recent plants (DiMichele & Phillips 1994) and indeed tissue compaction may also differ depending on the taxa concerned (Winston 1986). Exact analogies between modern peats and ancient coals may not, therefore, be found. There has been a tendency to imagine all coals of all ages are the same but clearly recent advances in our understanding of the vegetational composition demonstrates that this is clearly not the case (Collinson & Scott 1987a). Modern analogies are useful, as Lyell found, in visualising ancient deposits, but too much reliance on detailed interpretation of the Recent systems to interpret some of the older, for example the Upper Carboniferous, coals will lead to erroneous assumptions and results. One way to resolve these issues is through careful multidisciplinary work on plants, erect trees, coals and sediments etc. – the approach employed by Lyell and his colleagues.

Conclusions and future work

Personal field observations were important in shaping Lyell's view of the origin and significance coal and coal-bearing strata. In particular, his travels to several North American coal basins and especially his observations on the Joggins section, together with his observations on Recent peat-forming environments of the southern United States gave him a particular holistic approach to unravelling ancient wetland ecosystems. To Lyell, under-

standing the coal-forming vegetation, the investigation of animals and their traces, the origin of sediments and their rate of deposition, climate and weather were all part of the same problem to be explained. Lyell showed that careful field observation, combined with a knowledge of modern processes, was the key to interpreting the past.

What Lyell was unable to know was that the present is not always a completely reliable guide to interpreting the past. Lyell, and indeed many current researchers, did not realize that the peat–forming vegetation of today was quite different, both in terms of growth construction and reproductive strategy, from that of the Carboniferous. This, combined with the fact that atmospheric and climatic conditions were also different, means that detailed modern analogies are not possible for all but the most Recent coals.

I have shown three key areas that interested Lyell and which are also the subject of modern study. These are: the formation of coal and coal–bearing sequences; the role of fire in shaping coal forming ecosystems and the use of modern peats to interpret ancient coals. It is clear, as Lyell appreciated, that a multidisciplinary and interdisciplinary approach is needed to unravel these problems. Too often research has centred on only one aspect using only one tool. For example, coal measure sedimentologists have concentrated on clastic sediments between the coals, coal petrologists have concentrated on the coals and ignored the intervening sediments and both have often ignored the fauna, flora and their traces. Indeed, fossil forests atop coals provide an important link between these two (Calder *et al.* 1996).

If further progress is to be made in our understanding of coal measures systems then an integrated or holistic approach is necessary. This has been partly attempted only for a few sites (e.g. Willard *et al.* 1995, Calder *et al.* 1996) and the success of such an approach has been demonstrated on other terrestrial sequences such as the site for the current oldest reptile-like vertebrate from East Kirkton, Scotland (Rolfe *et al.* 1990, 1994 and subsequent papers). Unfortunately in this world of increasing specialisation, the ability, and indeed desire to undertake a more holistic approach, seems to be lacking both on the part of researchers and their granting agencies. Perhaps we should take a lesson from Sir Charles Lyell where the combination of careful field observations together with a theoretical framework helped him to promote the relatively new science of geology not only to other scientists but also to the public at large.

The work at Joggins with J. H. Calder was funded by NATO grant 940559. I thank K. de Souza for photographic assistance and L. Blything for drafting the diagrams.

References

ALLEN, P. 1974. The Wealden of the Weald – a new model. *Proceedings of the Geologists Association,* **86**, 389–437.

ALVIN, K. L. 1974. Leaf anatomy of *Weichselia* based on fusainised material. *Palaeontology,* **17**, 587–598.

BARTRAM, K. M. 1987. Lycopod succession in coals: an example from the Law Barnsly Sean (Westphalian B), Yorkshire, England. *In*: SCOTT, A. C. (ed.) *Coal and Coal-Bearing Strata: Recent Advances.* Geological Society, London, Special Publications, **32**, 187–199.

BENNETT, A. J. R. 1963. *Origin and Formation of Coal Seams. A Literature Survey.* Commonwealth Science Industrial Research Organisation Miscellaneous Report 239, Chatswood, N.S.W., 1–14.

BERNER, R. A. 1994. 3GeocarbII: a revised model of atmospheric CO$_2$ over Phanerozoic time. *American Journal of Science,* **291**, 56–91.

—— & CANFIELD, D. E. 1989. A new model for atmospheric oxygen over Phaneozoic time. *American Journal of Science,* **289**, 333–361.

BINNEY, E. W. 1846. On the Dukinfield and St. Helens *Sigillaria. Quarterly Journal of the Geological Society,* **2**, 392–400.

BROWN, R. 1846. On a group of fossil trees in the Sydney coalfield of Cape Breton. *Quarterly Journal of the Geological Society,* **2**, 393–396.

—— 1848. Description of an upright *Lepidodendron* with *Stigmaria* roots in the roof of the Sydney Main coal in the Island of Cape Breton. *Quarterly Journal of the Geological Society,* **4**, 46–50.

CALDER, J. H. 1994. The impact of climate change, tectonism and hydrology on the formation of Carboniferous tropical international mines: the Springfield coalfield, Cumberland Basin, Nova Scotia. *Palaeogeography, Palaeoclimatology, Palaeoecology,* **106**, 323–351.

—— & GIBLING, M. R. 1994. The Euramerican coal province: controls on Late Palaeozoic peat accumulation. *Palaeogeography, Palaeoclimatology, Palaeoecology,* **106**, 1–21.

——, ——, EBLE, C. F., SCOTT, A. C. & MACNEIL, D. J. 1996. The Westphalian D fossil lepidodendrid forest at Table Head, Sydney Basin, Nova Scotia: sedimentology, palaeoeocology and floral response to changing edaphic conditions. *International Journal of Coal Geology,* **31**, 277–313.

CAMERON, C. C., ESTERLE, J. S. & PALMER, C. A. 1989. The geology, botany and chemistry of selected peat forming environments from temperate and tropical latitudes. *International Journal of Coal Geology,* **12**, 105–156.

CARROLL, R. L. 1982. Sir William Dawson and the vertebrate fossils of the Joggins fauna. *In*: *Proceedings of the Third North American Paleontological Convention,* vol. 1, 71–75.

—— 1994. Evaluation of geological age and environmental factors in changing aspects of the terrestrial vertebrate fauna during the Carboniferous. *Transactions of the Royal Society of Edinburgh: Earth Sciences,* **84**, 427–431.

CECIL, C. B., DULONG, F. T., COBB, J. C. & SUPARDI 1993. *Allogenic and Autogenic Controls on Sedimentation in the Central Sumatra Basin as an Analogue for Pennsylvanian Coal-Bearing Strata in the Appalachian Basin.* Geological Society of America Special Paper 286, Boulder, Colorado, 3–22.

——, STANTON, R. W., NEUZIL, S. G., DULONG, F. T., RUPPERT, L. R. & PIERCE, B. S. 1985. Palaeoclimate controls on Late Paleozoic sedimentation and peat formation in the central Appalachian Basin (USA). *International Journal of Coal Geology,* **5**, 195–230.

CHALONER, W. G. 1989. Fossil charcoal as an indicator of palaeoatmospheric oxygen level. *Journal of the Geological Society of London,* **146**, 171–174.

—— & MCELWAIN, J. 1997. The fossil plant record and global climatic change. *Review of Palaeobotany and Palynology,* **95**, 73–82.

CLYMO, R. S. 1987. Rainwater–fed peat as a precursor of coal. *In*: SCOTT, A. C. (ed.) *Coal and Coal-Bearing Strata: Recent Advances.* Geological Society, London, Special Publications, **32**, 17–23.

COBB, J. C. & CECIL, C. B. (eds) 1993. *Modern and Ancient Coal Forming Environments.* Geological Society of America Special Paper 286, 1–198.

COHEN, A. D. 1974. Petrography and palaeoecology of Holocene peats from the Okefenokee swamp–marsh complex of Georgia. *Journal of Sedimentary Petrology,* **44**, 716–726.

—— & STACK, E. M. 1996. Some observations regarding the potential effects of draining of tropical peat deposits on the composition of coal beds. *International Journal of Coal Geology,* **29**, 39–65.

——, CASAGRANDE, D. J., ANDREJKO, M. & BEST, R. (eds) 1984. *The Okefenokee Swamp: Its Natural History, Geology and Geochemistry.* Wetland Surveys.

——, RAYMOND R., RAMUREZ, A., MORALES, Z. & PONCE, F. 1989. The Changuinda peat (coal) deposit of northwestern Panama: a tropical back-barrier coal-forming environment. *International Journal of Coal Geology,* **12**, 157–192.

COLLINSON, M. E. & SCOTT, A. C. 1987a. Implications of vegetational change through the geological record on models for coal-forming environments. *In*: SCOTT, A. C. (ed.) *Coal and Coal-Bearing Strata: Recent Advances.* Geological Society, London, Special Publications, **32**, 67–85.

—— & —— 1987b. Factors controlling the organisation of ancient plant communities. *In*: GEE, H. R. & GILLER, N. S. (eds) *Organisation of Communities, Past and Present.* British Ecological Society, London: Blackwell Scientific Publications, Oxford, 399–420.

COPE, M. J. 1993. A preliminary study of charcoalified plant fossils from the Middle Jurassic Scalby Formation of North Yorkshire. *Special Papers in Palaeontology,* **49**, 101–111.

—— & CHALONER, W. G. 1980. Fossil charcoal as evidence of past atmospheric composition. *Nature,* **283**, 647–649.

DAPPLES, E. C. & HOPKINS, M. E. (eds) 1969. *Environments of Coal Deposition.* Geological

Society of America Special Paper 114, Boulder, Colorado.

DAWSON, J. W. 1853. On the coal measures of the South Joggins, Nova Scotia. *Quarterly Journal of the Geological Society,* **10**, 1–51.

—— 1859. On a terrestrial mollusk, a chilognathous myriapod and some new species of reptiles from the coal-formation of Nova Scotia. *Quarterly Journal of the Geological Society,* **16**, 268–277.

—— 1863. *Air-breathers of the Coal Period.* Dawson Brothers, Montreal.

—— 1882. On the results of recent explorations of erect trees containing animal remains in the coal formation of Nova Scotia. *Philosophical Transactions of the Royal Society of London,* **173**, 621–659.

DAWSON, R. (ed.) 1901. *Fifty Years of Work in Canada, Scientific and Educational, Being the Autobiographical Notes by Sir William Dawson.* Ballentine, Hanson & Company, London.

DEMKO, T. M. & GASTALDO, R. A. 1992. Paludal environments of the Mary Lee Coal Zone, Pottsville Formation, Alabama: stacked clastic swamps and peat mires. *International Journal of Coal Geology,* **20**, 23–47.

DiMICHELE, W. A. & DeMARIS, P. J. 1987. Structure and dynamics of a Pennsylvanian-age *Lepidodendron* forest: colonisers of a disturbed swamp habitat in the Herrin (No 6) coal of Illinois. *Palaios,* **2**, 146–157.

—— & PHILLIPS, T. L. 1994. Palaeobotanical and Palaeoecological constraints on models of peat formation in the Late Carboniferous of Euramerica. *Palaeogeography, Palaeoclimatology, Palaeoecology,* **106**, 39–90.

EBLE, C. F. & GRADY, W. C. 1993. *Palynological and Petrographic Characteristics of Two Middle Pennsylvanian Coal Beds and a Probable Modern Analogue.* Geological Society of America Special Paper 286, 119–138.

ESTERLE, J. S. & FERM, J. C. 1994. Special variability in modern tropical peat deposits from Sarawak, Malaysia and Sumatra, Indonesia: analogues for coal. *International Journal of Coal Geology,* **26**, 1–41.

FALCON-LANG, H. 1998. The impact of wild fire on an early Carboniferous coastal environment, North Mayo, Ireland. *Palaeogeography, Palaeoclimatology, Palaeoecology,* **139**, 121–138.

FISK, H. N. 1960. Recent Mississippi river sedimentation and peat accumulation. *In: Compte Rendu 4 Congress Advances étude Stratigraphic et Geologie du Carbonifere.* vol. 1, Ernest, Van Aelst, Maestricht, 181–199.

FRIIS, E. M. & SKARBY, A. 1981. Structurally preserved angiosperm flowers from the Upper Cretaceous of southern Sweden. *Nature,* **291**, 485–486.

GRADY, W. C., EBLE, C. F. & NEUZIL, S. G. 1993. *Brown Coal Maceral Distributions in a Modern Domed Tropical Indonesia Peat and a Comparison with Maceral Distributions in Middle Pennsylvanian - Age Appalachian Bituminous Coal Beds.* Geological Society of America Special Paper 286, Boulder, Colorado, 63–82.

HOWER, J. C., CALDER, J. H., EBLE, C. F., SCOTT, A. C., ROBERTSON, J. D. & BLANCHARD, L. J. 1996. Petrology, geochemistry and palynology of Joggins (Westphalian A) coals, Cumberland Basin, Nova Scotia. *Bulletin of the American Association of Petroleum Geologists,* **80**, 1525.

HUTTON, J. 1795. *Theory of the Earth with Proofs and Illustrations.* Creech, Edinburgh, vol. 1.

JONES, T. P. 1993. New morphological and chemical evidence for a wildfire origin for fusain from comparisons with modern charcoal. *Special Papers in Palaeontology,* **49**, 113–123.

—— & CHALONER, W. G. 1991. Fossil charcoal, its recognition and palaeoatmospheric significance. *Palaeogeography, Palaeoclimatology, Palaeoecology,* **97**, 39-50.

——, SCOTT, A. C. & MATTEY, D. P. 1993. Investigation of 'fusain transition fossils' from the Lower Carboniferous: comparisons with modern partially charred wood. *International Journal of Coal Geology,* **22**, 37–59.

KOSTERS, E. C., CHMURA, G. L. & BAILEY, A. 1987. Sedimentary and botanical factors influencing peat accummulation in the Mississippi Delta. *Journal of the Geological Society of London,* **144**, 423–434.

KIRWAN, R. 1799. On coal mines. *In: Geological Essays.* London, vol. 7, 290–340.

LOGAN, W. E. 1841. On the character of the beds of clay lying immediately below the coal seams of South Wales. *Proceedings of the Geological Society of London,* **3**, 487–492.

LYELL, C. 1830–1833. *Principles of Geology.* Murray, London.

—— 1841. Letter addressed to Dr Fitton, by Mr. Lyell, and dated Boston the 15 of October 1841. *Proceedings of the Geological Society of London,* **3**, 554–558.

—— 1843. On the upright fossil trees found at different levels in the coal strata of Cumberland, Nova Scotia. *Proceedings of the Geological Society of London,* **4**, 176–178.

—— 1845. *Travels in North America.* 2 vols, Murray, London.

—— 1846a. Notice on the coal-fields of Alabama; *Quarterly Journal of the Geological Society of London,* **2**, 278–282.

—— 1846b. Observations on the fossil plants of the coal Field of Tuscaloosa, Alabama. *American Journal of Science,* **2**, 228–233.

—— 1846c. On footmarks discovered in the coal-measures of Pennsylvania. *Quarterly Journal of the Geological Society of London,* **2**, 417–420.

—— 1846d. On the evidence of fossil footprints of a quadruped allied to the Cheirotherium, in the coal strata of Pennsylvania. *American Journal of Science,* **2**, 25–29.

—— 1847. On the structure and probable age of the coal-field of the James River, near Richmond, Virginia. *Quarterly Journal of the Geological Society of London,* **3**, 261–288.

—— 1849a. *A Second Visit to the United States of North America.* 2 vols. Murray, London.

—— 1849b. Notes on some recent footprints on the Red

Mud in Nova Scotia. *Quarterly Journal of the Geological Society of London,* **5**, 344.

—— 1853*a*. On the discovery of some fossil reptilian remains, and a land-shell in the interior of an erect fossil-tree in the coal measures of Nova Scotia, with remarks on the origin of coal-fields, and the time required for their formation. *American Journal of Science,* **16**, 33–41; *Proceedings of the Royal Institution,* **1**, 281–288; *Edinburgh New Philosophical Journal,* **55**, 215–225.

—— 1853*b*.*Principles of Geology.* 9th edn. Murray, London.

—— 1855*a*. *Travels in North America, Canada and Nova Scotia.* vol. 2. 2nd edn. Murray, London.

—— 1855*b*. *Manual of Elementary Geology.* 5th edn. Murray, London.

—— & DAWSON, J. W. 1853. On the remains of a reptile (*Dendrerpeton acadianum,* Wyman and Owen) and of a land shell discovered in the interior of an erect fossil tree in the coal measures of Nova Scotia. *Quarterly Journal of The Geological Society,* **9**, 58–63.

LYELL, K. M. 1881. *Life, Letters and Journals of Sir Charles Lyell, Bart.* 2 vols. Murray, London.

McCABE, P. J. 1987. Facies studies of coal and coal-bearing strata. *In:* SCOTT, A. C. (ed.) *Coal and Coal-Bearing Strata: Recent Advances.* Geological Society, London, Special Publications, **32**, 51–66.

MEYER, G. A., WELLS, S. G., BALLING, R. C., JR & JULL, A. J. 1992. Response of alluvial systems to fire and climate change in Yellowstone National Park. *Nature,* **357**, 147–150.

MOORE, P. D. 1987. Ecological and hydrological aspects of peat formation. *In:* SCOTT, A. C. (ed.) *Coal and Coal-Bearing Strata: Recent Advances.* Geological Society, London, Special Publications, **32**, 7–15.

MOORE, T. A. & HUBBERT, R. 1992. Petrographic and anatomical characteristics of plant material from two peat forming environments of Holocene and Miocene age, Kalimantan, Indonesia. *Review of Palaeobotany and Palynology,* **72**, 199–227.

MUCK, F. 1881. *Grundzüge und Ziele der Skeinkohlenchemical.* Verlay van Enil Strauss, Bonn.

NEUZIL, S. G., SUPARDI, CECIL, C. B., KANE, J. S. & SOEDJONA, K. 1993. Inorganic geochemistry of domed peat in Indonesia and its implication for the origin of mineral matter in coal. *Geological Society of America Special Paper,* **286**, 23–44.

NICHOLS, G. & JONES, T. P. 1992. Fusain in Carboniferous shallow marine sediments, Donegal, Ireland: the sedimentological effects of wildfire. *Sedimentology,* **39**, 487–502.

PARKINSON, J. 1804. *Organic Remains of a Former World. Vol. 1 The Vegetable Kingdom.* Sherwood, Neely & Jones, London.

ROBINSON, J. 1987. Phanerozoic O_2 variations, fire and terrestrial ecology. *Global and Planetary Change,* **1**, 223–240.

ROLFE, W. D. I., DURANT, G. P., BAIRD, W. J., CHAPLIN, C., PATON, R. L. & REEKEE, R. J. 1994. The East Kirkton Limestone; Visean, of West Lothian, Scotland, introduction and stratigraphy: *Transactions of the Royal Society of Edinburgh: Earth Sciences,* **84**, 177–188.

——, ——, FALLICK, A. E. ET AL. 1990. An early terrestrial biota preserved by Visean vuleamicity in Scotland. *In:* LOCKLEY, M. G. & RICE, A. (eds) *Volcanism and Fossil Biotas.* Geological Society of America Special Paper 244, Boulder, Colorado, 13–24.

ROLLINS, M. S., COHEN, A. D. & DURIG, J. R. 1993. Effects of fire on the chemical and petrographic composition of peat in the Snuggedy Swamp, South Carolina. *International Journal of Coal Geology,* **22**, 101–117.

RUNDEL, P. W. 1981. Fire as an ecological factor. *In:* LANGE, O. L. *et al.* (eds) *Physiological Plant Ecology 1. Response to the Physical environment.* Springer Verlag, Berlin, 501–538.

SCHOPF, J. M. 1975. Modes of fossil preservation. *Review of Palaeobotany and Palynology,* **20**, 27–53.

SCOTT, A. C. 1974. The earliest conifer. *Nature,* **251**, 707–708.

—— 1977. A review of the ecology of Upper Carboniferous plant assemblages, with new data from Strathclyde. *Palaeontology,* **20**, 447–473.

—— 1989*a*. Observations on the nature and origin of fusain. *International Journal of Coal Geology,* **12**, 443–475.

—— 1989*b*. Deltaic coals: an ecological and palaeobotanical perspective. *In:* WHATELEY, M. K. G. & PICKERING, K. T. (eds) *Deltas: Sites and Traps for Fossil Fuels.* Geological Society, London, Special Publications, **41**, 309–326.

——, BROWN, R., GALTIER, J. & MEYER-BERTHAUD, B. 1994. Fossil plants from the Viséan of East Kirkton, West Lothian, Scotland. *Transactions of the Royal Society of Edinburgh: Earth Sciences,* **4**, 249–260.

—— & CALDER, J. H. 1994. Carboniferous fossil forests. *Geology Today,* **10**, 213–217.

—— & JONES, T. P. 1991*a*. Microscopical observations of Recent and fossil charcoal. *Microscopy and Analysis,* **24**, 13–15.

—— & —— 1991*b*. Fossil charcoal: a plant fossil record preserved by fire. *Geology Today,* **7**, 214–216.

—— & —— 1994. The nature and influence of fire in carboniferous ecosystems. *Palaeogeography, Palaeoclimatology, Palaeoecology,* **106**, 91–112.

SPACKMAN, W., DOLSEN, C. P. & RIEGEL, W. 1966. Phytogenie organic sedimentary environments in the Everglades–mangrove complex. Pt 1. Evidence of a transgressing sea and its effects on environments of the Shark River area of southwestern Florida. *Palaeontographica B,* **117**, 135–152.

STAUB, J. R. & ESTERLE, J. S. 1992. Evidence for a tidally influenced Upper Carboniferous ombrogenous mine system: upper bench, Beckly Bed (Westphelian A), Southern West Virginia. *Journal of Sedimentary Petrology,* **62**, 411–428.

STEVENSON, J. J. 1911. The formation of coal beds. An historical summary of opinion from 1700 to the present time. *Proceedings of the Americal Plulosophical Society,* **50**, 1–116.

STOPES, M. C. 1919. On the four visible ingredients in banded bituminous coal: studies in the composition of coal. *Proceedings of the Royal Society of London,* ser. B, **90**, 470–487.

STUTZER, O. (ed.) 1929. Fusit. Vorkommen, Entstehung und Pratische Bedeutung der Faserkohle (fossil Holzkohle). *Schriften aus dem Gebiet der Brennstoff-Geologie,* **2**, 1–139.

STYAN, W. B. & BUSTIN, R. M. 1983. Petrology of some Fraser River delta peat deposits: coal maceral and microlithotype precursors in temperate–climate peats. *International Journal of Coal Geology,* **2**, 321–370.

SUPARDI, SUBEKTY, A. D. & NEUZIL, S. G. 1993. General geology and peat resources of the Siak Kanan and Bengkalis Island peat deposits, Sumatra, Indonesia. *Geological Society of America Special Paper,* **286**, 45–61.

SWANSON, F. J. 1981. Fire and geomorphological processes. *In*: MOONEY, H. A. *et al.* (eds) *Fire Regimes and Ecosystem Properties.* USDA Forest Service General Technical Report W0–26, 401–420.

TEICHMÜLLER, M. 1989. The genesis of coal from the view point of coal petrology. *International Journal of Coal Geology,* **12**, 1–87.

WILLARD, D. A., DIMICHELE, W. A., EGGERT, D. L., HOWER, J. C., REXROAD, C. B. & SCOTT, A. C. 1995. Palaeoecology of the Springfield Coal Member (Desmoinesian, Illinois Basin) near the Leslie Cemetry Paleochannel, southwestern Indiana. *International Journal of Coal Geology,* **27**, 59–98.

WINSTON, R. B. 1986. Characteristic features and compaction of plant tissues traced from permineralized peat to coal in Pennsylvanian coals (Desmoinsian) from the Illinois Basin. *International Journal of Coal Geology,* **6**, 21–41.

—— 1993. Reassessment of the evidence for primary fusinite and degrade fusinite. *Organic Geochemistry,* **20**, 209–221.

The Carboniferous evolution of Nova Scotia

J. H. CALDER

*Nova Scotia Department of Natural Resources, PO Box 698, Halifax, Nova Scotia,
Canada B3J 2T9*

Abstract: Nova Scotia during the Carboniferous lay at the heart of palaeoequatorial Euramerica in a broadly intermontane palaeoequatorial setting, the Maritimes–West-European province; to the west rose the orographic barrier imposed by the Appalachian Mountains, and to the south and east the Mauritanide–Hercynide belt. The geological affinity of Nova Scotia to Europe, reflected in elements of the Carboniferous flora and fauna, was mirrored in the evolution of geological thought even before the epochal visits of Sir Charles Lyell. The Maritimes Basin of eastern Canada, born of the Acadian–Caledonian orogeny that witnessed the suture of Iapetus in the Devonian, and shaped thereafter by the inexorable closing of Gondwana and Laurasia, comprises a near complete stratal sequence as great as 12 km thick which spans the Middle Devonian to the Lower Permian. Across the southern Maritimes Basin, in northern Nova Scotia, deep depocentres developed en echelon adjacent to a transform platelet boundary between terranes of Avalon and Gondwanan affinity. The subsequent history of the basins can be summarized as distension and rifting attended by bimodal volcanism waning through the Dinantian, with marked transpression in the Namurian and subsequent persistence of transcurrent movement linking Variscan deformation with Mauritainide–Appalachian convergence and Alleghenian thrusting. This Mid-Carboniferous event is pivotal in the Carboniferous evolution of Nova Scotia. Rapid subsidence adjacent to transcurrent faults in the early Westphalian was succeeded by thermal sag in the later Westphalian and ultimately by basin inversion and unroofing after the early Permian as equatorial Pangaea finally assembled and subsequently rifted again in the Triassic.

The component Carboniferous basins have provided Nova Scotia with its most important source of mineral and energy resources for three centuries. Their combined basin-fill sequence preserves an exceptional record of the Carboniferous terrestrial ecosystems of palaeoequatorial Euramerica, interrupted only in the mid–late Viséan by the widespread marine deposits of the hypersaline Windsor gulf; their fossil record is here compiled for the first time. Stratal cycles in the marine Windsor, schizohaline Mabou and coastal plain to piedmont coal measures 'cyclothems' record Nova Scotia's palaeogeographic evolution and progressively waning marine influence. The semiarid palaeoclimate of the late Dinantian grew abruptly more seasonally humid after the Namurian and gradually recurred by the Lower Permian, mimicking a general Euramerican trend. Generally more continental and seasonal conditions prevailed than in contemporary basins to the west of the Appalachians and, until the mid-Westphalian, to the east in Europe. Palaeogeographic, paleoflow and faunal trends point to the existence of a Mid-Euramerican Sea between the Maritimes and Europe which persisted through the Carboniferous. The faunal record suggests that cryptic expressions of its most landward transgressions can be recognized within the predominantly continental strata of Nova Scotia.

I never travelled in any country where my scientific pursuits seemed to be better understood, or were more zealously forwarded, than in Nova Scotia... (Lyell, 1845, pp. 229–230)

The geological evolution of Nova Scotia during the Carboniferous and the evolution of geological thought about the Carboniferous strata in Nova Scotia both record a strong affinity to western Europe. During the nineteenth century, the splendid coastal exposures of Carboniferous strata (Fig. 1) were proving grounds for the geological principles and philosophy of Sir Charles Lyell, especially as they pertained to the Carboniferous Period. The geological affinity of Nova Scotia to Europe, recorded in elements of the Carboniferous flora and fauna, reflects its palaeogeographic position during the Carboniferous at the heart of palaeoequatorial Euramerica, in proximity to western Europe. If not a Euramerican Rosetta Stone, Nova Scotia and the Maritimes certainly from the keystone in the bridge to understanding the Carboniferous evolution of North America and Europe (Lyell 1843a, 1845; Dawson 1888; Belt 1968a, 1969; Carroll *et al.* 1972; Bless *et al.* 1987; Allen & Dineley 1988; Leeder 1988a; McKerrow 1988; Calder & Gibling 1994, among many others).

In this paper, the Carboniferous evolution of

CALDER, J. H. 1998. The Carboniferous evolution of Nova Scotia. *In*: BLUNDELL, D. J. & SCOTT, A. C. (eds) *Lyell: the Past is the Key to the Present.* Geological Society, London, Special Publications, **143**, 261–302.

261

Fig. 1. The Carboniferous section at Joggins, of which Lyell, in a discussion of coal measures in his The Student's Elements of Geology (1871), wrote, 'But the finest example in the world of a natural exposure in a continuous section ten miles long, occurs in the sea-cliffs bordering a branch of the Bay of Fundy in Nova Scotia.' From a nineteenth-century wood block engraving in Dawson's Acadian Geology.

Nova Scotia is considered by linking its tectonic history, basin fill sequence, fossil record and palaeoclimate with those of Carboniferous Euramerica to the west in North America and to the east in Europe. The aim of the paper is not only to review, but also to relate the records of disparate geological disciplines and to consider, from the author's perspective, implications that emerge for long established views of the Carboniferous history of Nova Scotia. A detailed treatment of all stratigraphic units of the Maritimes Basin in Nova Scotia is neither intended nor possible in a paper of this length. The references in this paper, as well as the overview of Gibling (1995, in van de Poll *et al.* 1995) and the *Lexicon* of Williams *et al.* (1985) will provide the reader with access to further details of the Carboniferous stratigraphy and geology of Nova Scotia.

Geological setting: the Maritimes Basin in Euramerica

The Carboniferous strata of Nova Scotia record most of the history of the larger, Late Palaeozoic Maritimes Basin (Williams 1974) of New Brunswick, Nova Scotia, Prince Edward Island and Newfoundland (Fig. 2). The Late Palaeozoic strata of the Maritimes Basin span the Middle Devonian (Dawson 1862; McGregor 1977; Forbes *et al.* 1979) through early Permian (Dawson 1845, 1891; Barss *et al.* 1963) with remarkably few gaps. The Maritimes Basin is a complex of predominantly

northeasterly trending intermontane basins, once variously interconnected and now, as then, defined by intervening massifs of the Avalon, Grenville and Meguma terranes. The basin was born of the Devonian (Emsian) Acadian orogeny (Poole 1967), contemporary of the latest stage of the Caledonian orogeny, both of which record final closure of the Iapetus Ocean (McKerrow 1988). The Carboniferous evolution of the Maritimes Basin bears witness to the nativity of Pangaea as Gondwana and numerous platelets of suspect terrane collided with Laurasia and the Old Red Continent, manifested in the Hercynian and Alleghenian orogenies (Schenk 1981; Rast 1988). The evolution of the Maritimes Basin during the late Devonian and Dinantian records extension (McCutcheon & Robinson 1987; Bradley 1982; Hamblin & Rust 1989) most pronounced between the Lubec-Bellisle, Cobequid and Hollow faults, an area that has been termed the Maritimes Rift (Belt 1969; van de Poll *et al.* 1995). This was suceeded in the Silesian by transpression and transtension in a renewed orogenic phase (Plint & van de Poll 1984; Nance 1987; Waldron *et al.* 1989; Yeo & Ruixiang 1987) and broadly across the basin by thermal sag (Bradley 1982) and ultimately in the Permo-Triassic, by inversion (Ryan & Zentilli 1993).

Nova Scotia and the Maritimes Basin in the Carboniferous lay within palaeoequatorial Euramerica, drifting northwards from a palaeo-latitude of 12 degrees south to cross the equator by the beginning of the Permian (Scotese & McKerrow 1990). Generally considered a northern

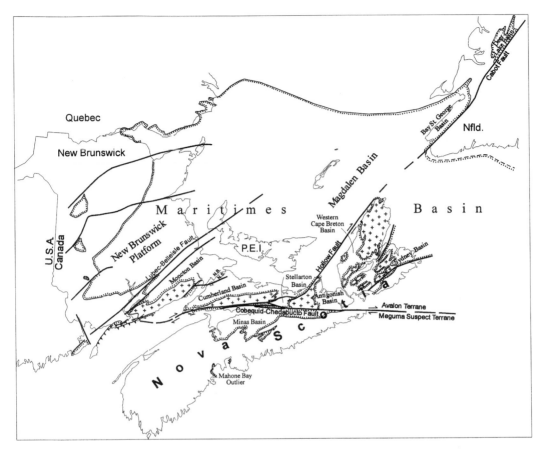

Fig. 2. Areas underlain by Middle Devonian - Permo-Carboniferous strata of the southern Maritimes Basin in Nova Scotia, and neighbouring provinces, with component depositional basins and intrabasinal massifs; modified after Gibling *et al.* (1992) and Gibling (1995).

part of the Appalachian orogenic belt, the Maritimes Basin lay situated at the palaeosoutheastern margin of the Appalachians in a palaeogeographic region distinct from the Appalachian Basin to the west. The mountain range posed an orographic climate barrier, drainage divide and phytogeographic barrier to biotic exchange between these two areas. No such *land* barrier existed to the east, however, and Nova Scotia can be included with Britain and western Europe in a broadly intermontane palaeogeographic region of tropical palaeolatitude lying to the east of the Appalachians, north of the Mauritanides and west of the Urals, and traversed by the Acadian–Hercynide upland belt, here called the Maritimes–West-European Province, modified after Leeder (1987), analogous to the Equatorial-Low-Latitude–Acadia phytogeographic unit of Rowley *et al.* (1985) and long ago known to Dawson (1888).

The evolution of geological thought on the Carboniferous of Nova Scotia

The first comprehensive account of the Carboniferous geology and mineral resources of Nova Scotia is that of Richard Brown (Fig. 3) in 1829, although this contribution in the past, if not overlooked altogether, has been attributed to Thomas Chandler Haliburton, in whose book it appears. Brown was employed by the London-based General Mining Association as manager of coal-mining operations in the Sydney coalfield, Cape Breton, from 1827 to 1864. 'This experienced observer', as he was respectfully described by Lyell (1845, p. 206), described most eloquently that *Stigmariae* are in fact the rootstock of lepidodendrid trees (Brown 1846; 1848), which are so splendidly exposed as fossil forests in the coastal sections of Nova Scotia (Lyell 1843b; Brown 1846, Dawson 1855, Calder *et al.* 1996; Scott & Calder

(a)   (b)   (c)   (d)

Fig. 3. Nineteenth century contemporaries of Lyell who contributed to the early understanding of the Carboniferous of Nova Scotia. (a) Richard Brown (1805–1882); (b) Abraham Gesner (1797–1864); (c) Sir William E. Logan (1798–1875); (d) Sir J. William Dawson (1820–1899).

1994; Scott this volume). Seemingly inocuous, this discovery had implications for the the origin of coal, the recognition of cyclic sedimentation and for reconsideration of the Deluge.

The stratigraphic nomenclature employed by Brown and his contemporaries of the nineteenth century immediately discloses two important facts: first, that they were both inclined and able to interpret the Carboniferous strata of Nova Scotia in terms of the geology of Europe, and in particular Britain, with which they were familiar; and second, the rock record permitted such a comparison to be drawn. Brown (1829) applied to the Carbon-iferous strata of Nova Scotia the then recently published stratigraphic nomenclature for Britain of Coneybeare and Phillips (1822). In contrast, the early geological account of Jackson and Alger (1829) was largely anecdotal and geographical in nature. The early stratigraphic interpretation of Brown is striking in its similarity to current subdivisions (Table 1). Current stratigraphic nom-enclature has been adopted largely from that employed by Dawson (1878), who drew upon the work in the coalfields of his contemporaries McOuat (1874) and Robb (1874) at the Geological Survey of Canada, and from the subsequently evolved nomenclature of Walter A. Bell (1929, 1944) and Edward Belt (1964).

The words of Sir William Dawson in his opus *Acadian Geology*, which he dedicated to Lyell, are testimony to the influence of Lyell and the esteem in which he was held in Nova Scotia, especially by Dawson:

The year 1842 forms an epoch in the history of geology in Nova Scotia. In that year Sir Charles Lyell visited the province, and carefully examined some of the more difficult features of its geological structure, which had baffled or misled previous inquirers. Sir Charles also performed the valuable service of placing in communication with each other, and with the geologists of Great Britain, the inquirers already at work on the geology of the province, and of stimulating their activity, and directing it into the most profitable channels. (Dawson 1868, p. 8)

In the following year (1843), Sir William Logan undertook the longstanding bed-by-bed description of the Joggins section, the first field project of the newly formed Geological Survey of Canada (Logan 1845).

Lyell gained much in return through the mutual respect between him and his collaborators Brown and Dawson. Many of his observations on the origin of the Joggins section (Fig. 4) clearly build upon those of Brown (1829). The 1845 geology map that accompanied his *Travels in North America* incorporated the stratigraphy of Dawson (1845) with respect to the age of the widespread Permo-Carboniferous red beds of Prince Edward Island and northern Nova Scotia, and is an admir-able precursor to the current map of the Maritimes Basin (Fig. 2). The problematic stratigraphic position of gypsum was resolved by Lyell (1843a, c) through discussions with Brown and Dawson (Lyell 1845, p. 206; Dawson 1847). Abraham Gesner, avid promoter of Nova Scotian geology and resources (Gesner 1836, 1843) and later inventor of a process to distil kerosene from coal and oil shale, strongly disagreed with this interpretation and sought from the Geological Society Lyell's censure (Lyell 1845). In response to Gesner's criticism, Lyell systematically detailed his case for the stratigraphic position of the Dinantian gypsum and limestone (1845, pp. 208–218), which has been accepted since that time.

Table 1. *The evolution of stratigraphic nomenclature for the Carboniferous basin-fill of Nova Scotia, with special reference to the nineteenth century*

Brown (1829)	Gesner (1836)	Gesner (1843)	Lyell (1843)	Dawson 1855	Dawson 1878++	Bell (1944)	Belt, 1964; Ryan et al. (1991)		
New Red Sandstone (& Gypsum)	New Red or Saliferous Sandstone	New Red Sandstone (& Gypsum)	Upper Carboniferous Division+	Upper Coal Formation	Permo-Carboniferous Series		Pictou Group	Lower Permian	
	Calcareous or Marine Deposit				Upper or Newer Coal Formation			S	Silesian
Coal Measures	———	Coal Measures	'Coal Measures'	Middle Coal Formation ('Coal Measures Proper')	Middle Coal Formation	Pictou, Morien, Stellarton, Cumberland	Cumberland Group	W	
'Millstone Grit'	Coal Measures			Millstone Grit Series (Formation)	Millstone Grit Series	Riversdale	? hiatus	N	Carboniferous
						? hiatus	Mabou Group		
						Canso			
Carboniferous Limestone	Old Mountain or Carboniferous Limestone	Old Red Sandstone or Devonian	Lower Carboniferous or Gypsiferous Series	Carboniferous Limestone (Lr. Carb. Limestone & Marine Fm.)	Windsor Series or Lr. Carb. Limestone & Gypsum Beds	Windsor	Windsor	V	Dinantion
						?	?		
Old Red Sandstone	Old Red Sandstone			Lower Coal Measures	Horton Series or Lower Carboniferous Coal Measures	Horton	Horton Group	T	
						?	?		
'Greywacke'	'Slate'						D / C Fountain Lake Group & Equivalents	Upper Devonian	

+ after Dawson
++ also McOuat (1874) and Robb (1874)

Fig. 18.

Section of the cliffs of the South Joggins, near Minudie, Nova Scotia.

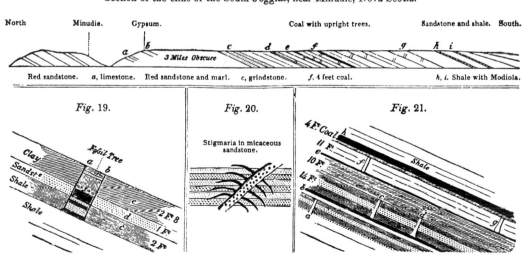

Fig. 4. Lyell's illlustrations of the Joggins section, from his *Travels in North America* (1845), reprinted in part in *The Student's Elements of Geology* (1871). Illustrated are the stratigraphic relationship of gypsum and coal measures (1871, fig. 18) and the nature and occurrence of fossil lepidodendrid trees, including: the nature of their casting (1871, fig. 19); *Stigmaria*, yet to be shown conclusively to be a lepidodendrid rootstock (1871, fig. 20); and the disposition of erect trees within the inclined strata, which was taken as evidence that the strata had been tilted subsequent to their deposition.

In 1852 Dawson and Lyell discovered, either through serendipity (Dawson 1868) or a careful search strategy (Lyell & Dawson 1853), the remarkable occurrence of tetrapods and land snails within the casts of erect lepidodendrid trees at Joggins. Dawson continued the search alone thereafter; in all, over 100 specimens comprising at least 11 tetrapod and five terrestrial invertebrate taxa (Appendix A) have been discovered in the trees (Carroll *et al.* 1972), the great majority by Dawson (1878, 1894). Perhaps the most famous of these, *Hylonomus lyelli* (Dawson 1860), for well over a century was the earliest known reptile. The strange circumstance of the tree stump fauna long has been favoured to have come about as the result of pitfall (Dawson 1878; Carroll *et al.* 1972); however, the weight of evidence including current research by the author and A. C. Scott (see Scott this volume) suggests otherwise; denning is seen by the author as a more probable scenario.

In the late nineteenth and early twentieth centuries, field relationships in the Carboniferous basins were mapped in precise detail across the Province of Nova Scotia by the Geological Survey of Canada. The resulting maps, at the scale of one inch to the mile, and accompanying reports are the valued legacy of Hugh Fletcher (1875–1909).

Bell (1929, 1940, 1944) erected series for the Carboniferous strata of Nova Scotia largely on the basis of the macroflora (Appendix B) and bivalve fauna (see Appendix A), which he correlated with equivalent European stages. The series of Bell, which established the age relationship of coal-bearing strata within the disjunct Carboniferous basins, subsequently were adopted as lithostratigraphic groups (Table 1). The diachronous nature of the lithostratigraphic units of the Maritimes Basin was illustrated by the application of miospore biostratigraphy (Belt 1964; Hacquebard *et al.* 1960; Barss *et al.* 1963). The problems inherent in the adoption of a lithostratigraphy born of biostratigraphy have been acknowledged by virtually all subsequent stratigraphers. This inherent stratigraphic problem has been accommodated in part by the growing practice of assigning diachronous coal measures within the disjunct coal basins to the Cumberland Group and succeeding red beds to the Pictou Group, as proposed by Ryan and colleagues (1991).

Because the biostratigraphy of the Carboniferous of the Maritimes Basin is rooted in the terrestrial fossil record except during the mid to late Viséan (Fig. 5), the effects of provincialism and palaeogeography can be particularly problematic in achieving precise correlations with stage boundaries based on marine fauna elsewhere in Euramerica. Depauperate miospore floras attend the red beds of Nova Scotia; in the lower Mabou

Group, for example, the impoverished flora has been ascribed to playa conditions (Neves & Belt 1970). The exceptional thickness of the component basin-fills diffuses the record of first and last appearances of miospores and the introduction of hinterland floras is problematic (Dolby, pers. comm. 1997). The scarcity of recognized tonsteins (Lyons *et al.* 1994) further stymies the use of absolute radiometric dates which otherwise could be employed to assist in the resolution of chronostratigraphy.

Evolution of the Carboniferous basin-fill sequence in Nova Scotia

The major basinal depocentres of the southern Maritimes Basin in Nova Scotia (Fig. 2), modified from Gibling (1995), from west to east, are: (1) Cumberland Basin; (2) Minas Basin (including Windsor–Shubenacadie, Musquodoboit–Mahone-Bay and depocentres along the southern Cobequids); (3) Stellarton Gap and Basin; (4) Antigonish Basin; (5) Western Cape Breton, at the margin of the submarine Gulf of St. Lawrence; (6) Central Cape Breton, including Glengarry, Loch Lomond and the Salmon River; and (7) Sydney Basin. Each of these component basins in turn comprises smaller depocentres, in part reflecting the anastomosing fault configuration generated by the transcurrent fault systems of the Maritimes Basin. The accrued Carboniferous fill of these component basins may reach 12 km in thickness (Belt 1968b). The reader is referred to Bell (1929, 1940, 1944, 1960), Belt (1965), Ryan *et al.* (1991), Williams *et al.* (1985) and Gibling (1995) for comprehensive details of their stratigraphy. The component formations of the six main lithostratigraphic groups are given in Table 2.

Beginnings: Middle – Upper Devonian

The formation of depocentres within the Maritimes Basin (Williams 1974) during the Late Palaeozoic was initiated at the close of the Acadian orogeny (Poole 1967), following the Caledonian orogeny in western Europe, which together record the final closure of the Iapetus Ocean (McKerrow 1988). The earliest basin-fill, recorded in the McAdam Lake Formation on Cape Breton Island, has been assigned a Lower or Middle Devonian age on the basis of its *Arthrostigma–Psilophyton* macroflora (Bell & Goranson 1938) but is better constrained to the latest Emsian to early Eifelian age on the strength of palynomorphs (McGregor 1977). The strata of the McAdam Lake Formation provide the earliest record of coal formation and sapropelic lacustrine deposits ('oil shales' according to Gilpin

1899, but unreplicated by Smith & Naylor 1990) in the evolution of the Maritimes Basin in Nova Scotia. The lacustrine deposits and basin margin conglomerates co-occur with felsic volcanics (Bell & Goranson 1938). Within the Murphy Brook Formation in the Cobequid Highlands, a flora has been reported and provisionally identified as the primitive tracheophytes *Taeniocrada* and *Drepanophycus* co-occuring with axes and sporangia 'probably referable to the genus *Psilophyton*' (Forbes *et al.* 1979), which has been taken to represent a similar age (Donohoe & Wallace 1982).

The subsequent Devonian basin-fill, which is assigned to the Fountain Lake Group and its equivalent, reaches nearly 3 km in thickness (Williams *et al.* 1985) and is consistent with continental rift facies: bimodal basalt–rhyolite volcanic suites, extensive basin margin conglomerates and basinal lacustrine deposits. Volcanics of this loosely defined group occur adjacent to major transform faults: the Cobequid and Hollow Faults of northern Nova Scotia and correlative faults in southwest Newfoundland, and along the Lubec–Belleisle Fault in southern New Brunswick (Blanchard *et al.* 1984) (Fig. 2). This facies assemblage historically has been interpreted as Devono-Carboniferous on the basis of rather limited and equivocal biostratigraphic data. The lacustrine strata have yielded late Devonian to early Carboniferous (Donohoe & Wallace 1982) and Tournaisian miospores (Blanchard *et al.* 1984) and, on Cape Breton Island, compressions of the Devonian progymnosperm *Archaeopteris* (Kasper *et al.* 1988). Recent U/Pb and biostratigraphic dating increasingly constrain these rocks to the pre-Carboniferous, from Middle to Late Devonian age (Barr *et al.* 1995; Martel *et al.* 1993; Dunning *et al.* 1997). The suggestion of Martel and colleagues (1993) to assign this early basin-fill to an expanded Horton Group (see below) as yet has not been widely adopted.

Volcanics of the Fountain Lake Group are inferred to be extrusive equivalents of high-level plutons emplaced along the Cobequid Fault (Pe-Piper *et al.* 1989, 1991, 1996). They co-occur with dyke swarms and are consistent with reflect within-plate crustal extension (Pe-Piper *et al.* 1989). Uranium/lead dating of rhyolites best constrains Devono-Carboniferous volcanism in Nova Scotia to the Middle Devonian to late Tournaisian / early Viséan, wherein four main episodes are recognized (Dunning *et al.* 1997). Along the trend of the Cobequid Fault system, volcanics were extruded during the Middle Devonian (385–389 Ma). In western Cape Breton, bimodal basalt–rhyolite suites of the Fisset Brook Formation in Cape Breton terminated in the Late Devonian (373±4

Ma; Barr *et al.* 1995). Extensive volcanism and related plutonism at the Devono-Carboniferous boundary (362–356 Ma) is recorded along the Kirkhill–Rockland-Brook Fault system of the Cobequid Highlands and associated faults in western Cape Breton, in association with transpression (Dunning *et al.* 1997). A subsequent phase of Mid-Dinantian volcanism is discussed below.

Dinantian

Latest Devonian to Tournaisian

In the latest Devonian and throughout the Tournaisian (Martel *et al.* 1993), predominantly continental alluvium was deposited across all tectono-stratigraphic terranes of both Avalonian and Meguma affinity, confirming the earlier assembly of the Northern Appalachian orogen (Hamblin & Rust 1989). Half-grabens developed adjacent to the Cobequid Fault system in northern Nova Scotia (Martel & Gibling 1991, 1996), the Hollow Fault in western Cape Breton (Hamblin & Rust 1989) and its extension in southwest Newfoundland (Miller *et al.* 1990) and probably within the Moncton Basin of southern New Brunswick as well. These half-graben segments record an early distensional phase of Maritimes Basin development consistent with the rifting proposed by Belt (1968a). Similar half-graben basins developed contemporaneously in the Dinantian of Britain (Leeder 1987, 1988a). This suggests a prevalence of regional distension across central and eastern Euramerica as plates reorganized between the Caledonian–Acadian and Alleghenian–Hercynian orogenies (Hamblin & Rust 1989); in Nova Scotia, basins adjacent to the Cobequid Fault system may have developed as collapse structures in response to thrusting that attended forceful intrusion of Devono-Carboniferous granites (Piper 1994). Mafic intrusions and basaltic flows of late Tournaisian to early Viséan age are scattered widely along the Cobequid Fault zone and from southern New Brunswick through Prince Edward Island to western Cape Breton (Dunning *et al.* 1997) (Fig. 2).

The basin-fill of the half grabens, assigned to the Horton Group (Dawson 1873, Bell 1929) of latest Devonian (Famennian) to latest Tournaisian (T_3) age (Utting *et al.* 1989; Martel *et al.* 1993), ranges from 600 m (Martel & Gibling 1996) or more (1100–1500 m; Bell 1960) in the type area of the Minas Basin to as much as 3000 m in western Cape Breton (Hamblin & Rust 1989). Characteristically, it comprises marginal thick extrabasinal conglomerates (Murphy *et al.* 1994) and a tripartite basinal stratigraphy of alluvial strata above and below intervening lacustrine beds (Hamblin & Rust 1989; Martel & Gibling 1996) (Fig. 5). The lacustrine component has been inferred by these authors to represent a period of accelerated subsidence during which the basins were underfilled. Sediment-starved organic-rich lakes accumulated oil shales (Smith & Naylor 1990). Coarsening upward sedimentary cycles have been ascribed to tectonism (Martel & Gibling 1991), but the lacustrine rocks, which record the effects of storm conditions (Martel & Gibling 1991) doubtless bear witness to climatic cyclicity, yet to be described.

The *Lepidodendropsis corrugatum* compression flora of the Horton Group (Bell 1960) was a cosmopolitan flora of the Tournaisian Old Red Continent (Jongmans 1952; Chaloner & Lacey 1973). The Tournaisian fauna of the Horton Group (see Appendix A) are represented in the aquatic realm by a restricted invertebrate record of ostracods, conchostracans and xiphosurans, and by a more diverse vertebrate fauna of palaeoniscid fish, crossopterygians, dipnoi, gyracanthids and symmoriid and probable other sharks. Recent re-evaluation of microfauna of the Horton Bluff Formation, long held to be solely lacustrine, has revealed the presence of a marginal marine ostracod fauna within profundal to lagoonal beds of the basal Blue Beach Member, including species of the western European genera *Copelandella* and *Carbonita* and the more cosmopolitan *Shemonaella* (Tibert 1996). The elasmobranch shark *Stethacanthus* (Bell 1929, p. 35) is known from the famous Carboniferous marine localities of Bear Gulch, Montana and Bearsden, Scotland.

Table 2. *Formations of the Carboniferous basin-fill of Nova Scotia[a]*

[a] Modified after Gibling (1995). Absolute age dates after Cowie & Bassett (1989).

The Tournaisian strata of Nova Scotia nonetheless are prevailingly continental (Belt 1968b) in contrast to the widespread marine beds of western Europe (Leeder 1988a) and afford an important locality for establishing the early evolution of terrestrial vertebrates, recorded in both trackways and osteological remains. Skeletal remains from Horton Bluffs include an anthracosauran reptiliomorph, comprising the earliest tetrapod record in both western Euramerica and the western hemisphere exclusive of Greenland (Carroll *et al.* 1972) and the only known Tournaisian tetrapod locality (Milner *et al.* 1986; Carroll 1992; Ahlberg & Milner 1994).

Viséan

Whereas the uppermost Horton strata dated are latest Tournaisian and the oldest dated succeeding strata of the Windsor Group (Lyell 1843a, c; Bell 1929) are of mid-Viséan (V_2 to early V_3) age, the possibility exists of a hiatus during part of the early (V_1) to middle (V_2) Viséan (Utting *et al.* 1989) (Fig. 5). Although the gap could be apparent given the paucity of age data from red beds of this interval (Utting *et al.* 1989), the regional stratigraphic relationship of the two groups suggests otherwise: the isochronous base of the Windsor overlies Horton rocks of variable Tournaisian age, and nowhere has an unequivocally

continuous record been documented between the two groups (P. S. Giles, pers. comm. 1997).

During the mid-Viséan, marine waters breached the Maritimes rift valley from the east, rapidly transgressing as far as the New Brunswick Platform (Fig. 2) (Bell 1929; Mamet 1970; Howie 1984). The body of water known as the Windsor Sea was a restricted, hypersaline tropical gulf near the scale of the Caspian Sea (Schenk *et al.* 1994). The sea opened to the northeast (Schenk 1969; pers. comm. 1997) of present-day Nova Scotia and east to southeast across the Meguma terrane (Mamet 1970; Giles 1981a; Howie 1984). There, in the east, an equatorial Phoibic seaway (McKerrow & Ziegler 1972), inferred to have been westward circulating (Allen & Dinely 1988), existed between the continental margins of the Old Red Continent (Laurussia) and Gondwana. At this time, marine conditions prevailed across much of tropical Euramerica (Bell 1929; Bless *et al.* 1987; Rast 1988), and the rapid incursion of the sea into the Maritimes rift valley may have occurred as the seaway became compressed by the convergence of Gondwana and the Old Red Continent; alternatively, regional distension across the Maritimes–Western-European province may have ushered in the sea. The marine carbonate–sulphate strata intertongue towards basin margins locally with red conglomerates of the Grantmire Formation (Schenk 1969) and in New Brunswick with conglomerates

Table 2. *Key.*

FORMATIONS

Pictou Group:

B – Balfron
BC – Broad Cove
CJ – Cape John
T – Tatamagouche

Cumberland Group:

BB – Big Barren
BPt – Boss Point
D – Delaney
EBk – Emery Brook
GV – Glengarry Valley
HYls – Henry Island
I – Inverness
J – Joggins
Mg – Malagash
MM – Mabou Mines
NGC – New Glasgow
    Conglomerate
P – Parrsboro*
PH – Port Hood
RR – Ragged Reef
SHM – Springhill Mines
St – Stellarton
SV – Scotch Village
SVM – Silver Mine

Mabou Group:

CD – Cape Dauphin
Cl – Claremont*
H – Hastings
L – Londonderry
M – Middleborough
McKL – McKeigan Lake
PQ – Pomquet
PtE – Point Edward
Sh – Shepody
WB – West Bay

Windsor Group:

AD – Addington
BV – Bridgeville
CC – Carroll's Corner
CV – Churchville
E – Enon
GO – Green Oaks
GR – Gays River
HBk – Holmes Brook
HH – Hartshorn
HIs – Hood Island
KH – Kempt Head
LKBk – Lime Kiln Brook
LL – Loch Lomond
Mc – Macumber
MCk – Miller Creek
McD – MacDonald Road
MUR – Murphy Road

MRd – Meadows Road
PM – Pugwash Mine
Sw – Stewiacke
SR – Sydney River
U – Uist
WQ – White Quarry
WRd – Woodbine Road
WW – Wentworth

Horton Group:

A – Ainslie
C – Creignish
Ch – Cheverie
CS – Coldstream
DBk – Diamond Brook
F – Falls
C – Grantmire
HB – Horton Bluff
S – Strathlorne
WBk – Wilkie Brook

Fountain Lake et al.

BBk – Byers Brook
FBk – Fisset Brook
FL – Fountain Lake
MBk – Murphy Brook
McABk – McArras Brook
McAL – McAdam Lake

of the Hopewell Cape Formation during the Mid Viséan (McCutcheon 1981). In southwestern Cape Breton, small gabbroic plutons and possible sills are coeval with the Windsor Group (U/Pb: 339 ± 2 Ma) and are inferred to reflect continued extension (Barr et al. 1994).

The Windsor Sea in its early history subsequent to its breach of the Maritimes Rift has been called Loch Macumber (Schenk et al. 1994); the widespread, basal Macumber Formation records relatively deep water with profundal laminated carbonate and mudrock associated with turbidite and debris flow deposits. The deeper basinal waters shoaled laterally, where formed lime mud reefs or hydrothermal tufa mounds of the Gays River Formation. Subsequent to the accumulation of evaporite deposits hundreds of metres in thickness, the Windsor seafloor became exposed subaerially (Schenk et al. 1994) with playa flats and subsequent karstification (Boehner 1986). This transgressive–regressive cycle (Subzone B of Bell 1929; Cycle 1 of Giles 1981a) defines a stratigraphic sequence (Schenk et al. 1994). Four subsequent sequences in the upper Windsor Group record repeated rapid transgression and protracted withdrawal from the Maritimes rift valley during the mid and late Viséan (Bell 1929; Schenk 1969) (Fig. 5). They comprise cyclic stratal successions of carbonate, thin red claystone, sulphate and thick red claystone that have been interpreted as algal carbonate strandlines shoaling to supratidal, dolomitizing salt flats with saline lakes and lagoons (Schenk 1969). Upon these sequences are superimposed numerous parasequence cycles recording lesser regressions (Giles 1981a; Schenk et al. 1994). The total thickness of Windsor strata is difficult to determine in areas of diapirism, but in the Minas Basin area is in the order of 800–900 m (Boehner 1986).

The transgressive–regressive sequences of the Windsor correlate approximately with the mid-Arundian to Brigantian of Britain and Belgium (Giles 1981b), represented by mid-mesothem 3–6 of Ramsbottom (1977). The driving mehanism of these sequences has been inferred to have been glacioeustacy (Giles 1981b), possibly caused by the waxing and waning of the developing Gondwanan ice cap (Crowell 1978; Veevers & Powell 1987; Gonzalez-Bonorino & Eyles 1995). The Windsor transgressive–regressive cycles mimic the strong asymmetry of glacial ice-cap meltout and growth evidenced by the oxygen isotope record from Pleistocene ice (Chappell & Shackleton 1986). The accumulation of thick evaporite deposits must also have contributed to progressive shoaling (Schenk et al. 1994).

The Viséan marine macrofauna record in Nova Scotia (Appendix A) is most prolific in the deeper-water Macumber limestone of the B subzone (Bell 1929, pp. 66–68), which also contains an undescribed fish fauna (R. G. Moore pers. comm. 1997). The subsequent faunal record derives primarily from carbonate units (Moore 1967), where the low diversity of bottom dwellers reflects adverse biotic conditions imposed by the shallow muddy carbonate sea bottom. The resulting meagre crinoid and echinoderm fauna contrasts with that of interior North America (Bell 1929). Suspension feeders such as the articulate brachiopod productids and Composita dominated, which together with pelecypods and gastropods, comprised a mollusc–brachiopod fauna (Bell 1929). The abundant pelecypod fauna is akin to that of northern Britain, where similar substrate conditions to those in Nova Scotia are inferred. The coral fauna, primarily Rugosa and Tabulata, is strikingly less abundant, albeit with notable exceptions such as in the Musquoboboit sub-basin (Boehner 1977), than in coeval sandy shelf areas of southern Britain. Taken together, these faunal elements suggest unstable substrate and salinity and restricted circulation within the Windsor gulf (Bell 1929; McKerrow 1978).

Restricted circulation is recorded as well in the increasing fractation of evaporitic brines from carbonates through calcium sulphate to salts northwestward across the Maritimes Basin (Howie 1984). This fractation suggests a palaeogeographic restriction of the Windsor Sea, either by shoaling or by straits, from deeper waters in western Europe, where carbonates prevail. The existence of straits above a basement low on the Grand Banks east of Newfoundland has been suggested (Geldsetzer 1978). The faunal record has been interpreted to indicate that normal marine conditions were approached only by the upper Viséan (Bell 1929). Corals overlain by thick sulphate and halite deposits in the upper Windsor Group, however, indicate that unstable salinities prevailed throughout the history of Windsor Group regressions (P. S. Giles, pers. comm. 1997).

The mollusc–brachiopod fauna of Nova Scotia show strong affinity with that of northwestern Europe, and therefore were interpreted by Bell (1929) before the theory of continental drift as common residents of a 'proto-North Atlantic', whereas in contrast, the Mississippian marine fauna of interior North America show but scant evidence of 'distant migratory connexions'. Benthonic foraminifers of the upper Viséan of the Maritimes Basin and eastern Newfoundland shelf, however, have affinity to North American fauna rather than the Tethyan fauna of western Europe, which suggests that a deep ocean barrier between Tethys and the Windsor Sea existed at this time (Jansa et al. 1978; Jansa & Mamet 1984). Evidence in support of the persistence of such a Mid-Euramerican Sea in the Silesian is discussed subsequently in this paper.

The ultimate withdrawal of the Windsor Sea from the Maritimes Basin by the late Viséan may have been a consequence of a major phase of Gondwanan glaciation and a consequent fall in global sea level. (Gibling 1995; Veevers & Powell 1987). The withdrawal pre-dates, however, the Namurian maxima inferred for the Gondwanan ice cap (Gonzalez-Bonorino & Eyles 1995) and is inconsistent with documented major global regressions (Veevers & Powell 1987), including the Mississippian–Pennsylvanian boundary event (Saunders & Ramsbottom 1986). At the same time across Euramerica, the converging circumequatorial continents heralded a Mid-Carboniferous episode of thrusting, transpression and inversion, discussed below, which had potential to effect a similar result.

Following withdrawal of the Windsor Sea, a semiarid climate persisted through the late Viséan as lake margins of the Hastings Formation, basal Mabou Group, contracted and desiccated, depositing cyclic 'cementstone' carbonate and red beds (Belt 1968b) in a schizohaline, inland mimic of the marine cycles of the earlier Viséan (Crawford 1995; cf. Leeder 1992). The predominance of grey beds and growth lines of the pelecypod *Carbonicola,* which span up to 10 years (E. S. Belt, pers. comm. 1997) has been cited as evidence in support of the persistence of deeper parts of the Mabou lakes; the ecology of *Carbonicola,* oft stated as 'brackish', is key to this argument.

Halokinesis, with flow into kilometre-scale diapirs and salt anticlines adjacent to basement fault blocks, deformed much of the Windsor Group and locally, the subsequent Permo-Carboniferous basin-fill (Bell 1944, 1958; Howie 1988; Boehner 1992; Brown *et al.* 1996). Extensional detachment faulting of possible basin-wide extent within the Windsor Group (Lynch & Giles 1995) and major salt movements (Bell 1944; Boehner 1992) were initiated early in the subsequent depositional history of the basins, possibly as early as the late Namurian, subsequent to deposition of the Mabou Group (Lynch & Giles 1995) and the Mid Carboniferous event discussed below. The halokinetic history of the Maritimes Basin fill as a whole stands in marked contrast to the Carboniferous of North America and western Europe.

Silesian

Namurian

The Namurian set the stage for the most profound changes in basin evolution during the Carboniferous in Nova Scotia, events that can be linked to the evolution of Euramerica. The Mabou Group (Belt 1964, superseding the Canso Group of Bell 1944, with revisions below) succeeds the Windsor Group in the late Viséan and persists through the Namurian A/B (Neves & Belt 1970). The age of the uppermost Mabou strata has proved to be especially problematic (see below); similar uncertainties in age attend red-bed-dominated strata within the Carboniferous basin-fill (Fig. 6; see also Neves & Belt 1970). The group attains a thickness of at least 3000 m in Nova Scotia (Belt 1965; Lynch & Giles 1995), where typically it is fine grained (Belt 1965; 1968b). Widespread grey beds of the Hastings and equivalent formations first were deposited throughout most basinal depocentres in Nova Scotia, and persisted adjacent to the Cobequid Fault in the Minas Basin, succeeded regionally by red mudrock-dominated strata of the Pomquet Formation (Belt, 1965, fig. 5). The younger grey strata of the Parrsboro Formation are assigned herein to the Cumberland Group as redefined by Ryan *et al.* (1991), as similarly advocated for the Emery Brook Formation (Giles 1995), formerly of the Mabou Group (Belt 1965).

The early to mid-Namurian aquatic fauna is typified by eocarid and conchostracan crustaceans and by sarcopterygian fishes, ctenacanth sharks and acanthodians represented by *Gyracanthus* (Fig. 6a) which has a sparse record elsewhere in North American Euramerica (Baird 1978). The absence of open marine fauna, however, underscores the relatively landward or inland position of the Mabou beds. The eocarid *Pseudotealliocaris* (Fig. 6c) and *Anthracophausia* fauna of the lower Mabou Group (Copeland 1957) have been interpreted elsewhere to be representatives of a nearshore marine community (Schram 1981). Within the lower Mabou occur also the ostracods *Shemonaella (Paraparchites) scotoburdigalensis* and *Beyrichiopsis* sp., which are indicative of a nearshore environment and which are found also in the Dinantian of Scotland (Copeland 1957; Tibert 1996). It is possible, therefore, that maximum transgressions from the retreating Windsor Sea continued for some time to influence the basinal waters of the Maritimes during deposition of the succeeding strata, the Mabou Group. The basal Hastings Formation may be the best candidate for recording a change from marine to lacustrine deposition (Crawford 1995). Continental deposition to which the Mabou Group historically has been ascribed (Belt 1968b), probably predominated, especially in the upper strata (Copeland 1957), but the ecology of the aquatic fauna should be studied further in order to ascertain whether transgressions still were still experienced from the withdrawing sea. The terrestrial vertebrate fauna of the late Viséan–Namurian is best recorded in the exceptional trackway record of the West Bay Formation of the Minas Basin (Carroll *et al.* 1972; Sarjeant & Mossman 1978).

The Mid-Carboniferous break/event in Nova Scotia: Mabou–Cumberland contact

The most profound changes in basin evolution and palaeoclimate during the Carboniferous evolution of Nova Scotia occurred during the mid-Namurian between the deposition of the Mabou Group and the disconformable to unconformable Cumberland

Group (Fig. 5). Namurian (K/Ar: 329 ± 11 Ma) deformation and overthrusting of plutons emplaced adjacent to the terrane-bounding Cobequid Fault in the early Dinantian have been interpreted to be a consequence of transpression or convergence of the Meguma and Avalon terranes along the Cobequid Fault (Waldron *et al.* 1989; Pe-Piper *et al.* 1991). Dextral transpression is inferred to have inverted the Cobequid and Caledonia massifs adjacent to the Cumberland Basin (Fig. 2), and strata of Namurian B age were later cannibalized by early Westphalian deposystems of the basin, as witnessed by reworked spores in the Westphalian strata (Dolby *in*

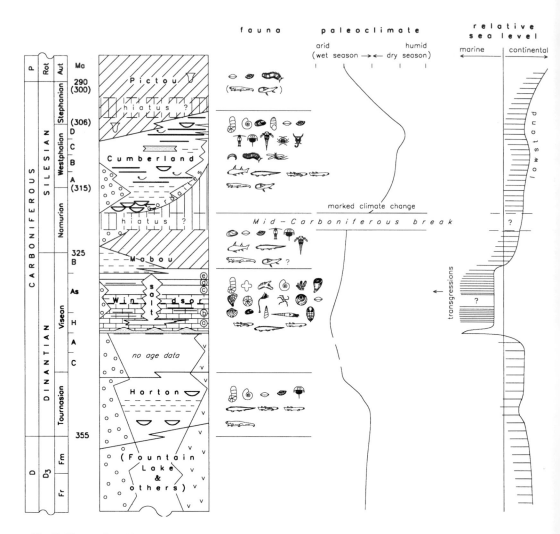

Fig. 5. The stratigraphic column for the Carboniferous basin-fill of the southern Maritimes Basin in Nova Scotia, with representative faunal groups and inferred palaeoclimate and sea-level curves. Sea-level curve for the mid to late Viséan modified after Giles (1981*b*); absolute age dates after Cowie and Bassett (1989), and Hess and Lippolt (1986) (in parentheses).

press). An intra-Namurian disconformity may be recorded in the miospore palynostratigraphy (Utting, pers. comm. 1997), but its placement remains equivocal at present.

This regional Namurian event in Nova Scotia

deformed to varying degree the entire Dinantian basin-fill of the Maritimes (Hamblin & Rust 1989) and defines a change in basin evolution from extensional to transtensional and transpressive (Gibling 1995). Supporting evidence of both the

Fig. 6. Selected fossils from the aquatic realm of the Carboniferous basin-fill of Nova Scotia. (a) *Gyracanthus cf. duplicatus,* NSMNH No. FGM.998.GF.1, Joggins Formation, Cumberland Group (Westphalian A); (b) *Pygocephalus (Anthrapalaemon) dubius,* dorsal view of carapace, Hypotype, GSC No. 12821, Joggins Formation, Cumberland Group (Westphalian A); (c) *Pseudotealliocaris (Tealliocaris) belli,* dorsal view of type specimen, Holotype, GSC No. 10381, West Bay Formation, Mabou Group (late Viséan – Namurian); (d) *Euproöps amiae,* ventral view of abdominal segments, Hypotype, GSC No. 12808a, Sydney Mines Formation, Cumberland Group *sensu latu* (Westphalian D).

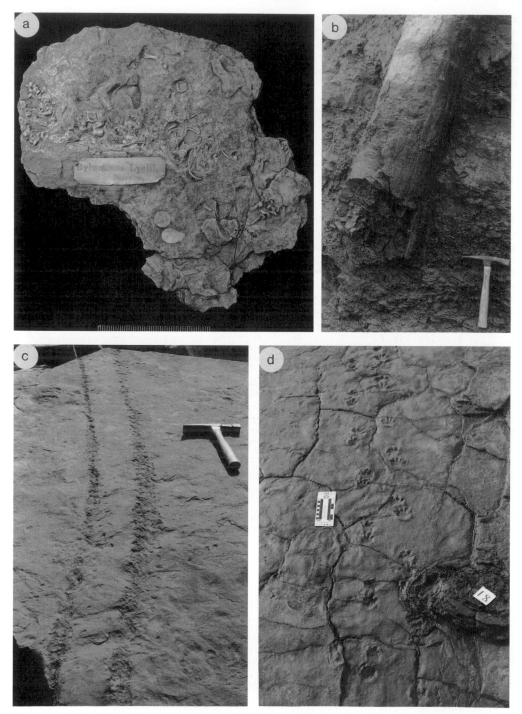

Fig. 7. Selected fossils from the terrestrial realm of the Permo-Carboniferous basin-fill of Nova Scotia. (a) *Hylonomus lyelli,* Holotype, NHM R4168, Joggins Formation (Westphalian A); (b) erect cast of a lycopsid tree, Joggins Formation, Cumberland Group (Westphalian A); (c) *Diplichnites sp.*, ichnogenus ascribed to *Arthropleura*, Joggins Formation, Cumberland Group (Wetphalian A); (d) *Amphisauropus latus* trackway, ascribed to a seymouriamorph cotylosaur, impressed in dessication-cracked redbeds adjacent walchian conifer stump casts, Cape John Formation, Cumberland Group (Carboniferous–Permian boundary beds).

timing and locus of deformation is found in the structural deformation (Hamblin & Rust 1989) and elevated thermal maturity of the Horton (Utting & Hamblin 1991) and Mabou Groups adjacent to the Cobequid Fault zone in the northern Minas Basin and in the Stellarton Gap, relative to Westphalian strata above, which generally are unaffected by the thermal event. A Mid-Carboniferous event, the 'Maritimes Disturbance' (Poole 1967), has long been recognized although poorly constrained and loosely defined. Its timing and significance as described herein are now becoming more fully understood.

This late Namurian event in Nova Scotia links the Variscan and Alleghanian orogenies (Nance 1987; Gibling 1995), with dextral transpression accomodating Alleghanian thrusting in the central and southern Appalachians (Quinlan & Beaumont 1984). The event coincides approximately with the Mississippian–Pennsylvanian boundary of continental North America (see Rehill 1996, fig. 4.3), which is marked by a widespread and pronounced unconformity across the Appalachian Basin (White 1891; Ettensohn & Chesnut 1989), and in shelf areas globally (Saunders & Ramsbottom 1986). This unconformity has been ascribed to a global eustatic event (Donaldson et al. 1985), which resulted in extinctions in the marine realm and marked evolutionary changes (Saunders & Ramsbottom 1986). In the Appalachian Basin, the unconformity coincides with early stages of the Alleghanian orogeny and has been ascribed to continental flexure (peripheral bulge according to Quinlan & Beaumont 1984) caused by continental convergence; foreland areas of greater subsidence cratonward of continental promontories do not record the unconformity (Ettensohn & Chesnut 1989). Evidence for a Mid to Late Namurian hiatus is found as far west as the Porcupine Basin of the western Irish shelf, which represents the westernmost edge of Carboniferous Europe (Tate & Dobson 1989).

The late Namurian and early Westphalian response to the events of the Namurian is recorded in a change in dominant depositional system from lacustrine to fluvial, the 'Coarse Fluvial Facies' of Belt (1964, 1965), now referred to the Cumberland Group (Ryan et al. 1991; redefined after Bell 1944). Across the Maritimes Basin an unparalleled cosmopolitan alluvial facies north of the Cobequid Fault dominantly comprising multistorey sandstones up to 90 m thick, the Boss Point Formation and equivalents, records response to seasonal rainfall and erosion of inverted and uplifted source areas southwest of the Maritimes Basin (Browne & Plint 1994). The distinctive composition of these mature Boss Point sandstones favoured both their use as grindstones and building stones during the nineteenth century and their assignment to the Millstone Grit (Brown 1829).

The establishment of these widespread river systems was preceded by deposition in north-western Nova Scotia and southern New Brunswick of red extrabasinal conglomerate and grit with reduction haloes, assigned to the Claremont and equivalent Enragé formations, which bespeak locally sourced deposition before the palaeoclimate changeover. Although currently included in the Cumberland Group (Ryan et al. 1991), their lithology, fossil record and the possibility of an intervening mid-Namurian hiatus point to their affinity with the Mabou Group. The sandy Boss Point river systems incorporated pedogenic calcrete, inferred to have been eroded from vertisols and aridosols (Browne & Plint 1994), although in some cases these may represent truncated soils (F. W. Chandler, pers. comm. 1997). Great cordaite trees, characteristically found in a permineralized state (*Dadoxylon acadianum*; Dawson 1868), were consumed from the inverted hinterland or from riparian sites and are found from the type section north of Joggins in the Cumberland Basin to the Port Hood section of the Western Cape Breton Basin. A simultaneous, marked introduction of the saccate cordaite pollen *Florinites* accompanies these macroflora (Dolby 1991; in press). A permineralized cordaite flora occurs above the Cumberland–Mabou unconformity in the Sydney Basin in strata as young as the mid-Westphalian South Bar Formation (R. Chisholm, pers. comm. 1997).

Bell (1944, p. 24) concluded that in the continental record of Nova Scotia, 'The top of the Canso [Mabou] group marks the most pronounced palaeontological break in the sequence of Canadian maritime Carboniferous floras.' This break is mirrored in the miospore record (Dolby pers. comm. 1997), but Copeland (1957) observed no such abrupt change in the arthropod record.

Westphalian

A period of rapid basinal subsidence unsurpassed in Euramerica, wherein 1000 m of strata may represent as little as one million years (Calder 1994), was generated in the southern and central Maritimes Basin by transcurrent wrench faulting, from the Late Namurian through Duckmantian/Westphalian B persisting locally through the Bolsovian/Westphalian C to early D. The dextral transpression of the Meguma against the Avalon terrane to the north (Fig. 2) resulted in inversion of the Caledonia massif in New Brunswick and the Cobequid massif north of the Cobequid Fault (Waldron et al. 1989). In southeastern New

Brunswick (Fig. 2), this dextral movement generated west-directed thrusting during the early to mid-Westphalian (Plint & van de Poll 1984), and consequent reworking of Devono-Carboniferous sediment in alluvial fans of the Cumberland Group. Concurrent with both these changes in palaeoclimate and basin tectonics, coal measures of Late Namurian to Cantabrian age were deposited diachronously in virtually all depocentres. The Cumberland Group coal measures range in thickness from less than 2 km in the Sydney Basin (Bell 1938) to 2.7 km in the Stellarton Basin (Hacquebard 1972) and 4 km in the Cumberland Basin, where they are associated with extensive alluvial fan deposits (Calder 1994).

Subsequent to the widespread deposition of the Boss Point and equivalent sandstones, rapidly subsiding depocentres bordered by alluvial fans developed in the latest Namurian and early Westphalian in basins adjacent the Cobequid and Hollow Faults. The early fine-grained fill of these basins, exposed at Joggins in the Cumberland Basin, the Parrsboro shore of the Minas Basin and along the Port Hood section of western Cape Breton Island, shows a remarkably similar lithology and fauna, dominated by mudrock alternating with metre-scale sheet sandsone bodies, with thin humic to sapropelic coals, some of which are overlain by basin-wide persistent black, organic-rich and bivalve-bearing shale or limestone. These fossiliferous beds yield a pelecypod–ostracod bivalve, eocarid pygocephalomorph and syncarid crustacean and varied, disarticulated fish fauna (Appendix A). The Emery Brook Formation bears biostratigraphic and lithostratigraphic affinity to these strata as well, which supports its reassignment from the Mabou to the Cumberland Group (Giles 1995).

During the early Westphalian, regional basin inversion (Waldron et al. 1989) led to a widespread diastem in the Duckmantian/Westphalian B of the Maritimes Basin (Bell 1944), but in the Cumberland Basin transpression along the Cobequid Fault created positive uplift of the neighbouring massifs and transtension of the Athol Syncline, which consequently became the main depocentre of this time in the Maritimes Basin. Before continuing their preferred northeastward course (Gibling et al. 1992), through-flowing rivers were diverted southward into the Cumberland Basin (Browne & Plint 1994; S. J. Davies & M. R. Gibling pers. comm. 1997). Rheotrophic mires developed in distributary settings and in areas of groundwater recharge along the northern Cobequid alluvial fan piedmont, under a humid climate prohibitively seasonal for the widespread development of raised mires (Calder 1994).

At this time, dextral strike-slip of the parallel Hollow and Cobequid Faults in the Stellarton Gap between the Cobequid and Antigonish massifs initiated an equally rapidly subsiding pull-apart basin (Yeo & Ruixiang 1987). Lacustrine sedimention predominated in the pull-apart through the Westphalian C/Bolsovian and early D (Bell 1940; Naylor et al. 1989; Yeo & Ruixiang 1987), with thick, rheotrophic peat accumulation resulting in bituminous coal beds up to 13.4 m thick (Hacquebard & Donaldson 1969; Calder 1979) and organic-rich cannel shales in deeper water during periods of underfilling (Smith & Naylor 1990; Smith et al. 1991). Elsewhere, red beds with calcrete-bearing vertisols (Chandler 1997) developed widely.

By the Westphalian D, diminished subsidence rates attended the thermal sag experienced earlier in the Pennine and neighbouring basins of the British Isles, although sporadic thermal events continued along the Avalon–Meguma terrane boundary at least until the late Westphalian (K/Ar: 303 ± 11 Ma; Waldron et al. 1989). Widespread peat accumulated across the Sydney Basin (Hacquebard & Donaldson 1969; Marchioni et al. 1994) and beneath the Gulf of St Lawrence (Hacquebard 1986; Grant 1994; Rehill 1996) on near coastal plains where the effects of glacioeustacy were felt in the cyclothemic alternation of peat formation and palaeovalley incision of red beds (Gibling & Bird 1994). Fauna of the coal roof strata include the xiphosuran *Euproöps amiae* (Fig. 6d), which is exclusive to the strata of the eastern Sydney Basin (Copeland 1957) and agglutinated foraminifera, which show an easterly gradient from upper to lower estuarine environments (Wightman et al. 1993). The oppportune 'coal window' for peat accumulation (Calder 1994; Calder & Gibling 1994) and succeeding changeover to red beds was broadly diachronous from west to east across the Maritimes Basin in Nova Scotia, with the exception of widespread early fluviolacustrine deposits coeval with the basal Joggins coals. By the Stephanian, the last vestiges of mires and hydromorphic gleysols had all but given way to the deposition of red beds across the Maritimes Basin.

Stephanian to Lower Permian

In the Westphalian D to Lower Permian (Barss & Hacquebard 1967), continental red beds of the Pictou Group (Bell 1944; redefined by Ryan et al. 1991) were deposited widely across the Maritimes Basin during a period of regional thermal sag and growing aridity. The Pictou red beds reach 1650 m in thickness in the Cumberland Basin (Ryan et al. 1991) and 3000 m northward in the Gulf of St Lawrence (van de Poll et al. 1995). By the end of the Cantabrian (Zodrow & Cleal 1985), lepido-dendrid-based wetland ecosystems, last witnessed in Nova Scotia by the fossil forest at Cranberrry Head, Sydney Basin, collapsed and were replaced by a mesic to xeric flora, recorded in the compression flora by taxa such as *Pecopteris arborescens*, *Cordaites* sp. and *Walchia* sp.

The paucity of fossil fauna and flora from the latest Carboniferous basin-fill of the Maritimes can be ascribed to loss of habitat, particularly for wetland and aquatic biota, and also for environments conducive to the preservation of fossils. This has proved troublesome in defining the chronostratigraphy of the Stephanian to Permian red beds; indeed Bell (1944), who defined the age relationships of the Carboniferous of Nova Scotia primarily on the basis of macroflora, considered the basin-fill in Nova Scotia to be no younger than Westphalian D. The miospore record (Barss & Hacquebard 1967) indicates that the youngest red beds of the Maritimes Basin reach into the Permian. Abundant *Vittitina* from the Cape John Formation lend support to an Early Permian age for the uppermost red beds in Nova Scotia (Dolby 1991). The possibility of a mid-Stephanian hiatus within the red beds of Euramerica, although suggested in Nova Scotia (Wagner & Lyons 1997), has yet to be recognized definitively here and will be challenging to ascertain because of the dearth of biostratigraphic data for the red beds of this time interval. Paraconformities between late Silesian formations in the Tatamagouche Syncline of the eastern Cumberland Basin (Ryan *et al.* 1991) warrant closer inspection in search of an intra-Stephanian hiatus.

Eloquent testimony to the age and environment of the uppermost red beds of the Maritimes Basin fill in Nova Scotia has recently been discovered in red beds of the Cape John Formation of the Pictou Group, Cumberland Basin (Calder *et al.* in press). At Brule, on the Northumberland Strait, the only known walchian conifer forest is preserved within thinly bedded, mud-draped silty sandstones with pervasive desiccation cracks, infilling a monsoon-fed dryland river bed. Impressed within the red beds (Fig. 7d) is a prolific record of vertebrate trackways that bear close affinity to the lower Rotliegend ichnofauna of western Europe, including *Amphisauropus latus, A. imminutus, Batrachichnus delicatulus* aff. *B. (Anthichnium) salamandroides, Varanopus microdactylus* and *Dimetropus nicolasi* (Calder *et al.* in press). Included are ichnotaxa that are rare or excluded west of the Appalachian Mountains, suggesting that the Appalachians continued to pose a barrier to terrestrial species exchange between North America and the Maritime–West-European province until the end of the Carboniferous.

The final assembly of Pangaea brought about basin inversion regionally both in the Maritimes Basin (Ryan & Zentilli 1993) and in basins of western Europe (Leeder 1988a). Fission track data suggest unroofing of 1500–4000 m of strata from the Maritimes Basin during the time interval 280–200 Ma (Ryan & Zentilli 1993), and a pronounced unconformity exists between Carboniferous and Triassic strata of the Mesozoic Fundy Rift along the Minas Basin.

The Carboniferous fossil record of Nova Scotia

One of the most notable aspects of the Carboniferous palaeontology of Nova Scotia is its affinity to western Europe (Bell 1929, 1944; Baird 1978; Zodrow & Vasey 1986), at least since the Acadian orogeny and closing of the Iapetus Ocean (Nowlan & Neuman 1991). The faunal record of the Carboniferous basin-fill in Nova Scotia (Appendix A) is compiled herein, in part from unpublished data, for the first time since the pioneering work of Lyell's contemporary of the nineteenth century, Sir William Dawson. The faunal list will provide the reader with details of the taxonomic groups discussed in the paper and hopefully will induce subsequent researchers to make further comparisons with the palaeontological record of Euramerica in North America and Europe. The chronostratigraphy of the Carboniferous of Nova Scotia was developed foremost on the basis of floral biostratigraphy, but with notable exceptions (Zodrow & Cleal 1985; Zodrow & Vasey 1986) has been underutilized in recent decades. The macroflora taxonomy, although in need of revision, similarly is compiled here for the first time in recognition of its fundamental importance to the chronostratigraphy of the Nova Scotian Carboniferous (Appendix B). Systematic study of the revised macrofloral taxonomy, such as that undertaken for the later Westphalian and Stephanian by Zodrow & Cleal (1985) is required to realize their potential contribution to Maritimes Basin and Euramerican correlation (Wagner & Lyons 1997).

The terrestrial environments that attended the lower sea levels of the Maritimes Basin favoured the evolution and migration of early tetrapods. The Nova Scotia record is particularly significant during the Tournaisian, when tetrapods are virtually unknown elsewhere in the world (Carroll 1992; Ahlberg & Milner 1994). Seasonality and fluctuating water levels may have contributed to the evolution of terrestrial vertebrates, and early amniotes in particular. The record in Nova Scotia of terrestrial vertebrate evolution during the Carboniferous, both osteological (Carroll *et al.* 1972) and ichnological (Sarjeant & Mossman 1978; Calder *et al.* submitted), may be the most complete of equatorial Euramerica. The fossil record in Nova Scotia of tetrapods spans the earliest Carboniferous to earliest Permian, interrupted only by the marine conditions of the mid to late Viséan, and here too the regressive shorelines may yet prove to be productive. Conspicuously absent from this predominantly terrestrial record, however, are the aquatic lepospondyls (Carroll *et al.* 1972).

To the vertebrate record can be added a fuller account of invertebrate and plant life during this formative period in the evolution of terrestrial ecosystems. Fossil forests exposed on the sea coast range from earliest Carboniferous *Lepidodendropsis* stands of the Horton Group to a late North American example of a lepidodendrid forest of the Cumberland Group coal measures in the Cantabrian and the only known example in the world of the succeeding xeric conifer *Walchia*, at the Permo-Carboniferous boundary.

Aquatic fauna: how 'nonmarine' ?

Apart fom the marine fauna of the Viséan, virtually all other aquatic fauna of the Carboniferous in Nova Scotia historically have been described as nonmarine, which, as asserted in this paper, is a too restrictive generalization. The term 'nonmarine' fails to describe the spectrum from marine to inland aquatic communities. Aquatic invertebrate taxa of equivocal affinity are found among the agglutinated foraminifera, spirorbids, limulids, ostracods, eocarid crustaceans and pelecypods. The crustacean fauna, including the eocarid *Pseudotealliocaris–Anthracophausia* and ostracod *Shemonaella–Beyrichiopsis* communities of the lower Mabou Group, belie the consistency of 'nonmarine' conditions. The pygocephalomorph–syncarid fauna of the Cumberland Group coal measures, typified by the widely occurring *Pygocephalus dubius*, (Fig. 6b) which occurs also in the coal measures of Britain (Copeland 1957), may represent specialization of a more inland crustacean community relative to the nearshore *Pseudotealliocaris–Anthracophausia* (Fig. 6c) community of the Mabou Group (Schram 1981). It has beeen suggested (Brooks 1962) that *Pygocephalus* was anadramous, migrating from marine to estuarine and inland environments to spawn.

Similar comments apply to the 'nonmarine' bivalve taxa, wherein an ecological gradient from near marine to inland freshwater is likely. Apart from *Carbonicola* of the Windsor and Mabou

Groups, however, the taxa found in Nova Scotia historically have been designated restrictively as 'nonmarine'. The apparent contradiction that certain of these are associated with near-marine environments in western Europe has been accomodated by invoking ecological adaptation to the lower-salinity waters inferred for Nova Scotia (Bell 1944; Vasey 1984). An exception is the study of the Joggins section by Duff & Walton (1973), who concluded that 'in the light of European studies, *Curvirimula* and *Naiadites* could suggest a salinity nearer the "marine" rather than the "fresh" end of the spectrum'. A similar expression of marine influence in associated beds has been suggested for a *Kouphichnium–Cochlichnus–Treptichnus* invertebrate trace fossil assemblage (Archer *et al.* 1995). The pseudoplanktonic, hence widespread, pelecypod *Curvirimula* of the early Silesian (Vasey, 1984) is associated with near-marine faunas in transgressive sequences of the Appalachian Basin and British Isles (Rogers 1985), and *Anthraconaia* similarily may record marine inluence in the Westphalian D (Vasey 1984). The geochemistry of aragonitic *Naiadites* shells from the Joggins Formation, which shows both high Sr/Ca and Mg and a $^{87}Sr/^{86}Sr$ signature of >0.7093, led Brand (1994) to conclude a freshwater habitat although aspects of the geochemistry are equivocal. Within the Maritimes–Western-European province, the nearshore to nonmarine pelecypod taxa exhibit striking similarities (Eagar 1961), with *Carbonicola* and *Naiadites* being endemic genera (Vasey 1984).

The fish fauna of Nova Scotia (see Appendix A) – comprising acanthodians, chondrichthyes, palaeonisciformes and sarcopterygians – is inferred to represent a restricted freshwater habitat and is similarly represented in western Europe (Allen & Dineley 1988). The fauna includes the palaeoniscids *Rhadinichthys* and *Elonichthy*s from the Tournaisian, the acanthodian *Gyracanthus* (Fig. 6a) and the dipnoid *Sagenodus*, common in Europe but rare in North America (Baird 1978; Allen & Dinely ibid.); in Nova Scotia. *Sagenodus* lungfish scales are more common than the literature would suggest, however, preserved most often within bivalve-bearing organic-rich limestones.

Similar comments about nearshore marine to inland palaeoenvironments apply to the fish fauna. *Rhabdoderma*, for example, has been shown to have spawned in estuaries (Schultze 1985). Preliminary isotope data from fish fossils derived from bivalve-bearing organic-rich limestones of the Cumberland Group (H. Falcon-Lang pers. comm. 1997) show $^{87}Sr/^{86}Sr$ signatures consistent with estuarine salinities. These range from 0.7097772 ± 300 (palaeoniscid fish scale) to 0.710338 ± 83 (xenacanth shark tooth, *Diplodus*).

It can be generally stated that the palaeoecology of the faunal groups has been underutilized in modern times in interpretations of the basin-fill. The search for the inland expression of transgressive events in stratal groups historically deemed perennially 'nonmarine', in particular, will benefit from closer scrutiny of the faunal record. Ultimately, the faunal and floral records bear witness to the palaeogeographic and phytogeographic evolution of Euramerica and the relationship between Europe and North America during the Carboniferous.

The Carboniferous palaeoclimate of Nova Scotia

The arid–humid–arid cycle of Late Palaeozoic Euramerican palaeoclimate (Bless *et al.* 1987, fig. 4.2) is recorded in the basin-fill of the Maritimes Basin (Fig. 5). The semiaridity of the Dinantian and early Namurian was approached once again by the early Permian. Maximum humidity is recorded during the Westphalian, followed by a sharp decline in the Stephanian, a trend that mimics broadly that deduced for Euramerica on the basis of the floral record (Phillips & Peppers 1984). The most significant departure from the palaeoclimate inferred for Euramerica west of the Appalachians and east of the Mid-Euramerican Sea is the greater seasonal distribution of rainfall inferred for the Maritimes Basin during the early Silesian. The preliminary Carboniferous palaeoclimate curve for the Maritimes Basin, here published for the first time (Fig. 5), is intended to serve as a model to be tested and refined in the course of future research.

During the Tournaisian, grey lacustrine beds are widespread. The lakes are inferred to have ranged in water depth from relatively deep in the type Horton Bluffs Formation, where hummocky cross-stratification indicates water depths below wave base (Martel & Gibling 1991), to shallow where on, Cape Breton Island for example, playa lake beds exhibit calcrete and desiccation cracks (Hamblin & Rust 1989); algal stromatolites are developed locally. Oil shales of the Horton Bluff and neighbouring Albert Mines formations occur within the *Vallatisporites vallutus* spore zone (Utting 1987), a flora representative of the subtropical arid belt of southern Euramerica (Van der Zwan 1981). The conchostracan fauna that occurs in strata of Tournaisian, early Namurian and Stephanian to early Permian age is similar to extant 'clam shrimp' fauna that can withstand prolonged periods of desiccation, opportunistically inhabiting water-stressed environments where ephemeral ponds develop in prevailing dryland settings (Tasch 1969). The Viséan has long been held to represent a

semiarid palaeoclimate (Bell 1929; Schenk 1967a, b) in which strandline algal stromatolite carbonates, playa salt flats, halite, anhydrite and potash formed. Laminated carbonate of the Macumber Formation, basal Windsor Group, has been ascribed to semiarid seasonality (Schenk *et al.* 1994). Considerable evidence exists for the persistence of semiaridity during the late Viséan following retreat of the Windsor Sea: playa lakes, deep desiccation cracks, gypsum casts, calcareous 'cementstone' beds (Belt et al. 1967; Neves & Belt 1970; McCabe & Schenk 1982; Crawford 1995) and a conchostracan fauna (Copeland 1957). The palaeoclimate may have moderated by the early Namurian (Crawford 1995), during which time the developing monsoon is said to have been recorded in vertisols of the Hastings Formation (Chandler 1995).

The dramatic floral changeover across the Mid-Namurian divide reflects a marked climate change from semiarid to subhumid or seasonal humid (Fig. 5), similarly recorded in the palaeoclimate of the Central Appalachian Basin (Cecil *et al.* 1985). This changeover in macroflora (Bell 1944) similarly is recorded in the miospore record by the entry of saccate *Florinites* prepollen and *Lycospora* (G. Dolby, pers. comm. 1997; J. Utting pers. comm. 1997). Care must be excercised in the interpretation of the basin-fill record across this time interval, given its coincidence with the marked change in tectonic regime from distensional to transpressional at the Mid-Carboniferous event, which may correlate with the Mississippian–Pennnsylvanian unconformity. The attendant inversion of massifs and neighbouring basinal areas has potential to introduce to the basin elements of hinterland flora (*cf.* Chaloner 1958), such as the saccate *Florinites,* which accompanied the abundant permineralized cordaite logs of the Boss Point Formation. Indeed, the hinterland flora includes bisaccate pollens of Gondwanan affinity (Dolby *in press*), and sand grain microtextures have even been cited as evidence of local mountain glaciation (D'Orsay & van de Poll 1985). The increase in *Lycospora*, however, reflects the widespread development of lepidodendrid peat-forming wetland ecosystems in the lowlands. Densopores *sensu latu* are relatively rare in the Carboniferous of the Maritimes Basin in Nova Scotia (Neves & Belt 1970; Dolby in press, pers. comm. 1997), in marked contrast to their abundance in coal-bearing strata in the Appalachian Basin (Eble 1996) and in Europe (Butterworth 1966). Smith (1962) and Butterworth (1966) attributed the paucity of densospores to a precipitation deficit, which is consistent with their relative scarcity in the Maritimes Basin.

Seasonality prevailed even during the most humid periods when mires persisted, represented by rheotrophic to mesotrophic coal beds (Calder

1994; Calder *et al.* 1996), and climate shifts were experienced in the range of the Crowell–Milankovitch band at which time seasonality was more pronounced (Calder 1994), to the point that calcrete was developed (Tandon & Gibling 1994). The trophic status of coal can be used to infer the relative degree of dry seasonality within a humid climate: solely rain-fed, ombrotrophic, raised mires require year-round distribution of precipitation, whereas rheotrophic peats can be sustained through a short dry season by groundwater. The development of calcrete during the Westphalian D, when the climate and glacioeustatic cycles also favoured the development of mesotrophic coals (Tandon & Gibling 1994), inferred to tolerate only moderate dry seasons, serves to illustrate the point that palaeoclimate history should always be considered in terms of climate maxima and minima (see Cecil 1990; Perlmutter & Matthews 1989). Calcrete occurs in red beds coeval with coals from the Langsettian/Westphalian A (Chandler 1995) to D (Tandon & Gibling 1994; Chandler 1995, 1997).

Evidence of Westphalian seasonality (Calder 1979; 1994; Calder *et al.* 1996; Chandler 1995, 1997) perhaps is recorded most eloquently in the rhythmic alternation of siliciclastic and organic-rich laminae in lacustrine mudrocks (Kalkreuth *et al.* 1990) and in the ubiquitous fusain clasts throughout the 13.4 m thick Foord seam and other coal beds, all of the Stellarton Basin. The fossil charcoal bears witness to a humid palaeoclimate with a pronounced dry season with recurring wildfires ignited by lightning strikes, as in modern continental regions of the tropics (Lottes & Ziegler 1994). Vertisols with calcrete in coeval red beds of the Malagash Formation (Chandler 1997) bear similar palaeoclimatic testimony in areas of better drainage (Naylor *et al.* 1989). This record of seasonality contrasts with inferences of a non-seasonal humid climate drawn from the palaeobotanical record elsewhere in Euramerica (Chaloner & Creber 1975; Phillips *et al.* 1985). Seasonality notwithstanding, across the Mid-Namurian divide, the Westphalian clearly records an abrupt shift to a more humid palaeoclimate (Fig. 4), suggesting a marked change in oceanic or orographically influenced circulation patterns. The humid seasonal palaeoclimate of the Westphalian Maritimes Basin was a harbinger of the developing Pangaean monsoon (Rowley *et al.* 1985; Broadhurst 1988; Parrish 1992). The timing of monsoonal circulation in the Maritimes in comparison with other regions of Euramerica has implications for modelling the effects of the equatorial Appalachian–Caledonian, Mauritanide and Hercynide mountain belts on climate circulation models for Euramerica (Parrish 1992; Rowley *et al.* 1985).

The Stephanian to early Permian trend of increasing aridity is mirrored across Euramerica, with notable wetland refugia persisting in the Hercynides and Cantabrians of western Europe and in China (DiMichele & Phillips 1994). The Late Carboniferous palaeoclimate change in Euramerica to more arid conditions has been ascribed to northward continental drift from an equatorial rainy belt into a tropical climate belt (Rowley *et al.* 1985; Bless *et al.* 1987, Cecil 1990, Calder & Gibling 1994), and locally to development of orographic rainshadows (Besly 1988). The assembly of Pangaea into a cross-equatorial landmass and closure of latitudinal seaways may have had even greater palaeoclimatic impact (Rowley *et al.* 1985; Parrish 1992). Orographic effects of the rising assembly of the Pangaean plates and consequent basin inversion and subsequent unroofing in the Maritimes–West-European province and global fall in world ocean levels during the Permian also contributed to the trend. The latest vestiges of peat accumulation and hydromorphic gleysols are found in the Cantabrian (Zodrow & Cleal 1985) fossil lepidodendrid forest atop the Point Aconi seam of the Sydney Basin. Succeeding calcrete-bearing vertisols (Chandler pers. comm. 1997), monsoonally fed dryland river systems and xeric walchian flora of the Cape John Formation, Pictou Group at the Permo-Carboniferous boundary (Fig. 7d; Calder *et al.* in press) serve to frame the end Carboniferous palaeoclimate of Nova Scotia and the Maritimes Basin.

Carboniferous mineral and energy deposits

For more than two centuries, and since the first recorded export of minerals from Canada, from the coal mines of Cape Breton to the Boston colonies in 1720 (Brown 1871), the Carboniferous strata have proved to be the most important source of metallic, industrial and energy minerals in Nova Scotia. The mineral resources of the Carboniferous strata can be linked directly or indirectly to the genesis of the Carboniferous basin-fill. They can be categorized by their mode of occurrence within the basin-fill: (1) primary sedimentary (stratal) deposits (Dinantian salt, limestone and rehydrated gypsum; Silesian rheotrophic to mesotrophic coal and oil shale; Carboniferous marine to terrestrial hydrocarbon source rocks), which have proved to be the most productive of the province's resources; (2) strata-bound deposits (lead, zinc, copper, silver; liquid and gaseous hydrocarbons, including coal bed methane); (3) infracontact and intracontact deposits, commonly at redox boundaries (copper, silver, uranium); (4) fracture-fill deposits (barium, lead, zinc, silver, iron); and (5) physically derived (palaeoplacer) deposits (gold). The disposition and types of mineral deposits in the basin-fill are shown in Fig. 8.

The thermal evolution and burial history of the Maritimes Basin have determined the generation of petroleum and base metal deposits. The source of heat flow has variously been ascribed to underlying plutons, depth of burial or elevated heat flow inherent within the Maritimes Basin (see Sangster et al. 1998). Lead–zinc–barium deposits, which occur as strata-bound, replacement and fracture fillings, are inferred to have been sourced from basinal brine expulsion during the Carboniferous; apatite fission track analysis places the expulsion as earlier than 280 Ma (Ryan & Zentilli 1993). Transpression and inversion associated with the Mid-Carboniferous event in Nova Scotia may have initiated the movement of basinal brine. The source of these brines remains contentious. Geochemical analysis of mine waters in the collieries of the Sydney Basin provides evidence that marine evaporative brines of probable Windsor Group affinity are present at depth in Carboniferous strata (Martel et al. 1997). Metallogenic models for the continental formations of the Maritimes Basin which were developed on the strength of consistently terrestrial conditions (van de Poll 1978; Sangster & Vaillancourt 1990) may warrant

reconsideration in light of the possible cryptic inland signatures of marine transgressions suggested here. Continental red bed copper–silver–uranium deposits in late Westphalian–Stephanian strata of the Cumberland Basin have been ascribed to epigenetic reddening associated with unroofing and aridity in the Permo-Triassic (Ryan & Boehner 1994). As asserted earlier, however, the red beds of the Pictou Group are inferred to record increasing aridity in the latest Carboniferous and early Permian, which alternatively raises the possibility of earlier diagenetic processes, for example in rheotrophic 'copper bogs' (Chandler 1997).

Bituminous coal deposits of the Cumberland Group of Westphalian A–C age developed as areally restricted rheotrophic mires nourished through the dry season by supplemental groundwater flow (Calder 1994) generated at piedmont margins (Springhill coalfield: Calder 1994), in distributary (Joggins coalfield: Calder 1994) and lacustrine (Stellarton Basin: Calder 1979; Naylor et al. 1989; Waldron 1996) settings during regional transgression and transtension. Subsequent mire development in the Westphalian D to Cantabrian may have attained a mesotrophic status if only through the increased, hence insular, area of the peatlands on coastal plains during thermal sag (Hacquebard & Donaldson 1969; Gibling & Bird 1994; Marchioni et al. 1994; Calder et al. 1996).

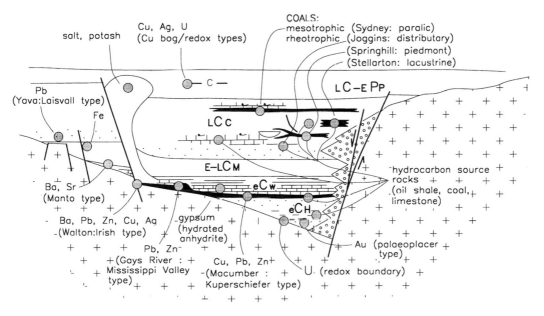

Fig. 8. Schematic representation of the occurrence of mineral and energy deposits in the Carboniferous basin-fill of Nova Scotia. Base metals disposition modified from Ryan and Boehner (1994).

The latter coal deposits constitute the main economic seams of Nova Scotia, mined in collieries of the Sydney Basin.

With few exceptions (Utting & Hamblin 1991), the strata of the Maritimes Basin at surface lie everywhere within the oil or gas windows, with R_0 (vitrinite reflectance) at surface ranging from 0.4 where suppressed by liquid hydrocarbons (Mukhophadyay et al. 1991) to >2 (Hacquebard & Donaldson 1970; Ryan & Boehner 1994). Oil seeps occur at present (Bell 1958; Short 1986), even though the apatite fission track record indicates that the oil window was attained early in the basin's history, before 250 Ma (Grist et al. 1995). Kinematic research into coal bed methane generation indicates that gas desorption in coal beds of R_0 <0.9 may be impeded by micropore blockage by earlier generated oil, but enhanced above that maturity by cracking of the oil (Mukhopadhyay et al. 1993). In all basins, rank increases with depth of burial (Hacquebard & Donaldson 1970). Hydrocarbon source rocks include sapropelic shales of the MacAdam Lake Formation and Horton Group, widespread organic-rich carbonate laminites of the Windsor Group, and sapropelic shales, sapropelic and humic coals, and basin-wide, organic-rich bivalve-bearing limestones and shales, all of the Cumberland Group. New models of hydrocarbon generation in Nova Scotia should incorporate the different thermal and structural histories of Horton, Windsor and Mabou strata before the Mid-Carboniferous event, and Cumberland and Pictou strata thereafter, and should consider possible cryptic marine transgressions within groups traditionally described as 'nonmarine'.

Gypsum, as rehydrated anhydrite (Bell 1929; D. Shearman pers. comm. 1997), forms thick, areally extensive deposits across much of the Maritimes Basin. The open-pit operations are the largest in the world (Adams 1991). Halite, which together with potash is unrepresented in outcrop due to solubility in the present humid climate, was discovered in Nova Scotia only in 1917, at Malagash, Cumberland Basin (Bell 1929). This deposit in the Cumberland Basin subsequently became the location of Canada's first underground salt mine (Bell 1929 1944); mining continues in the basin from salt-cored halokinetic anticlines that parallel basement faults (Boehner 1986).

The Carboniferous history of Nova Scotia: implications for Euramerican studies

The Carboniferous history of Nova Scotia can be understood fully only by considering it in the context of the regional history of Euramerica (Fig. 9); the same observation can be made for Europe and for North America, and in these cases, the story as recorded in Nova Scotia is significant. Salient points of the palaeogeography and evolution of Euramerica arising from this study of the Carboniferous evolution of Nova Scotia are discussed below.

Implications for the tectonic history of circum-Atlantic Euramerica

The Carboniferous evolution of Nova Scotia serves to corroborate or weigh against tectonic interpretations for the Appalachians of the United States and for western Europe. The Namurian transpressive event and subsequent changeover to a transcurrent regime in the Maritimes records the Variscan orogeny in the west of the Maritimes–West-European province and is consistent with Alleghenian thrusting in the Appalachians (Nance 1987; Gibling 1995; Rehill 1996). This Mid-Carboniferous event in Nova Scotia is coeval with the Mississippian–Pennsylvanian unconformity west of the Appalachian Mountains. The transpression in Nova Scotia at this time lends support to the role of tectonic inversion in the development of the unconformity (Ettensohn & Chesnut 1989) rather than to eustasy alone (Saunders & Ramsbottom 1986). These events in Euramerica and coincident inversion and volcanism in Gondwana (Gonzalez-Bonorino & Eyles 1995) mark the global convergence of plates and assembly of Pangaea.

In Nova Scotia, the distensional period of rifting had ended by the Westphalian and so does not accord with widespread extension and rifting across the Maritimes–West-European province subsequently in the Silesian (Hazeldine 1984). Rather, it lends credence to continued linkage of transcurrent tectonism in western Europe with that in the Maritimes (Matte 1986; Leeder 1988a) as plates continued to reorganize in eastern Euramerica to accommodate Gondwana.

Silesian Appalachian drainage divide

Palaeoflow compilations from the Appalachian Basin (Archer & Greb 1995) and from the Maritimes Basin (Gibling et al. 1992) have determined that prevailing drainage was to the east across the Maritimes Basin during the Silesian and to the west through the Appalachian Basin. The Appalachian Basin study, however, infers that drainage was sourced in part in the Maritimes basinal area, which clearly could not have been the case, with the possible exception of during the

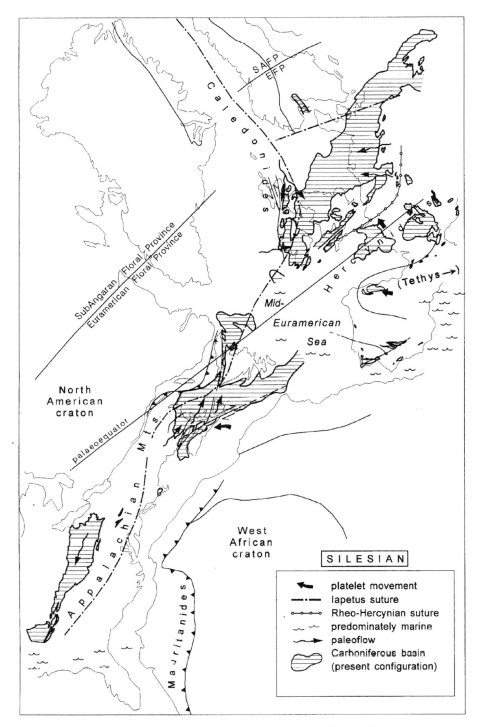

Fig. 9. Nova Scotia in the context of equatorial Euramerica during the Carboniferous (Silesian). Base map is that used by Hazeldine (1984) and Williams (1984). Palaeoflow after Archer and Greb (1995) for the Appalcahian Basin, Gibling *et al.* (1992) for the Maritimes Basin, and Leeder (1988*b*, 1992) for western Europe. Western European Carboniferous Basin outline after Maynard *et al.* (1997).

period of inversion attending the Mid-Carboniferous event. The Maritimes Basin lay to the palaeosoutheast of the orogen, whereas the Appalachian Basin lay to the palaeonorthwest (Fig. 9). Opposing palaeoflow directions suggest the existence of an Appalchian Divide during the Silesian (Gibling et al. 1992; Rehill 1996) with reversed polarity of basins about the position of the New York Promontory / Pennsylvania Embayment.

The Mid-Euramerican Sea

Circumstantial evidence points to the existence of a Mid-Euramerican Sea between the Maritimes Basin and western Europe during the Carboniferous (Fig. 9). The presence of an intervening deep oceanic realm which acted as a barrier to benthonic species exchange in the Viséan was hypothesized by Jansa et al. (1978) and Jansa & Mamet (1984). The invertebrate macrofauna (Bell 1929) and ostracode fauna (Dewey 1989), however, show affinity to western European fauna of similar environments, and conodonts such as Taphrognathus transatlanticus permit cross Maritimes–West-European correlation (Von Bitter & Austin 1984). The Silesian pelecypod fauna of Nova Scotia demonstrates a close affinity to western Europe, suggesting a palaeooceanic link between the two areas and an Appalachian barrier to the west (Vasey 1984). It has been proposed that the Mid-Euramerican Sea persisted through the Dinantian and Silesian and may have been one of the last remaining seas between the closing continents of Gondwana and the Old Red Continent, representing either a part of the Phoibic Ocean (McKerrow & Ziegler 1972) or Proto-Tethys (Leeder 1988b) or their vestige (W. S. McKerrow, pers. comm. 1997). Marine connections with the Boreal Sea may have been possible via rifted basins in the area of Svalbard and Greenland (Stemmerik et al. 1991; Rehill 1996), although faunal and palaeogeographic trends favour a cross-Maritimes-West European connection. Less certain is whether the ocean was connected with the southern Appalachian Basin by a seaway between the converging African and North American cratons, which would have become increasingly restricted if not closed during Alleghenian–Mauritainide orogenesis.

River drainage through the Maritimes Basin during the Silesian period of transcurrent movement followed the northeast structural grain of intervening basins and massifs to a common destination with the system draining southwestward across the Pennine Basin (Leeder 1988b) – inferred to have been the Mid-Euramerican Sea. Palaeoflow from the Maritimes Basin earlier had been inferred to have fed into the Rheo-Hercynian Ocean (Gibling et al. 1992); however, the Rheic purportedly had closed by that time (Leeder 1988a; Maynard et al. 1997). A thick succession of Dinantian and Silesian strata in the Porcupine Basin west of Ireland at the westernmost edge of the European plate contains marine strata of Westphalian A age and a record of marine transgressions through the Stephanian (Tate & Dobson 1989). This evidence of a Mid-Euramerican Sea is supported by recurring marine transgressions from the southwest across the British Isles until the Westphalian C (Ramsbottom 1977), and by westerly derived marine influence across the Western European Carboniferous Basin, from Poland and Germany to the British Isles (Maynard et al. 1997).

Phytogeographic implications for Euramerican correlations

Although the stage boundaries of the Carboniferous, with the notable exception of the Westphalian D, have been established largely on the strength of goniatite faunal zones, these seldom coincide with major floral breaks (Wagner 1984). In the Maritimes Basin where goniatites are absent, this situation is exacerbated by phytogeogeographic differences from other Euramerican regions. The recognition of the Namurian–Westphalian boundary in Nova Scotia is a case in point. Few floral extinctions or appearances occur across this boundary, which is defined by the basal Westphalian Gastrioceras subcrenatum goniatite band in Europe (Ramsbottom et al. 1978). Index miospores used to recognize the base of the Westphalian elsewhere in Euramerica (Clayton et al. 1977; Maynard et al. 1997) include densospores sensu latu (including the related crassicingulate genera Densosporites, Cristatisporites and Radiizonites; DiMichele & Phillips 1994) produced by the subarboreous lycopsid Bodeodendron (Sporangiostrobus) (Wagner 1989), and Endosporites produced by the diminutive lycopsid Chaloneria (Polysporia). The subarboreous lycopsids, and densospores and their parent plant in particular, are rare in the fossil record of Nova Scotia and may have been environmentally excluded, as discussed earlier. The biostratigraphy of the prevailingly continental Maritimes Basin in Nova Scotia, perhaps more so than elsewhere in Euramerica, must be interpreted in the light of its regional palaeoenvironment and provincialism. The faunal record, even with its affinity to western Europe, nonetheless requires similar interpretation, specifically with respect to the aquatic spectrum from open marine to nearshore and inland communities.

Conclusions

Doubtless, we are biased in our perceptions of Carboniferous Euramerica by the configuration of the modern world and by geopolitical boundaries. The Carboniferous of Nova Scotia, though adjacent to the northern Appalachian orogen, nonetheless represents part of the Maritime–West-European palaeogeographic province of tropical Euramerica, albeit with an intervening Carboniferous sea. It shares many attributes with Europe, including elements of its tectonic history and fossil fauna and flora, as described more than a century ago by Lyell (1845 and others) and by Dawson (1888 and earlier works). The Carboniferous history in Nova Scotia of distension, transpression, thermal sag and inversion links the tectonic history of the Alleghanian orogeny in North America and the Variscan–Hercynian orogeny in western Europe. The Mid-Carboniferous Break marks the most profound period of change in the evolution of the Carboniferous of Nova Scotia; the tectonic and palaeoclimatic changes experienced across this equivocally dated interval are linked to global plate tectonics, and in particular, to Appalachian orogenesis.

The palaeoclimate record of the Silesian stands Nova Scotia apart from the nonseasonal humid palaeoclimate inferred for Euramerica and suggests that Nova Scotia may have experienced seasonal rainshadow effects, presumably as a consequence of its intermontane palaeogeography. The record of cycles of duration, largely within the Crowell–Milankovitch band, is superimposed on the longer-term evolution from marine to continental strata (Fig. 6). Carbonate–sulfate cycles of the Viséan probably represent semiarid marine equivalents of seasonally humid continental coal-bearing cyclothemic stratal successions of Westphalian coastal plains (Gibling & Bird 1994) and alluvial valley piedmonts (Calder 1994), as hypothesized by Schenk (1969 p. 1060). Intermediate between these end members are the predominantly continental, schizohaline late Viséan to mid-Namurian cycles of the Mabou Group (Belt 1968b; Crawford 1995), in which the marine record may yet be discerned. All are primarily allocyclic, driven most probably by orbitally induced climate change as it affected global glaciocustacy (Veevers & Powell 1987) and local climate change (Calder 1994), although tectonic causes have long been proposed for the 'nonmarine' strata.

Of the great river systems that drained the Maritimes–West-European province during the Silesian, two drained eastwards through the interconnected Maritimes basins and southwestward across the Pennine Basin towards a common destination, the Mid-Euramerican Sea. The existence of such a sea during the Carboniferous has attendant implications for cross-Maritimes–West-European correlations of sea-level change.

Although open marine conditions occurred only during the mid to late Viséan, the signature of transgression and regression should be variably discernible throughout much of the Carboniferous basin-fill. Traditional interpretation of the basin-fill simply in terms of marine or 'nonmarine' is too restrictive. It is better to consider the palaeogeographic gradient from open marine to nearshore and inland, which may be identified through careful reconsideration of the faunal record, even within the prevailingly continental, inland strata of Nova Scotia. A particular challenge, however, will be to discern brackish water caused by evaporation and contraction of lakes from the most inland expressions of a transgressing sea. This approach will have implications for modelling not only the basin-fill of the Maritimes but also its mineral and energy deposits. Strata-bound base metal and hydrocarbon models that have had to assume consistently continental conditions for groups other than the Windsor in particular stand to profit from such a re-examination of basin modelling. The Mid-Carboniferous Break, which links the tectonic evolution of Euramerica to the west and east in the Namurian, must also be considered in the development of hydrocarbon and base metal models for Nova Scotia.

To achieve these ends, the collaboration of European and North American geologists with those of Nova Scotia, so profitable during the time of Sir Charles Lyell, is required. In so doing, we will continue to loosen Lyell's 'Gordian knot' as it pertains to Euramerican Carboniferous geology.

The concepts of this paper draw on the distillation of published research by many Carboniferous workers, and discussions during the past 20 years with my colleagues in Nova Scotia and abroad. The research for this paper has been supported by the Nova Scotia Department of Natural Resources. Graham Dolby and John Utting were most open in sharing their wealth of experience in the palynostratigraphy of Nova Scotia. The faunal compilation would not have been possible without the generous and open collaboration of Andrew Milner and Reg Moore. Jean Dougherty and David Lewis are thanked for graciously supplying photographs of specimens in the collections of the GSC and NHM, respectively. The *Gyracanthus* spine (Fig. 6a) was discovered in 1997 by Brian Hebert. The incisive reviews by E. S. Belt and M. R. Gibling and generous insightful discussions with Bob Boehner, Fred Chandler, Sarah Davies, Howard Falcon-Lang, Peter Giles, David Piper, Paul Schenk, John Waldron and Erwin Zodrow provided welcome improvement to the breadth of the paper. Many thanks to Patricia Fraser and Janet Webster for drafting the figures, and to Tracey Lenfesty for her literature searches. I am grateful for the invitation and financial support of the convenors of the Lyell Bicentennial Meeting, Professors Derek Blundell and Andrew Scott, to undertake this comprehensive assessment of the Carboniferous evolution of Nova Scotia and I thank the many participants at the Lyell meeting from whom I received welcome input; to the likes of Sir Charles Lyell and Sir J. William Dawson, we owe much. Finally, to my uncle, John Lynton Martin, my thanks for your wonderful gift those years ago of *Acadian Geology*.

References

ADAMS, G. C. 1991. *Gypsum and Anhydrite Resources in Nova Scotia*. Nova Scotia Department of Natural Resources Economic Geology Series 91-1, Halifax.

AHLBERG, P. E. & MILNER, A. R. 1994. The origin and early diversification of tetrapods. *Nature*, **368**, 507–514.

ALLEN, K. C. & DINELEY, D. L. 1988. Mid-Devonian to mid-Permian floral and faunal regions and provinces. *In*: HARRIS, A. L. & FETTES, D. J. (eds) *The Caledonian–Appalachian Orogen*. Geological Society, London, Special Publications, **38**, 531–548.

ARCHER, A. W. & GREB, S. F. 1995. An Amazon-scale drainage system in the Early Pennsylvanian of Central North America. *Journal of Geology*, **10**, 611–628.

——, CALDER, J. H., GIBLING, M. R., NAYLOR, R. D., REID, D. R. & WIGHTMAN, W. G. 1995. Invertebrate trace fossils and agglutinated foraminifera as indicators of marine influences within the classic Carboniferous section at Joggins, Nova Scotia, Canada. *Canadian Journal of Earth Sciences*, **32**, 2027–2039.

BAIRD, D. 1978. *Studies on Carboniferous freshwater fishes*. American Museum Novitates, No. 2641, New York, 1–22.

BARR, S. M., GRAMMATIKOPOULOS, A. L. & DUNNING, G. R. 1994. Early Carboniferous gabbro and basalt in the St. Peters area, southern Cape Breton Island, Nova Scotia. *Atlantic Geology*, **30**, 247–258.

BARR, S. M., MacDONALD, A. S., ARNOTT, A. M. & DUNNING, G. R. 1995. Field relations, structure and geochemistry of the Fisset Brook Formation in the Lake Ainslie–Gillanders Mountain area, central Cape Breton Island, Nova Scotia. *Atlantic Geology*, **31**, 127–139.

BARSS, M. S. & HACQUEBARD, P. A. 1967. Age and the stratigraphy of of the Pictou Group in the Maritime Provinces as revealed by fossil spores. *In*: NEALE, E. R. W. & WILLIAMS, H. (eds) *Collected Papers on Geology of the Atlantic Region*. Geological Association of Canada, Special Paper 4, 267–282.

——, —— & HOWIE, R. D. 1963. *Palynology and Stratigraphy of Some Upper Pennsylvanian and Permian Rocks of the Maritime Provinces*. Geological Survey of Canada Paper, Ottawa, 63–3.

BELL, W. A. 1929. *Horton-Windsor District, Nova Scotia*. Geological Survey of Canada, Ottawa, Memoir 155.

—— 1938. *Fossil Flora of the Sydney Coalfield*. Geological Survey of Canada, Ottawa, Memoir 215.

—— 1940. *The Pictou Coalfield, Nova Scotia*. Geological Survey of Canada, Ottawa, Memoir 225.

—— 1944. *Carboniferous Rocks and Fossil Floras of Nova Scotia*. Geological Survey of Canada, Ottawa, Memoir 238.

—— 1958. *Possibilities for Occurrence of Petroleum Reservoirs in Nova Scotia*. Nova Scotia Department of Mines, Halifax.

—— 1960. *Mississippian Horton Group of Type Windsor–Horton District, Nova Scotia*. Geological Survey of Canada, Ottawa, Memoir 314.

BELL, W. A. & GORANSON, E. A. 1938. *Sydney Sheet (West Half), Cape Breton and Victoria Counties, Nova Scotia*. Geological Survey of Canada Map 360A.

BELT, E. S. 1964. Revision of Nova Scotia Middle Carboniferous units. *American Journal of Science*, **262**, 653–673.

—— 1965. Stratigraphy and palaeogeography of Mabou Group and related middle Carboniferous facies, Nova Scotia, Canada. *Geological Society of America Bulletin*, **76**, 777–802.

—— 1968a. Post-Acadian rifts and related facies, eastern Canada. *In*: ZEN, E-AN, WHITE, W. S., HADLEY, J. B. & THOMPSON, J. B. (eds) *Studies of Appalachian Geology: Northern and Maritime*. Wiley-Interscience, New York, 95–113.

—— 1968b. Carboniferous continental sedimentation, Atlantic Provinces, Canada. *In*: KLEIN, G. DE V. (ed.) *Late Paleozoic and Mesozoic Continental Sedimentation, Northeastern North America*. Geological Society of America Special Paper 106, Boulder, 127–176.

—— 1969. Newfoundland Carboniferous stratigraphy and its relation to the Maritimes and Ireland. *In*: KAY, M. (ed.) *North Atlantic Geology and Continental Drift*. American Association of Petroleum Geologists, Memoir 12, Tulsa, 734–753.

——, FRESHNEY, E. C. & READ, W. A. 1967. Sedimentology of Carboniferous cementstone facies, British Isles and eastern Canada. *Journal of Geology*, **75**, 711–721.

BESLY, B. M. 1988. Palaeographic implications of late Westphalian to early Permian red-beds, Central England. *In*: BESLY, B. M. & KELLING, G. (eds) *Sedimentation in a Synorogenic Complex: The Upper Carboniferous of Northwest Europe*. Blackie, Glasgow & London, 200–221.

BLANCHARD, M. -C., JAMISEON, R. A. & MORE, E. B. 1984. Late Devonian–Early Carboniferous volcanism in western Cape Breton Island, Nova Scotia. *Canadian Journal of Earth Sciences*, **21**, 762–774.

BLESS, M. J. M., BOUCKAERT, J. & PAPROTH, E. 1987. Fossil assemblages and depositional environments: limits to stratigraphical correlations. *In*: MILLER, J., ADAMS, A. E. & WRIGHT, V. P. (eds) *European Dinantian Environments*. John Wiley & Sons, New York, 61–73.

BOEHNER, R. C. 1977. *The Lower Carboniferous Stratigraphy of the Musquodoboit Valley, Central Nova Scotia*. MSc thesis, Acadia University, Nova Scotia.

—— 1986. *Salt and Potash Resources in Nova Scotia*. Nova Scotia Department of Mines and Energy, Halifax, Bulletin.

—— 1992. An overview of the role of Windsor Group

evaporites in the structural devlopment of Carboniferous Basins in Nova Scotia. *In*: MacDonald, D. R. (ed.) *Mines and Minerals Branch Report of Activities, 1991*. Nova Scotia Department of Natural Resources Report 92-1, Halifax, 39–56.

Bradley, D. C. 1982. Subsidence in Late Paleozoic basins in the northern Appalachians. *Tectonics*, **1**, 107–123.

Brand, U. 1994. Continental hydrology and climatology of the Carboniferous Joggins Formation (lower Cumberland Group) at Joggins, Nova Scotia: evidence from the geochemistry of bivalves. *Paleogeography, Palaeoclimatology, Palaeoecology*, **106**, 307–321.

Broadhurst, F. M. 1988. Seasons and tides in the Westphalian. *In*: Besly, B. M. & Kelling, G. (eds) *Sedimentation in a Synorogenic Complex: The Upper Carboniferous of Northwest Europe*. Blackie, Glasgow & London, 264–272.

Brooks, H. K. 1962. The Paleozoic Eumalacostraca of North America. *American Paleontology Bulletin*, **44**, 163–338.

Brown, R. 1829. Geology and mineralogy [of Nova Scotia]. *In*: Haliburton, T. C. (ed.) *An Historical and Statistical Acount of Nova Scotia*. Joseph Howe, Halifax, vol. 2, section 3, 414–453.

—— 1846. On a group of fossil trees in the Sydney coalficld of Nova Scotia. *Journal of the Geological Society of London*, **2**, 393–396.

—— 1848. Description of an upright Lepidodendron with stigmaria roots in the roof of the Sydney Main coal in the Island of Cape Breton. *Journal of the Geological Society of London*, **5**, 46–50.

—— 1871. *The coal fields and coal trade of the Island of Cape Breton*. Sampson Law, Marston, Law & Searle, London.

Brown, J. P., Davison, I., Alsop, I. & Gibling, M. R. 1996. Deformation related to Carboniferous salt tectonics, western Cape Breton, Nova Scotia. *Atlantic Geology*, **32**, 68.

Browne, G. H. & Plint, A. G. 1994. Alternating braidplain and lacustrine deposition in a strike-slip setting: the Pennsylvanian Boss Point Formation of the Cumberland Basin, Maritime Canada. *Journal of Sedimentary Research*, **B64**, 40–59.

Butterworth, M. A. 1966. The distribution of Densospores. *The Palaeobotanist*, **15**, 16–28.

Calder, J. H. 1979. *Effects of Subsidence and Depositional Environment on the Formation of Lithotypes in a Hypautochthonous Coal of the Pictou Coalfield*. Nova Scotia Department of Mines Paper 79-6, Halifax.

—— 1994. The impact of climate change, tectonism and basin hydrology on the formation of Carboniferous intermontane mires: the Springhill coalfield, Cumberland Basin, Nova Scotia. *Palaeogeography, Palaeoclimatology, Palaeoecology*, **106**, 323–351.

—— & Gibling, M. R. 1994. The Euramerican coal province: controls on Late Paleozoic peat accumulation. *Palaeogeography, Palaeoclimatology, Palaeoecology*, **106**, 1–21.

——, ——, Eble, C. F., Scott, A. C. & MacNeil, D. J. 1996. The Westphalian D fossil lepidodendrid forest

at Table Head, Sydney Basin, Nova Scotia: sedimentology, paleoecology and floral response to changing edaphic conditions. *International Journal of Coal Geology*, **31**, 277–313.

——, Van Allen, H. E. K., Adams, K. S. & Grantham, R. G. 1995. Tetrapod trackways in a fossil Walchia forest: a new discovery from the early Permian of Nova Scotia. *Atlantic Geology*, **31**, 42.

——, ——, Brown, G. & Hunt, A. P. In press. A fossil forest of walchian conifers and the trackway record of its tetrapod community from the Permo-Carboniferous of Nova Scotia. *Palaios*.

Calver, M. A. 1969. Westphalian of Britain. *Sixième Congrés International de la Stratigraphie et de la Géologie Carbonifère, Sheffield*. vol. 1, Ernest van Aelst, Maastricht, 233–254.

Carroll, R.L. 1964. The earliest reptiles. *Journal of the Linnaean Society (Zoology)*, **45**, 61–83.

—— 1992. The primary radiation of terrestrial vertebrates. *Annual Review of Earth and Planetary Sciences*, **20**, 45–84.

——, Belt, E. S., Dineley, D. L., Baird, D. & McGregor, D. C. 1972. *Vertebrate paleontology of Eastern Canada*. Guidebook, Excursion A59, 24th International Geological Congress, Montreal, Ottawa.

Cecil, C. B. 1990. Paleoclimate controls on stratigraphic repetition of chemical and siliciclastic rocks. *Geology*, **18**, 533–536.

——, Stanton, R. W., Neuzil, S. G., Dulong, F. T., Ruppert, L. F. & Pierce, B. S. 1985. Paleoclimate controls on Late Paleozoic sedimentation and peat formation in the central Appalachian Basin. *International Journal of Coal Geology*, **5**, 195–230.

Chaloner, W .G. 1958. The Carboniferous upland flora. *Geological Magazine*, **95**, 261–262.

—— & Creber, G. T. 1975. Do fossil plants give a climatic signal? *Journal of the Geological Society*, **147**, 343–350.

—— & Lacey, W. S. 1973. *The Distribution of Late Palaeozoic Floras*. Special Paper in Palaeontology, **12**, 271–289.

Chandler, F. W. 1995. Geological mapping in the Stellarton Gap (NTS 11E/7,9,10,15), a status report. *Atlantic Geology*, **31**, 43.

—— 1997. The Canfield Creek copper deposit, Nova Scotia – a late Carboniferous cupriferous bog deposit: implications for exploration for redbed copper in Carboniferous clastics in Nova Scotia. *In*: *Current Research 1997-D*, Geologial Survey of Canada, Ottawa, 35–42.

Chappell, J. & Shackelton, N. J. 1986. Oxygen isotopes and sea level. *Nature*, **324**, 137–140.

Clayton, G., Coquel, R., Doubinger, J., Guenin, K. J., Loboziak, S., Owens, B. & Streel, M. 1977. Carboniferous miospores of western Europe: illustration and zonation. *Mededelingen van's Rijks Geologische Dienst*, **19**, 1–71.

Conybeare, W. D. & Phillips, W. 1822. *Outlines of the Geology of England and Wales*. William Phillips, London.

Copeland, M. J. 1957. *The arthropod fauna of the Upper Carboniferous rocks of the Maritime Provinces*. Geological Survey of Canada Memoir 286, Ottawa.

COWIE, J. W. & BASSETT, M. G. 1989. International Union of Gological Sciences 1989 global stratigraphic chart with geochronometric and magnetostratigraphic calibration. *Episodes*, **12** (Suppl.).

CRAWFORD, T. L. 1995. Carbonates and associated sedimentary rocks of the Upper Viséan to Namurian Mabou Group, Cape Breton Island, Nova Scotia. *Atlantic Geology*, **31**, 167–182.

CROWELL, J. C. 1978. Gondwanan glaciation, cyclothems, continental positioning and climate change. *American Journal of Science*, **278**, 1345–1372.

DAWSON, J. W. 1845. On the Newer Coal Formation of the eastern part of Nova Scotia. *Journal of the Geological Society of London*, **1**, 322.

—— 1847. The gypsum of Nova Scotia. *Proceedings of the Academy of Natural Science, Philadelphia*, **3**, 270–274.

—— 1847. *Quarterly Journal of the Geological Society of London*, **4**, 59.

—— 1855. *Acadian Geology: an Account of the Geological structure and Mineral Resources of Nova Scotia, and Portions of the Neighbouring Province of British America.* Oliver & Boyd, Edinburgh.

—— 1868. *Acadian Geology. The Geological Structure, Organic Remains, and Mineral Resources of Nova Scotia, New Brunswick and Prince Edward Island.* 2nd edn. Macmillan, London.

—— 1878. *Acadian Geology. Supplement to the Second Edition, Containing Additional Facts as to the Geological Structure, Fossil Remains, and Mineral Resources of Nova Scotia, New Brunswick and Prince Edward Island.* Macmillan, London.

—— 1891. *The Geology of Nova Scotia, New Brunswick and Prince Edward Island or Acadian Geology.* Oliver & Boyd, Edinburgh.

—— 1860. On a terrestrial mollusk, a millepede, and new reptiles, from the Coal Formation of Nova Scotia. *Quarterly Journal of the Geological Society of London*, **16**.

—— 1862. On the flora of the Devonian period in N.-E. America. *Journal of the Geological Society*, **18**, 296.

—— 1873. *Report on the Fossil Plants of the Lower Carboniferous and Millstone Grit Formations of Canada.* Geological Survey of Canada, Separate Report 430.

—— 1888. On the Eozoic and Palaeozoic rocks of the Atlantic coast of Canada, in comparison with those of Western Europe and of the interior of America. *Quarterly Journal of the Geological Society of London*, **176**, 797–817.

—— 1894. Synopsis of the air-breathing animals of the Palaeozoic in Canada, up to 1894. *Transactions of the Royal Society of Canada*, **4**, 71–88.

DEWEY, C. 1989. Lower Carboniferous ostracodes from the Maritimes Basin of eastern Canada: a review. *Atlantic Geology*, **25**, 63–71.

DIMICHELE, W. A. & PHILLIPS, T. L. 1994. Paleobotanical and paleoecological constraints on models of peat formation in the Late Carboniferous of Euramerica. *Palaeogeography, Palaeoclimatology, Palaeoecology*, **106**, 39–90.

DONOHOE, H. V., JR & WALLACE, P. 1982. *Geological Map of the Cobequid Highlands, Colchster, Cumberland & Pictou Counties, Nova Scotia.* Nova Scotia Department of Mines and Energy Map 82-8, sheet 3, Halifax.

DOLBY, G. 1991. *The Palynology of the Western Cumberland Basin, Nova Scotia.* Nova Scotia Department of Mines and Energy Open File Report 91-006.

—— In press. *The Palynology of the Cumberland Group (Upper Carboniferous) in the Western Part of the Cumberland Basin, Nova Scotia.* Geological Survey of Canada Paper, Ottawa.

DONALDSON, A. C., RENTON, J. J. & PRESLEY, M. W. 1985. Pennsylvanian deposystems and paleoclimates of the Appalachians. *International Journal of Coal Geology*, **5**, 167–193.

D'ORSAY, A. M. & VAN DE POLL, H. W. 1985. Quartz-grain surface textures: evidence for middle Carboniferous glacial sediment input to the Parrsboro Formation of Nova Scotia. *Geology*, **13**, 285–287.

DUFF, P. MCL. D. & WALTON, E. K. 1973. Carboniferous sediments at Joggins, Nova Scotia. *In: Seventh International Congress on Carboniferous Stratigraphy and Geology, Compte Rendu.* Vol. 2, Geologisches Landesamt Nordrhein-Westfalen, Krefeld, 365–379.

DUNNING, G., PIPER, D. J. W., GILES, P .S., PE-PIPER, G. & BARR, S. M. 1997. Chronology of early phases of rifting of the Devonian–Carboniferous Magdalen Basin in Nova Scotia from U/Pb dating of rhyolites. *In: Geological Association of Canada / Mineralogical Association of Canada Annual Meeting, Ottawa*, **22**, A42.

EAGAR R. M. C. 1961. A summary of the results of recent work on the palaeoecology of Carboniferous non-marine lamellibranchs. *In: Comptes Rendus du Congrès d'Avances Etudes de la Stratigraphie et Géologie du Carbonifére, Heerlen, 1958.* Vol. 1, 137–149.

EBLE, C. F. 1996. Paleoecology of Pennsylvanian coal beds in the Appalachian Basin. *In*: JANSONIUS, J. & MCGREGOR, D. C. (eds) *Palynology: Principles and Applications.* American Association of Stratigraphic Palynologists Foundation. vol. 3, 1143–1156.

ETTENSOHN, F. R. & CHESNUT, D. J. 1989. The nature and probable origin of the Mississippian-Pennsylvanian unconformity in the eastern United States. *In: XIe Congrès International de Stratigraphie et de Géologie du Carbonifère, Beijing, 1987, Compte Rendu.* Vol. 4, Nanjing University Press, 145–159.

FLETCHER, H. 1875–1909. Geological Survey of Canada, Reports of Progress, with accompanying maps, scale 1 inch = 1 mile.

FORBES, W. H., KASPER, A. E., DONOHOE, H. V. & WALLACE, P. 1979. A new Devonian flora from Nova Scotia. *In*: MACGILLIVARY, J. M. & MCMULLIN, K. A. (eds) *Mineral Resources Division Report of Activities 1978.* Nova Scotia Department of Mines Report 79-1, Halifax.

GARDINER, B. G. 1966. *Catalogue of Canadian Fossil Fishes.* Royal Ontario Museum Life Sciences Contribution No. 68, University of Toronto.

GELDSETZER, H. H. J. 1978. *The Windsor Group in Atlantic Canada – An Update.* Geological Survey of Canada, Paper 78-1C, Ottawa, 43–48.

GESNER, A. 1836. *Remarks on the Geology and Mineralogy of Nova Scotia.* Gossip & Coade, Halifax.

—— 1843. A geological map of Nova Scotia, with an accompanying memoir. *Proceedings of the Geological Society of London*, **4**, 186–190.

GIBLING, M. R. 1995. Upper Paleozoic rocks, Nova Scotia. *In*: WILLIAMS, H. (ed.) *Geology of the Appalachian–Caledonian Orogen in Canada and Greenland.* Geological Survey of Canada, **6**, 493–523.

—— & BIRD, D. J. 1994. Late Carboniferous cyclothems and alluvial palaeovalleys in the Sydney Basin, Nova Scotia. *Geological Society of America Bulletin*, **106**, 105–117.

——, CALDER, J .H., RYAN, R. J., VAN DE POLL, H. W. & YEO, G. M. 1992. Late Carboniferous and Early Permian drainage patterns in Atlantic Canada. *Canadian Journal of Earth Sciences*, **29**, 338–352.

GILES, P. 1981a. *The Windsor Group of the Mahone Bay Area, Nova Scotia.* Nova Scotia Department of Mines and Energy Paper 81-3, Halifax.

—— 1981b. Major *Transgressive-Regressive Cycles in Middle to Late Visean Rocks of Nova Scotia.* Nova Scotia Department of Mines and Energy Paper 81-2.

—— 1995. Namurian and early Westphalian stratigraphy of western and southwestern Cape Breton Island. *Atlantic Geology*, **32**, 72.

GILPIN, E., JR 1899. *The Minerals of Nova Scotia, Canada.* Queen's Printer.

GONZALEZ-BONORINO, G. & EYLES, N. 1995. Inverse relation between ice extent and the late Paleozoic glacial record of Gondwana. *Geology*, **23**, 1015–1018.

GRANT, A. C. 1994. Aspects of seismic character and extent of Upper Carboniferous coal measures, Gulf of St. Lawrence and Sydney basins. *Palaeogeography, Palaeoclimatology, Palaeoecology*, **106**, 271–285.

GRIST, A. M., RYAN, R. J. & ZENTILLI, M. 1995. The thermal evolution and timing of hydrocarbon generation in the Maritimes Basin of eastern Canada: evidence from apatitie fission track data. *Bulletin of Canadian Petroleum Geology*, **43**, 145–155.

HACQUEBARD, P. A. 1972. The Carboniferous of eastern Canada. *In*: *7th International Congress on Carboniferous Stratigraphy and Geology.* Vol. 1, Geologisches Landesamt Nordrhein-Westfalen, Krefeld, 69 90.

—— & DONALDSON, J. R. 1964. Stratigraphy and palynology of the Upper Carboniferous coal measures in the Cumberland Basin of Nova Scotia. *In*: *Cinquième Congrès International de la Stratigraphie et de la Géologie du Carbonifère.* Vol. 3, 1157–1169.

—— & —— 1969. Carboniferous coal deposition associated with flood-plain and limnic environments in Nova Scotia. *In*: DAPPLES, E. C. & HOPKINS, M. E. (eds) *Environments of Coal Deposition.* Geological Society of America Special Paper 114, Boulder, 143–191.

—— & —— 1970. Coal metamorphism and hydrocarbon potential in the Upper Paleozoic of the Atlantic Provinces. *Canadian Journal of Earth Sciences*, **7**, 119–163.

—— 1986. The Gulf of St. Lawrence Carboniferous Basin: the largest coalfield of eastern Canada. *Canadian Institute of Mining and Metallurgy Bulletin*, **79**, 67–78.

——, BARSS, M. S. & DONALDSON, J. R. 1960. Distribution and stratigraphic significance of small spore genera in Upper Carboniferous of the Maritime Provinces of Canada. *In*: *Fourth International Carboniferous Stratigraphy and Geology Congress, Heerlen.* Vol. 1, 237–245.

HAMBLIN, A. P. & RUST, B. R. 1989. Tectono-sedimentary analysis of alternate-polarity half-graben basin-fill sucessions: Late Devonian–Early Carboniferous Horton Group, Cape Breton Island, Nova Scotia. *Basin Research*, **2**, 239–255.

HAZELDINE, 1984. Carboniferous North Atlantic palaeogeography: stratigraphic evidence of rifting, not megashear or subduction. *Geological Magazine*, **121**, 443 463.

HESS, J. C. & LIPPOLT, H. J. 1986. $^{40}Ar/^{39}Ar$ ages of tonstein and tuff sanidines: new calibration points for the improvement of the Upper Carboniferous time scale. *Chemical Geology (Isotope Geoscience Section)*, **59**, 143–154.

HOWIE, R. D. 1984. Carboniferous evaporites in Atlantic Canada. *In*: GORDON, MACK, JR (ed.) *Neuvième Congrès International de Stratigraphie et de Géologie du Carbonifère, 1979, Compte Rendu.* Vol. 3, Southern Illinois University Press, Carbondale, 131–142.

—— 1988. *Upper Paleozoic Evaporites of Southeastern Canada.* Geological Survey of Canada, Ottawa, Bulletin 380.

JACKSON, C. T. & ALGER, F. 1829. A description of the mineralogy and geology of a part of Nova Scotia. *American Journal of Science and Arts*, **15**, 132–160.

JANSA, L. F. & MAMET, B. (1984) Offshore Viséan of Estern Canada: paleogeographic and plate tectonic implications. *In*: *Neuvième Congrès International de Stratigraphie et de Géologie du Carbonifère, 1979, Compte Rendu.* Vol. 3, Southern Illinois University Press, Carbondale, 205–214.

——, —— & ROUX, A. 1978. Viséan limestones from the Newfoundland shelf. *Canadian Journal of Earth Sciences*, **15**, 1422–1436.

JOHNSON, G. A. L. 1973. Closing of the European Sea in Western Europe. *In*: TARLING, D. H. & RUNCORN, S. K. (eds) *Implications of Continental Drift to the Earth Sciences.* Academic Press, London & New York, vol. 2, 843–850.

JONGMANS, W. J. 1952. Some problems of Carboniferous stratigraphy. *In*: *Troisième Congrès d'Avances Etudes de la Stratigraphie et Géologie du Carbonifère.* Vol. 1, 295–306.

KALKREUTH, W. D., NAYLOR, R. D., PRATT, K. & SMITH, W. D. 1990. Flourescence properties of alginite-rich oil shales from the Stellarton Basin, Canada. *Fuel*, **69**, 139–144.

KASPER, A. E., FORBES, W. H. & JAMIESON, R. A. 1988. Plant fossils from the Fisset Brook Formation of Cape Breton Island, Nova Scotia. *Maritime Sediments and Atlantic Geology*, **24**, 199.

LEEDER, M. R. 1987. Tectonic and palaeogeographic models for Lower Carboniferous Europe. *In European Dinantian Environments, In* : MILLER, J., ADAMS, A. E. & WRIGHT, V. P. (eds) John Wiley & Sons, Chichester, 1–20.

—— 1988a. Recent developments in Carboniferous geology: a critical review with implications for the British Isles and N.W. Europe. *Proceedings of the Geologists' Association*, **99**, 73–100.

—— 1988b. Devono-Carboniferous river systems and sediment dispersal from the orogenic belts and cratons of NW Europe. *In*: HARRIS, A. L. & FETTES, D. J. (eds) *The Caledonian–Appalachian Orogen*, Geological Society, London, Special Publications, **38**, 549–558.

—— 1992. Dinantian. *In*: DUFF, P. MCL. D. & SMITH, A. J. (eds) *Geology of England and Wales*. Geological Society, London, 207–238.

LOGAN, W. E. 1845. A section of the Nova Scotia coal measures as developed at Joggins on the Bay of Fundy, in descending order, from the neighbourhood of the west Ragged Reef to Minudie, reduced to vertical thickness. *In: Geological Survey of Canada Report of Progress for 1843.* Geological Survey of Canada, Ottawa, Appendix, 92–153.

LOTTES, A. L. & ZIEGLER, A. M. 1994. World peat occurrence and the seasonality of climate and vegetation. *Palaeogeography, Palaeoclimatology, Palaeoecology*, **106**, 23–37.

LYELL, C. 1843a. On the coal-formation of Nova Scotia, and on the age and relative position of the gypsum and accompanying marine limestones. *Proceedings of the Geological Society of London*, **4**, 184–186.

—— 1843b. On upright fossil trees in the coal strata of Cumberland, Nova Scotia. *Silliman's Journal*, **45**, 353.

—— 1843c. Coal formations, gypsum and marine limestones of Nova Scotia. *Silliman's Journal*, **45**, 356.

—— 1845. *Travels in North America; with Geological Observations on the United States, Canada, and Nova Scotia*. Murray, London, vol. 2.

—— 1871. *The Student's Elements of Geology*. Murray, London.

—— & DAWSON, J. W. 1853. On the remains of a reptile (*Dendrerpeton acadianum* Wyman and Owen), and of a land shell discovered in the interior of an erect fossil tree in the coal measures of Nova Scotia. *Quarterly Journal of the Geological Society of London*, **9**, 58–63.

LYNCH, G. & GILES, P. S. 1995. The Ainslie Detachment: a regional flat-lying extensional fault in the Carboniferous evaporitic Maritimes Basin of Nova Scotia, Canada. *Canadian Journal of Earth Sciences*, **33**, 169–181.

LYONS, P. C., SPEARS, D. A., OUTERBRIDGE, W. F., CONGDON, R. D. & EVANS, H. T., JR 1994. Euramerican tonsteins: overview, magmatic origin and depositional–tectonic implications. *Palaeo-*

geography, Palaeoclimatology, Palaeoecology, **106**, 23–47.

——, ZODROW, E. L., MILLAY, M. A., DOLBY, G., GILLIS, K. S. & CROSS, A. T. 1997. Coal-ball floras of Maritime Canada and palynology of the Foord seam: palaeobotanical and palaeoecological implications. *Review of Palaeobotany and Palynology*, **95**, 31–50.

MCCABE, P .J. & SCHENK, P. E. 1982. From sabkha to coal swamp: the Carboniferous sediments of Nova Scotia and southern New Brunswick. *11th International Congress on Sedimentology, Field Excursion Guidebook* 4A. International Association of Sedimentologists.

MCCUTCHEON, S. R. 1981. *Stratigraphy and Paleogeography of the Windsor Group in Southern New Brunswick.* New Brunswick Department of Natural Resources, Mineral Resources Division, Geological Surveys Branch, Frederichton, Open File Report 81-31.

—— & ROBINSON, P. T. 1987. Geological constraints on the genesis of the Maritimes Basin, Atlantic Canada. *In*: BEAUMONT, C. & TANKARD, A.J. (eds.), *Sedimentary Basins and Basin-Forming Mechanisms.* Canadian Society of Petroleum Geology, Calgary, Memoir 12.

MCGREGOR, D. C. 1977. *Spores from the McAdam Lake Formation, in Samples Submitted by W. Potter (Nova Scotia Dept. of Mines) from Two Boreholes in Cape Breton Island (NTS 11K/1)*. Geological Survey of Canada, Eastern Paleontology Section, report no. F1-12-1977-DCM.

MCKERROW, W. S. 1978. *The Ecology of Fossils*, Duckworth, London.

—— 1988. Wenlock to Givetian deformation in the British Isles and the Canadian Appalachians. *In*: HARRIS, A. L. & FETTES, D. J. (eds) *The Caledonian–Appalachian Orogen.* Geological Society, London, Special Publications, **38**, 437–448.

—— & ZIEGLER, A. M. 1972. Palaeozoic Oceans. *Nature Physical Science*, **240**, 92–94.

MCOUAT, W. 1874. *Report on a Portion of the Coalfield of Cumberland County, Nova Scotia.* Geological Survey of Canada, Ottawa. Report of progress for 1873-1874, 161–170.

MAMET, B. L. 1970. *Carbonate Microfacies of the Windsor Group (Carboniferous), Nova Scotia and New Brunswick.* Geological Survey of Canada, Ottawa, Paper 70-21.

MARCHIONI, D., KALKREUTH, W., UTTING, J. & FOWLER, M. 1994. Petrographical, palynological and geochemical analyses of the Hub and Harbour seams, Sydney Coalfield, Nova Scotia, Canada – implications for facies development. *In*: CALDER, J. H. & GIBLING, M. R. (eds) The Euramerican coal province: controls on tropical peat accumulation in the Paleozoic. *Palaeogeography, Palaeoclimatology, Palaeoecology*, **106**, 241–270.

MARTEL, A. T. & GIBLING, M. R. 1991. *Wave-Dominated Lacustrine Facies and Tectonically Controlled Cyclicity in the Lower Carboniferous Horton Bluff Formation, Nova Scotia, Canada.* International Association of Sedimentologists, Special Publication 13, Blackwell, Oxford, 223–243.

—— & —— 1996. Stratigraphy and tectonic history of the Upper Devonian to Lower Carboniferous Horton Bluff Formation, Nova Scotia. *Atlantic Geology*, **32**, 13–38.

——, KENNEDY, A. & GIBLING, M. R. 1997. Saline brines of the Sydney Basin: origin as evaporative Windsor residues? *Atlantic Geology*, **33**, 69–70.

——, MCGREGOR, D. C. & UTTING, J. 1993. Stratigraphic significance of Upper Devonian and Lower Carboniferous miospores from the type area of the Horton Group, Nova Scotia. *Canadian Journal of Earth Sciences*, **30**, 1091–1098.

MASSON, A. G. & RUST, B. R. 1984. Freshwater shark teeth as paleoenvironmental indicators in the Upper Pennsylvanian Morien Group of the Sydney Basin, Nova Scotia. *Canadian Journal of Earth Sciences*, **21**, 1151–1155.

MATTE, 1986. Tectonics and plate tectonic model for the Variscan Belt of Europe. *Tectonophysics*, **126**, 329–377.

MAYNARD, J. R., HOFMAN, W., DUNAY, R. E., BENTHAM, P. N., DEAN, K. P. & WATSON, I. 1997. The Carboniferous of western Europe: the development of a petroleum system. *Petroeum Geoscience*, **3**, 97–115.

MILLER, H. G., KILFOIL, G. J. & PEAVY, S. T. 1990. An integrated geophysical interpretation of the Carboniferous Bay St. George Subbasin. *Canadian Petroleum Geology Bulletin*, **38**, 320–331.

MILNER, A. 1987. The Westphalian tetrapod fauna; some aspects of its geography and ecology. *Journal of the Geological Society, London*, **144**, 495–506.

——, SMITHSON, T. R., MILNER, A. C., COATES, M. I. & ROLFE, W. D. I. 1986. The search for early tetrapods. *Modern Geology*, **10**, 1–28.

MOORE, R. G. 1967. Lithostratigraphic units in the upper part of the Windsor Group, Minas Sub-basin, Nova Scotia. *In*: NEALE, E. R. W. & WILLIAMS, H. (eds) *Collected Papers on Geology of the Atlantic Region*. Geological Association of Canada, Toronto, Special Paper 4, 245–266.

—— & RYAN, R. J. 1976. *Guide to the Invertebrate Fauna of the Windsor Group in Atlantic Canada*. Nova Scotia Department of Mines, Halifax, Paper 76-5.

MUKHOPADHYAY, P. K., HATCHER, P. & CALDER, J. H. 1991. Hydrocarbon generation of coal and coaly shale from fluvio-deltaic and deltaic environments of Nova Scotia and Texas. *Organic Geochemistry*, **17**, 765–783.

——, CALDER, J. H. & HATCHER, P. G. 1993. Geological and physicochemical constraints on methane and C_6 hydrocarbon generating capabilities and quality of Carboniferous coals, Cumberland Basin, Nova Scotia, Canada. *In*: *Proceedings of the Tenth Annual International Pittsburgh Coal Conference*, University of Pittsburgh.

MURPHY, J. B., STOKES, T. R., MEAGHER, C. & MOSHER, S. J. 1994. *Geology of the eastern St. Mary's Basin, Central Mainland Nova Scotia*. Geological Survey of Canada, Ottawa, Current Research 1994-D, 95–102.

NANCE, R. D. 1987. Dextral transpression and Late Carboniferous sedimentation in the Fundy coastal zone of southern New Brunswick. *In*: BEAUMONT, C. & TANKARD, A. J. (eds) *Sedimentary Basins and Basin-Forming Mechanisms*. Canadian Society of Petroleum Geology, Calgary, Memoir 12, 363–377.

NAYLOR, R. D., KALKREUTH, W., SMITH, W. D. & YEO, G. M. 1989. *Stratigraphy, sedimentology and depositional environments of the coal-bearing Stellarton Formation, Nova Scotia*. Geological Survey of Canada, Ottawa, Paper 89-8, 2–13.

NEVES, R. & BELT, E. S. 1970. Namurian and Viséan spores from Nova Scotia, Britain and Spain. *In*: *Sixième Congrès International du Stratigraphie et Géologie Carbonifère, Sheffield, 1967, Compte Rendu*. Vol. 3, Ernest van Aelst, Maastricht, 1233–1248.

NOWLAN, G. S. & NEUMAN, R. B. 1991. Paleontological contributions to Paleozoic paleogeographic and tectonic reconstructions. *In*: WILLIAMS, H. (ed.) *Geology of the Appalachian–Caledonian Orogen in Canada and Greenland*. Geology of Canada, Geological Survey of Canada, Ottawa, vol. 6, 817–842.

PARRISH, J. T. 1992. Climate of the Supercontinent Pangaea. *Journal of Geology*, **101**, 215–233.

PE-PIPER, G., CORMIER, R. F. & PIPER, D. J. W. 1989. The age and significance of Carboniferous plutons of the western Cobequid Highlands, Nova Scotia. *Canadian Journal of Earth Sciences*, **26**, 1297–1307.

——, PIPER, D. J. W. & CLERK, S. B. 1991. Persistent mafic igneous activity in an A-type granite pluton, Cobequid Highlands, Nova Scotia. *Canadian Journal of Earth Sciences*, **28**, 1058–1072.

PERLMUTTER, M. A. & MATTHEWS, M. D. 1989. Global cyclostratigraphy – a model. *In*: CROSS, T. A. (ed.) *Quantitative Dynamic Stratigraphy*. Prentice Hall, Englewood Cliffs, NJ, 233–260.

PHILLIPS, W. & PEPPERS, R. A. 1984. Changing patterns of Pennsylvanian coal-swamp vegetation and implications of climatic control on coal occurence. *International Journal of Coal Geology*, **3**, 205–255.

PHILLIPS, T. L., PEPPERS, R. A. & DIMICHELE, W. A. 1985. Stratigraphic and interregional changes in Pennsylvanian coal-swamp vegetation: environmental influences. *International Journal of Coal Geology*, **5**, 43–109.

PIPER, D. J. 1994. Late Devonian–earliest Carboniferous basin formation and relationship to plutonism, Cobequid Highlands, Nova Scotia. *In*: *Current Research 1994-D*. Geological Survey of Canada, Ottawa, 109–112.

PIPER, D. J. W., PE-PIPER, G. & PASS, D. J. 1996. The stratigraphy and geochemistry of late Devonian to early Carboniferous volcanic rocks of the northern Chignecto peninsula, Cobequid Highlands, Nova Scotia. *Atlantic Geology*, **32**, 39–52.

PLINT, A. G. & VAN DE POLL, H. W. 1984. Structural and sedimentary history of the Quaco Head area, southern New Brunswick. *Canadian Journal of Earth Sciences*, **21**, 753–761.

POOLE, W. H. 1967. Tectonic evolution of the Appalachian region of Canada. *In*: NEALE, E. R. W. & WILLIAMS, H. (eds) *Geology of the Atlantic*

Region. Geological Association of Canada Special Paper No. 4, Toronto, 9–51.

QUINLAN, G. M. & BEAUMONT, C. 1984. Appalachian thrusting, lithospheric flexure and the Paleozoic stratigraphy of the eastern interior of North America. *Canadian Journal of Earth Sciences*, **21**, 973–996.

RAMSBOTTOM, W. H. C. 1977. Major cycles of transgression and regression (mesothems) in the Namurian. *Proceedings of the Yorkshire Geological Society*, **41**, 261–291.

——, CALVER, M. A., EAGAR, R. M. C., HODSON, F., HOLLIDAY, D. W., STUBBLEFIELD, C. J. & WILSON, R. B. 1978. *A Correlation of Silesian Rocks in the British Isles*. Geological Society, Special Report. 10, London.

RAST, N. 1988. Tectonic implications of the timing of the Variscan orogeny. *In*: HARRIS, A. L. & FETTES, D. J. (eds) *The Caledonian–Appalachian Orogen*. Geological Society, London, Special Publications, **38**, 585–595.

REHILL, T. A. 1996. *Late Carboniferous nonmarine sequence stratigraphy and petroleum geology of the Central Maritimes Basin*. PhD thesis, Dalhousie University, Halifax.

ROBB, C. 1874. *Report on Exploration and Surveys in Cape Breton, Nova Scotia*. Geological Survey of Canada, Ottawa, Report of progress for 1873-1874, 171–191.

ROGERS, M. J. 1985. Non-marine pelecypods from the Riversdale Group (Upper Carboniferous), Nova Scotia. *In*: GORDON, MACK, JR (ed.) *Neuvième Congrès International de Stratigraphie et de Géologie du Carbonifère, 1979, Compte Rendu*. Vol. 5, Southern Illinois University Press, Carbondale, 261–270.

ROWLEY, D. B., RAYMOND, A., TOTMAN PARRISH, J., LOTTES, A L., SCOTESE, C. R. & ZIEGLER, A. M. 1985. Carboniferous paleogeographic, phytogeographic and paleoclimatic reconstructions. *International Journal of Coal Geology*, **5**, 7–42.

RYAN, R. J. & BOEHNER, R. C. 1994. *Geology of the Cumberland Basin, Cumberland, Colchester and Pictou Counties, Nova Scotia*. Nova Scotia Department of Natural Resources, Mines and Energy Branches Memoir 10, Halifax.

—— & ZENTILLI, M. 1993. Allocyclic and thermochronological constraints on the evolution of the Maritimes Basin of eastern Canada. *Atlantic Geology*, **29**, 187–197.

——, BOEHNER, R. C. & CALDER, J. H. 1991. Lithostratigraphic revision of the Upper Carboniferous to Lower Permian strata in the Cumberland Basin, Nova Scotia and the regional implications for the Maritimes Basin in Atlantic Canada. *Canadian Society of Petroleum Geologists Bulletin*, **39**, 289–314.

SANGSTER, D. R. & VAILLANCOURT, P. D. 1990. Geology of the Yava sandstone–lead deposit, Cape Breton Island, Nova Scotia, Canada. *In*: SANGSTER, A. L. (ed.) *Mineral Deposit Studies in Nova Scotia*. Geological Survey of Canada, Ottawa, Paper 90-8, vol. 1, 203–244.

——, SAVARD, M. M. & KONTAK, D. J. In press. An integrated model for mineralization of the lower Windsor carbonates of Nova Scotia. *In*: SANGSTER, D. R. & SAVARD, M. M. (eds) Zinc–lead mineralization and basinal brine movement, Lower Windsor Group (Viséan), Nova Scotia, Canada. *Economic Geology*, **3**.

SARJEANT, W. A. S. & MOSSMAN, D. J. 1978. Vertebrate footprints from the Carboniferous sediments of Nova Scotia: a historical review and description of newly discovered forms. *Palaeogeography, Palaeoclimatology, Palaeoecology*, **23**, 279–306.

SAUNDERS, W. B. & RAMSBOTTOM, W. H. C. 1986. The mid-Carboniferous eustatic event. *Geology*, **14**, 208–212.

SCHENK, P. E. 1967a. The significance of algal stromatolites to palaeoenvirinmental and chronostratigraphic interpretations of the Windsorian stage (Missipian), Maritime Provinces. *In*: NEALE, E. R. W. & WILLIAMS, H. (eds) *Geology of the Atlantic Region*, Geological Asociation of Canada Special Paper No. 4, Toronto, 229–243.

—— 1967b. The Macumber Formation of the Maritime Provinces, Canada – a Mississippian analogue to Recent strandline carbonates of the Persian Gulf. *Journal of Sedimentary Petrology*, **37**, 65–376.

—— 1969. Carbonate–sulphate–redbed facies and cyclic sedimentation of the Windsorian stage (Middle Carboniferous), Maritime Provinces. *Canadian Journal of Earth Sciences*, **6**, 107–1066.

—— 1981. The Meguma Zone of Nova Scotia – a remnant of western Europe, South America or Africa? *In*: KERR, J. W. & FERGUSSON, A. J. (eds) *Geology of the Atlantic Borderlands*. Canadian Society of Petroleum Geologists Memoir 7, Calgary, 119–148.

——, MATSUMOTO, R. & VON BITTER, P. H. 1994. Loch Macumber (early Carboniferous) of Atlantic Canada. *Journal of Paleolimnology*, **11**, 151–172.

SCHRAM, F. R. 1981. Late Paleozoic Crustacean communities. *Journal of Paleontology*, **55**, 126–137.

SCHULTZE, H. -P. 1985. Reproduction and spawning sites of *Rhabdoderma* (Pisces, Osteichthyes, Actinistia) in Pennsylvanian deposits of Illinois, USA. *In*: GORDON, MACK, JR (ed.) *Neuvième Congrès International de Stratigraphie et de Géologie du Carbonifère, 1979, Compte Rendu*. Vol. 5, Southern Illinois University Press, Carbondale, 326–330.

SCOTESE, C. R. & MCKERROW, W. S. 1990. Revised world maps and introduction. *In*: MCKERROW, W. S. & SCOTESE, C. R. (eds) *Palaeozoic Palaeogeography and Biogeography*. Geological Society, Memoir 12, London, 1-21.

SCOTT, A. C. 1998. The legacy of Charles Lyell: advances in our knowledge of coal and coal-bearing strata. *This volume*.

—— & CALDER, J. H. 1994. Carboniferous fossil forests. *Geology Today*, **10**, 213–217.

SHORT, G. 1986. *Surface Petroleum Shows Onshore Nova Scotia*. Nova Scotia Department of Mines and Energy Information Series 11, Halifax.

SMITH, A. H. V. 1962. The palaeoecology of Carboniferous peats based on the miospores and petrography of bituminous coals. *Proceedings of the Yorkshire Geological Society*, **33**, 423–474.

SMITH, W. D. & NAYLOR, R. D. 1990. *Oil Shale Resources of Nova Scotia*. Nova Scotia Department of Mines and Energy, Economic Geology Series 90-93, Halifax.

——, ST PETER, C. J., NAYLOR, R. D., MUKHOPADHYAY, P. K., KALKREUTH, W. D., BALL, F. D. & MACAULEY, G. 1991. Composition and depositional environment of major eastern Canadian oil shales. *International Journal of Coal Geology*, **19**, 385–438.

STEMMERIK, L., VIGRAN, J. O. & PASECKI, S. 1991. Dating of Late Paleozoic rifting events in the North Atlantic: new biostrtigraphic data for the uppermost Devonian and Carboniferous of East Greenland. *Geology*, **19**, 218–221.

TANDON, S. K. & GIBLING, M. R. 1994. Calcrete and coal in late Carboniferous cyclothems of Nova Scotia, Canada: climate and sea-level changes linked. *Geology*, **22**, 755–758.

TASCH, P. 1969. Branchiopoda. *In*: MOORE, R. C. (ed.) *Treatise on Invertebrate Paleontology, Part R, Arthropoda 4*, vol. 1, R128–191.

TATE, M. P. & DOBSON, M. R. 1989. Pre-Mesozoic geology of the western and north-western Irish continental shelf. *Journal of the Geological Society of London*, **146**, 229–240.

TIBERT, N. E. 1996. *A Paleoecological Interpretation for the Ostracodes and Agglutinated Foraminifera from the Earliest Carboniferous Marginal Marine Horton Bluff Formation, Blue Beach Member, Nova Scotia, Canada*. MSc thesis, Dalhousie University, Halifax.

UTTING, J. 1987. *Palynology of the Lower Carboniferous Windsor Group and Windsor-Canso Boundary Beds of Nova Scotia, and Their Equivalants in Quebec, New Brunswick and Newfoundland*. Geological Survey of Canada, Ottawa, Bulletin 374.

—— & HAMBLIN, A. P. 1991. Thermal maturity of the Lower Carboniferous Horton Group, Nova Scotia. *International Journal of Coal Geology*, **19**, 439–456.

——, KEPPIE, J. D. & GILES, P. S. 1989. Palynology and age of the Lower Carboniferous Horton Group, Nova Scotia. *Contributions to Canadian Palaeontology, Geological Survey of Canada Bulletin*, **396**, 117–143.

VAN DE POLL, H. W. 1978. Paleoclimate control and stratigraphic limits of syn sedimentary mineral occurrences in Mississippian–Early Pennsylvanian strata of eastern Canada. *Economic Geology*, **73**, 1069–1081.

——, GIBLING, M. R. & HYDE, R. S. 1995. Introduction: Upper Paleozoic rocks. *In*: WILLIAMS, H. (ed.) *Geology of the Appalachian-Caledonian Orogen in Canada and Greenland*. Geology of Canada, Geological Survey of Canada, vol. 6, Ottawa, 449–455.

VAN DER ZWAN, C. J. 1981. Palynology, phytogeography and climate of Lower Carboniferous. *Palaeogeography, Palaeoclimatology, Palaeoecology*, **33**, 279–310.

VASEY, G. M. 1984. *Westphalian macrofaunas in Nova Scotia: palaeoecology and correlation*. PhD thesis, University of Strathclyde, Glasgow.

VEEVERS, J. J. & POWELL, C. McA. 1987. Late Paleozoic glacial episodes in Gondwanaland reflected in transgressive–regressive depositional sequences in Euramerica. *Geological Society of America Bulletin*, **98**, 475–487.

VON BITTER, P. 1976. Paleoecology and distribution of Windsor Group (Visean–?early Namurian) conodonts, Port Hood Island, Nova Scotia, Canada. *In*: BARNES, C. R. (ed.) *Conodont Paleoecology*. Geological Association of Canada, Special Paper 15, Waterloo, 225–241.

VON BITTER, P. H. & AUSTIN, R. L. 1984. The Dinantian *Taphrognathus transatlanticus* conodont range zone of Great Britain and Atlantic Canada. *Palaeontology*, **27**, 95–111.

WAGNER, R. H. 1984. Megafloral zones of the Carboniferous. *In*: GORDON. MACK, JR (ed.) *Neuvième Congrès International de Stratigraphie et de Géologie du Carbonifère*. Vol. 2, Southern Illinios University Press, Carbondale, 109–134.

—— 1989. A Late Stephanian forest swamp with *Sporangiostrobus* fossilized by volcanic ash fall in the Puertollano Basin, central Spain. *International Journal of Coal Geology*, **12**, 523–552.

—— & LYONS, P. C. 1997. A critical analysis of the higher Pennsylvanian megaflora of the Appalachian region. *Review of Palaeobotany and Palynology*, **95**, 255–283.

WALDRON, J. W. F. 1996. *Differential Subsidence and Tectonic Control of Sedimentation in the Stellarton Basin, Pictou Coalfield, Nova Scotia*. Geological Survey of Canada, Ottawa, Current Research 1996-E, 261–268.

——, PIPER, D. J. W. & PE-PIPER, G. 1989. Deformation of the Cape Chignecto Pluton, Cobequid Highlands, Nova Scotia: thrusting at the Meguma-Avalon boundary. *Atlantic Geology*, **25**, 51–62.

WHITE, I. C. 1891. *Stratigraphy of the Bituminous Coal Field of Pennsylvania, Ohio and West Virginia*. United States Geological Survey Bulletin No. 65.

WIGHTMAN, W. G., SCOTT, D. B., MEDIOLI, F. S. & GIBLING, M. R. 1993. Carboniferous marsh oraminifera from coal-bearing strata at the Sydney Basin, Nova Scotia: a new tool for identifying paralic coal-forming environments. *Geology*, **21**, 631–634.

WILLIAMS, E. P. 1974. Geology and petroleum possibilities in and around Gulf of St. Lawrence. *American Association of Petroleum Geologists Bulletin*, **58**, 117–155.

WILLIAMS, H. 1984. Miogeosynclines and suspect terranes of the Caledonian-Appalachian orogen: tectonic patterns in the North Atlantic region. *Canadian Journal of Earth Sciences*, **21**, 887–901.

WILLIAMS, G. L., FYFFE, L. R., WARDLE, R. J., COLMAN-SADD, S. P. & BOEHNER, R. C. 1985. *Lexicon of Canadian Stratigraphy, Volume VI, Atlantic Region*. Canadian Society of Petroleum Geologists, Calgary.

YEO, G. M. & RUIXIANG GAO, 1987. Stellarton Graben: an Upper Carboniferous pull-apart basin in Northern Nova Scotia. *In*: BEAUMONT, C. & TANKARD, A. J. (eds) *Sedimentary Basins and Basin-forming*

Mechanisms. Canadian Society of Petroleum Geology, Calgary, Memoir 12, 299–309.

ZODROW, E. L. & CLEAL, C. J. 1985. Phyto- and chronostratigraphical correlations between the late Pennsylvanian Morien Group (Sydney, Nova Scotia) and the Silesian Pennant Measures (South Wales). *Canadian Journal of Earth Sciences*, **22**, 1465–1473.

—— & GAO ZHIFENG 1991. *Leeites oblongifolios* nov. gen. et sp., (Sphenophyllaean, Carboniferous),

Sydney Coalfield, Nova Scotia, Canada. *Palaeontographica Abst. B*, **223**, 61–80.

—— & MCCANDLISH, K. 1980. *Upper Carboniferous Fossil flora of Nova Scotia in the Collections of the Nova Scotia Museum with Special Reference to the Sydney Coalfield*. Nova Scotia Museum, Halifax.

—— & VASEY, G. M. 1986. Mabou Mines section: biostratigraphy and correlation (Pennsylvanian Pictou Group, Nova Scotia, Canada). *Journal of Paleontology*, **60**, 208–232.

Appendix A. The Carboniferous fauna of Nova Scotia

Systematic taxonomy for many of these groups in the fossil record is problematic and many variations exist on the structure of the compilation presented here. Stratigraphic occurrence: Horton Group [H]; Windsor group [W]; Mabou Group [M], with formations: Point Edward [Mpe], Grand Etang [Mge], West Bay [Mwb]; Cumberland Group, *sensu* Ryan *et al.* (1991) [C], with formations: Joggins [Cj], Parrsboro [Cp], Port Hood [Cph], Springhill Mines [Csp], Stellarton [Cs], Mabou Mines [Cmm], Sydney Mines [Csm]; Pictou Group red beds [P], with formation: Cape John [Pcj]. Primary sources: Baird (1978); Bell (1929, 1944, 1960); Carroll *et al.* (1972); Copeland (1957); Dawson (1855 and others); Gardiner (1966); J. Kukulova-Peck (pers. comm. 1994); Mamet (1970); Masson & Rust (1984); A. Milner (pers. comm. 1994, 1995); R.G. Moore (unpublished data, pers. comm. 1997); Moore & Ryan (1976); A. C. Scott (pers. comm. 1996); Tibert (1996); Von Bitter (1976); Wightman *et al.* (1993); Zodrow & Hitchcock (unpublished, 1995); and Calder (this study).

Phylum Protista (single-celled organisms)
Class Sarcodina
  Order Foraminifera
  Suborder Textularidae (agglutinated)
    Trochammina sp. [H, Cc, Csm]
    Ammobaculites sp. [H, Cc, Csm]
    Ammodiscus sp. [H,W]
    Ammotium sp. [H, Cc, Csm]
  Suborder Miliolina
    Cornuspira sp. [W]
  Suborder Textulariina
    Ammovertella sp. [W]
    Pseudoammodiscus volgensis [W]
    Trepeilopsis sp. [W]
  Suborder Fusulinina
    Archaediscus sp. [W]
    A. krestovnikovi [W]
    A. koktjubensis [W]
    A. infantis [W]
    A. aff. *A. moelleri* [W]
    A. aff. *A. chernoussovensis* [W]
    Archaesphaera (Vicinesphaera) sp. [W]
    Asteroarchaediscus baschkiricus [W]
    Biseriammina? windsorensis [W]
    Brunsia sp. [W]
    Climacammina aff. *C. patula* [W]
    C. aff. *C. prisca* [W]
    Cribostomum sp. [W]
    Earlandia aff. *E. clavatula* [W]
    E. aff. *E. elegans* [W]
    E. aff. *E. vulgaris* [W]
    Endothyra bowmani [W]
    E. obsoleta [W]
    E. excentralis [W]
    E. aff. *E. prisca* [W]
    E. af. E. similis [W]
    Endothyranella sp. [W]
    Endothyranopsis compressa [W]
    E. crassa [W]
    E. sphaerica [W]
    Eoendothyranopsis aff. *E. pressa – E. rara* [W]
    E. aff. *E. ermakiensis* [W]
    Eostaffela? discoidea [W]
    E. radiata [W]
    Globoendothyra globulus [W]
    ?Haplophragmella sp. [W]
    ?Haplophragmina sp. [W]
    Irregularina sp. [W]
    Mikhailovella sp. [W]
    Neoarchaediscus grandis [W]
    Neoarchaediscus sp. [W]
    N. incertus [W]

    N. parvus regularis [W]
    Palaeotextularia consobrina [W]
    P. aff. *P. longiseptata* [W]
    P. asper [W]
    P. dagmarae [W]
    Parathurammina sp. [W]
    Pseudoendothyra aff. *P. ornata* [W]
    Pseudoglomospira spp. [W]
    P. infinitesima [W]
    ?Saccaminopsis sp. [W]
    Tetrataxis aff. *T. angusta* [W]
    T. aff. *T. conica* [W]
    T. aff. *T. eominima* [W]
    T. aff. *T. maxima* [W]
    Tuberitina sp. [W]
  Incertae sedis
    Calcisphaera sp. [W]
    C. laevis [W]
    C. pachysphaerica [W]
    Diplosphaerina sp. [W]
    Koskinobigerina sp. [W]
    Koskinotextularia sp. [W]
    Planospirodiscus gregorii [W]
    P. minimus [W]
    Radiosphaera sp. [W]
    Zellerina spp. [W]
    Zellerina discoidea [W]
Class Arcellinida (thecamoebians)
  Order Arcellinida
    cf.Centropxsis sp. [H, Csm]
    ?Difflugia sp. [H]
    cf. Nebela sp. [Csm]

Phylum Conodonta
    Apatognathus? spp. [W]
    Bispathodus spp. [W]
    Cavusgnathus windsorensis [W]
    C. aff. *C. regularis – C. unicornis* [W]
    Ellisonia spp. [W]
    Gnathodus bilineatus [W]
    G. girtyi intermedius [W]
    G. scotiaensis [W]
    Hindeodus cristulus [W]
    H. parva [W]
    Kladognathus tenuis [W]
    Mestognathus spp. [W]
    Spathognathus campbelli [W]
    S. cristatus [W]
    S. scitulus [W]
    S. sp. [W]
    Taphrognathus sp. [W]
    T. transatlanticus [W]

Vogelgnathus campbelli [W]
V. pesaquidi [W]
V. dhindsai [W]
V. postcampbelli [W]

Phylum Annelida (segmented worms)
Class Polychaeta
    Serpula annulata [W]
    S. caperatus [W]
    S. hartii [W]
    Spirorbis avonensis [H]
    S. carbonarius [C]
    ?S. arietina [C]
Scolecodonts
    Anisocerasites spp. [W]
    Arabellites spp. [W]
    Diopatraites spp. [W]
    Eunicites spp. [W]
    Lumbriconcereites spp. [W]
    Staurocephalites spp. [W]
    Stauronereisites spp. [W]
    Ungulites spp. [W]

Phylum Porifera
    Belemnospongia sp. [W]

Phylum Cnidaria
Class Anthozoa (corals)
    Chaetetes spp. [W]
  Order Rugosa
    Bothrophyllum sp. [W]
    Caninia juddi [W]
    Corwenia sp. [W]
    Clisiophyllum (Dibunophyllum) billingsi [W]
    Dibunophyllum lambii [W]
    Diphyphyllum sp. [W]
    Enniskillenia enniskilleni
      (=*Zaphrenitis minas*) [W]
    Lonsdaleia floriformis [W]
    L. pictoense [W]
    Koninckophyllum (Lophophylum) avonensis [W]
    K. (Lophophyllum) interruptum [W]
    K. cf. *O⊖*[W]
    Lithostrotion pauciradiale [W]
    L. proliferum [W]
    L. aff. *scoticum* [W]
    Thysanophyllum cf. *orientale* [W]
  Order Tabulata
    Cladochonus sp. [W]
    Pseudoromingeria sp. [W]
    Syringopora sp. [W]

?Phylum Conulariida
    Paraconularia (Conularia) planicostata [W]
    P. sorrocula [W]

Phylum Bryozoa
Class Gymnolaemata
  Incertae sedis
    Paleocrisidia (Nodosinella) priscilla [W]
  Order Cryptostomata
    Rhombopora sp. [W]
    Streblotrypa biformata [W]
  Order Fenestrata

Fenestrellina (Fenestella) yelli [W]
Polypora schucherti [W]
Septopora primitiva [W]
Order Trepostomata
    Anisotrypa sp. [W]
    Tabulipora acadica [W]
    Stenopora sp. [W]
Order Halloprina
    Batostomaella abrupta [W]
    B. exilis [W]
Order Ctenostomata
    Eliasopora spp. [W]

Phylum Echinodermata
Class Echinoidea (sea urchins)
    Cravenechinus spp. [W]
    Proterocidaris spp. [W]
    ?Archaeocidaris spp. [W]
Class Crinoidea (crinoids)
    indet. columnals, plates [W]
Class Ophiuroidea (brittle stars)
    2 genera, 3 species [W]

Phylum Brachiopoda
Class Inarticulata
  Order Acrotretida
    Orbiculoidea limata [W]
    Crania cincta [W]
    C. brookfieldensis [W]
Class Articulata
  Order Terebratulida
    Beecheria davidsoni [W]
    B. v. latum [W]
    B. v. mesaplanum [W]
    B. v. milviformis [W]
    B. (Dielasma) sp. [W]
    Cranaena tumida [W]
    Hartella dielasmoidea [W]
    H. parva [W]
    Romingerina anna [W]
    Tornquistia polita (=*Chonetes politus*) [W]
  Order Strophomenida
    Avonia spinocardinata [W]
    Buxtonia cogmagunensis [W]
    Diaphragmus (Productus) avonensis [W]
    D. tenuicostiformis [W]
    Dictyoclostus (Productus) subfasciculatus [W]
    Echinoconchus exigunus
      (=*Pustula exigua*) [W]
    Ovatia (Productus) dawsoni [W]
    O. (Productus) lyelli [W]
    O. (Productus) semicubicula [W]
    Protoniella baddeckensis [W]
    P. beedii [W]
    Rugosochonetes aff. *hindi* [W]
    Schellwienella kennetcookensis [W]
    Schuchertella pictoense [W]
    Semiplanus aff. *latissimus* [W]
    Spinulicosta (Productella) baddeckensis [W]
  Order Rhynchonellida
    Allorhyncus hartti [W]
    A. macra [W]
    A. ramosum [W]
    Camarotoechia acadiansis [W]

C. atlantica [W]
Pugnax dawsonianus [W]
P. dawsonianus v. magdalena [W]
Pugnoides sp. [W]
Streptorhyncus cf. *minutum* [W]
Order Spiriferida
  Ambocoelia acadica [W]
  Composita dawsoni [W]
  C. obligata [W]
  C. strigata [W]
  C. windsorensis [W]
  Gigantoproductus giganteus [W]
  Martinia galatea [W]
  M. thetis [W]
  Punctospirifer (Spiriferina) octoplicata [W]
  P. (Spiriferina) verneuli [W]
  Spirifer adonis [W]
  S. nox [W]

Phylum Mollusca
Class Pelecypoda (bivalves: clams)
  ?*Carbonicola angulata* [Mpe, wb]
  ??*Carbonicola bradorica* [W, Mh, pq]
  Carbonita sp. [Csm]
Order Paleoconcha
  Edmondia hartti [W]
  E. rudis [W]
  Sanguinolites niobe [W]
  S. parvus [W]
  S. striatogranulatus [W]
Order Taxodonta
  Grammatodon (Parallelidon) dawsoni [W]
  G. (Parallelidon) hartingi [W]
Order Schizodonta
  Schizodus chevierensis [W]
  S. depressus [W]
Order Dysodonta (Mytilacea)
  Anthraconauta phillipsii [Csm]
  Aviculopecten lyelli [W]
  A. lyelliformis [W]
  A. subquadratus [W]
  Bakewellia shubenacadiensis [W]
  Curvirimula ?ovalis [Cp]
  C. sp [C]
  Leptodesma acadica [W]
  L. borealis [W]
  L. dawsoni [W]
  Lithophaga (Lithophagus) poolii [W]
  Modiolus dawsoni [W]
  M. hartti [W]
  Naiadites carbonarius [Cp]
  N. longus [C]
  Pteronites gayensis [W]
  Spathella insecta [W]
  Streblopteria (Pseudamusium) debertianum [W]
  S. (Pseudamusium) simplex [W]
Order Heterodonta
  Cypricardella acadica [W]
  Scaldia fletcheri [W]
  S. fundiensis [W]
Class Gastropoda (snails)
Subclass Amphigastropoda
Order Bellerophontida
  Bellerophon sp. [W]

Bucanopsis beedii [W]
Subclass Prosobranchia
Order Archaeogastropoda
  Aclisi(n?)a acutula [W]
  Anematina (Holopea) cf. *proutana* [W]
  Cyclonema ?subangulatu [W]
  Euphemites cf. *urei* [W]
  Euomaphalus exortivus [W]
  Mourlonia sp. [W]
  Murchisonia gypsea [W]
  Naticopsis hartii [W]
  N. howi [W]
  Platyschisma ?dubium [W]
  Pseudophorus (Flemingia) dispersa [W]
  P. minuta [W]
  Stegocoelia abrupta [W]
  S. compactoidea [W]
  Straparollus minutus [W]
  Worthenia longi [W]
Order 'Caenogastropoda'
  Bulimorpha maxneri [W]
  Pseudozygopleura (Zygopleura) ?cara [W]
Subclass Pulmonata (Stylommatophora) (land snails)
Order Orthurethra
  Dendropupa vetusta [Cj]
  Pupa bigsbii [Cj]
Incertae sedis
  Zonites priscus [Cj]
Class Cephalopoda
Subclass Nautiloidea
Order Nautilida
  Diodoceras avonensis [W]
  Stroboceras hartti [W]
Order Michelinoceroidea/Orthocerida
  Campyloceras cf. *C. unguis* [W]
  Hemidolorthoceras belli [W]
  H. windsorensis [W]
  Kionoceras spp. [W]
  Michelinoceras (Orthoceras) vindobonense [W]
  Mitorthoceras sp. [W]
  Mooreoceras aff. *M. hindei* [W]
  Pseudorthoceras knoxense [W]
Order Oncocerida
  Poterioceras sp. [W]

Phylum Arthropoda (jointed-leg invertebrates)
SuperClass Trilobitomorpha
Class Trilobita
Order Opisthoparia/Polymerida
  Paladin (Phillipsia) eichwaldi [W]
Superclass Crustacea
Class Ostracoda
Order Palaeocopida
  Amphisites sp. aff. *A. centronotus* [W]
  Beryichiopsis cornuta [W]
  B. ophota [W, M]
  Copelandella (Hollinella) novascotica [H,W]
  Glyptopleura parvacostata [W]
  G. elephanta [W]
  'Gortanella' sp. [W]
  Kirkbya novascotica [W]
  Paraparchites gibbus [W]
  P. inornatus [W]
  P. okeni [W]

P. scotoburdigalensis [W]
Pseudoparaparchites ensigner [W]
Sansabella carbonaria [C]
S. reversa [C]
Youngiella sp. [H]
Order Podocopida
   Acrat(a)ia acuta [W]
   Bairdia brevis [W]
   B. pruniseminata [H]
   Bairdiacypris quartziana [W]
   B. striatiformis [H]
   Bythocypris aequalis [W]
   Carbonita agnes [C]
   C. altilis [C]
   C. elongata [C]
   C. fabulina [M, C, c, s, m, sm, P?]
   C. inflata [Cm, sm, Ppei]
   C. pungens [C]
   Carbonita rankiniana
     (=*Candona salteriana*) [H,C]
   C. scalpellus (=*C.* cf. *subula*) [H, Csm, P]
   C. secans [C]
   Chamisella suborbiculata [W]
   Chamishaella sp. [H]
   Gutschickia ninevehensis [C]
   G. bretonensis [C]
   Healdianella sp. [W]
   Hilboldtina evelinae [C]
   H rugulosa [C]
   Monoceratina youngiana [W]
   ?Paraparchites okeni [M]
   P. sp. aff. *P. kelletae* [W]
   Paraparchites sp. [H]
   Shemonaella (Paraparchites) scotoburdigalensis
     (=*Limnoprimitia hortenensis*) [H,W]
   S. tatei [H]
   Sishaella moreyi [W]
   Sulcella levisulcata [W]
Order Platycopida
   Cavellina ?lovatica [H]
   Geisina sp. [H]
Order Myodocopida
   ?Cypridina acadica [W]
   Polycope spinula [W]
Class Branchiopoda
Order Conchostraca (clam shrimps)
   Paraleaia leidyi (=*Leaia leidyi*) [M]
   Leaia baentschiana [M, ?C]
   L. tricarinata [M,C]
   L. silurica [M,Cl,p]
   L. (Eoleaia) laevicostata [H]
   L. (Eoloeaia) leaiaformis [H]
   L. laevis [C-P]
   L. acutangularis [M]
   L. acutilirata [M]
   L. magnacostata [Cp]
   L. elongata [C]
   L. sp. [H]
   L. sp. [P]
   Monoliolophus (conemaughensis)
     unicostatus [C]
   Asmussia alta [H, Cl, s]
   A. tenalla [C, P]
   Cyzicus (Euestheria) belli [H]

   C. (Euestheria) dawsoni [M]
   C. (Euestheria) lirella [H]
   C. (Euestheria) raymondi [?M]
   ?C. (Lioestheria) simoni [?M]
   C. (Lioestheria) striata [M,Cs]
   Cycletherioides blackstonensis [M]
   Palaeolimnadiopsis pruvoti [?C]
?Order Notostraca
   Lynceites cansoensis [M]
Class Malacostraca (soft-shelled: crayfish etc.)
Superorder Eocarida
Order Pygocephalomorpha
   Pygocephalus (Anthrapalaemon) dubius (hillianus)
     (=*Diplostylus dawsoni*) [Cj, p, spm, ph, mm]
   P. cooperi [?Mwb, Cp]
   Pseudotealliocaris caudafimbriata [Mm]
   P. belli (=*Tealliocaris barathrota*) [Mm, wb]
Order Eocaridacea
   Anthracophausia sp. cf. *dunsiana* [M]
?Superorder Syncarida
   Paleocaris cf. *typus* [Cs]
   Incertae sedis
   Dithyrocaris glabradoides [M]
Superclass Chelicerata (pincer-bearing)
Class Merostomata
Order Xiphosurida (sword-tailed: horseshoe crabs etc.)
   Belinurus reginae [Mwb, Cp, rv]
   B. grandaevus [Mwb, C]
   Euproops cf. *danae* [Mmar, ?C]
   E. amiae [C?s,sm]
   E. sp. [H]
Subclass Eurypterida (wing-like legs)
   Eurypterus brsadorensis [Csm]
   indet. cuticle, cf. *Hibbertopterus /Mycterops* [Cj]
   Dunsopterus sp. [Mge]
Class Arachnida (spiders, scorpions)
Order Anthracomartida
   Coryphomartus triangularis [?Cj]
Order Phrynichida (whip spiders)
   Graeophonus carbonarius
     (=*Libellula carbonaria*) [Cj, c, sm]
Order Scorpionida
   Eoscorpius sp. [?Cs]
   indet. cuticle [Cj]
Superclass Myriapoda
Class Diplopoda (millipedes)
Order Eurysterna/Spirobolida
   Xyloiulus (Xylobius) sigillariae [Cj]
   Incertae sedis
   Archiulus xylobioides [Cj]
Order Amynilyspedida
   Amynilyspes springhillensis [Cj, sp]
Incertae sedis
   Arthropleurida[1]
   Arthropleura sp. [Cj, Pcj]
Superclass Hexapoda
Class Insecta
Subclass Pterygota (winged insects)
Infraclass Palaeoptera
Order Megasecoptera
   Megasecoptera incertae familiae [C]
   Brodioptera cumberlandensis [C]
   B. amiae [?M]
Order Protodonata

Meganeura sp. [?Csm]
Order Palaeodictyoptera
   Palaeodictyoptera incertae sedis [?Csm]
Infraclass Neoptera
  Order Blattaria
    Archimylacris ?acadica [Cs]
    Archimylacris moriensis [Csm]
    A. sp. [Cs]
    Hemimylacris sp. [Cs]
    'Blattoidea' carri [Csm]
    'Blattoidea' schchertiana [Csm]
    ?Phylloblatta sp. [Cs]
  ?Order Mixotermitoidea
    Geroneura wilsoni [Cl]

Phylum Chordata
Subphylum Vertebrata
Superclass Pisces
Class Acanthodii
  Incertae sedis
  Gyracanthidae
    Gyracanthus duplicatus [H, M, Cj, p, ph]
    G. magnificus [M]
Class Chondrichtyes (cartilaginous fishes)
  (sharks)
  Order Symmoriida
    Stethacanthus sp. [H]
  Order Xenacanthida
    Xenacanthus acinaces [Cs]
    X. penetrans [Cs]
    X. sp. [Cj, s, sm]
    Oracanthus sp. [W]
    Orthacanthus sp. [Csm]
  Order Ctenacanthiformes
    Ctenacanthus sp. [?H, M, C]
  Order Petalodontiformes (?ray-like fishes)
    Ctenoptychius cristatus [Cj]
  Incertae sedis
    Ageleodus (Calloppristodus) pectinatus [Cj]
Class Osteichthyes (bony fishes)
  Subclass Actinoptcrygii (ray-finned fishes)
  Order Palaeonisciformes
    'Acrolepis hortonensis' [H]
    Amblypterus sp. [Csm]
    Canobius modulus [H]
    Elonichthys sp. [H]
    ?Gyrolepis sp.[Csm]
    Haplolepis cf. *corrugata* [C]
    H. (Parahaplolepis) canadensis [Cp]
    Palaeoniscus sp. [Csm]
    indet. palaeoniscid [W, Csm]
  Subclass Sarcoptcrygii (lobe fins)
    [Order Actinistia]
  Order Crossopterygii
    Megalichthys hibberti
      (=*Psammodus bretonensis*) [?Cph]
    Megalichthys sp. [Mpe, Cj, p, sm]
    Rhadinichthys sp. [H, Cj]
    Rhizodopsis/ Strepsodus sp. [H, Mpe, ge, Cj, sm]
    Rhizodopsis (Strepsodus) dawsoni [Cj]

Rhizodus hardingi [H, Cs]
    R. lancifer [H, Cs]
  Suborder Coelacanthini
    Coelacanthus sp. [?Cm]
    ?Rhabdoderma sp. [Csm]
    (indet. scales) [Cj]
  Superfamily Dipnoi (lungfish)
    Ctenodus sp. [?H]
    Ctenodus cristatus [Mge]
    C. murchisoni [Csm]
    Monongahela stenodonta [Csm]
    Sagenodus cristatus [Cj]
    S. sp [Mpe, ge, Csm]
    S. plicatus nomen vanum [Cj]
Superclass Tetrapoda
  Incertae sedis, nomen vanum
    Novascoticus (Hylonomus) multidens [Cj]
    Hylonomus ociedentatus
      (= *Smilerpeton aciedentatum*) [Cj]
    Hylonomus wymani [Cj]
Class Amphibia
  Subclass Batrachomorpha (true amphibians)
  Order Temnospondyli
    Temnospondyli *incertae sedis* [Cp]
    Dendrerpeton acadianum
      (=*D. oweni, Platystegos loricatum,*
      Dendryazousa dikella) [Cj]
    Dendrepeton helogenes
      (=*Dendrysekos helogenes*) [Cj]
    Denderpeton sp. nov. [Cj]
    Spathicephalus pereger [Mpe]
    Cochleosauridae *incertae sedis* [Cj, Csm]
  Order Microsauria (Lysoropha)
    Archerpeton anthracos [Cj]
    Asaphestera (Hylerpeton) intermedium [Cj]
    Hylerpeton dawsoni
      (= *Amblydon problematicum)* [Cj]
    Leiocephalikon problematicum
      (?=*Trachystegos megalodon*) [Cj]
    ?Ricnodon sp. [Cj]
  Subclass Reptiliomorpha
  Order Anthracosauria (Embolomeri)
    Baphetes planiceps [Cs]
    'B'. minor [Cj]
    Calligenethlon watsoni [Cj]
    Carbonoherpeton sp. [Csm]
    'Pholiderpeton' bretonense [Mpe]
    Seymouriamorpha *incertae sedis* [H]
    (*'Eosaurus acadiensis'*= exotic ichthyosaur?)
Class Reptilia
  Subclass Anapsida
  Order Captorhinomorpha (protorothyrids and
  captorhinids)
    Romeriscus periallus [Cph]
    Hylonomus lyelli
      (=*Fritschia curtidentata*) [Cj]
    Paleothyris acadiana [Csm]
  Subclass Synapsida
  Order Pelycosauria
    Protoclepsydrops haplous [Cj]

Appendix B. Carboniferous fossil flora of Nova Scotia

Principal sources: Bell (1929, 1940, 1944), Calder *et al.* (1996), W. G. Chaloner (pers. comm. 1995), Lyons *et al.* (1997), A. C. Scott (pers. comm. 1994–1996), Zodrow & Gao (1991), Zodrow & Cleal (1985), Zodrow & McCandlish (1980); and Calder (this study). Stratigraphic occurrence: H = Horton Group; W = Windsor Group; M = Mabou Group; C = Cumberland Group *sensu stricto*; CR = Riversdale Series of Cumberland Group; CS = Stellarton Series of Cumberland Group; CM = Morien Series of Pictou Group. 'r' denotes rare occurrence.

Kingdom Plantae

Division Bryophyta
Class Hepaticaea (liverworts)
  Marchenites sp. [CM]

Division Tracheophyta
Class Lycopsida (club mosses and relatives)
  Order Protolepidodendrales
    Lepidodendropsis corrugatum [H]
  Order Lepidodendrales (trees with stigmarian rootstock)
    Stem genera
      Asolanus camptolaenia [CM]
      *Bothrodendron cf. *minutifolium* [CM]
      *B. cf. *punctatum* [C]
      Diaphorodendron (Lepidodendron)
      scleroticum [CM]
      Lepidocarpon lomaxii [CS]
      Lepidodendron aculeatum [C, CM]
      L. bretonense [CM]
      L. dawsoni [CM]
      L. dichotomum var. *bretonensis* [C, CS, CM]
      L. jaraczewskii [C]
      L. lanceolatum [CR, C, CM]
      L. lycopodioides [CM]
      L. obovatum [C]
      L. ophiurus [CS, CM]
      L. praelanceolatum [M]
      L. pictoense [CR, CM]
      L. rimosum [C]
      L. wortheni [C, CM]
      Lepidophloios laricinus [CR, CM]
      Paralycopodites brevifolius [CS]
        (=*Ulodendron majus* [C, CM])
      Sigillaria boblayi [CM]
      S. cf. *brardi* [CM]
      S. elegans [C]
      S. cf. *elegans* [CM]
      S. fundiensis [C]
      S. laevigata [C, CM]
      S. lorwayana [CM]
      S. mammillaris [C]
      S. ovata [CS]
      S. reticulata? [C]
      S. scutellata [C]
      S. tessellata [CM]
      S. tessellata var. *eminens* [CM]
    Root genera
      S tigmaria ficoides [CR, C, CS, CM]
    Foliar genera
      Cyperites sp.
        (=*Lepidophylloides* sp.) [C, CM]
    Reproductive organ genera
      Lepidophyllum (Lepidostrobophyllum)
      fimbriatum [H]
      Lepidostrobophyllum acuminatum [CS]
      L. fletcheri [C]
      L. cf. *jenneyi* [CM]

L. lanceolatum [CR, C, CS, CM]
L. majus [C]
L. cf. *mintoensis* [CM]
L. moyseyi [CM]
L. triangulare [CM]
Lepidostrobus variabilis [CR, CM]
L. hydei [M]
L. olryi [CR, C]
Sellaginites gutbieri [CM]
Sigillariostrobus? crépini [CM]
Class Sphenopsida
  Order Sphenophyllales
    Foliar genera
      Leeites oblongifolis [CM]
      Sphenophyllum cuneifolium [CR, C, CM]
      S. emarginatum [CM]
      S. majus [CM]
      S. myriophyllum [rCM]
      *S. oblongifolium forma *trizygia* [CM]
      S. trichomatosum [CM]
  Order Equisetales
    Stem genera
      Asterocalamites scrobiculatus [H, M]
      Calamites carinatus [CM]
      C. cisti [C]
      C. (Mesocalamites) cistiiformis [M]
      C. discifer [CM]
      C. extensus [CR]
      C. multiramis [CM]
      C. paniculata [C]
      C. ramosus [CR, C, CM]
      C. suckowi [CR, C, CM]
      C. undulatus [CR, CM]
      C. of *varians* group [C]
      C. waldenburgensis [CM]
      Calamostachys germanica [C, CM]
      C. of *varians* group [C]
      C. paniculata [C]
      Nematophyllum sp. [H]
    Root genera
      *Pinnularia sp. [CR, C, CM]
    Foliar genera
      Annularia acicularis [CR, C]
      A. aculeata [CR, C]
      A. asteris [C]
      A. latifolia [C]
      A. mucronata [CM]
      A. radiata [C, CM]
      A. sphenophylloides [CM]
      A. stellata [rC, CM]
      A. stellata forma *longifolia* [C]
      Asterophyllites charaeformis [CR, C, CS, ?CM]
      A. equisetiformis [?M, CR, C, CM]
      A. grandis [CR, C]
      A. longifolius forma *striata* [C]
    Reproductive organ genera
      Calamostachys of. *A. aculeta* [CR, C]

Calamostachys of *A. charaeformis* [CR, C]
Calamostachys of *A. grandis* [CR, C]
C. germanica [C]
C. superba [CM]
C. paniculata [C]
C. tuberculata [CM]
Macrostachya hauchecorni [CM]
M. ifundibuliformis [CM]
Palaeostachya elongata [C, CM]
P. striata? [C]
Class Filicopsida (ferns and relatives)
　Order Filicales
　Foliar genera
　　Adiantites adiantoides [C]
　　A. bondi [CM]
　　A. oblongifolius [CR]
　　A. obtusus [C]
　　A. pooli [CS]
　　A. tenuifolius [H]
　　Botryopteris tridentata [CS]
　　Corynepteris sternbergi [CM]
　　C. winslovii [CM]
　　C. sp. [CR]
　　Oligocarpia brongniarti [C, CM]
　　O. missouriensis [CM]
　　O. sp. cf. gutbieri [CM]
　　Renaultia gracilis [C]
　　R. hydei [CR]
　　Sphenopteris (Oligocarpia?) crenatodentata [CM]
　　S. (Renaultia) rotundifolia [C]
　　S. (Zeilleria) hymenophylloides [C]
　　S. (Zeilleria) sp. [C]
　　Zeilleria avoldensis [CM]
　　Z. delicatula [CM]
　　Z. frenzli [C, rCM]
　　Z. schaumburg-lippeana [C]
　　Senftenbergia sp. [CM]
　Order Marattiales (tree ferns)
　Stem genus
　　Caulopteris sp. [C, CM]
　Foliar genera
　　Eupecopteris (Asterotheca) cyathea [CM]
　　E. (Senftenbergia) obtusa [CM]
　　Lobatopteris (Asterotheca) miltoni [CM]
　　L. (Pecopteris) vestita [jc: CM]
　　Pecopteris (Asterotheca) acadica [CM]
　　P. clarkii [CM]
　　P. cf. densifolia [CM]
　　P. (Asterotheca) hemitelioides [CM]
　　P. herdii [CMb]
　　P. (Asterotheca) miltoni [CM]
　　P. (Senftenbergia) pennaeformis [CM]
　　P. pilosa [C]
　　P. plumosa forma *crenata* [C]
　　P. plumosa forma *dentata* [CM]
　　P. sterzeliformis [CM]
　　P. sp [C]
　　P. (Ptychocarpus) unitus [CM]
　　Pecopteridium sullivantii [CM]
　Reproductive organ genera
　　Asterotheca cf. *abbreviata* [CM]
　　A. daubreei [CM]
　　A. herdi [CM]
　　A. oreopteridia [CM]
　　A. robbi [CM]

Scolecopteris sp. [CS]
Order Zygopteridales
　Alloiopteris (Corynepteris) almaensis [CR]
　A. (Corynepteris) coralloides [CR]
　A. (Corynepteris) major [CM]
　A. (Corynepteris) sternbergi [M?, C, CM]
Class Progymnospermopsida
　Order *incertae*
　　Rhacopteris robusta [W]
Class Gymnospermopsida (plants with naked seeds)
　Order Pteridospermales (seed ferns)
　Stem genus
　　Medullosa sp. [C]
　Foliar genera
　　Alethopteris davreuxi [CM]
　　A. decurrens [CR, C]
　　A. friedeli [CM]
　　A. grandini [CS, CM]
　　A. hartti [CR]
　　A. lonchitica [CR, C, CM]
　　A. scalariformis [CM]
　　A. serli [CM]
　　A. (Megalethopteris) hartii [CR]
　　A. valida [CM]
　　Aneimites acadica [H]
　　Callipteridium sullivanti [CM]
　　Dicksonites pluckeneti [CM]
　　Eusphenopteris neuropteroides
　　　(= *Sphenopteris squamosa*) [CM]
　　E. (Sphenopteris) nummularia forma *dilatata* [C]
　　E.(Sphenopteris) obtusiloba [CR, C]
　　E. (Sphenopteris) striata [CM]
　　E. (Sphenopteris) trifoliolata [CR]
　　Eremopteris artemisiaefolia [CS, CM]
　　Fortopteris (Mariopteris) latifolia [CM]
　　Karinopteris (Mariopteris) acuta [CR]
　　K.? (Mariopteris) grandepinnata [C]
　　K. (M.) soubeirani [CM]
　　Linopteris bunburii [CM]
　　L. muensteri [CM]
　　L. neuropteroides var. *major* [CM]
　　L. obliqua [CM]
　　L. obliqua var. *bunburii* [CM]
　　Lonchopteris eschweileriana [eCM]
　　Mariopteris comata [C]
　　M. disjuncta [C]
　　M. hirsuta [CM]
　　M. nervosa [rC, CM]
　　M. sphenopteroides [CM]
　　M. tenuifolia [C, CM]
　　M. tenuis [CM]
　　M. sp. [C]
　　Megalopteris kellyi [Cng]
　　Neuralethopteris kosmanni? [CR]
　　N. schlehani forma *rectinervis* [CR, C]
　　N. smithsii [CR]
　　N. sp. [C]
　　Neurocardiopteris barlowi [CR]
　　Neuropteris aculeata [CM]
　　N. crenulata [CM]
　　N. (Mixoneura) flexuosa [CM]
　　N. gigantea [C]
　　N. heterophylla [CM]
　　N. macrophylla [CM]
　　N. (Mixoneura) obliqua [C]

N. (Mixoneura) ovata [CM]
N. pseudogigantea [C, CM]
N. rarinervis [CM]
N. scheuchzeri [CM]
N. tenuifolia [C, CM]
Odontopteris cantabrica [CM]
O. minor [CM]
O. schlotheimii [CM]
O. subcuneata [CM]
Paripteris spp. [CM]
Pseudomariopteris (M.) *ribeyroni* [CM]
Rhodea laqueata [CR]
R. cf. *sparsa* [CR]
R. wilsoni [C]
Sphenopteridium crassum? [M]
S. dawsoni [M]
S. macconochei? [H]
S. sp. [H]
Sphenopteris (Diplotmema) furcatum [CR, C, CM]
S. (D.) geniculatum var. *erectum* [CM]
S. (D.) patentissimum [H]
S. (D.) zobeli? [CM]
Telangium cf. *affine* [M]
T. bretonensis [W]
T.? potieri [CM]
Triphyllopteris minor [H]
Triphyllopteris virginiana [H]
Incertae sedis
   Reticulopteris muensteri [CM]
Reproductive organ genera
   Heterangium sp. [CS]
   *Holcospermum sp. [C]
   Polypterocarpus sp. [C]
   *Trigonocarpus parkinsoni [C]
   T. praetextus [CR]
   Whittleseya brevifolia [CR, C]
   W. desiderata [CR]
Filicales or Pteridospermales *incertae sedis*
  Foliar genera
    Sphenopteris aculeata [CM]
    S. amoenaeformis? [CR]
    S. brittsii [CM]
    S. cantiana [CM]
    S. cuneoliformis [CR]
    S. deltiformis [CR, C]
    S. dixoni [C]
    S. fletcheri [C]
    S. goniopteroides [CM]
    S. (Renaultia) gracilis [C]
    S. haliburtoni [CS]
    S. cf. *hoeninghausi?* [CM]
    S. licens [CR]
    S. lineata [CR]
    S. missouriensis? [CM]
    S. moriensis [CM]
    S. mixta [C]
    S. moyseyi [C]
    S. oxfordensis [CR]
    S. philipensis [CR]
    S. polyphylla [CR]
    S. pseudo-furcata [CR]
    S. rhomboidea [CR, C]
    S. (Renaultia?) schatzlarensis [CR, C]

S. spiniformis [CM]
S. spinulosa [CM]
S. spp. [C]
S. stipulataeformis [C]
S. sulcata [CR]
S. cf. *suspecta* [CM]
S. strigosa [H]
S. valida [C]
S. (Diplotmema) whitii [rCM]
S. (Rhodea) wilsoni [C]
Reproductive organ genera
   Crossotheca boulayi [CS]
   C. communis [CM]
   C. compacta [CM]
   C. denticulata [CM]
Incertae sedis
   Hymenophyllites quadridactylites [CM]
   Hymenotheca broadheadi [CM]
   H. bronni [CM]
   H. dathei [CM]
   Sphenopteris (Hymenophyllites) bronni [CM]
   (?= *Boweria schatzlerensis* [C, CM])
Order Cordaitales
Stem genus (pith cast)
   Artisia transversa [CR, C, CM]
Stem genus (permineralized)
   Cordaixylon sp. [CR, C, CM]
Foliar genera
Cordaites principalis [CR, C, CM]
Reproductive organ genera
   Cardiocarpon carinatum [CM]
   Cordaianthus devonicus [C]
   C. pitcairniae [CS]
   C. spinosus [CM]
   Cordaicarpus dawsoni [C, CM]
   Samaropsis ampullacea [CM]
   S. baileyi [C, (Cng)]
   S. cornuta [CR, C, CM]
   S. crampii [C]
   S. ingens [C]
   S. wilsoni [C]
Order Voltziales
Form genus
   Walchia sp.[P]
Gymnopermopsida *incertae sedis*
   Gymnocladus salisburyi [C]
Seeds *incertae sedis*
   Carpolithus tenellus [H]
   Rhabdocarpus sp. [C]
   Radiospermum sp.[CM]
Order Dicranophyllales
   Dicranophyllum glabrum [C*]
   ?*Dicranophyllum* sp. [P]
Incertae sedis
   Nematophyllum sp. [H]
Incertae sedis
   Acitheca polymorpha [CM]
   Dactylotheca plumosa [CM]
   Desmopteris elongata (cf. *D. longifolia*) [CM]
   Sporangites acuminata [C]
   Tetrameridium caducum [CM]
   Volkmannia? sp. [CR]

* First report of taxon.

Sequence stratigraphy: a revolution without a cause?

R. C. L. WILSON

Department of Earth Sciences, The Open University, Milton Keynes MK7 6AA, UK

Abstract: The paper presents personal reflections on the origins and utility of sequence stratigraphic models. These focus on two questions: (1) does sequence stratigraphy represent a revolution in our understanding of the stratigraphic record, and (2) does it provide a new means of global correlation? The first question is answered in the affirmative, at least insofar as sequence stratigraphy enables us to integrate a wide range of data and interpretations across a huge range of spatial and temporal scales. The recognition of the importance of stratal surfaces has led to a greater understanding of the response by sedimentary to climatic, tectonic and eustatic changes. But it has yet to be shown that eustatic signals can be detected unequivocally in the stratigraphic record. Therefore this 'new global stratigraphy', based on the premise that sequence boundaries are primarily controlled by eustatic changes, is not yet a reality. Testing this hypothesis is beyond the resolution of current biostratigraphic and chronostratigraphic techniques.

When I was invited to contribute to the Lyell bicentenary meeting, I was asked to talk about 'Sequence stratigraphy and sea-level change'. It was hardly surprising that 'sea-level change' was included in the title, because over 150 years after Lyell addressed the sea-level controversy, we are still trying to unravel tectonic and eustatic signals from the sedimentary record. But sequence stratigraphy is not just about sea-level change. It identifies genetic packages of strata bounded by time-related physical surfaces: unconformities and their correlative conformities, and surfaces caused by flooding events. Peter Vail and the 'Exxon school' claimed that eustatic sea-level changes are the dominant control on stratal geometries and facies distributions within them. For readers not familiar with it, the sequence stratigraphic approach is summarised in Fig. 1. It develops a simplified version of a frequently used diagram which is often referred to as the 'Exxon slug'.

To some geologists, the sequence stratigraphic approach heralded the possibility of the 'new global stratigraphy' whereby stratigraphic surfaces caused by globally synchronous eustatic sea-level changes could be used as a means of global correlation. However, others regarded it as dressing up regressions, transgressions etc. in unnecessary new terminology.

This paper presents personal reflections on the sequence stratigraphic approach gained both as a teacher and researcher. It focuses on the questions explicitly and implicitly stated in the title, but does not attempt to provide a thorough description or comprehensive critical review. Brief introductions to the subject are given by Wilson (1992) and Christie-Blick and Driscoll (1995). Textbooks by Miall (1996) and Emery and Myers (1996) provide in-depth coverage, with Miall's text offering much constructive criticism. The best introduction from the Exxon school is the colourful volume on siliciclastic stratigraphy (Van Wagoner *et al.* 1990). Reflective articles by Posamentier and Weimer (1993) and Walker (1990) are well worth reading.

Method, problems and doctrine

The beginnings

The roots of sequence stratigraphy lie in the recognition, some 40 years ago, of packages of strata bounded by continent-wide unconformities. Technological advances that enabled the petroleum exploration industry to move offshore played a key role in stimulating stratigraphers to take this 'big view'. High-quality marine seismic data enabled stratigraphic architecture to be determined at the basin and continental margin scales. Out of this grew the conceptual framework that enables the integration of a range of scales of stratigraphic information from a single laminae to a first order stratigraphic sequence.

I first learnt about seismic and sequence stratigraphy through a workshop run by Esso UK soon after the 'old testament' was published (Payton 1977). They ran another in the 1980s, around the time of the appearance of the 'new testament' (Wilgus *et al.* 1988). The first meeting

WILSON, R. C. L. 1998. Sequence stratigraphy: a revolution without a cause?.
In: BLUNDELL, D. J. & SCOTT, A. C. (eds) *Lyell: the Past is the Key to the Present.* Geological Society, London, Special Publications, **143**, 303–314.

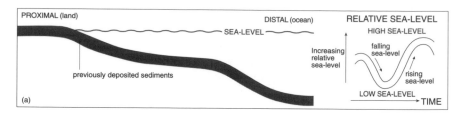

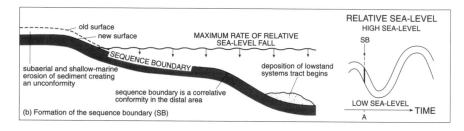

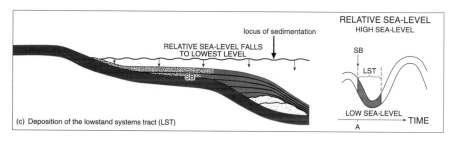

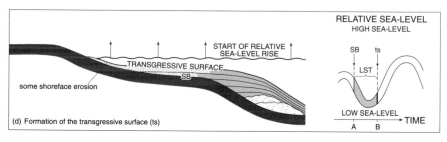

was, of course, seismically oriented, and the second one focused more on the outcrop and well log scales.

At the first meeting there was some robust discussion about a number of problems and contentious issues, which was hardly surprising with people such as Drummond Matthews and Tony Hallam present. A few of these problems are discussed below.

Are seismic reflections really time lines?
The examples given by the Exxon group at the Esso workshop (e.g. Vail, Todd & Sangree 1977, figs 3–6) were not entirely convincing. A diagram for

which I find more use with students is given in Fig. 2 – but this is not *proof* of the proposition!

How can eustatic signals be determined from coastal onlap charts?
Exxon charts have a characteristic saw-tooth shape with abrupt initial falls followed by gradual rises. The early derivative sea-level curves reflected such asymmetry. In the early 1980s revised smoother curves were published, as by then it had been recognised that the landward encroachment of sediments across older strata (onlap) along continental margins, as seen on seismic sections, involved marine *and* fluvial sediments. The latter were deposited during periods when coastlines

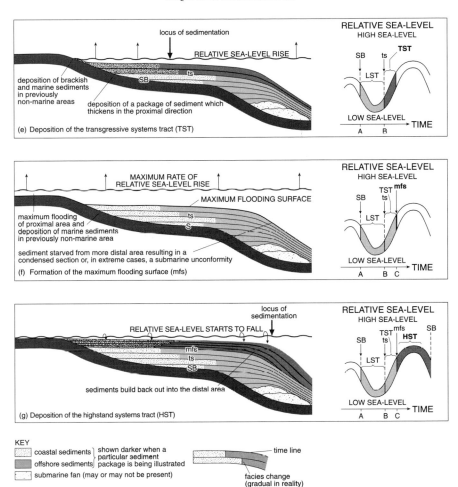

Fig. 1. Sketch cross-sections illustrating the basic concepts and terminology of sequence stratigraphy. Sketches (a) to (g) illustrate the stages in the development of one depositional sequence during one cycle of relative sea-level change. Slightly modified from Skelton *et al.* (1997).

moved seaward, yet on seismic sections landward onlap continued.

How can sea-level curves for individual locations yield a means of global correlation?
This was, and still is, a minefield. Different approaches, and different authors, produce different curves for the same period of time (Fig. 3). Twenty years after the first global curves were published, all the evidence on which a series of Exxon school curves are based have still not been published although some is contained in the appendix of Haq *et al.* (1988). Despite such shaky foundations (and others which are discussed by Miall 1996) many geologists felt that the contents of the old and

new testaments, and especially the famous – or infamous – Haq curve (Haq *et al.* 1988), heralded the start of a new era of stratigraphic studies. Perhaps a means of global chronostratigraphic correlation was within reach? This euphoria probably resulted from the fact that, to many people, traditional stratigraphy had slipped into the doldrums. Many with an historical geology bent had taken refuge in sedimentology and found that the sequence stratigraphic approach widened their horizons – in space and time – as they attempted to analyse basin-wide data sets and interpretations.

The historical development of sequence stratigraphy was significantly different from the way Lyell's ideas were introduced to, and eventually

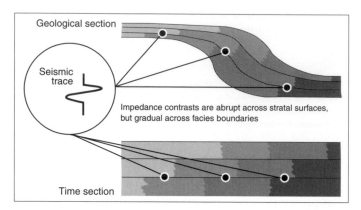

Fig. 2. A cartoon to explain why seismic reflections are time lines, and do not follow lithostratigraphic boundaries. Changes in tone density depict facies changes; these are relatively abrupt across time lines, but gradual parallel to them.

accepted by, the scientific community, or for that matter the exposure to peer review of later conceptual advances. Seismic stratigraphy came out of the corporate closet in the late 1970s, about ten years after it had been routinely applied during petroleum exploration activities. The birth and early childhood of sequence stratigraphy were not subjected to open debate as were Lyell's observations and interpretations. Perhaps there were robust debates within Exxon, but they failed to ensure that the method was distanced from eustatic explanations when the subject entered the public domain in 1977. Unfortunately, this led to judgements about the value of the sequence stratigraphic approach being clouded by the claim

that it provides a new means of *global* correlation based on the supposed eustatic signals. So method became tainted by doctrine. Lyell avoided this trap, as discussed by Jim Secord during his Linnean Society lecture at the Bicentenary Meeting (see also Secord 1997).

Therefore it is important to distinguish the analytical framework of sequence stratigraphy from debates about eustatic signals and the 'new global stratigraphy'. With this in mind, this paper addresses two questions contained within its title: (1) does sequence stratigraphy represent a revolution in our understanding of the stratigraphic record, and (2) does it enhance our ability to determine the causes (i.e. the relative contributions made by eustasy,

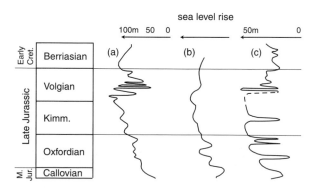

Fig. 3. Late Jurassic sea-level curves compared. (a) Haq *et al.* (1988) based on a compilation of coastal onlap and sequence stratigraphic studies. (b): Hallam (1988) based on 'stage by stage analysis of the areal spread of seas over the continents as inferred from palaeographic studies' with shorter-term cyclic changes inferred from facies analyses which enabled 'the recognition of extensive shallowing and deepening events in epicontinental marine sequences'. (c) Sahagian *et al.* (1996) based on the sedimentary record of repeated flooding and exposure of the very low gradient ramp of the Russian Platform.

tectonics and depositional processes) of relative sea-level change at single locations and across basins?

A *revolution?*

What are we looking for in a revolution? I have neither the space nor the expertise to discuss at length ideas about scientific revolutions, so I will quote just two views. Kuhn (1970) used the concepts of 'normal' and 'extraordinary' science to explain what he considered to be the nature of scientific revolutions. *Normal science* is like puzzle solving; it is research firmly based on one or more past scientific achievements that a particular discipline acknowledges for a time as supplying the foundation for its continued practice. *Extraordinary science* is preceded by a period of uncertainty, when investigators may divide into different schools of thought, so that there is no generally accepted consensus. Extraordinary science begins when a new paradigm takes over from a previously held one: a new consensus then prevails and the 'revolution' has begun. New ideas suddenly enable a whole range of previously puzzling phenomena to be explained, and so a rigorous 'mopping-up' operation commences. The emergence of plate tectonics fits this definition of a scientific revolution very well.

Not all historians of science agree with Kuhn, maintaining that it is too simplistic to define just two distinct kinds of scientific progress (normal and extraordinary). Jevons (1973) wrote:

Contributions form a continuous spectrum, within which really major changes are rare, and minor ones most frequent, but with intermediate ones occurring with intermediate frequency.

Did the advent of sequence stratigraphy represent a revolution in historical geology, or a minor or intermediate step forward? Consider this question in the context of the following definitions:

The subdivision of basin fills into genetic packages bounded by unconformities and their correlative conformities. (Emery & Myers (1996).

The predictable succession of physical stratigraphic units including sequences, systems tracts and parasequences. These depositional units are defined on the basis of internal stratal 'geometries' and evolve in response to changes in shelfal accommodation space. (Vail, pers. comm. 1992).

Notice that the definition given by Peter Vail in a short course avoids the dreaded word 'eustasy'. 'Shelfal accommodation space' is used instead, which is perhaps a tacit recognition of the difficulty (many geologists would say impossibility) of unravelling from the rock record the relative contributions of eustasy, tectonics and sedimentary processes in creating and filling the space available for sediments to fill (Fig. 4). Both the definitions quoted above focus on method, but 'explanation' still figures in the Vail definition. So has sequence stratigraphy enabled 'a whole range of previously puzzling phenomena to be explained', or is it just another step forward in stratigraphic procedures and understanding? We do not have a global model – a cause – that integrates a variety of observations and interpretations in a manner similar to that achieved by plate tectonics. The Vail definition quoted above clearly recognises this by leaving out the link with eustasy. So have we witnessed a revolutionary change in method and approach? I will consider this question first and then return to the issue of 'causes'.

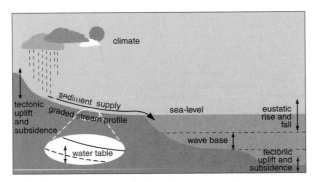

Fig. 4. Sketch cross section across a basin margin showing the controls that determine the space available for sediment to accumulate. The primary controls are eustatic, tectonic and climatic. They determine sediment supply, and the development of equilibrium profiles in erosional and depositional systems that are linked to base levels related to sea level, the water table and wave base.

Building on the shoulders of others

Table 1 is a chronological list of geologists, and the concepts they proposed, that are embraced by sequence stratigraphy. These are the foundations that underpin sequence stratigraphy, starting with Hutton and Lyell. The younger roots of sequence stratigraphy lie in our fascination with cyclic sequences, and the recognition by Larry Sloss of stratal assemblages bounded by continent wide unconformities (Sloss *et al.* 1949; Sloss 1963). Sloss also contributed to the evolution of sequence stratigraphy through his supervisory role when Peter Vail was a graduate student. We should not forget other contributions, such as Rich's (1951) depositional topography (fondoform, clinoform, undaform) which was the forerunner of the Exxon 'slug', Wheeler's (1958) chronostratigraphic charts that were (and still are) used to dissect the Exxon 'slug' in time, and Busch's (1971) genetic increments of strata and genetic sequences of strata, the definitions of which are similar to the parasequences and parasequence sets of the Exxon school.

The ability to take the 'big view', as was done by Sloss and Rich, was significantly enhanced by technological advances that enabled stratigraphic architecture at the basin and continental margin scales to be determined using high quality marine seismic data. Out of this has grown the conceptual framework, albeit with a continental margin bias, that enables the integration of a range of scales of stratigraphic information (see below).

A new paradigm or an effective catalyst?

Sequence stratigraphy represents a new paradigm in geology. (Miall 1996)

Sequence stratigraphy is often regarded as a relatively new science, evolving in the 1970s from seismic stratigraphy. In fact sequence stratigraphy has its roots in the centuries-old controversies over the origin of cyclic sedimentation and eustatic versus tectonic controls on sea level. (Emery & Myers 1996)

These two views represent end members of a range of opinion about the significance of the sequence stratigraphic approach. Miall's view is perhaps a little surprising, given that he has been such a strong critic of the 'new global stratigraphy' claims of the Exxon school. However his accolade is for the method, not the doctrine of eustatic control.

Few stratigraphers would deny that sequence stratigraphy has reinvigorated stratigraphic research over the past 25 years. At the very least, it has enabled us to look at old data with new eyes, and take a basin scale (and larger) view when accumulating and interpreting new data. This has resulted in the publication of a multitude of papers and special volumes on the subject. The burgeoning sequence stratigraphic literature consists of a spectrum of papers from those applying the new nomenclature in a predominantly descriptive way, to others who use the concept to come to a better understanding of basin dynamics. For an example

Table 1. *The foundations of sequence stratigraphy*

Year	Author(s)	Concepts
1788	Hutton	Unconformities separating cycles of uplift, erosion and deposition
1835	Lyell	Sea-level change and uplift of land
1839	Sedgwick & Murchison	Unconformities as physical boundaries of geological Periods
1840	Agassiz	Glacial theory
1842	Maclaren	Glacioeustasy
1894	Walther	Law of succession of facies
1898/1909	Chamberlin	Diastrophic control of stratigraphy by sea-level changes
1906	Suess	Onlap and offlap attributed to eustasy
1913	Grabau	Recognised interrelationship of subsidence and sediment supply
1924	Stille	Global unconformities caused by tectonism and resulting eustatic effects
1935	Wanless & Shephard	Carboniferous cyclothems and glacioeustacy
1940	Grabau	~30 Ma 'rhythm of the ages': heat flow and transgressions/regressions
1949	Sloss *et al.*	First proposal of concept of stratigraphic sequences
1951	Rich	Depositional topography: genetically related associations of sediment types.
1958	Teichert	Facies sequence
1958	Wheeler	Chronostratigraphic charts
1963	Sloss	Major continent-wide unconformity-bound sequences related to orogenies
1965	De Raaf *et al.*	Vertical profiles (facies successions)
1971	Busch	Genetic increments (GIS) and sequences of strata (GSS)

of the utility of the method one has only to look at the extent to which papers presented at successive meetings on the petroleum geology of NW Europe (Woodland 1975: Illing & Hobson 1981; Brooks & Glennie 1987; Parker 1993) show the degree to which geologists and managements of companies have progressively espoused the new approach. However, a mountain of paper does not signal a revolution in stratigraphy.

Sequence stratigraphy does not provide an all embracing global causal model. It is a catalyst, or perhaps an up-to-date toolkit, designed to utilize and integrate a wide range of data sets and explore more deeply spatial and temporal stratal relationships. *Integration and interrelationships* are the essence of sequence stratigraphy, based on the well established traditions of biostratigraphy, facies analysis and chemostratigraphy. It provides a framework with which to integrate observations and interpretations derived from different data sets, from outcrops through basins to crustal plates, and through time (Figs 4 & 5). It has forced us to think more deeply about how space for sediments to accumulate is created, about the significance of stratal surfaces and units, and about the hierarchical structure of sedimentary successions and the factors that shape their architecture.

The difficulty of manipulating so much data, and the complexity of considering possible controls, inevitably led to the development and application of modelling studies. Not surprisingly, there are optimistic and pessimistic views about their value, as the following quotations from the same volume (Dott 1992) show:

Mathematical modelling in the late 1970s ... allowed us to relate sea level, subsidence, and sediment supply to produce the curves on our global cycle chart, which implies a major role for eustasy throughout Panerozoic time. (Vail 1992)

A wide range of geological characteristics places limits on the tectonism and eustasy. This allows the application of a family of reasonable tectonic and eustatic models to explain basin history ... models can generate complex basinal sequences with high fidelity using plausible inputs. Thus assumptions heaped on assumptions work. (Kendall *et al.* 1992)

Using modelling studies, can we really unravel the effect of a multitude of controls on the architecture of a given basin fill? They help us understand a multitude of possibilities and rule out unlikely combinations of possible controls, but they cannot separate and quantify the effects of different controls on real stratigraphic patterns which we observe. This does not mean that stratigraphic modelling is worthless: it makes us aware that

similar patterns may be produced by several different combinations of controlling factors. Modelling aids our understanding of what rates of processes are unrealistic or impossible, and what combination of rates are plausible (Fig. 5(c)). Models can also be used to predict facies distributions ahead of drilling in partially explored basins (e.g. Lawrence *et al.* 1990).

Modelling studies demonstrate the likelihood that the combination of eustatic changes and differential subsidence across rift basins can produce sequence boundaries and depositional systems tracts differing in age across a relatively small area (e.g. Gawthorpe *et al.* 1994). Christie-Blick and Driscoll (1995) highlighted the difficulty of predicting the timing of the sedimentary response to external 'drivers'. They stated:

if the phase relation between the eustatic signal and the resulting stratigraphic record varies from one place to another, then the synchrony or lack thereof of observed stratigraphic events may prove to be less useful then previously thought as a criterion for distinguishing eustasy from other controls on sedimentation.

The new global stratigraphy

Mission impossible?

The identification of facies and their environmental interpretation is crucial to the application of sequence stratigraphic concepts. As this approach is, to say the least, not entirely objective, it is not surprising that different workers may arrive at different interpretations, as demonstrated by the examples from the Upper Jurassic shown in Fig. 6. So can sequence boundaries and other stratal surfaces offer a means of global correlation because they are generated by eustatic changes of sea level?

As already discussed, the hypothesis that sequence boundaries and flooding-related surfaces signal global changes in sea level is fundamentally flawed because we are unable to isolate the eustatic signals from those caused by tectonic and sedimentary processes. This is not the only difficulty. Notwithstanding the problem that different workers may not be able to agree on sequence stratigraphic interpretations of the same succession, *proving* that stratal surfaces are globally synchronous is an almost impossible task within the resolution of the current geological timescale. This is particularly difficult in successions deposited before the current icehouse period began, and there is still a long way to go even within this interval of time.

Miall (1994) pointed out the problems of correlating stratal surfaces between basins: the

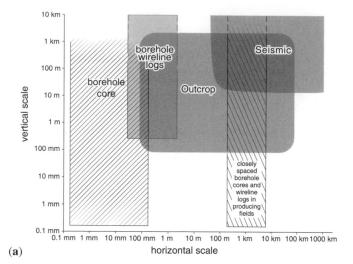

(a)

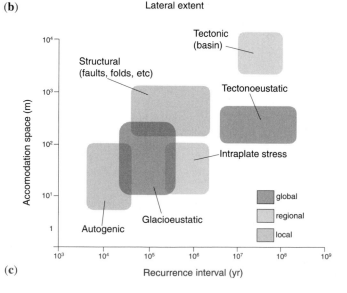

(b)

(c)

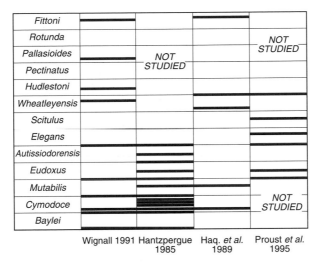

	Wignall 1991	Hantzpergue 1985	Haq. *et al.* 1989	Proust *et al.* 1995
Fittoni				
Rotunda				*NOT STUDIED*
Pallasioides		*NOT STUDIED*		
Pectinatus				
Hudlestoni				
Wheatleyensis				
Scitulus				
Elegans				
Autissiodorensis				
Eudoxus				
Mutabilis				
Cymodoce				*NOT STUDIED*
Baylei				

Fig. 6. The difficulty of reaching agreement on the location of sequence boundaries: alternative interpretations of the Kimmeridgian *sensu anglico* of northwestern Europe (from Proust *et al.* 1995). Sequence boundaries are shown as thick lines.

inability to map them directly from place to place means that they must be correlated using biostratigraphy or exceptionally chemostratigraphy or magnetostratigraphy. He showed how the limitations of time resolution involved in the various steps necessary to date and correlate stratigraphic events ranges between a few hundred thousand and several millions years. Thus error may be heaped on error so that the error range in biostratigraphic resolution is often greater than the duration of stratigraphic sequences.

The existence of cycles occurring at different frequencies and amplitudes adds to the problems of recognising global signals (Fig. 7), because different depositional settings may record different frequency events. So there is a danger that a lower order event in one place may be correlated with a higher order one in another, and so on.

Optimists argue that if key events as indicated by sequence boundaries and transgressive/flooding surfaces do turn up in the same zonal positions in many places around the world, then there is a possibility that the global signal is strong enough to overcome regional variability in tectonic subsidence and uplift.

The future

Despite all the difficulties of global correlation, there can be little doubt that our profession will pursue cycles in the geological record for a long time to come – just as it has done in the past. A promising line of investigation is to determine the extent to which global geochemical signatures correlate with flooding events identified using stratigraphic and sedimentological features. Studies of Cretaceous pelagic limestones suggest that 'within certain limits, the correlation of the ∂^{13}C profile with published sea-level curves may be extremely close' and that 'such a relationship could well relate to shelf-sea area governing the global burial rate of organic carbon' (Jenkyns 1996). During the present icehouse period, ∂^{18}O variations are a proxy for glacioeustatic sea-level changes, and in some areas these have been shown to be correlated with sequence boundaries (e.g. Carter *et al.* 1991; Naish & Kemp 1997). Notably, the oxygen isotope studies of Browning *et al.* (1996) showed that ice-volume changes occurred across sequence boundaries beneath the New Jersey coastal plain after 42 Ma (when Antarctic ice sheets

Fig. 5. Sequence stratigraphy: its role in promoting the integration of different data sets, and as a framework for exploring the interrelationships in space and time of the controls on the creation of sediment accommodation and the resultant stratal units. (a) Sequence stratigraphy provides a framework for the integration of data sets collected across a range of scales, shown on a log–log plot. (b) The hierarchy of scale of stratal units in thickness and time, shown on a log–log plot. (c) Estimates of the ranges of sediment accommodation space versus the time periods over which various processes operate to produce such space, shown on a log–log plot (Dickinson *et al.* 1994).

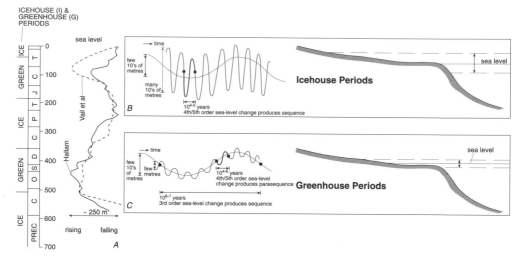

Fig. 7. Changing frequencies and amplitudes of eustatic change. (A) The 'first order' curves of Hallam (1977) and Vail *et al.* (1977) show sea-level peaks in the Early Palaeozoic and Cretaceous, separated by a low in the Permo-Triassic. The smaller scale fluctuations on the Hallam curve represent second order changes. Left side of (B) and (C) (from Tucker 1993): cartoons showing the possible changes in the amplitude of third–fifth-order sea-level changes during icehouse and greenhouse periods. During icehouse periods, higher frequency and amplitude Milankovich-related changes dominate over lower frequency and amplitude third order changes. During greenhouse times, the third-order changes probably had higher amplitudes than Milankovich-related changes, and so are more likely to generate sequences. Right side of (B) and (C): cartoons (not to scale) showing differences in mean elevation and amplitudes of sea levels during icehouse and greenhouse periods. Lower mean sea level during icehouse times are accompanied by high amplitude short-term changes so that the upper parts of continental slopes, continental shelves and coastal plains are subjected to frequent flooding and exposure. In contrast, high mean sea level during greenhouse periods result in extensive epeiric seas covering platforms and low angle ramps, so that relatively small changes in sea level cause extensive flooding or emergence over large areas.

began to form), but no link was revealed between $\partial^{18}O$ variations and sequence boundaries before this time.

There are limitations to the application of studies of recent sedimentary systems and Neogene basin architecture in order to understand the controls on sequence development during greenhouse periods. This is because during icehouse periods depositional systems may not reach equilibrium during the relatively short time spanned by a single higher order cycle of sea-level change (Fig. 7B). They were more likely to have done so during greenhouse periods. Would storm processes be more or less significant than they are today in controlling architecture during greenhouse times? During greenhouse times, the amplitudes of global sea-level changes were probably much smaller than those of icehouse times, and they occurred over longer periods of time: this must have had a significant effect on rates of sediment supply. So does sequence architecture change between greenhouse and icehouse times as suggested by Read *et al.* (1991)? These and many other questions remain to be explored.

Conclusion

Few geologists would deny that sequence stratigraphy has reinvigorated stratigraphic research over the past 25 years. At the very least, it has enabled us to look at old data with new eyes, and take a basin scale (and larger) view when acquiring and interpreting new data. This has resulted in the publication of a plethora of papers and special volumes on the subject. But does this signal a revolution in stratigraphy?

Judgements about the value of the sequence stratigraphic approach are often clouded by arguments concerning whether it provides a new method for *global* correlation because sequences were largely the result of eustatic sea-level changes. Therefore it is important to distinguish the analytical framework and predictive capability of sequence stratigraphy from debates about eustatic signals and the 'new global stratigraphy'.

It is probably unlikely that high resolution sequence stratigraphic studies will ever, on their own, succeed in proving that sequence boundaries can be used as globally synchronous stratigraphic markers, especially during greenhouse periods. In the absence of a significant improvement in the resolution of dating methods, the ultimate test will be to show whether these surfaces are related to geochemical signals of global change.

The quotation below provides a strong reminder about the need to objectively record the nature of the stratigraphic record, and then consider the processes that were responsible for what we observe.

> Perhaps Lyell's greatest single contribution was to separate study of the geological record – geology in its historical sense – from the study of processes that produced this record. Sutton (1975)

We should not forget the pitfalls of using the genetic stratigraphy route to interpreting the rock record, for there is always the risk that method will slip into doctrine.

I am indebted to Peter Vail and his Exxon colleagues for stimulating my interest in sequence stratigraphy. An unpublished perspective on sequence stratigraphy by Gerald Salisbury provided food for thought a year before this paper was written. The comments of Angela Coe and two anonymous reviewers are gratefully acknowledged. I thank Angela Coe and the Open University for permission to reproduce Fig. 1. The contributions of Yvonne Englefield and Denise Swann (word processing) and Andrew Lloyd and John Taylor (figure drafting) are gratefully acknowledged.

References

AGASSIZ, L. 1840. *Études sur les Glaciers*. Privately published, Neuchatel.

BROOKS, J. & GLENNIE, K. W. (eds) 1987. *Petroleum geology of North-West Europe,* Graham & Trotman, London, vols 1–2.

BROWNING, J. V., MILLER, K. G. & PAK, D. K. 1996. Global implications of Lower to Middle Eocene sequence boundaries on the New Jersey coastal plain: the icehouse cometh. *Geology,* **24,** 639–642.

BUSCH, D. A. 1971. Genetic units in delta prospecting. *American Association of Petroleum Geologists Bulletin,* **55,** 1137–1154.

CARTER, R. M., ABBOTT, S. T., FULTHORPE, C. S., HAYWICK, D. W, & HENDERSON, R. A. 1991. Applications of global sea-level and sequence stratigraphic models in southern hemisphere Neogene strata from New Zealand. *In:* MACDONALD, D. I. M. (ed.) *Sedimentation, Tectonics and Eustasy: Sea-Level Changes at Active Margins.* International Association of Sedimentologists Special Publication **12,** 41–65.

CHAMBERLAIN, T. C. 1898. The ulterior basis of time divisions and the classification of geologic history. *Journal of Geology,* **6,** 449–462.

—— 1909. Diastrophism as the ultimate basis of correlation. *Journal of Geology,* **17,** 685–693.

CHRISTIE–BLICK, N. & DRISCOLL, N. W. 1995. Sequence stratigraphy. *Annual Reviews of Earth and Planetary Science,* **23,** 451–478.

DICKINSON, W. R., SOREHEGAN, G. S. & GILES., K. A. 1994. Glacioeustatic origin of Permo-Carboniferous stratigraphic cycles; evidence from the southern Cordilleran foreland region. *In:* DENNISON, J. M. & ETTENSOHN F. R. (eds) *Tectonic and Eustatic Controls on Sedimentary Cycles.* Concepts in Sedimentology and Palaeontology 4, Society of Sedimentary Geology, 25–34.

DOTT, R. H., Jr 1992. *An Introduction to the Ups and Downs of Eustasy.* Geological Society of America Memoir 180, 17–24.

EMERY, D. & MYERS, K. (eds) 1996. *Sequence Stratigraphy.* Blackwell Science, Oxford.

GAWTHORPE, R. L., FRASER, A. J. & COLLIER, R. E. LL. 1994. Sequence stratigraphy in active extensional basins: implications for the interpretation of ancient basin fills. *Marine and Petroleum Geology,* **11,** 642–648.

GRABAU, A. W. 1913. *Principles of Stratigraphy.* Seiler, New York.

—— 1940. *The Rhythm of the Ages.* Henri Vetch, Peking.

HAQ, B. V., HARDENBOL, J. & VAIL, P. R. 1988. Mesozoic and Cenozoic chronostratigraphy and cycles of sea-level change. *In:* WILGUS, C. K., POSAMENTIER, H., VAN WAGONER, J., ROSS, C. & KENDALL, C. St. C. (eds) *Sea-Level Changes: an Integrated Approach.* Society of Economic Palaeontologists and Mineralogists Special Publication 42, 71–108.

HALLAM, A. 1977. Secular changes in marine inundation of USSR and North America through the Phanerozoic. *Nature,* **269,** 762–772.

—— 1988. A re-evaluation of Jurassic eustasy in the light of new data and the revised Exxon curve. *In:* WILGUS, C. K., HASTINGS, B. S., KENDALL, C. G. St. C., POSAMENTIER, H. W., ROSS, C. A., VAN WAGONER, J. C. (eds) *Sea-Level Changes: an Integrated Approach.* Society of Economic Palaeontologists and Mineralogists Special Publication 42, 261–273.

ILLING, L. V. & HOBSON, G. D. (eds) 1981. *Petroleum Geology of the Continental Shelf of North-West Europe.* Heyden & Son, Institute of Petroleum, London.

JENKYNS, H. C. 1996. Relative sea-level change and carbon isotopes: data from the Upper Jurassic (Oxfordian) of central and southern Europe. *Terra Nova,* **8,** 75–85.

JEVONS, F. R. 1973. *Science Observed.* Allen & Unwin, London.

KENDALL, C. G. St. C., MOORE, P. & WHITTLE, G. 1992. A challenge: is it possible to determine eustasy and does it matter? *In:* DOTT, R. H., Jr (ed.) *Eustasy: the Historical Ups and Downs of a Major Geological Concept.* Geological Society of America Memoir 180, 93–107.

KUHN, T. S. 1970. *The Structure of Scientific Revolutions.* 2nd edn. University of Chicago Press, Chicago.

LAWRENCE, D. T., DOYLE, M. & AIGNER, T. 1990. Stratigraphic simulation of sedimentary basins: concepts and calibration. *American Association of Petroleum Geologists Bulletin,* **74,** 273–295.

LYELL, C. 1835. The Bakerian Lecture – On the proofs of a gradual rising of the land in certain parts of Sweden. *Philosophical Transactions of the Royal Society of London*, **125**, 1–38.

MacLAREN, C. 1842. Art XVI – The glacial theory of Prof Agassiz. Reprinted in *American Journal of Science*, **42**, 346–365.

MIALL, A. D. 1994. Sequence stratigraphy and chronostratigraphy: problems of definition and precision in correlation, and their implications for global eustasy. *Geoscience Canada*, **21**, 1–26.

—— 1996. *The Geology of Stratigraphic Sequences*. Springer-Verlag, Berlin.

NAISH, T & KEMP, P. J. J. 1997. Sequence stratigraphy of sixth-order (41 k.y.) Pliocene–Pleistocene cyclothems, Wanganui Basin, New Zealand. A case for regressive systems tract. *Geological Society of America Bulletin*, **109**, 978–999.

PARKER, J. R. (ed.) 1993. *Petroleum Geology of North-West Europe: Proceedings of the 4th Conference*. Geological Society, London, vols 1–2.

PAYTON, C. E. (ed.) 1997. *Seismic Stratigraphy – Applications to Hydrocarbon Exploration*. American Association of Petroleum Geologists Memoir 26.

POSAMENTIER, H. W. & WEIMER, P. 1993. Siliciclastic sequence stratigraphy and petroleum exploration – where to from here? *American Association of Petroleum Geologists Bulletin*, **77**, 731–742.

PROUST J. N., DECONNICK, J. F., GEYSSANT, J. R., HERBIN, J. P. & VIDIER, J. P. 1995. Sequence analytical approach to the Upper Kimmeridgian – Lower Tithonian storm-dominated ramp deposits of the Boulonnais (Northern France). A landward time-equivalent to offshore marine source rocks. *Geologische Rundshau*, **84**, 255–271.

READ, J. F., OSLEGER, D. A. & ELRICK, M. E. 1991. Two-dimensional modelling of carbonate ramp sequences and component cycles. *In*: FRANSEEN, E. K., WATNEY, W. L., KENDALL, C. G. ST. C., & ROSS, W. (eds) *Sedimentary modelling: Computer Simulations and Methods for Improved Parameter Definition*. Kansas Geological Survey Bulletin 233, 473–488.

RICH, J. L. 1951. Three critical environments of deposition, and criteria for recognition of rocks deposited in each of them. *Bulletin, Geological Society of America*, **62**, 1–20.

SAHAGIAN, D., PINOVS, O., OLFERIEV, A. & ZAKHAROV, V. 1996. Eustatic curve for the Middle Jurassic – Cretaceous based on Russian Platform and Siberian stratigraphy: zonal resolution. *American Association of Petroleum Geologists Bulletin*, **80**, 1433–1458.

SECORD, J. A. 1997. Introduction. *In*: LYELL, C. *Principles of Geology* (abridged). Penguin, London.

SEDGWICK, A. & MURCHISON, R. J. 1839. On the classification of the older rocks of Devon and Cornwall. *Proceedings of the Geological Society of London*, **3**, 121-123.

SKELTON, P., TURNER, C., WILSON, C., HYDEN, F. & COE, A. 1997. *Siliciclastic Sediments and Environments: Study Commentary for Block 1, S338 Sedimentary Processes and Basin Analysis*. Open University, Milton Keynes.

SLOSS, L. L. 1963. Sequences in the cratonic interior of North America. *Geological Society of America Bulletin*, **74**, 93–114.

——, KRUMBEIN, W. C. & DAPPLES, E. C. 1949. *Integrated Facies analysis*. Geological Society of America Memoir 39, 91–124.

STILLE, H. 1924. *Grundfragen der vergleichenden Tektonik*. Borntraeger, Berlin.

SUESS, E. 1906. *The Face of Earth*. Clarendon, Oxford.

SUTTON, J. 1975. Charles Lyell and the liberation of geology. *New Scientist*, (20 February), 442–445.

TEICHERT, C. 1958. Concepts of facies. *American Association of Petroleum Geologists Bulletin*, **42**, 2718–2744.

TUCKER, M. E. 1993. Carbonate diagenesis and sequence stratigraphy. *In*: WRIGHT, V. P. (ed.) *Sedimentology Review 1*. Blackwell Scientific Publications, Oxford, 51–72.

VAIL, P. R. 1992. The evolution of seismic stratigraphy and the global sea-level curve. *In*: DOTT, R. H., Jr (ed). *Eustasy: The Historical Ups and Downs of a Major Geological Concept*. Geological Society of America Memoir 180, 83–91.

——, MITCHUM, R. M. & THOMPSON, S. 1977. Seismic stratigraphy and global changes of sea level. *In*: *Seismic Stratigraphy – Applications to Hydrocarbon Exploration*. American Association of Petroleum Geologists Memoir 26, 83–97

——, TODD, R. G. & SANGREE, J. B. 1977. Seismic stratigraphy and global changes of sea level, part 5: chronostratigraphic significance of seismic reflections. *In*: PAYTON, C. E. (ed.) 1997. *Seismic Stratigraphy – Applications to Hydrocarbon Exploration*. American Association of Petroleum Geologists Memoir 26, 99–116.

VAN WAGONER, J. C., MITCHUM, R. M., CAMPION, K. M. & RAHMANIAN, V. D. 1990. *Siliciclastic Sequence Stratigraphy in Well Logs, Cores, and Outcrops*. Methods in Exploration Series, No. 7, American Association of Petroleum Geologists.

WALKER, R. G. 1990. Facies modelling and sequence stratigraphy. *Journal of Sedimentary Petrology*, **60**, 777–786.

WALTHER, J. 1894. *Einleitung in die Geologie als Historische Wissenschaft*. Fischer Verlag, Jena, 535–1055.

WHEELER, H. E. 1958. Time-stratigraphy. *American Association of Petroleum Geologists Bulletin* **77**, 1208–1218.

WILGUS, C. K., POSAMENTIER, H., VAN WAGONNER, J., ROSS, C. & KENDALL, C. G. ST C. (eds) 1988. *Sea-Level Changes: an Integrated Approach*. Special Publication 42, Society of Economic Palaeontologists and Mineralogists.

WILSON, R. C. L. 1992. Sequence stratigraphy. *In*: BROWN, G. C., HAWKESWORTH, C. J. & WILSON, R. C. L. (eds) *Understanding the Earth*. Cambridge University Press, Cambridge, 388–414.

WOODLAND, A. (ed.) 1975. *Petroleum and the Continental Shelf of North-west Europe, Vol. 1, Geology*. Applied Science Publishers, Institute of Petroleum, London.

Extrusions of Hormuz salt in Iran

CHRISTOPHER J. TALBOT

Hans Ramberg Tectonic Laboratory, Institute of Earth Sciences,
Uppsala University, S-752 36 Uppsala, Sweden

Abstract: This work illustrates that Lyell's approach to inferring geological processes from field observations can be improved by using field measurements of their rates to scale dynamic models. A controversy in the 1970s about whether crystalline rock salt can flow over the surface echoed an earlier controversy resolved with Lyell's help about whether ice could flow and carry exotic blocks. This work argues that the external shapes, internal fabrics and structures and rates of flow of current salt extrusions in the Zagros Mountains are keys not only to understanding past and future extrusion of salt, but also to the extrusion of metamorphic cores of orogens.

Salt is shown to extrude first in hemispherical domes, which later spread to the shapes of viscous fountains until they are isolated from their source. They then rapidly assume the shapes of viscous droplets, which they maintain until they degrade to heaps of residual soils. Salt emerges as a linear viscous Bingham fluid, and strain rate hardens downslope to a power-law fluid ($n = 3$) beneath a thickening carapace of brittle dilated salt.

The velocities of salt constrained in emergent diapirs by measurements of extruding salt sheets in Iran are remarkably rapid compared with rates estimated for buried equivalents elsewhere in or under which it is planned to store nuclear waste or to extract hydrocarbons.

Because halite (NaCl) is the product of evaporation exceeding the supply of sea water, beds of rock salt often occur near the base of sedimentary sequences that accumulate in new basins. Diagenesis can render evaporated salt sequences essentially solid only tens of metres beneath the evaporative surface (Lowenstein & Hardie 1985). As a result, beds of rock salt soon attain a density close to the 2200 kg m^{-3} that they maintain as they are buried by clastic sediments that usually compact to greater densities beyond a depth of about 1 km. Superposed clastic sediments can sink into the salt, leaving salt bodies near the depositional surface which become the crests of salt diapirs downbuilt by flow during progressive burial of the surrounding source layer. Gravity can also upbuild diapirs into or through prehalokinetic overburdens that are ductile (Ramberg 1981) or brittle (Schultz-Ela *et al.* 1993). Only recently has it been appreciated that the buoyancy of planar salt layers is sufficiently low that most overburdens are so stiff that they have to be faulted and thinned before reactive diapirs rise into or extrude through them (Vendeville & Jackson 1992a, b; Jackson & Vendeville 1994).

As well as being buoyant, buried salt is also weak enough to act as a lubricant along decollements beneath basin sequences undergoing lateral extension and/or shortening. Thrust prisms moving over salt are wide with gentle taper angles incorporating more upright box folds with shorter hinges – compared to adjoining thrust wedges not underlain by salt, which are narrower and have a steeper taper angle with constant structural vergence (Davis & Engelder 1987). Laterally unsupported shelf sequence may collapse *en masse* downslope into basins opened over salt deposited at the rifting stage, as in the Gulf of Mexico (Wu *et al.* 1990), Atlantic Ocean (Liro & Coen 1995) and Red Sea (Heaton *et al.* 1995). Gravity sliding and spreading of open continental margins mimics the thin-skin effects of ocean-closure orogeny (Talbot 1992). Progradation of the cover squeezes the substrate into deep swells which break through the cover as giant nappes that laterally extend beneath the upper slopes as they spread towards or over fold belts of autochthonous cover rucked up beneath them (Wu *et al.* 1990).

The first workers who described the many majestic extrusions of Hormuz salt in the Zagros Mountains of Iran (Fig 1) appear to have appreciated that the salt in many of them is still flowing (De Böckh *et al.* 1929, Harrison 1930). However, optimists advocating the storage of radioactive waste in salt in the 1970s (e.g. Gera 1972), knowing that the rock mechanics of the time could not account for salt flow at the surface (Carter & Hansen 1983), interpreted extruded salt as having flowed over the surface only when it was hot, like lavas (Gussow 1966). This argument was rebutted

TALBOT, C. J. 1998. Extrusions of Hormuz salt in Iran. *In*: BLUNDELL, D. J. & SCOTT, A. C. (eds)
Lyell: the Past is the Key to the Present. Geological Society, London, Special Publications, **143**, 315–334.

315

by good 'present is the key to past' arguments. Tall extrusions of salt must still be active because some avalanche episodically, and supply must exceed loss because many still rise a kilometre above the surrounding plains despite an annual rainfall capable of dissolving them in a few tens of thousands of years (Kent 1966). Such rebuttals were followed by remote (Wenkert 1979) and then field (Talbot & Rogers 1980) measurements demonstrating that some of the Zagros salt extrusions are still active. Measurements suggesting that one of these salt mountains flowed faster when it rained (Talbot & Rogers 1980) were presented as a natural demonstration in bulk crystalline salt of the long-known Joffé effect, in which water softens single crystals of halite (Kleinhamns 1914). Field measurements prompted laboratory experiments (Urai *et al.* 1986) which led to the recognition that water-assisted dynamic recrystalization can account for the low-temperature flow of salt. Arguments over the mobility of salt in the 1970s mirrored an earlier controversy that raged over the mobility of ice in the Alps, in which Lyell played a major role (see Boylan this volume, p. 157).

During studies of many of the 130 or so salt intrusions emergent in the Zagros, Player (1969) made similar inferences for salt as those made earlier by Lyell for ice. Thus Player used spreads of distinctive multicoloured insoluble Hormuz rocks among the honey-coloured country rocks to infer the timings and magnitudes of past extrusions of salt and thereby constrain rates of post-Triassic salt intrusion (1969; in Edgell 1996).

The remainder of this work treats the Zagros as a unique laboratory where salt extrusions not only mimic past and present salt extrusions elsewhere, but also represent smaller and faster natural analogue models of the gravity spreading of other tectonites extruded as piles of nappes in orogens. This review of salt extrusions is based on visits to at least 45 examples since 1993 as part of the research programme on '*Zagros halokinesis*' run jointly by geologists from the Institute of Earth Sciences at Uppsala University, Sweden, and the Institute for Earth Sciences at the Geological Survey of Iran. By November 1997, field parties had spent five field seasons of a month each making new surveys and documenting the displacements, apparent to theodolites stationed on country rock, of hundreds of markers on ten mountains of Hormuz salt.

A brief introduction to the Hormuz salt and its history is followed by descriptions of the external shapes of salt extrusions at different stages and their internal fabrics and structures, both ductile and brittle. Recent measurements of the rates of one of the salt extrusions are used to scale a non-dimensional flow model and to constrain its viscosity and current extrusion velocity.

The Hormuz Formation

Beds of salt in the Hormuz sequence are estimated to have an original total thickness of about 1 km (Kent 1958) and are divisible into two parts by red beds (probably of the Middle Cambrian Lalun Formation). Neoproterozoic salt consists of beds of multicoloured halite interlayered with beds of anhydrite, magnesium carbonates and fetid limestone containing stromatolites. This cyclic sequence probably accumulated in shallow Proto-Tethyan basins along the tropical margin of Gondwana from Iraq to Australia (Talbot & Alavi 1996). Apart from dispersed anhydrite and blocks of bimodal volcanic rocks, Cambrian salt is more pure and uniform and probably accumulated in graben that trended north–south, along the local grain of the underlying Pan-African (900–500 Ma) crystalline basement. The salt was buried by clastic sediments in the Palaeozoic and platform carbonates from Triassic until Miocene times. An overburden 2.5–3 km thick by Jurassic times

Fig. 1. Photographs of profiles of salt extrusions at different stages in the Zagros. Young bun-shaped post-Zagros domes of Hormuz salt, (a) extruding through a collar of Cretaceous carbonates at Kuh-e-Gach (28° 37′ N, 52° 18′ E), reaches 956 m above sea level, but only about 400 m above the surrounding molasse plain (b) extruding through a dome of Eocene carbonates at Kuh-e-Charhal (dome 67, 28° 00′ N, 54° 58′ E) seen from the west; (c) Kuh-e-Charhal seen from the north. Salt fountains breaching anticlines at: (d) Angoreh (Dome 9, 27° 20′ N, 55° 55′ E) from the southwst; (e) Kuh-e-Namak (Dashti, 28° 16′ N, 51° 42′ E) from the northwest; and (f) Darbhast (dome 49, 26° 57′ N, 54° 08′ E) from the east-northeast. Salt fountains rising through molasse plains: (g) the 'fried egg' (dome 80, 28° 00′ N, 54° 56′ E) from east-southeast; (h) Kuh-e-namak (Feroosabad, 28° 45′ N, 51° 22′ E), from the south-southeast showing a dynamic fountainhead sinking into the plateau of a salt droplet. The southern reaches of namakiers at (i) Kuh-e-Namak (Feroosabad) and (j) Kuh-e-Jahani (28° 37′ N, 52° 25′ E) also have the profiles of viscous droplets. Shrunken salt droplets: (k) at Kuh-e-Talkeh (Dome 48, 27° 03′ N, 54° 14′ E) from the west; and (l) Kuh-e-Bachoon (28° 57′ N, 52° 06′ E) from the south. Snouts of advancing namakiers: (m) Kuh-e-Jahani and (n) Kuh-e-Namak (Feroosabad). (o) Satellite image of the 'fried egg' dome 80, courtesy of Exxon. (p) Last remnants of salt cloaked by Hormuz soils at Roxana (28° 40′ N, 52° 15′ E). (q) Heaps of Hormuz soils bury a melange chimney at Bogdana (28° 25′ N, 52° 16′ E).

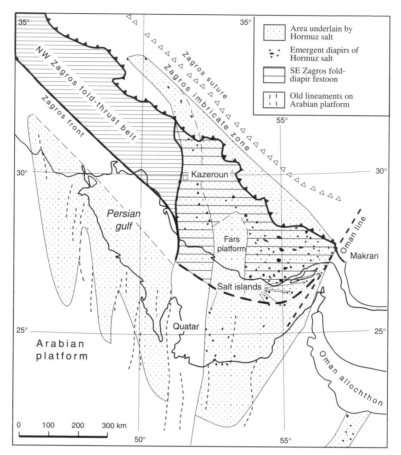

Fig. 2. Map of the Persian Gulf showing components of the Zagros Mountains and areas where Hormuz salt occurs between the cover and basement with its old faults.

appears to have been sufficiently stiff to constrain the underlying salt for approximately 300 Ma. This delay in the movement of salt long buried to a buoyant state was probably responsible for the general acceptance that rock salt had a yield strength equivalent to 3 km of overburden (Gera 1972), until it was appreciated that salt pierced prehalokinetic overburden only 350 m thick in the interior basins of the US gulf coast (Rosenkrans & Marr 1967) and only 610 m thick in the North Sea (Brunstrom & Walmsley 1969).

Hormuz salt in and around what is now the Persian Gulf (Fig. 2) began flowing very slowly into deep conformable pillows with north and north-northeast trends from late Jurassic to early Cretaceous times (Edgell 1996). The development of salt pillows was in response to a wave of instability travelling eastward along the northern margin of Gondwana which riffled fault blocks in

the basement like piano keys as Neo-Tethys subducted beneath central Iran (Talbot & Alavi 1996). Neo-Tethys began to close like a zip fastener pulled from the west in the Late Cretaceous. Since then the front of the Zagros Mountains has been propagating southwestwards away from the suture in central Iran and driving the Persian Gulf in front of it as a foreland basin. Several faults in the deep basement have reactivated as transverse faults since the suturing. Two of these, the Kazaroun and Mangarak fault zones, are defining an incipient syntaxis along the western boundary of Hormuz salt with significant thickness (Talbot & Alavi 1996). The Zagros has accumulated serial structures with styles that depend on the nature of its basal decollement. To the northwest of the Kazaroun fault zone, where any Hormuz salt has negligible thickness, the northwestern Zagros fold-thrust belt (Fig. 1) is a steep and narrow wedge characterized by long thrusts and thrust antiforms all verging southwestwards. East of the Kazaroun fault zone, the southeastern Zagros fold-diapir festoon (Fig. 1) is twice the width with short upright box folds punctured by salt diapirs behind a much gentler front which has advanced twice as fast over Hormuz salt.

Diapirs of Hormuz salt rose from many of the deep pillows in three main phases: Lower Cretaceous, Eocene to Miocene, and Recent (Player; 1969, in Edgell 1996) and some surfaced in what was still the foreland far beyond the front of the southeastern Zagros fold-diapir festoon. Much of the Hormuz salt squeezed to the surface from depth then may have been recycled north-westward as the Fars (Miocene) evaporites which later acted as a high level detachment and the source of a completely different suite of diapirs in the north-western Zagros fold-thrust belt (O'Brien 1957).

On arrival of the fold-thrust front, large pre-Zagros pillows of Hormuz salt still at depth east of the Kazaroun fault zone suddenly deflated to feed syn-Zagros diapirs with smaller diameters. These are locally flanked by salt withdrawal basins full of Bhaktiari or Recent molasse noticeably thicker than surrounding equivalents. Salt still remaining at depth behind the Zagros front now rises in post-Zagros diapirs along a few conjugate strike-slip faults (Fürst 1976).

Many diapirs now onshore near the gulf coast were emergent by Miocene times when they were buried by soft sediments; their subsequent reactivation led to them piercing upturned collars of superposed cover. However, the Persian Gulf is a shallow foreland basin being pushed over a low desert and not like the Gulf of Mexico, the depocentre of the mighty Mississippi. Instead of giant sheets of Jurassic salt squeezed to the edge of the prograding shelf being repeatedly reactivated to downbuild new generations of diapirs under the Gulf of Mexico, most of the Hormuz sequence is extruded subaerially until the deep source is locally exhausted or isolated by Zagros shortening. The dissolution of emergent salt then results in melange chimneys of Hormuz residues resembling those made by salt of similar age in the Amadeus and Adelaide Basins in Australia (Talbot & Alavi 1996). The planforms of heaps of soils that mark dissolved syn-Zagros or post-Zagros diapirs are rhombic or round but those of pre-Zagros diapirs are smeared toward long and narrow multicoloured stripes along thrusts (Kent 1979).

Every stage in the recycling of Hormuz salt still exists in the Gulf region. Pre-Zagros pillows and diapirs are still deep beneath the surface south of the gulf but emergent in Oman. Syn-Zagros salt islands mark the Zagros front in the Gulf itself and post-Zagros diapirs already extrude along strike-slip faults behind the front among the wasted and deformed remnants of earlier extrusions.

Ductile profiles of salt extrusions in the Zagros

Blind diapirs and young extrusive domes

Player (1969) constrained the rates of rise of subsurface (blind) diapirs in the Zagros to $0.3-2$ mm a^{-1} (in Edgell 1996) and Ala (1974) first outlined the various stages through which Zagros salt extrusives evolve. There are so many Kuh-e-Namaks (mountains of salt) and Gach (sodium sulphate) in the Zagros that identification is usually aided by bracketing their name with the district or nearest city or by quoting a (still incomplete) numbering system begun by de Böckh and colleagues (1929).

Topographic salt domes

Simple Zagros anticlines of carbonate beds rise with the smooth rounded profiles of buckles but are soon carved by rivers and slope failures to skylines that are blocky on scales of hundreds of metres. By contrast, extrusions of salt rapidly develop and long maintain the smooth rounded outlines of flows of another crystalline fluid, ice. However, whereas the profiles of ice sheets are shaped by gravity acting on a budget of ice supplied as snow from above and lost mainly by basal melting (along with some river erosion *etc.*), extrusions of salt are shaped by gravity acting on salt supplied from below and lost by dissolution from above.

Before they surface on the plains, still-blind diapirs deflect the drainage pattern by doming first unconsolidated gravels and, after these have been eroded, their carbonate overburden. Syn-Zagros diapirs emerge at the crests of anticlinal mountains, in the intervening plains of molasse, or along faults anywhere in-between (Kent 1970).

The imminent emergence of blind diapirs in mountain massifs is heralded by the exposure of almost circular domes in the Phanerozoic cover which contrast with the elongation of adjoining thrust anticlines. Examples abound within a few anticlines of the shore of the southeastern Zagros fold-diapir festoon. The salt is first exposed as dike-like masses rising through residual soils in the core of a dome of deep country rocks as reported for dome 44, 9 km east of Bastak by Player (1969; in Edgell 1996). A simple hemispherical or bun-shaped topographic dome of clean rock salt then rises above the local scenery (Figs 1(a)–(c) and 3(a)–(c)). An example at Charhal reaches approximately 2.5 km above sea level but protrudes only perhaps 300 m above an orifice that breaches the crest of a dome of carbonates. Viewed from the west, this magnificent dome appears to rise proud of all its country rocks (Figs 1(b) and 3(b)) but, seen from the north (Figs 1(c) and 3(c)), a nearby peak of country rock is about the same height.

Salt fountains and namakiers

Young domes of salt probably start rising into the gulf or sky with the rhombic planform of the gap opened along a fault in the country rocks. However, the rising salt cannot support its own weight and soon overflows the edges of its orifice. Domes are soon surrounded by a plateau of salt flowing in unconstrained directions over the surrounding scenery so that mature salt extrusions develop the profiles of viscous fountains (Figs 1(d)–(h) and 3(d)–(h); see also Talbot & Jarvis 1984; Talbot 1993). Narrow flanges occasionally spread along

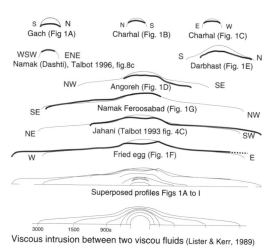

Superposed profiles Figs 1A to I

Viscous intrusion between two viscou fluids (Lister & Kerr, 1989)

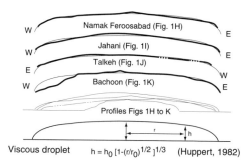

Profiles Figs 1H to K

Viscous droplet $h = h_0 [1-(r/r_0)^{1/2}]^{1/3}$ (Huppert, 1982)

Fig. 3. Profiles of Zagros salt extrusions traced from Fig. 1 with the same identification letters (but with two extra examples) compared to profiles of a viscous fluid spreading along a fluid interface (Lister & Kerr 1989) or a droplet of constant volume spreading over a non-slip surface (Huppert 1982; McKenzie et al. 1992).

the crests of anticlines but most extrusive salt flows down one or more fault gullies or uptilted river valleys and/or out over the surrounding plains to form sheets of allochthonous salt, known as 'namakiers' (from the Farsi: namak = salt), which are analogues of valley or piedmont glaciers. Altogether there are perhaps 20 namakiers still extant onshore in the Zagros. Some are still advancing, others are essentially steady state, and still others are in retreat behind terminal moraines. Past examples are represented by spreads of Hormuz material in the stratigraphic record (Fig. 4(a)) or by blocks of Hormuz materials too large to be carried by current streams spread over alluvial plains darkened by finer-grained Hormuz clasts.

The shapes, sizes, and rates of flow of individual salt extrusions reflect mainly the rate at which the supply of salt through their orifice exceeds surface dissolution by rain, and the topographies of the scenery over which they spread. The surfaces of namakiers that flow rapidly down the steep flanks of Zagros anticlines tend to be steep (15 degrees) and comparatively thin (Figs 1(e), (f) and 5(e), (f)). Piedmont namakiers that spread slowly over horizontal plains are characteristically about 40 m thick beneath gentle (less than 1 degree) top slopes (Fig. 1(g)–(j)). Most salt extrusions rising on high ground are asymmetric and feed only one namakier. Two clear examples are a Kuh-e-Gach (dome 68, near Lar) and Burkh (dome 48), which rise on adjacent anticlines and flow southwards and northwards, respectively, across the flat floor of the same west–east valley and would have met were they not offset by 7 km. By contrast, extrusions on the plains are more symmetric, and dome 80 resembles a fried egg with a summit dome encircled by a continuous namakier approximately 20 m thick divided by a river bed draining south (Figs 1(g), (o) and 3(g)).

Spreading salt droplets

When the supply of diapiric salt to a salt fountain is exhausted or isolated from its deep source by thrusting, the fountainhead sinks into and thickens the surrounding salt plateau. No longer supplied from below, the extruded salt mass thereafter merely spreads with a volume slowly diminishing by erosion. Together with the profiles of namakiers supplied from out of section, profiles of such salt mountains fit the non-dimensional, mathematically modelled profiles of viscous droplets (Huppert 1982, see Figs 1(i), (k) and 3(i), (k)). They tend to maintain these profiles even as they waste beneath cloaks of insoluble cap soils (Figs 1(l) and 3(l)). Mathematical models of viscous fountains and droplets do not constrain the profiles of their distal terminations very well. However, the snouts of

Fig. 4. (a) Conglomerate of black and green Hormuz debris (approximately 50 cm thick) in cream-coloured Miocene sandstones, northwest of Kuh-e-Bam (dome 79). (b) Fresh sink hole near the summit of Kuh-e-Namak (Feroosabad). (c) Funnel-shaped sink holes on the eastern plateau of Kuh-e-Namak (Feroosabad). (d) Badland soils on the northern margin of Ku-e-Bachoon. (e) Compound dissolution collapse hollow about 30 m deep, western Jahani. (f) Efflorescence along a stream on Kuh-e-Namak (Dashti). (g) Salt in a dissolution hole near the summit of Kuh-e-Namak (Feroosabad). (h) Cave (6m high) in the northern namakier of Kuh-e-Namak (Dashti).

namakiers advancing over their own debris have the same steep bulbous profiles of advancing ice flows (Fig. 1(m), (n)), and viscous models of advancing flows (Ramberg 1981, figs 9.10 and 9.11). By contrast, the distal margins of clean wasting namakiers have the same feather edges (Fig. 6(b)) as clean retreating ice. Submarine namakiers have not been reported as spreading

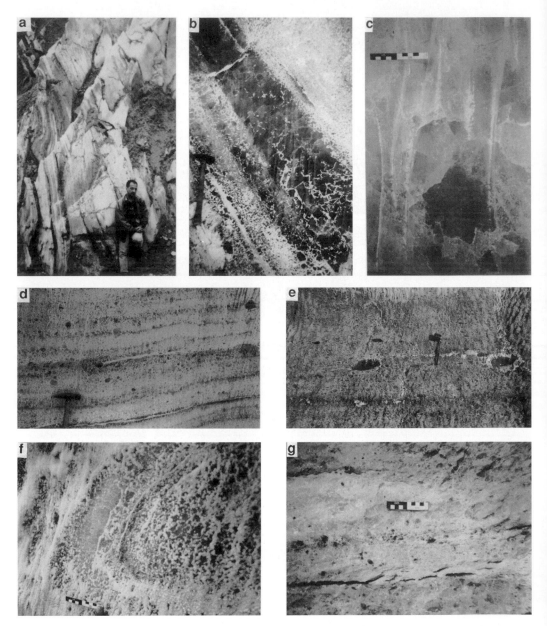

Fig. 5. Internal fabrics. (a) Steep layers (bedding) of course-grained gneissose halite with upright flow folds and boudin of pegmatite inherited from the long journey to the surface exposed in the orifice of Kuh-e-Bachoon. (b) Pegmatite of halite near the summit of Kuh-e-Namak (Dashti). (c) Close-up of halite pegmatite in the orifice of Kuh-e-Bachoon. (d) Relict megacrysts of halite dispersed in fine-grained halite are usually spherical, here on the southern slopes of Kuh-e-Namak (Dashti) but become elliptical (e) where salt flow is impeded a few metres downstream. (f) A gneissose grain shape fabric axial planar to a recumbent fold in the snout of Kuh-e-Namak (Feroosabad). (g) Gneissose halite above and below a thin mylonite in a loose block of salt at Roxana.

Fig. 6. Internal structures, mainly of Kuh-e-Namak (Dashti). (a) Gentle recumbent folds on the southern face (about 200 m high) of the summit dome. Such gentle flow folds amplify and close to piles of nappes downstream. (b) Flow folds amplifying (right to left) to distal thrusts in the lower slopes just behind the feather edge of the southern namakier; the thickness of salt visible is about 30 m. (c) Inclined folds in salt flowing right to left over a basal obstruction in the upper slopes of the northern namakier at Kuh-e-Namak (Dashti), rucksack in right foreground for scale. (d) The main recumbent fold behind the snout of the northern namakier at Kuh-e-Namak (Dashti) simulates a tank track. (e) Recumbent folds about 100 m upstream of the last obstruction encountered by the northern namakier at Kuh-e-Namak (Dashti). (f) The sequence seen in the lower half of (e) is here thinned and repeated five times 20 m short of the limestone ridge. (g) Upright folds (foreground) and the tank track (background) near the northern snout of Kuh-e-Namak (Dashti). (h) Thrusts in distal basal mylonitic salt, Kuh-e-Namak (Bandar Abbas). (i) Salt at the snout of the northern namakier at Kuh-e-Namak (Dashti); the thickness of salt visible is approximately 3 m.

from islands of extrusive Hormuz salt out in the Persian Gulf, although this seems likely (see later).

The huge sheets of allochthonous salt beneath the Gulf of Mexico were introduced to the scientific literature as intrusions of Jurassic salt into Tertiary sediments (Nelson & Fairchild 1989). However, rather than having the profiles of laccoliths under plastic roofs, profiles of the tops of salt bodies beneath the Gulf of Mexico are so like those of the subaerial salt extrusions in the Zagros (Talbot 1993) that the picture accepted now is that many, if not most, originated as submarine namakiers beneath thin veneers of muds and were subsequently buried, with or without reactivation (e.g. Fletcher *et al.* 1995).

Salt extrusions as models of orogenic extrusion.

Speculative sketches by recent workers (e.g. Fig. 7(A), after Anderson *et al.* 1992, and (B), after Grant 1992) suggest that the metamorphic cores of ororgens extrude from the sutures between

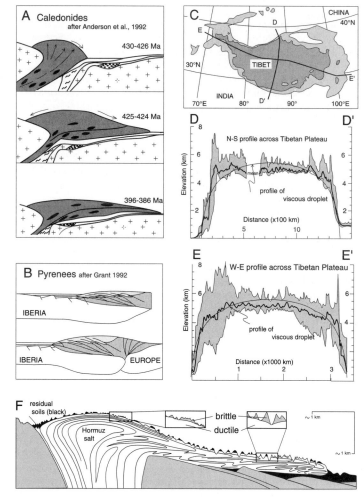

Fig. 7. (A) West–east sketch profiles of the Scandinavian Caledonides interpreted as two-dimensional viscous extrusions evolving from semi-circular through fountain to droplet profiles (after Anderson *et al.* 1992). (B) South–north sketch profiles of two stages of the Pyrenees extruding from the suture between Iberia and Europe and gravity spreading over an irregular substrate (from Grant 1992). (C) Locations of (D) north–south and (E) west–east topographic profiles across the Tibetan Plateau (from Fielding *et al.* 1994) with suitably scaled profiles of a droplet of viscous fluid (Huppert 1982; see bottom diagram of Fig. 3, p. 320) superposed to show close fit. (F) Cartoon profile of Hormuz salt extruding from an orifice in a Zagros anticline and gravity spreading beneath a brittle carapace thickening downstream.

converging continents with profiles that begin as simple bulges before they gravity spread as piles of nappes and eventually assume profiles reminiscent of spreading viscous droplets like the extrusions of salt described here (Figs 1 and 3). I know of no orogenic profiles that are simple bulges or have dynamic fountainheads. However, north–south and west–east profiles acros the Tibetan Plateau (Fig. 7(D), (E)) fit the profile of a viscous droplet spreading over a non-slip surface (Huppert 1982) sufficiently well to suggest that the profiles through which salt extrusions evolve may well be considered as smaller and faster natural analogues of orogenic extrusion. The cartoon profile of a 'typical' salt extrusion (Fig. 7(F)) emphasizes that not only does the external topography of salt extrusions model orogens; the fabrics, structures and brittle carapaces described in later sections also model orogenic nappe piles. But first, spatial dimensions and velocities will be added to a non-dimensional model of a salt extrusion to constrain the viscosity of the Hormuz salt and the rate at which it extrudes in the Zagros.

Flow rates

Markers painted along lines of sight monitored on Kuh-e-Namak (Dashti) for two months early and late in 1977 established that salt still damp from rain flowed over a metre in two days (Talbot & Rogers 1980). By our next visit to this mountain, in June 1994, markers painted on the salt were no longer recognizable and the 1977 position of only one could still be reconstructed with reasonable confidence. This had moved over 50 cm downslope in two days in January 1977 but in the next 17 years moved downslope at a rate that averaged only ≈ 4 cm a^{-1}. We renewed the 1977 markers and added many more which survived the excessive rains of the winter of 1994/1995. A new marker spray-painted in November 1995 on the position mentioned above showed insignificant movement in November 1996 but had moved 180 cm downstream during a particularly wet November in 1997. This particular marker therefore moved over 3 m in 20 years but the rate of 15 cm a^{-1} averaged over 20 years occurred in two increments of 50 cm and 180, the second of which occurred within a year, and the first within two days. Salt extrusions can apparently surge, but whereas glaciers probably surge over wet bases, namakiers appear to surge within an hour of their tops being dampened by light rain (e.g. 5 mm during a day: Talbot & Rogers 1980).

The displacements of 42 natural markers on the salt of Kuh-e-Jahani (mountain of the universe) apparent to three theodolite stations between 1994 and 1996 are about to be reported (Talbot et al., in preparation). The general pattern of displacements

on this, currently one of the largest salt fountains in Iran (Fig. 2(j)), demonstrates that salt is extruding vigorously and spreading down a gentle route through the modern scenery. Differential displacements indicate that the lip of the summit dome is sinking and spreading downslope at approximately 4 m a^{-1} while the middle reaches are flowing at approximately 2 m a^{-1}. These are much the same velocities as measured in valley glaciers of about the same size.

The most startling of all the apparent displacements on Jahani are a 24 m outward and 62 cm downward motion of one marker over an 18 month period. This extraordinary rapid motion, at around 1 m per month, occurred by avalanching of salt destabilized by ductile flow behind a location where steep river cliffs 400 m high have been cut almost into the core of Kuh-e-Jahani (Talbot et al., in preparation).

The dimensions and velocities constrained at Kuh-e-Jahani have been used to scale numerical models (Sergei Medvedev, in preparation) of a viscous fountain extruding over a flat plain while its top surface loses salt to uniform dissolution at rates of 2–3 cm a^{-1}. The resulting dimensionalization suggests that salt rises out of the centre of its bedrock orifice at between 0.5–3 m a^{-1} and gravity spreads with viscosities in the range of 10^{16}–10^{17} Pa s. This rate is consistent with modern formulae for the mechanics of rock salt (e.g. Cristescu & Hunsche 1993). It exceeds the extrusion rate of 17 cm a^{-1} budgeted by Talbot and Jarvis (1984) with poorer constraint in both time and space for another salt fountain 100 km further south. These are the sort of velocities necessary to advance the fronts of salt sheets downslope over the floor of the Gulf of Mexico at 1.5–275 mm a^{-1} in Miocene to Pleistocene times (Wu et al. 1990).

Talbot and Jarvis (1984) interpreted the planform of bedding retarded along the lateral margins of a particular salt fountain as mapping the velocity profiles. They then compared these planforms with those of channel flows in linear and nonlinear fluids and thereby constrained the power law exponent of the distal salt to $n = 3$, a value lower than the $n = 5$ to 7 determined for most rock salts in the laboratory (Cristescu & Hunsche 1993). Applying the same approach to aerial photos of salt layering along the main stream of Kuh-e-Jahani leads to the conclusion that the salt there is essentially Newtonian ($n = 1$) where it leaves the summit dome, but that $n \sim 3$ near the snout.

Salt extrusions as tectonic pressure gauges

Salt extrusions act as tectonic pressure gauges when their heights are related to their levels of neutral buoyancy. In practice, the heights that salt extrusions reach relative to the overburden surrounding them are useful qualitative guides to tectonic overpressures (Talbot 1992; see also Davison *et al.* 1996). Onshore, the most vigorous current salt extrusions are those rising up gaps pulled apart at releasing side-steps along transpressional strike-slip transfer faults trending across the grain of the Zagros folds and thrusts (Talbot & Alavi 1996). Instead of pull-apart basins being filled by clastic sediments fed from above, they are filled by salt squeezed from below. These extrusions may reach nearly 2.5 km above the surrounding plains but even the most impressive onshore examples are no higher than one or more of the surrounding mountain peaks (Fig. 1(b), (c)). This implies that, however the extruding diapir originated, the height multiplied by density of the salt column equals the height multiplied by the density of the overburden sitting on its deep source layer; any friction between the salt and its country rock appears to be negligible. Thus all the current onshore salt extrusions in the Zagros appear to be driven by gravity alone. The only salt extrusions that currently rise above the maximum heights of the surrounding country rocks are those forming salt islands along the front of the southeastern Zagros fold-diapir festoon offshore in the Persian Gulf (Fig. 2). Several of these islands (e.g. domes 17 and 19–22) can provisionally be interpreted as the fountainheads of submarine salt fountains if the surrounding shoals are largely floored by salt. Taking the level that the salt reaches above the surrounding sea bed as an approximation of the overpressure due to tectonic shortening, this reaches about 4 MPa around dome 20.

Internal ductile fabrics and structures

Internal salt fabrics

Most salt first emerges into the sky of the Zagros with a coarse-grained (5–10 mm) gneissose fabric. Layers of salt with different grain size, colours and inclusions are interpreted as bedding (Figs 5 and 6) modified during its 3–10 km horizontal transport along the deep source layer and 3–12 km vertical journey up the diapiric stem to the surface. Layers of extruding salt deform at similar rates to each other but faster than their inclusions (which are all Hormuz in age: Gansser 1992; Weinberg 1993). Early joints across carbonate beds tend to have been separated by extending flow before they surface, and the gneissose salt emerges containing occasional pegmatites of halite recrystallized to grain sizes of decimetres (Fig. 5(b), (c)) in the strain shadows between such boudins. These pegmatites may have started over a metre thick but have usually thinned to a few decimetres as they

extended to lengths of tens of metres. Proximal coarse-grained fabrics in salt generally decrease in grain size downslope (Fig. 5).

Slow-moving, thick namakiers tend to retain gneissose fabrics (Fig. 5(d)–(f)) between mylonites that are decimetres thick (Fig. 5(g)). In relatively rapid steep and thin namakiers, relict megacrysts of the coarse emergent salt may remain a few centimetres across but decrease in number as they are increasingly dispersed downstream in a fine-grained groundmass of increasingly mylonitic halite. All the megacrysts are circular in most exposed sections (Fig. 5 (d)), but where salt flow is impeded they are elliptical with long axes either parallel to the layering (Fig. 5(e)) or axial planar to flow folds (Fig. 5(f) and Talbot 1979). Such foliations are emphasized by the orientations of dispersed needles of anhydrite or flakes of hematite.

Ramberg (1975) explained the results of competent inclusions rotating in bulk flows by a process since called 'supershear'. The aspect ratios of competent inclusions increase by progressive addition of strain increments as they rotate towards the plane of bulk shear, reach maxima as they rotate through that plane, and thereafter decrease by retrogressive strain increments to minima perpendicular to that plane (see e.g. Ramsay & Huber 1983, p. 130). Talbot & Jackson (1987) argued that each of the grains in essentially monomineralic salt rotates by supershear because it is competent relative to it surroundings of grains plus boundaries. Most grains are likely to start this oscillation in shape with different orientations and shapes. Talbot and Jackson (1987) went on to illustrate how supershear can account for statistical orientations of populations of halite grains defining flow foliations. Coarse grains of similar size have low aspect ratios (less than 2:1) and define gneissose foliations with grain sizes that are probably close to steady state. Every grain emphasizes and diminishes the bulk foliation twice during every complete rotation. Strain that accumulates in progressive increments is subsequently removed or 'forgotten' by a special (dynamic) category of annealing where the strain increments are retrogressive. Halite megacrysts that decrease in size and define foliations in mylonitic salt elongate to aspect ratios as high as 5:1 before parting along one of their basal cleavages to subrounded smaller grains (Talbot 1979). This constant replacement of the most deformed megacrysts by several smaller 'undeformed' grains (another mechanism of dynamic annealing) leads to constant renewal of the foliation. As a consequence, the fabric of mylonitic salt has such a short strain memory that the foliation can change direction over distances of a metre or so (Talbot 1979).

Fig. 8. The brittle carapace, mainly of Kuh-e-Namak (Dashti). (a) Joints with dips of 90 degrees, 45 degrees and zero (±10 degrees) define the trapezoidal profile of this river valley. (b, c) Joints with dips of 45 degrees etched by dissolution, some infilled by infloresence. (d) Master joints above the last obstruction to the northern namakier; about 15 m thickness of salt visible. (e) Polygonal joints in a vertical river cliff, western base of Kuh-e-Angoreh (dome 9). (f) Vertical layers of halite with a rounded block of Neoproterozoic carbonate, summit of Kuh-e-Namak (Dashti). (g) Flaked slab of salt bulging away from a steep face on Kuh-e-Gach. (h) Joints and a master joint with orientations relating to the ductile flow fabric.

In effect, flow foliations in both gneissose and mylonitic salt map the flow lines of the latest increment of bulk flow. Where the flow lines in crystalline fluids merely deflect, constantly regenerating gneissose foliation can remain parallel to the layering so that foliation and layering can undulate together (as in inverse 'folds' between boudins). Where the flow is unsteady because of changes in velocity or boundary conditions in either space or time, new flow lines can develop across passive layers so that, in salt slowed behind an obstruction, the flow foliation regenerates parallel to the axial surfaces of new flow folds in the layering (Talbot 1979, 1992). Where the salt accelerates beyond the obstruction, the flow lines converge, the layering rotates to parallel the flow foliation, and the inherited flow folds tighten to become inconspicuous. Similar processes are likely in other rocks with gneissose fabrics, particularly if they are monominerallic and nearly molten, like ice.

Internal salt structures

Whereas grain-shape fabrics in flowing rocks can have short strain memories, bedding is annealed only by melting. As a result, extrusive salt emerges with sheath folds (Fig. 5(a)) and possible curtain folds inherited from its long journey to the surface (Talbot & Jackson 1987). However, these folds are seldom conspicuous because the extruding steep salt layers turn in short distances to parallel the gentle top and bottom boundaries of the extrusion (Fig. 7(F)). Bedding remains conformable with the outer surfaces of young extrusive domes until gentle recumbent folds in steep bedding extruding from a steep-walled orifice signal gravity spreading of the unsupported lower flanks of the dome. Two or three recumbent folds in the flanks of summit domes also record the beginning of outward and downward flow of mature salt fountains (Fig. 6(a)). Traced downslope, these similar flow folds amplify and tighten about axial surfaces that curve to dip downslope (Fig. 6(b)). The general effect is that of Pennine-type Alpine nappes extruding from steep root zones – a cascade of flow folds tightening as they spread downslope.

The first-order structure of thin namakiers that have spread rapidly down steep slopes is a single sheath fold. Like the axial profiles of Helvetic-type Alpine nappes, downstream profiles of these salt flows resemble tank tracks. Beds in the upper limbs parallel the upper (driving) slopes and although they can locally be thinner than the lower limb (Fig. 6(d)), they usually thicken as they roll over the snout, which advances over relatively stationary lower limbs smeared into finer-grained mylonites a few decimetres thick along slides parallel to the no-slip bottom boundary (Talbot 1979; Talbot & Jackson 1987, fig. 6).

Competence contrasts between clean and impure layers tend to increase downslope. Thus pink salt layers at the base of Kuh-e-Namak (Bandar Abbas) display both shortening (Fig 6(h)) and extending faults a few tens of metres apart downstream of where the salt begins to flow over its own fluvially reworked moraine. Small pebbles appear to be entrained within basal salt flowing over moraines or fluvial plains elsewhere, implying that flowing salt can ruck up soft substrates and incorporate inclusions.

Successive generations of flow folds and slides are picked up where flow planes diverge in salt slowing and thickening upstream of fixed bedrock obstructions. Each generation usually tightens so much where the flow accelerates downstream that the folds are usually cryptic before the next train of flow folds is superposed at the next bedrock obstruction. Refolding is therefore uncommon, and signs of the superposition of several generations of folds in mylonites rapidly carried to distant namakier snouts are usually subtle and recognizable only from repetition of the colour banding and a few isoclinal fold closures (Fig. 6(i)). Fast salt nappes tend to pile one above the other (Fig. 7(F)), slow salt nappes separated by thrusts along repetitions of red beds usually tend to stack one behind the other (Talbot et al. in preparation).

Bedding in basal salt is usually precisely parallel to planar basal contacts on dip slopes of limestone. However, distinctive trains of flow folds and slides develop where the velocity of the flowing salt changes sufficiently for the flow paths to cross the passive bedding. The foliation regenerating along paths of unsteady flow parallels the axial surfaces of new flow folds. These paths tend to diverge upstream of bedrock scarps and converge where the folds tighten down the dip slope beyond (Talbot 1979). The sweep of new folds in steep belts up and over (or around) irregularities fixed in their substrate and the convergence of their limbs in flat belts beyond simulate structures often attributed to late folds in piles of orogenic nappes. Like profiles of their top free surfaces (Fig. 3), the fabrics (Fig. 5) and structures (Fig. 6) within namakiers model gravity spreading of orogens extruded over irregular substrates (Fig. 7(F)).

Like orogenic nappes, namakiers do not appear to erode their substrates like sheets of ice. This may be because inclusions in Hormuz salt are weak and not concentrated along the bottom boundary as harder blocks can be concentrated along the base of flowing ice. Instead of an abrasive lower boundary maintained by basal melting and abrasion, inclusions tend to stay dispersed within flowing salt or to be concentrated by dissolution along its top surface. As a result, namakiers tend to smooth their effective bottom and side boundaries by shearing past irregularities infilled by complex melanges of static salt (Talbot 1979).

Brittle skins and wasting

Destressing joints and dilation

Like ice and other tectonites, crystalline salt can also fracture as it flows. However, namakiers are not crossed by gaping crevasses like accelerating or decelerating ice sheets. Instead, extrusive salt is more like orogenic tectonites; fractures in the brittle carapace remain essentially closed over still-ductile deeper levels.

Fractured surfaces of crystalline salt in Zagros salt extrusions can be hidden beneath insoluble soils, or exposed on gentle clean-washed slopes, in cliffs along rivers or around sink holes. The nearest equivalents in salt of the open crevasses in ice are bergschrunds that gape behind rare avalanching salt cliffs. Steep fracture zones can develop in salt flowing over convex upward breaks in slope but these fractures are sufficiently tight that temporary streams flow over them without effect. Joints (as distinct from thrusts and slides, which generally subparallel foliation and/or bedding) provide the most general evidence for a surficial skin embrittled by dilation and stress relief to the free surfaces of salt extrusions (Fig. 8).

Three characteristic sets of joints tend to control the attitudes of free surfaces of exposed crystalline salt with layering of any orientation. Both the free surfaces and the stress-relief joints paralleling them have dips that are vertical, 45 degrees, and less often (beneath waterfalls eroded in salt) horizontal (± 10 degrees). As a result, profiles of stream valleys eroded in salt tend to be box shaped or trapezoidal (Fig. 8(a)) with flat bottoms flanked by slopes near 45 degrees (of etched crystalline salt rather than screes of broken salt) beneath vertical walls (Fig. 8(b), (c)). Strikes of faces and joints with dips near vertical and 45 degrees box the compass, not only throughout a salt mountain, but also along individual valleys incised by meandering rivers and around intervening salt plateaux or pedestals.

The patterns of fractures in salt extrusions in the Zagros are therefore essentially close to radially symmetric about the vertical on scales over hundreds of metres (controlled by gravity, salt viscosity and basal relief). However, they relate to local relief on scales less than hundreds of metres (controlled by stream erosion and dissolution collapse). This general pattern can be attributed to stress fields that remain isotropic as they increase with depth in a crystalline (Bingham) fluid that is viscous on scales of kilometres and years, but usually fails as a brittle solid on scales of hundreds of metres and days to months. Gravity dilates vertical joints that bisect shears with 45 degree dips indicating that the angle of internal friction of the salt is close to that of an ideal elastic solid – zero. Valley floors locally bulge above horizontal fractures opened by the relief of vertical loads and lateral spreading of the valley walls.

The only visible crystalline salt not fractured in Zagros salt extrusions is that exposed at the bottoms of active river channels. Exposed brittle relief that moves downstream must be carried by

still-ductile flow of the underlying salt, even where the salt beneath markers is likely to be only about ten metres thick. Combined, these two observations suggest that the thickness of the brittle carapace dilated by destressing around each salt extrusion is that of its surface relief, probably with upward bulges over unbroken salt beneath the divides between stream beds (inserts on Fig. 7(F)). If this is so, the embrittled upper levels thicken from metres on the youngest salt domes through tens of metres on vigorous namakiers, to about a hundred metres on static extrusions. Where the brittle–ductile transition zone reaches the base of a namakier, ductile salt may back up behind static brittle salt or even override it.

These field observations are consistent with (and explain) the results of sophisticated laboratory experiments on shearing cores of Zechstein salts from several diapirs in Germany. These find that rock salt begins to dilate (by developing microcracks) in conditions that are independent of its strength, load geometry (compression or extension), damage to failure, impurities or grain size (Hunsche 1997). The boundary between salt shearing with constant volume and dilation extrapolates close to zero shear stress at zero normal stress, the conditions expected after stress relief towards a free salt surface of any origin. Dilated rock salt with its fractures and microcracks can still creep; it is particularly vulnerable to changes in humidity or dampening by rain, when it can creep 40 times faster than non-dilated salt (Cristescu & Hunsche 1993). Such humidity assisted creep almost certainly accounts for a small proportion of slabs of salt loosened from vertical faces bulging outward (Fig. 8(g)) or bending or kinking under their own weight.

Shrinkage cracks

Apart from gravity and its relief, the other main force that affects surficial layers of extruded salt is probably thermal. Large planar surfaces of salt with steep (Fig. 8(e)) or gentle layering (see Talbot et al. 1996, fig. 17) commonly display patterns of polygonal joint columns that are a few metres across and perpendicular to free surfaces, with any of the three sets of orientations already mentioned. At about 1 vol% and 100°C, the thermal expansivity of salt is about an order of magnitude higher than that of most other rocks (Clark 1966). The polygonal joint patterns are therefore attributed to shrinkage when the temperature falls rapidly.

The length of a Zagros salt mountain 7 km across expands and shrinks about 7 mm as a result of 0.1 linear per cent strains during the 30°C diurnal temperature range common during the summer.

Survey stations 9 and 40 m on to the salt monitored every 15 minutes of the day (Talbot & Rogers 1980) moved downstream 8 mm and 4 cm within 15 and 30 minutes, respectively, of the sun rising on the other side of the mountain. When the salt was dry, these markers returned to their overnight positions as the dry elastic salt cooled at nightfall. Strain gauges bonded across joints found them to open at sunrise and partially close and reopen within minutes of clouds shadowing them (Talbot & Rogers 1980). Geologists sleeping on the summit salt overnight soon learn to dress before dawn because socks left dangling neatly over the edges of joints gaping the night before are gripped by salt teeth clenched by the rising sun.

The elastic shrinkage of a mountain of dry salt cooling in the dusk is more spectacular than its expansion at sunrise. This is because loose fragments fall deeper into joints opened silently by heat and no longer fit when they close on cooling. Over longer periods, joints propped and levered open by fallen fragments may be filled by salt infloresence. The occasional groans and clicks, and the fewer cracks like rifle shots followed by rattling falls, that record brittle salt failure during the day increase in number and intensity to an impressive crescendo every dusk. By contrast, salt still damp from rain is eerily silent in its continual flow and does not recover all its daily advance (Talbot & Rogers 1980).

A few fractures, hundreds of metres long and with widths of decimetres and spacings of the order of 50 m, are prominent across large exposures of distal salt (Figs 6(b) & 8(d)). The origins of such master joints are unclear. Most are planar but a particular example rises from the crest of a sharp ridge of bedrock and curves upward and downstream through the complete 30 m thickness of a particular extrusion (Fig. 8(d)). This master joint was attributed to a vertical planar fracture due to thermal contraction of the salt above a stress concentrator in its bedrock in the previous summer being distorted by flow during the following winter (Talbot 1979). However, this interpretation needs modification, for the same fracture had undergone insignificant change 17 years later. This fracture is now assumed to have formed when dilation first temporarily extended through the complete salt profile, and its curvature is attributed to subsequent flow of dilated salt dampened by rain over many seasons.

Wasting by dissolution and erosion

Rain in the Zagros is seasonal and a recent map (Sahab Geographic Institute, 1992) indicates mean average annual falls of 100–600 mm a^{-1} where salt

mountains survive. Some salt dissolved upstream is reprecipitated on salt downstream (Fig. 4(f)) but the vast majority of the rain falling on the salt drains off it and can be assumed to carry salt with it, mainly in solution. Estimates of the potential rate of loss based on theory (1.67 mm of salt per 10 mm of rain; Talbot & Jarvis 1984) have not yet been improved by field measurements.

Emerging salt begins to degrade as soon as it is exposed to rainfall. Most divides where rising salt diverges to flow downslope from the summits of salt fountains are blanketed by accumulations of insoluble components in the Hormuz sequence. Such soils cloak most but not all gentle salt surfaces (Fig. 7(F)). Unlike hard cap rocks cemented in reducing conditions below the water table in the US gulf coast or Germany, the soils accumulating on salt emerging in oxidizing conditions in the Zagros are uncemented and loose. On cyclic Neoproterozoic salt, such soils are mainly black or red with blocks of black carbonates. On the relatively pure Cambrian salt, they are usually uniform buff-coloured anhydrite with local red beds and dispersed blocks of greenstone. Such residual soils are usually more fertile than any developed on surrounding carbonates or alluvium and, on the plateau of Kuh-e-Namak (Feroosabad), are farmed in preference to the surroundings.

Where relatively clean salt is increasingly exposed, valley sides retreat to leave crenulated ridges, ribbed columns or sharp spires of clean salt. Blocks in the soil protect residual pillars of salt as the surrounding soils and salt are eroded (Fig. 9(a)). Salt pillars that have lost their capstones are sharpened by dissolution to serrated salt pinnacles and spikes (Fig. 9(b), (d)). Groups of tall pillars and pinnacles tend to bend in a variety of directions unrelated to attitudes of internal layering or likely wind direction (Fig. 9(c)); it is not clear whether this is due to differential dissolution or the weight of the capstones.

All exposed salt tends to develop downslope dissolution ribbing. Run-off tends to dissolve downslope channels about 5 cm wide with semi-circular cross-section separated by serrated knife-sharp ridges with cuspate profiles (Fig. 9(b), (e)).

Several generations of salt inflorescence infill most master joints (Fig. 8(d)) and polygonal joints (Fig. 8(e)). Fringes of halite efflorescence, a few fibres or hopper crystals thick, decorate the traces of most grain boundaries over dry surfaces of crystalline rock salt (eg. Fig. 5(b), (f)). Secondary halite is redeposited along dry stream courses. Efflorescent salt masses can reach thicknesses of metres around brine pools (Fig. 4(f)) and areas of hundreds of square metres over flood plains both on (Fig. 9(c), (h)) and off the salt. Such deposits start as skeletal efflorescent hopper crystals (Fig. 9(f))

Fig. 9. Erosive and blanketing effects on Kuh-e-Namak (Dashti). (a) Stone-capped salt pillars sharpen to (b) grooved pinnacles when they lose their caps; matchbox perched on the skyline for scale. (c) Both pillars and pinnacles can bend in directions unrelated to the attitude of internal bedding or the likely wind direction; rucksack beside closest column for scale. (d) General view of vertical joints, pillars and spires in about 25 m thickness of salt with subhorizontal layering. (e) View down on dissolution ribs draining from a crown. (f) Halite efflorescence on a brine fall. (g) Stalactite of halite with fibres a few millimetres long. (h) Coral-shaped halite efflorescence where brine surfaces on salt; note the miniature spires dissolved in reprecipitated salt in the background, beyond the boots for scale.

but may become hummocky with artichoke-like or coral-like growths of halite where subsurface brine evaporates (Fig. 9(h)), or spiked where secondary salt is etched.

Efflorescence can vary in character very rapidly in response to changes in microclimate. Surfaces of crystalline salt normally sprout only stubby blades of halite along fracture traces but, five minutes after the beginning of an evening breeze, whiskers consisting of stacks of microscopic halite cubes (Fig. 9(g)) have been seen to grow up to about 10 mm long and to build twisted and tangled fibrous mats over every visible surface.

The beginning of the end of a salt extrusion is when erosion starts to degrade the profiles of viscous droplets faster than the damage can be repaired by viscous spreading of ductile salt. Sink holes appear in plateaux of salt capped by soils. The floors of young sink holes are merely circles of the surroundings dropped several metres, complete with healthy trees (Fig. 4(b)). Sink holes surrounded by vertical salt cliffs reach depths of at least 10 m (Fig. 4(g)) and compound sink holes can reach diameters exceeding 100 metres (Fig. 4(e)). Most sink holes soon fill with soils shed from the surroundings but some develop into funnels (Fig. 4(c)) leading to cave systems in the salt beneath. Caves in salt typically have upright elliptical profiles with smooth walls (Fig. 4(h)) and many of their roofs are decorated by stalactites of halite. Caves originally connecting sink holes at depths estimated to exceed 50 m evolve into surface streams that flow along straight gorges or meander through salt plateaux. Waterfalls work back from knick points first developed where streams leave salt to flow over moraines or alluvial plains. However, where adjoining bedrock is higher than the salt, rivers flowing along the contact preferentially erode the salt. The initial rounded outlines of sink-holes are lost as they widen and merge into badland topographies of multicoloured soils (Fig. 4(d)).

Summary

The profiles of domes and fountains of extrusive Hormuz salt in the Zagros have been shown to match the profiles of growing extrusions of viscous fluids until they exhaust their deep source when they fit the profiles of droplets of viscous fluids spreading with diminishing volumes. If one neglects their surficial relief (which indicates the thickness of carapaces of embrittled salt dilated by fractures) the topographies of salt extrusions resemble those of other extrusions of ductile rocks with brittle carapaces. Piles of nappes of halitic gneisses separated by halitic mylonites in the Zagros are natural models of orogenic nappe piles.

Significant differences are that there are more salt extrusions active in the Zagros than there are active orogens on Earth and that salt extrusions are smaller, with velocities that are faster and more easily constrained.

Hormuz salt in the Persian Gulf region flowed very slowly into deep conformable pillows, one of which rose at a rate of 0.01 mm a^{-1} averaged over 102 Ma (Edgell 1996). Intruding salt diapirs that are still blind can rise at 0.3–2 mm a^{-1}, 30–200 times faster (Player 1969; in Edgell 1996). Once the roof has been pierced, measurements demonstrate that, at 50–200 cm a^{-1}, emergent salt can extrude over a hundred times faster than that. The top free surface of flowing salt can reach velocities down subaerial slopes of 4 m a^{-1} averaged over a year, and 50 cm per day^{-1} after rain (Talbot & Rogers 1980). These figures suggest that individual episodes can extrude, dissolve and recycle by dissolution 200 km^3 of deep salt source layer in about 50 000 years.

This simple fact may explain the dramatic disparity between salt flow veolocities measured in the Zagros and equivalent figures estimated by backstripping of salt diapirs elsewhere. Zirngast (1996) used volumes of excess sediment deposited in salt withdrawal basins around a diapir of Zechstein salt in Germany to constrain its flow rates. However, the geological constraint available to Ziirngast means that his estimate of maximum extrusion rate of only 0.145 mm a^{-1} between the Cenomanian and the Early Palaeocene had to be averaged over 39 Ma. This interval may have included brief episodes when German diapirs of Zechstein salt may have extruded as vigorous salt fountains a thousand times faster. Similarly, the fall and flow of cliffs of salt at Kuh-e-Jahani at about 1 m per month suggests that the closure of a Quaternary subglacial channel that trapped gravels in a now inclined slot reaching about 380 m below sea level in the crest of Gorliben dome may have involved flow much faster than the 0.033 mm a^{-1} constrained by Zirngast for the last 23 Ma (Talbot et al., in preparation).

As rock salt flows downslope over the surface, the linear Newtonian–Bingham fluid seen in summit domes hardens to a power-law fluid with $n = 3$, and increasing proportions of its vertical thickness dilate to a broken brittle solid downslope. As in all tectonites, grain-shape fabrics and macroscopic structures show that flow penetrates throughout coarse-grained masses that have flowed without interruption, but is increasingly partitioned into narrowing zones of decreasing grain sizes wherever unimpeded masses flow past obstructions. The highly irregular flow profiles exposed in innumerable Zagros extrusions (Fig. 6) are warnings to oil companies planning to develop production wells penetrating sheets of allochthonous salt beneath the Gulf of Mexico where flow profiles interpolated between a few side-wall cores (Diggs *et al.* 1997) look suspiciously smooth.

References

ALA, M. A. 1974. Salt diapirism in southern Iran. *American Association of Petroleum Geologists Bulletin*, **58**, 1758-1770.

ANDERSON , M. W., BARKER, A. J., BENNETT, D. G. & DALLMEYER, R. D. 1992. A tectonic model for Scandian terrane accretion in the northern Scandinavian Caledonides. *Journal of the Geological Society of London*, **149**, 727–742.

BOYLAN, P. J. 1998. Lyell and the dilemma of Quaternary glaciation. *This volume.*

BRUNSTROM, R. G. W. & WALMSLEY, P. J. 1969. Permian evaporites in the North Sea basin. *American Association of Petroleum Geologists Bulletin*, **53**, 870–888.

CARTER, N. J. & HANSEN, F. D. 1983. Creep of rocksalt. *Tectononphysics*, **92**, 275, 333.

CLARK, S. P. 1966. *Handbook of Physical Constants.* Geological Society of America Memoir 97, New York.

CRISTESCU, N. & HUNSCHE, U. 1993. A consitutive equation for salt. *In*: WITTE, W. (ed.) *Proceedings 7th International Congress Rock Mechanics, Aachen, Sept. 1991.* Balkema, Rotterdam, vol. 3, 1821–1830.

DAVIS, D. M. & ENGELDER, R. 1987. Thin-skinned deformation over salt. *In*: LERCHE. I. & O'BRIEN, J. J. (eds) *Dynamical Geology of Salt and Related Structures.* Academic Press, New York, 301–337.

DAVISON, I, BOSENCE, D., ALSOP, G. I., & AL-AAWAH, M. A. 1996. Deformation and sedimentation around active Miocene salt diapirs on the Tihama Plain, northwest Yemen. Geological Society, London, Special Publications, **100**, 23–39.

DE BÖCKH, H., LEES, G. M. & RICHARDSON, F. D. S. 1929. Contribution to the stratigraphy and tectonics of the Iranian ranges. *In*: GREGORY, J. W. (ed.) *The Structure of Asia.* Methuen, London, 58–176.

DIGGS, T. N., URAI, J. L. & CARTER, N. L. 1997. Rates of salt flow in salt sheets, Gulf of Mexico: quanitfying the risk of casing damage in subsalt playas. *Terra Nova*, **9**, abstract 11/4B29.

EDGELL, H. S. 1996. Salt tectonism in the Persian Gulf basin. Geological Society, London, Special Publications, **100**, 129–151.

FIELDING, E., ISACKS, B., BARAZANGI, M. & DUNCAN, C. 1994. How flat is Tibet? *Geology*, **22**, 163–167.

FLETCHER, R. C., HUDEC, M. R. & WATSON, I. A. 1995. *Salt Glacier and Composite Sediment–Salt Glacier Models for the Emplacement and Early Burial of Allochthonous Salt Sheets.* American Association of Petroleum Geologists, Memoir **65**, Tulsa, OK, 77–107.

FÜRST, M. 1976. Tektonic und diapirismus der österlichen Zagrosketten. *Zeitschrift der deutschen geologischen Gesellschaft*, **127**, 183–225

GANSSER, A. 1992. The enigma of the Persian dome inclusions. *Eclogae Geologica Helveticae*, **85**, 825–846.

GERA, F. 1972. Review of salt tectonics in relation to disposal of radioactive wastes in salt formation. *American Association of Petroleum Geologists Bulletin*, **83**, 3551–3574.

GRANT, N. T. 1992. Post emplacement extension within a thrust sheet from the central Pyrenees. *Journal of the Geological Society of London*, **149**, 775–792.

GUSSOW, W. C. 1966. Salt temperature: a fundamental factor in salt dome intrusion. *Nature*, **210**, 518–519.

HARRISON, J. V. 1930. The geology of some salt-plugs in Lariistan (southern Persia*). Quarterly Journal of the Geological Society of London*, **86**, 463–522.

HEATON, R. C., JACKSON, M. P. A., BARBAHMOUD, M. & NANI, A. S. O. 1995. *Superposed Neogene Extension, Contraction and Salt Canopy Emplacement in the Yemeni Red Sea.* American Association of Petroleum Geologists, Memoir **65**, Tulsa, OK, 333–351.

HUNSCHE, U. 1997. Determination of the dilatancy boundary and damage up to failure for four types of rock salt at different stress geometries. Proceedings 4th Conference on the Mechanical Behaviour of Salt, Montreal, June 1996.

—— 1998. Determination of the dilatancy boundary and damage up to failure for four types of rock salt at different stress geometries. *In*: AUBERTIN, M. & HARDY, H. R. (eds) *The Mechanical Behavior of Salt IV. Proceedings of the Fourth Conference (MECASALT IV), Montreal 1996.* TTP Trans. Tech. Publications, Clausthal, 163–174.

HUPPERT, H. E. 1982. The propagation of two-dimesional and axisymmetric viscous gravity currents over a rigid horizontal surface. *Journal of Fluid Mechanics*, **121**, 43–58.

JACKSON, M. P. A. & VENDEVILLE, B. C. 1994. Regional extension as a geologic trigger for diapirism. *Geological Society of America Bulletin*, **106**, 57–73.

KENT, P. E. 1958. Recent studies of south Persian salt plugs. *American Association of Petroleum Geologists Bulletin*, **422**, 2951–2972.

—— 1966. Salt temperature: a fundamental factor in salt dome intrusion: discussion. *Nature*, **211**, 1387–1388.

—— 1970. The salt plugs of the Persian Gulf region. *Transactions of the Leicester Literary and Philosophical Society*, **64**, 56–88.

—— 1979. The emergent Hormuz salt plugs of southern Iran. *Journal of Petroleum Geology*, **2**, 117–144.

KLEINHAMNS, K. 1914. Die abhängigkeit der plastizität des steinzalzes von umgebenden medium. *Zeitschrift für Physik*, **15**, 362–363.

LIRO, L. M. & COEN, R. 1995. *Salt Deformation History and Postsalt Structural Trends, Offshore Southern Gabon, West Africa.* American Association of Petroleum Geologists, Memoir **65**, Tulsa, OK, 323–331.

LISTER, J. R. & KERR, R. C. 1989. The propagation of two-dimensional and axisymmetric viscous gravity currents at a fluid interface. *Journal of Fluid Mechanics*, **203**, 215–249.

LOWENSTEIN, T. K. & HARDIE, L. A. (1985) Criteria for the recognition of salt pan evaporites. *Sedimentology*, **32**, 627–644.

MCKENZIE, D., FORD, P. G., LIU, F. & PETTENHILL, G. H. 1992. Pancakelike domes on Venus. *Journal of Geophysical Research*, **97**, 15 967–15 976.

NELSON, T. H. & FAIRCHILD, L. H. 1989. Emplacement and evolution of salt sills. *American Association of Petroleum Geologists Bulletin*, **73**, 395.

O'BRIEN, C. A. E. 1957. Salt diapirism in south Persia. *Geologie en Mijnbouw*, **19**, 357–376.

PLAYER, R. A. 1969. *Salt Plugs Study*. Iranian Oil Operating Companies, Geological and Exploration Division, Tehran, Report No. 1146.

RAMBERG, H. 1975. *Superposition of Homogeneous Strains and Progressive Deformation of Rocks*. Bulletin of the Geological Institutions of the University of Uppsala, New Series, no. 6, Uppsala.

—— 1981. *Gravity, Deformation and the Earth's Crust in Theory, Experiments and Geological Application*. Academic Press, London.

RAMSAY, J. G. & HUBER, M. I. 1983. *The Techniques of Modern Structural Geology, Volume 1: Strain Analysis*. Academic Press, London.

ROSENKRANS, R. R. & MARR, D. J. 1967. Modern seismic exploration of the Gulf coast Smackover trend. *Geophysics*, **32**, 184–206.

SAHAB GEOGRAPHIC INSTITUTE 1992. 1:10 million map of Iran, PO Box 11365-617, Tehran.

SCHULTZ-ELA, D. D. JACKSON, M. P. A. & VENDEVILLE, B. 1993. Mechanics of active salt diapirism. *Tectonophysics*, **228**, 275–312.

TALBOT, C. J. 1979. Fold trains in a glacier of salt in southern Iran. *Journal of Structural Geology*, **1**, 5–18.

—— 1992. Quo vadis tectonophysics? with a pinch of salt! *Journal of Geodynamics*, **16**, 1–20.

—— 1993. Spreading of salt structures in the Gulf of Mexico. *Tectonophysics*, **228**, 151–166

—— & ALAVI, M. 1996. *The Past of a Future Syntaxis Across the Zagros*. Geological Society, London, Special Publications, **100**, 89–109

—— & JACKSON, M. P. A. 1987. Internal kinematics of salt diapirs. *American Association of Petroleum Geologists Bulletin*, **71**, 1068–1093.

—— & JARVIS, R. J. 1984. Age, budget and dynamics of an active salt extrusion in Iran. *Journal of Structural Geology*, **6**, 521–533.

—— & ROGERS, E. A. 1980. Seasonal movements in a salt glacier in Iran. *Science*, **208**, 395–397.

——, STANLEY, W., SOUB, R. & AL-SADOUN, N. 1996. Epitaxial salt reefs and mushrooms in the southern Dead Sea. *Sedimentology*, **43**, 1025–1047.

URAI, J. L., SPIERS, C .J., ZWART, H. J. & LISTER, G. S. 1986. Weakening of rock salt by water during long term creep. *Nature*, **324**, 554–557.

VENDEVILLE, B. C. & JACKSON, M. P. A. 1992a. The rise of diapirs during thin-skinned extension. *Marine and Petroleum Geology*, **9**, 331–353.

—— & —— 1992b. The fall of diapirs during thin-skinned extension. *Marine and Petroleum Geology*, **9**, 354–371.

WEINBERG, R. F. 1993. The upward transport of inclusions in Newtonian and power-law salt diapirs. *Tectonophysics*, **228**, 141–150.

WENKERT, D. D. 1979. The flow of salt glaciers. *Geophysical Research Letters*, **6**, 523–526.

WU, S., BALLY, A. W. & CRAMEZ, C. 1990. Allochthonous salt, structure and stratigraphy of the north-eastern Gulf of Mexico. Part 11: structure. *Marine and Petroleum Geology*, **7**, 334–370.

ZIRNGAST, M. 1996. The development of the Gorliben salt dome (northwest Germany) based on qualitative analysis of peripheral sinks. Geological Society, London, Special Publications, **100**, 203–226.

Mount Etna: monitoring in the past, present and future

HAZEL RYMER[1], FABRIZIO FERRUCCI[2] & CORINNE A. LOCKE[3]

[1] Department of Earth Sciences, The Open University, Walton Hall, Milton Keynes, Bucks, MK7 6AA, UK

[2] Dip. di Scienze della Terra, Universita della Calabria, 87036 Arcavacata di Rende, Cosenza, Italy

[3] Department of Geology, The University of Auckland, Private Bag 92019, Auckland, New Zealand

Abstract: Mount Etna is important to the economy of eastern Sicily, with agriculture and summer and winter tourism providing employment for thousands of people. Although there are no permanent homes within 10 km of the summit, year-round human activities on the upper slopes are proliferating and the risks from even a small eruption are consequently magnified. The earliest form of monitoring at Etna, as for other volcanoes, was direct observation. Modern volcano monitoring at its most effective is a synergy between basic science and hazard assessment. A prerequisite to effective monitoring is an understanding of volcanic structure and history. Sir Charles Lyell was among the first to make systematic observations of Mount Etna and laid the foundation of more modern studies. A huge array of monitoring techniques has been tested on Etna; methods that have proved successful in monitoring and sometimes in predicting eruptions include observations of seismicity, ground deformation and microgravity. These, together with electromagnetic, magnetic, gas geochemistry and various remote sensing techniques have also provided key information on the volcanic plumbing system and the eruption process. Monitoring techniques were formerly based on the most easily measured phenomena; other effects were either not recorded or were treated as noise. Future progress will be enhanced by taking account of these more subtle or complex effects and by the more comprehensive acquisition and real-time analysis of continuous data sets over extended periods. Important monitoring techniques and strategies available both now and in the near future are reviewed here in the context of Etna. The need to develop a reliable scientific platform for routine and inexpensive volcano monitoring throughout the world is highlighted.

Why monitor volcanoes?

Volcano monitoring serves two key functions: it provides basic scientific data to develop our understanding of the structure and dynamics of volcanoes and is crucial for hazard assessment, eruption prediction and risk mitigation at times of volcanic unrest. Monitoring provides the means to address questions of vital interest to communities affected by impending eruptions, such as: When and where will the volcano erupt? Which areas are safe or dangerous? When will eruptions cease? Optimal interpretation of data from monitoring, especially for the purpose of prediction, depends critically on an adequate scientific understanding of volcano structure and processes, both in general and for each specific volcano. Thus, volcano monitoring at its most effective is a synergy between basic science and hazard assessment.

Historically, volcanic activity has been considered to be primarily of local interest because typically, with the exception of relatively infrequent large explosive events, eruptions seriously affect only a few square kilometres at most. As populations expand, however, more people are at risk from the direct or indirect impact of volcanic eruptions. In addition to the immediate hazards posed by eruptive products to life, property and food production, volcanic activity of any kind may have a significant effect on the economy. During minor eruptions tourism may increase and boost the local economy. The loss, however, of agricultural farm land, farmhouses, communication and service infrastructure (roads, ports, water supplies, electric cables, etc.) and permanent changes in the local ground water and drainage system can have severe economic consequences. A volcanic disaster on one side of the world can now have a significant economic impact on countries on the other. Insurance companies are particularly vulnerable because of this, but governments are also at risk. The National Plan for volcanic emergencies at Mount Vesuvius

RYMER, H., FERRUCCI, F. & LOCKE, C. A. 1998. Mount Etna: monitoring in the past, present and future. *In*: BLUNDELL, D. J. & SCOTT, A. C. (eds) *Lyell: the Past is the Key to the Present*. Geological Society, London, Special Publications, **143**, 335–347.

335

for example forecasts that around 700 000 people would need to be evacuated during a period of unrest preceding an explosive eruption such as the one that occurred in 1631. The cost of evacuation and resettlement would represent a significant fraction of Italy's GNP and the rest of the EU would certainly suffer economically from such an event.

Links between global climate change and volcanic eruptions have been postulated and there is good evidence for a relationship between large explosive ash injection to the stratosphere and unseasonal weather. Well documented examples include the 1815 eruption of Tambora (Indonesia) which was followed by the 'year without a summer' in the northern hemisphere, and more recently El Chichon (1982) and Pinatubo (1991) (described for example in Francis 1993). In each case there is some doubt as to the extent of climate change (caused by cooling due to increased high altitude aerosols) during these short explosive eruptions, but it is likely that longer-lived eruptions such as the 1783 lava eruption at Laki (8 months), have a more prolonged effect on local and global temperatures.

A more recent hazard posed by volcanoes is the effect of volcanic ash clouds on transportation. There have been more than 80 incidents since 1982 of jet aircraft encountering volcanic ash, with tens of millions of dollars worth of damage done on each occasion (e.g. Casadevall 1994). Millions of US dollars have also been lost to the aviation industry as a result of disruption when flights have been diverted or made emergency unscheduled landings. Catania international airport has been closed on several occasions in the last 20 years as a result of ash on the runway and the plume intersecting the flight path of planes. The Messina–Catania motorway was closed in 1995 when accumulating ash made the surface slippery and dangerous (J. B. Murray, pers. comm.).

It is therefore of interest to all nations that volcanoes are monitored so that eruptions can be predicted and mitigating action taken. Successful mitigation of the local effects of an eruption (such as lava flow diversion and community evacuation) requires a detailed understanding of the eruption process, for which monitoring data are a vital component. More widespread effects can be mitigated to some degree with early warning, for

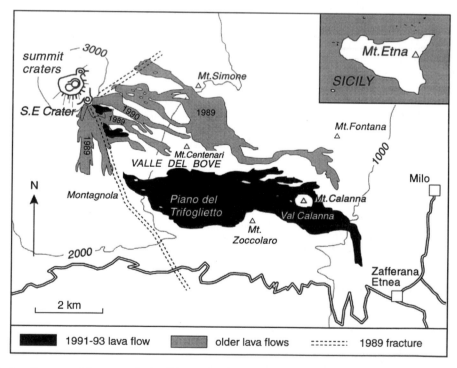

Fig. 1. Location map of the eastern flank of Mount Etna showing: (1) the eruptive fissure of 14–15 December 1991; (2) northeast and south-southeast trending fractures of 1989; (3) area covered by lava flows of the 1991–1993 eruption (after Calvari *et al.* 1994).

example allowing aircraft re-routing. Similarly, an appreciation of the climatic effects both locally and globally requires data from surveillance before, during and after eruptive activity.

Historical activity and observations at Etna

Volcanic activity in Sicily has been focused at Etna since the Mid-Pleistocene, resulting in the development of one of the largest active continental volcanoes in the world (Chester *et al.* 1985). The landscape of eastern Sicily is dominated by the edifice of the Etna volcano, which rises from sea level to over 3300 m. Its activity, which commenced some 230 000 years ago (Kieffer 1985), has been recorded for the last 3500 years and in considerable detail for the last 400 years. Activity at this basaltic volcano occurs both at the summit and on the flanks. Effusion rates from the summit are generally moderate and at least one of the several vents appears to be open to the feeder system at any one time (Chester *et al.* 1985). As well as construction, there are periods of collapse at the summit vents resulting in ash clouds. Flank eruptions tend to be confined to specific rift zones (McGuire & Pullen 1989) and generally have a higher effusion rate (Kieffer 1975) producing extensive lava flows and cinder cones. Other types of activity at this volcano include ash eruptions, pyroclastic flows and slope failure – which ranges from minor mud flows to major landslides. It may have been a series of such landslides that formed or catastrophically enlarged the Valle de Bove, an amphitheatre some 5 km across, 10 km long and up to 1 km deep on the eastern flank of Etna (Fig. 1). One of the largest and most destructive historical eruptions was a flank eruption in 1669 from near Nicolosi (at 800 m a.s.l.) which produced the cinder cone Monti Rossi and a 14 km long lava flow responsible for destroying part of the city of Catania.

Whilst most Greeks and Romans held supernatural beliefs about Etna, some notable exceptions such as Empedocles (492–432 BC) made observations of the eruptive phenomena which unfortunately do not survive. It was not until the sixteenth century that such natural phenomena began to be studied systematically. In the early nineteenth century Leopold von Buch (of Neptunist fame) developed his idea of 'craters of elevation', i.e. that 'volcanic mountains formed by upwarping of formerly horizontal beds of basalt as a result of pressures exerted by molten materials at depth' (von Buch 1818–1819), a hypothesis which he deduced from observations in the Canary Islands. Whilst Scrope (1825) argued for 'craters of eruption', Élie de Beaumont (1838) supported von Buch by suggesting that Etna lavas had been erupted in thin sheets on to a subhorizontal surface above sea level accumulating to a considerable thickness. These were thought to have then been uplifted along the line of the Valle de Bove.

Elie de Beaumont's work focused Sir Charles Lyell's mind on how volcanoes formed; Lyell visited Etna on a number of occasions and in 1858 The Royal Society published a monograph titled: *On the Structure of Lavas which have Consolidated on Steep Slopes; with Remarks on the Mode of Origin of Mount Etna and on the Theory of 'Craters of Elevation'*, in which he describes his own work and that of others and condemns the 'craters of elevation' hypothesis (Lyell 1858). From the morphology and flow characteristics of historic and ancient lava flows he deduced that they had solidified on steep slopes. He also determined from Valle de Bove exposures that there were two eruptive centres, the present summit (Mongibello) and the Trifoglietto centre to the east (Fig. 1). Lyell declared:

> ... we must abandon the elevation-crater hypothesis; for although one cone of eruption may envelop and bury another cone of eruption, it is impossible for a cone of upheaval to mantle round and overwhelm another cone of upheaval so as to reduce the whole mass to one conical mountain. (Lyell 1858)

Recent activity at Etna

In 1989, four months of strombolian activity at the summit craters was followed in September by an effusive eruption from the southeast crater and the opening of two fractures (Bertagnini *et al.* 1990). A non-eruptive south-southeast-trending fracture extended more than 7 km from the summit (Fig. 1), passed close to the headwall of the Valle del Bove and crossed the road between the Sapienza and Zafferana Etnea. The northeast-trending fracture extended about 2 km from the summit crater area and lava was erupted along much of its length. No further activity occurred until a small explosive eruption from the southeast crater in January 1990.

Activity on Etna was then unusually quiet until 14 December 1991 with the start of the most voluminous eruption since the end of the 1669 eruption. The activity was largely confined to a single bocca within the western wall of the Valle del Bove (2200 m a.s.l.) erupting 231×10^6 m^3 of lava at an average rate of 5.7 m^3 s^{-1} (Stevens *et al.* 1997). The flow was more than 8 km long and threatened the town of Zafferana Etnea when it was successfully diverted (Barberi & Villari 1994; Di Palma *et al.* 1994; Vassale 1994). The eruption ended on 31 March 1993, after 471 days.

Past and present monitoring on Etna

Volcano watchers have long ventured to the mouth of an erupting volcano and noted changing intensity of degassing, lava eruption, etc. New or increased hydrothermal activity, felt earthquakes and in some cases visible ground deformation (rifting or faulting), have been recognized for centuries as being associated with volcanic activity, but usually during or after the event. Until recently volcano monitoring has tended to be responsive in that resources were forthcoming only after the beginning of an eruption. In this review, the techniques applied during the 1989 and 1991–1993 Etna eruptions are taken as examples of volcano monitoring today. The high level of activity at Etna and its easy accessibilty have led to most available techniques for volcano monitoring and some involving mitigation being tested there.

Seismology

A century of seismological data testifies that seismic unrest in the form of earthquakes and tremor almost always precedes and/or accompanies volcanic unrest at all types of volcanoes. Seismic activity is considered to be the best indicator, and often a reliable short to mid term (days to weeks) predictor, of the level, type and evolution of volcanic activity (Ferrucci 1995).

Seismic sources at volcanoes are highly complex and involve the interaction of gas, melt and solid. The role of the melt and gas may be either (i) active, giving rise to pressurized intrusions of magma into pre-existing or newly formed zones of weakness, or to sustained vibration of the melt and host rocks; or (ii) passive, where brittle failures and the consequent stress readjustments modify the distribution of melts in the crust. Since volcanic media comprise dense systems of pores, fractures and faults at all scales, sudden modification of the local stress field may induce seismic failure independent of melt propagation. These factors contribute to the substantial degree of ambiguity in volcano eruption prediction on the basis of seismology alone.

The first instruments set up to record earthquakes at Etna were deployed in the early 1900s, however the first network of seismometers was not established on Etna until 1973. This network, comprising two to six stations recording the short period vertical component of seismic waves, was used to investigate eruption-related earthquakes and volcanic tremor (Cosentino *et al.* 1982). These studies showed that shallow (1–5 km) seismic activity usually preceded flank fissure eruptions by a few days. In contrast precursory seismic activity is not usually associated with summit eruptions, although they are often accompanied by tremor. Tremor is common at active volcanoes, but the process that sustains these low magnitude and frequency signals is poorly understood (Schick & Mugiono 1991). Analysis of seismicity at a large number of active volcanoes, however, has shown that in general an increase in the rate of low-frequency earthquakes gives an increased probability that tremor episodes may follow. Increased levels of tremor energy are consistent with an increased probability that an eruption may occur within weeks, days or even hours (Ferrucci 1995). Modern seismometers detect ground vibration in 3 dimensions over a range typically of 0.1–10 Hz. High-frequency events are usually associated with deep (several kilometres) fracture events, while low-frequency and tremor events are thought to be caused by forced resonance of fluids in shallow (a few kilometres) volcanic conduits.

A new seismic array was being developed and expanded on Etna by the Istituto Internazionale di Vulcanologia, Catania prior to the 1989 activity and data from up to 22 seismometers were available for analysis and comparison. Seismicity at Etna remained at background levels until July 1989 when a 10 km deep (b.s.l.) magnitude 2.9 event beneath the southeastern Valle del Bove and later a seismic swarm heralded the onset of renewed activity. A second swarm in August 1989 originated only 5 km b.s.l. No noticeable seismicity occurred subsequently until the beginning of eruptive activity in the southeast crater in September 1989 when a further swarm and large individual events some of magnitude >3.0 to the northwest of the summit craters occurred. These events may have been linked to the opening of the northeast-trending fracture (Bertagnini *et al.* 1990) that erupted a small amount of magma (Fig. 1), while several hundred shallow seismic events over a two-day period (1–2 October) accompanied the opening of the non-eruptive south-southeast-trending fracture (Fig. 2). Although the causative mechanism of the seismicity is unclear (Ferrucci 1990), a temporal relationship exists between the occurrence of the swarm and the propagation of the fracture beyond the scarp of the Valle del Bove. No definitive evidence for shallow magma beneath this fracture was found (see below). Volcanic tremor throughout the 1989 eruption correlated strongly with eruptive activity; in particular a sharp increase in tremor amplitude always accompanied the change in activity from weak Strombolian, to strong Strombolian, to lava fountaining, to lava flows (Ferrucci 1990).

There was no further noteworthy seismic activity observed from the end of the 1989 eruption until one month before a small explosive eruption from

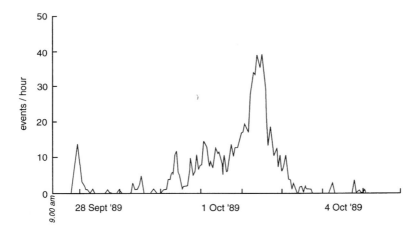

Fig. 2. Hourly frequency of events recorded at a station about 6 km southeast of the summit during the seismic swarm 27 September to 3 October 1989 (after Ferrucci 1990).

the northeast crater in January 1990. The brief explosive activity in 1990 was associated with significantly higher amplitude tremor than the 1989 eruptions.

Seismic activity on Etna ceased until January 1991. Earthquake swarms were located along the north-northwest–south-southeast trending structure along which magma later rose to erupt in December 1991. Focal-plane solutions and the hypocentral pattern for these data have led to the suggestion that activation of the intrusion at depth occurred about two months before the eruption (Patane *et al.* 1994). The foci of the events became shallow (1–4 km b.s.l.) only one week before the eruption began.

Ground deformation

In common with seismology, this is one of the oldest volcano monitoring techniques. Tilt and relative elevation have been monitored with high precision at active volcanoes for decades, but the disadvantage of the method has been that it is time-consuming and labour intensive. Also, in the case of precision levelling, the reference point is usually too close to the active area to be unambiguously stationary relative to the region of interest. The survey level of the past has been replaced in the last five years by a digital level that automatically reads a bar-coded staff. Theodolites have also become digitised and EDM (electronic distance measurement) using laser or infra-red light reflecting from a target prism or even rock faces has provided accurate horizontal, vertical and line-of-sight

distance measurements. During the 1990s EDM theodolites (total stations) became automated and deployed to make periodic measurements on fixed targets without the need for a survey crew.

The first ground-deformation studies in the summit area of Etna (using an EDM) suggested an open, cylindrical magma column beneath the summit craters (Wadge 1976). No evidence for high-level magma storage was found using detailed levelling traverses of the summit region (Murray & Guest 1982). Tiltmeter data indicate inflation/deflation cycles at elevations below 1800 m, consistent with deep magma movements. Whilst some localized inflation prior to eruption at the summit craters has been observed, subsidence at the summit has occurred prior to some flank eruptions coupled with inflation on the flanks (Murray 1990). In addition, ground deformation studies have provided evidence for slope instability and collapse into the Valle del Bove, stimulating interest in the mechanism of formation of this major feature (Murray *et al.* 1994; McGuire *et al.* 1990).

Murray *et al.* (1994) noted that the instability on the western wall of the Valle del Bove appears to have been at least partly responsible for the siting of the 1989 eruption. Their precision levelling data indicate relative elevation changes in the summit area of between −71 cm and +206 cm, consistent with the intrusion of a vertical dyke beneath the Valle del Leone and a 2 km long vertical dyke beneath the upper part of the 7 km long non-eruptive south-southeast fracture. Like other flank eruptions, it appears to have been fed by a near-vertical dyke but in this case there were two dykes radiating from the southeast crater. In the first, the magma travelled northeastwards and intersected the surface, erupting lava. In the second, the magma travelled south-southeastwards but did not reach the surface though it did cause fracturing. There is evidence for dyke intrusion up to 3 km from the summit along the south-southeast fracture (Murray *et al.* 1994), but the pattern of ground deformation at the distal end of the fracture was characteristic of an elastoplastic medium, and therefore was not linked with magma intrusion (Luongo *et al.* 1990).

Precise levelling and triangulation measurements were made before, during and after the 1991–1993 event (Murray 1994). Data from over 300 stations in the summit region of Etna revealed a narrow trough of more than 1 m of subsidence running south-southeast from the summit, connecting the southeast crater with the eruption site, flanked by two zones of inflation up to 37 cm to the east and 7 cm to the west (Fig. 3a). This is interpreted in terms of a syn-eruption dyke intrusion at a depth of 450 m to its top and 1100 m to its base (Murray 1994).

GPS

The global positioning system (GPS) is now widely used; by placing two or more GPS receivers either permanently or for the duration of a survey at locations of interest, the user can obtain x, y and z coordinates for each location with centimetre or even sub-centimetre precision horizontally and vertically. Traditional geodetic techniques are not suitable for continuous observations, but GPS provides this opportunity. Feasibility studies on the possibility of using permanently fixed receivers capable of automatically tracking satellites, recording and transmitting data for automatic processing and storage began on Etna in 1988 (Nunnari & Puglisi 1995). Data collected during the 1991–1993 eruption on a network of stations over the summit of the volcano extending down the flanks as far as Catania indicate that contraction of the whole edifice occurred during the first months of 1992. This has been modelled in terms of depressurisation within a source (magma storage region) 1.5–3.5 km b.s.l. The volume of magma

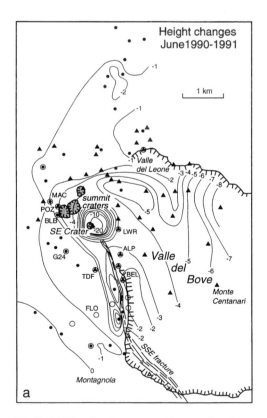
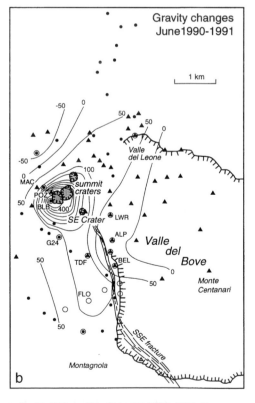

Fig. 3. (a) Elevation changes over the summit and upper eastern flank of Mount Etna between 1990–1991. Contour interval is 1 cm (after Rymer *et al.* 1995). (b) Gravity changes over the summit and upper eastern flank of Mount Etna between 1990–1991. Contour interval is 50 mGal (after Rymer *et al.* 1995).

erupted during the same period exceeds the volume loss within the modelled region, suggesting this relatively superficial storage region may be linked to a deeper feeder system (Nunnari & Puglisi 1995).

GPS technology has improved dramatically since 1993, and permanent stations have now been established at several key locations on and around the volcanic edifice. In addition, twice yearly campaigns now aim to occupy stations in the summit and flank areas; results have shown consistent radial expansion since 1992 consistent with relaxation of the edifice (Murray 1997).

Micro-gravity

Etna was one of the first volcanoes to be studied using micro-gravity (Sanderson 1982; Sanderson *et al.* 1983). The technique typically involves repeated high-precision surveys in which the relative value of gravity at a number of stations is determined. Repeat micro-gravity surveys are used to investigate subsurface mass changes; since their effect is sensitive to variations in the distance from mass anomalies, micro-gravity data must be corrected for height changes and hence these surveys must always be accompanied by a ground deformation survey.

A micro-gravity increase (approximately 20 mGal) attributed to a shallow (a few hundred metres depth) source was observed across the southernmost end of the south-southeast-trending non-eruptive fracture a few days after formation in October 1989 (Budetta *et al.* 1990). However, neither these data nor data from a network of micro-gravity stations around the volcano flank at an elevation of 1500–2000 m provided evidence in support of a magma intrusion associated with the fracture process.

The period June 1989 – November 1989 was characterised by a gravity increase in the summit region (up to 100 mGal) and was followed by a decrease of similar magnitude between November 1989 and June 1990. These data imply subsurface mass increases of 10^9–10^{10} kg followed by decreases, and are consistent with the eruption of 10^9 kg of pyroclastic material in January 1990 (Rymer *et al.* 1993). Between June 1990 and June 1991, although there was no surface eruptive activity, there were significant gravity changes (Fig. 3b) in the summit area; the changes are too large by an order of magnitude to be caused by the minor deflation observed over the same period (Murray 1994). The data have been interpreted in terms of a passive (aseismic) intrusion of magma into the pre-stressed region beneath the 1989 south-southeast-trending fracture (Rymer *et al.* 1993). The intrusion must have occurred some time between June 1990 and June 1991, several months after the formation of the surface fractures and several months before the onset of the 1991–1993 eruption. The intrusion appeared to be in the location of the dyke that fed the later eruption (Fig. 4). Relative micro-gravity values remained constant in the vicinity of the fracture zone after the end of the 1991-1993 eruption, but decreased and subsequently increased again by 1994 in the summit crater region. This suggests that intruded

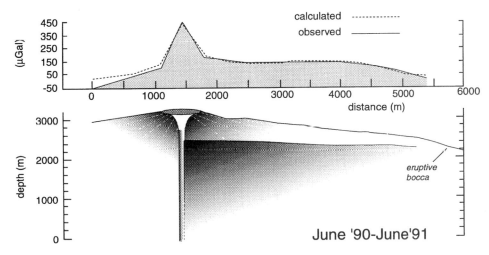

Fig. 4. Cross-section along a profile through the summit and along the south-southeast fracture zone showing a computed model of a 50 m diameter feeder pipe rising towards the summit and a 4 m wide dyke extending along the fracture zone, which can account for the gravity changes observed between June 1990 and June 1991.

magma solidified within the fracture zone, and the magma level within the summit feeder fell then rose again, which is consistent with the increased summit activity observed since 1994 (Rymer *et al.* 1995).

Tidal gravity

Micro-gravity data are corrected for the effects of Earth tides (up to about 200 mGal peak to peak, diurnal) before interpretation. A predicted Earth tide can be calculated for any location (eg. Brouke *et al.* 1972) which represents the vertical component of gravitational acceleration on a solid, uniform Earth but does not account for variations due to topography, rheological heterogeneities and ocean loading. Use of the predicted tide is satisfactory for many applications of micro-gravity (it is good to better than 8 μGal), but more accurate tidal corrections are possible if concurrent tidal observations are made or if tidal observations have already been made for at least three months and the phase lags and amplitudes of the key tidal components (Melchior 1978) have been determined for the specific location.

Mason *et al.* (1975) reported an anomalously large response to Earth tides at Etna for tidal observations accompanying a micro-gravity survey preceding an eruption, from which they suggested that the crust beneath Etna was elastically inhomogeneous because of the presence of subsurface magma bodies. It has subsequently been shown using combined tiltmeter and gravity data that this observed anomalous behaviour was probably a transient response related to ground deformation and eruptive activity at Etna and that there is no anomalous tidal response under normal quiescent conditions (Davis 1981).

Continuous tidal measurements were made at the astrophysical observatory (Serra la Nave) near the Sapienza on the southern flank of Etna some 7 km from the summit in the period 1992–1994, i.e. during and after the 1991–1993 eruption (d'Oreye *et al.* 1994). However the station was too far from the active region to detect a volcanic response in the tidal signal.

Currently there are continuously recording tidal gravimeters operating at Serra la Nave (for control) and at the Torre del Filosofo (TDF) near the summit as part of an EU funded initiative. Clearly although the spatial resolution of tidal gravity data is very limited compared with that provided by a micro-gravity survey, the temporal resolution is considerably improved (typically one observation per minute rather than several per year) which allows the rates of processes (such as dyke intrusion, magma emplacement/drainage, etc.) to be determined.

Self-potential

Several electromagnetic techniques have been used to investigate variations in electrical conductivity due to sub-surface volcanic structures. Self-potential (SP) measurements for example, made across the lower part of the south-southeast-trending fracture (Patella *et al.* 1990) after the end of the 1989 eruption with an electrode separation of 110 m over a profile length of 1.32 km showed two large positive anomalies up to approximately 100 mV in magnitude and several hundred metres in wavelength. These were interpreted in terms of shallow (near-surface) fluid movements in response to magma intrusion (Patella *et al.* 1990). These data therefore suggest that magma was intruded at depth beneath the south-southeast-trending fracture during the 1989 active period although there was no eruption south of the summit. The SP anomaly was attributed to the heating of water in saturated rocks by volcanic gases from the intrusion. Regular SP observations (approximately monthly) between October 1989 and April 1993 (Di Maio & Patella 1994) showed that the anomaly gradually decreased which was considered to be consistent with such a heating process.

Gas geochemistry

Magmatic intrusions can influence the concentration of soil gases. For a period of several days after the formation of the south-southeast-trending fracture, H_2, CO_2, CO and ^{222}Rn variations were monitored at the distal end of the fracture in an attempt to determine whether or not a magmatic intrusion was responsible. Continuous CO_2 flux monitoring revealed a dramatic increase from about 500 ppm to 2000 ppm about one week after the appearance of the fracture, but the flux returned to normal values after one day (Carapezza *et al.* 1990). Although the ^{222}Rn activity was five times higher in the fractured region than in the adjacent unfractured areas, Chiodini *et al.* (1990) were unable to prove a link between the data and degassing of a magmatic body because their observations were in the range previously measured in Etnean soil gases (Seidel & Monnin 1984).

The 1991 eruption was preceded by a relative decrease in CO_2 flux across the distal part of the SSE trending 1989 fracture, suggesting that as magma rises, degassing preferentially occurs through the main conduits in the summit area (Badalamenti *et al.* 1994). Some rapid fluctuations of soil gas fluxes across the south-southeast-trending 1989 fracture also occurred during the 1991–1993 eruption. These results suggest that soil gas monitoring along the flanks of Etna can be useful for detecting early phases of volcanic unrest, while monitoring of degassing from the summit or fractures can provide early indications of near-surface magmatic intrusion (Badalamenti *et al.* 1994).

COSPEC

The most widely used non-satellite based instrument for remote sensing of volcanic plume compositions is the correlation spectrometer (COSPEC). The relative absorption of solar uv light by a volcanic plume is analysed along with plume width and wind speed to give an estimate of the flux of SO_2 molecules. Relatively low SO_2 flux has been identified previously as a precursor to eruptions on Etna (Malinconico 1979).

A slow increase in SO_2 flux was observed from August 1989, with a transition from a low, 1000 metric tonnes per day (t d^{-1}) to medium values (7000 t d^{-1}) by mid September (Caltabiano & Romano 1990). During the second half of September and coinciding with the opening of the eruptive northeast-trending fissure in the Valle del Leone, SO_2 flux increased to an average of 15 000 t d^{-1} and peaked at 23 000 t d^{-1}. The flux decreased after this time and was at a low of 850 t d^{-1} just five weeks before the 1991 eruption began. During the 1991–1993 Etna eruption, COSPEC measurements were taken regularly – sometimes as much as twice weekly (Bruno *et al.* 1994). Comparison of data from the summit craters and the SSE trending 1989 fracture during the first few months of the eruption showed an anti-correlation which may indicate that whilst the magma is degassing significantly from the summit craters, the magma from the fissure is relatively volatile poor and vice versa (Fig. 5). It was not often possible to compare the flux from the summit and fissure sites as the plumes were usually quickly mixed by the wind. The average SO_2 flux during the eruption was 5800 t d^{-1}, but in April 1992 it dropped to 980 t d^{-1}; this may have presaged the introduction of new magma to recharge the volcano feeder system (Bruno *et al.* 1994).

Remote sensing

Fumarole and ground temperature increases have been found to precede volcanic eruptions. Changes in the radiant energy flux, or even the ground temperature of a volcano can be measured remotely via satellite or airplane and can in certain circumstances provide eruption precursor information (Rothery *et al.* 1995). Remote sensing is a fast-growing field of volcano monitoring as more data and software become readily available. Data from the thematic mapper (TM) carried by the Landsat satellites and the advanced very high resolution radiometer (AVHRR) carried by NOAA satellites have been used to test the viability of the technique at Etna (Bonneville & Gouze 1992).

Airborne TIMS data (Thermal Infrared Multispectral Scanner) were collected in the summit and southeast flank regions immediately after the cessation of eruptive activity in 1989 (Bianchi *et al.* 1990). The temperature distribution in lava flows identified using this technique agreed well with 'ground truth' observations made on the ground, but no thermal anomaly was detected along the south-southeast-trending fracture. This supports the conclusion based on other data that the opening of at least the lower end of this fracture was not driven by the shallow intrusion of a dyke.

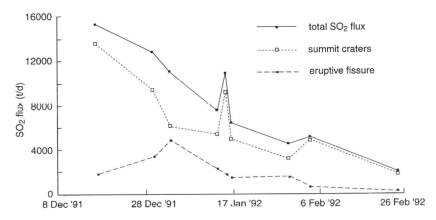

Fig. 5. Variations in SO_2 flux from the summit craters and eruptive fissure together with the total flux between December 1991 and February 1992 (after Bruno *et al.* 1994).

The possibility of real-time volcano monitoring using remote sensing came closer during analysis of the 1991–1993 lava flow field using AVHRR data. Ground truth data and lava flow maps were used to confirm remote sensing estimates of lava flow volume, active lava area, thermal flux and effusion rates (Harris et al. 1997). Although this technique does not have an application for eruption prediction, it could provide frequent information on the progress of an inaccessible lava flow, the sites of secondary breakouts and assessment of the associated hazards.

SAR interferometry

Synthetic aperture radar satellites illuminate the Earth's surface with microwave radiation and can be used for mapping topography to high precision (Francis et al. 1996). Provided that a reliable digital elevation model (DEM) is available, images obtained on ascending and descending orbits of the host satellite can be obtained. Interferograms are obtained by combining two such images and removing the effects of topography. Images for Etna during the waning stages of the 1991–1993 eruption indicated possible volcano-wide deflation (Massonnet et al. 1995). There are concerns, however, about the validity of this model (J. B. Murray, pers. comm.) as almost simultaneous ground surveys with GPS do not show the deflation, and there seems to be a strong correlation between apparent elevation change and topography.

Volcano monitoring in the future

The focus of Lyell's studies of Mt Etna was an understanding of the nature and evolution of volcanoes which has formed the foundation of more modern studies. These studies have provided the basis of the science of monitoring volcanic activity which is now an increasing imperative as both the density of population and air traffic increases in volcanic regions. Whilst monitoring solely for the purpose of risk mitigation may reduce loss of life or property to a degree, its effectiveness can be greatly improved through the concurrent development of scientific methods and enhanced understanding of volcanic processes. Given the average interval between eruptions at many volcanoes, volcano monitoring tends not to be a high spending priority of governments, yet major volcanic eruptions can be national (and international) social and economic disasters.

Future developments in volcanic monitoring will depend on many factors including technological advances, improvements in analytical and interpretation methods, the ability to successfully integrate multiple datasets and having detailed knowledge of volcanic structure and processes. The continued development of robust, inexpensive and effective technologies for monitoring volcanoes is a vital and sometimes urgent requirement for volcano-affected countries. Advances in electronics in recent decades have helped progress towards cheap expendable sensors for example, thus reducing the inherent financial risk of losing expensive monitoring equipment and greatly benefiting the monitoring of volcanoes in developing countries.

The importance of a good knowledge of the nature of background activity at each particular volcano cannot be overemphasized; if what is 'normal' is not known then the 'abnormal' cannot readily be recognized. Most of the world's potentially active volcanoes, however, are not currently monitored. Furthermore, since no reliable laws of prediction can be determined from the data collected by volcanologists to date (Scarpa & Tilling 1996) multiple integrated monitoring methods are needed at each volcano.

Monitoring techniques were, of course, originally based on the most easily measured phenomena; other phenomena were either not measured, or filtered from the signal as noise. Future progress will be made by the recording and analysis of these more subtle or complex effects using a variety of methods. For example, it is now recognized that three component and broad-band seismometers provide vital information on shallow magmatic processes and should be incorporated into any volcano-seismological network. In the case of micro-gravity monitoring, the effects of the Earth tides are subtracted from the data before analysis and modelling. Advances in technology and analytical methods, however, will allow modelling of the response of both the volcanic edifice and the plumbing system to tidal effects, from which further insight could be gained into eruption precursors and processes at depth.

Traditionally, most types of data (e.g. for gravity or ground deformation) have been recorded periodically at a number of stations. Data collection in such cases is labour intensive, time consuming and hence expensive and provide only snapshots of the real picture. Ideally, continuously recorded data at numerous stations should be telemetered to an observatory at which initial analysis could take place automatically, in real time, as has been achieved at some seismic networks. Recent advances in telemetry (Lesage et al. 1995) make such systems more viable and once established would be less labour intensive and, in times of eruption, safer.

GPS and SAR have the potential to revolutionise ground deformation studies given the large areas that can be covered simultaneously. Since these methods give information of broad-scale changes they would be particularly effective in determining the significant edifice-wide inflation that might be expected to precede a dramatic change in eruptive style. COSPEC data are likely to become more reliable and consistent as more instrumentation becomes available and data collection becomes automated (Andres & Rose 1995). Other promising techniques for remotely investigating volcanic gases, e.g. GASPEC (Williams & Dick 1997) which measures CO_2 flux, and Fourier transform infra-red (FTIR) spectroscopy which measures the ratio of various gas species (Mori et al. 1993), are currently being developed. For example, variations in the SO_2:HCL and HCl:SO_2:SiF_4 ratios have been measured using FTIR (Francis et al. 1995, 1996). Multispectral infrared instruments to be carried by the first Earth Observing System satellite, due for launch in 1998, will greatly expand the capabilities and accessibility of spaceborne remote sensing for both thermal and gas studies (Francis et al. 1996).

Potential developments in analysis and interpretation include improvements to automatic detection and processing (ADP) of seismic data such that the large volume of data generated by an intense seismic swarm can be processed and that real events can be more reliably distinguished from noise. The effectiveness of ADP depends on the density and geometry of the array compared with the extent and depth of the active zones; data from the 1989 and 1991–1993 events on Etna suggest an array density of 1 seismometer per km^2 is required to resolve events associated with shallow magmatic intrusion.

The Integrated Mobile Volcano Monitoring System developed by the USGS for use in volcanic crises (Murray et al. 1996) was used to very good effect in the 1991 Pinatubo eruption, however, portable seismometers do not constitute an adequate alternative to permanent recording stations (Ferrucci 1995). Whilst they provide satisfactory monitoring during volcanic events they have no role in background surveillance or in recording the onset of seismic activity.

Advances in interpretation will require more detailed knowledge of the structural complexities and heterogeneities typical of volcanoes in order to develop adequate models . Recent advances in non-linear, high-resolution three-dimensional seismic tomography, for example, may allow the detailed interpretation of broad-band seismic data. Such broad-band data with a period greater than a few seconds are considered to be associated with mass transport (Chouet 1996) and are currently the subject of much research into their causative mechanisms

Although seismology will remain the most important monitoring tool for the foreseeable future, the integration of seismic data with results from other methods, e.g. geodesy, gravity, electrical, gas and temperature flux etc. will be far more powerful (McNutt 1996). Smart systems using artificial neural networks (ANNs) are now being developed to integrate the vast quantities of data resulting from technological developments, to produce the simplest best-fitting models consistent with as much of the data as possible (Cristaldi et al. 1997). Such systems, however, are only as good as the input data, hence the need for detailed structural models and background information.

Important questions in contemporary volcanology include: What are the processes that trigger eruptions? Can the timing and location of an eruption be predicted? Is it possible to determine in advance the nature of an eruption and its likely duration? These issues present pressing problems both to basic science and to affected societies; they are being studied on Mount Etna and other active volcanoes using a combination of geophysical and geochemical techniques.

The priority now in volcanic monitoring is the acquisition of more data and the development of rapid and reliable methods to collate, analyse and interpret them. We do have some basic understanding of the mechanisms and processes operating within the Etna volcano but we now need to expand this knowledge by integrating the experience gained from decades of independent researches. As well as technological and scientific advances, effective communication and interaction between volcanologists, civil authorities and the affected populace are required to improve the mitigation of volcanic disasters.

We are very grateful to John Cassidy, John B. Murray, Peter W. Francis, Dave A. Rothery and two anonymous reviewers for valuable comments.

References

ANDRES, R. J. & ROSE, W. I. 1995. Remote sensing spectroscopy of volcanic plumes and clouds. In: McGUIRE, W. J., KILBURN, C. R. J. & MURRAY, J. (eds) Monitoring Active Volcanoes. UCL Press, London, 301–314.

BADALAMENTI, B., CAPASSO, M., CARAPEZZA, M. L. ET AL. 1994. Soil gases investigations during the 1991–1993 Etna eruption. Acta Vulcanologica, 4, 135–142.

BARBERI, F. & VILLARI, L. 1994. Volcano monitoring and civil protection problems during the 1991–1993 Etna eruption. Acta Vulcanologica, 4, 157–166.

BIANCHI, R., CASSACCHIA, R., PICCHIOTTI, A. & SALVATORI, R. 1990. Airborne thermal IR survey. In: BARBERI, F., BERTAGNINI, A. & LANDI, P. (eds) Mt Etna: the 1989 Eruption. Giardini, Italy, 69–72.

BERTAGNINI, A., CALVARI, S., COLTELLI, M., LANDI, P., POMPILO, M. & SCRIBANO, V. 1990. The 1989 eruptive sequence. *In*: BARBERI, F., BERTAGNINI, A. & LANDI, P. (eds) *Mt Etna: The 1989 Eruption*. Giardini, Italy, 10–22.

BONNEVILLE, A. & GOUZE, P. 1992. Thermal survey of Mount Etna volcano from space. *Geophysical Research Letters*, **19**(7), 725–728.

BROUKE, R. A., ZURN, W. E. & SLICHTER, L. B. 1972. Lunar tidal acceleration on a rigid Earth. *Geophysical Monograph Series, American Geophysical Union*, **16**, 319–324.

BRUNO, N., CALTABIANO, T., GRASSO, M. F., PORTO, M. & ROMANO, R. 1994. SO_2 flux from Mt. Etna volcano during the 1991–1993 eruption: correlations and considerations. *Acta Vulcanologica*, **4**, 143–147.

BUDETTA, G., GRIMALDI, M. & LUONGO, G. 1990. Gravity variations. *In*: BARBERI, F., BERTAGNINI, A. & LANDI, P. (eds) *Mt. Etna: The 1989 Eruption*. Giardini, Italy, 56–57.

CALTABIANO, T. & ROMANO, R. 1990. Soil gas geochemistry (c) COSPEC. *In*: BARBERI, F., BERTAGNINI, A. & LANDI, P. (eds) *Mt. Etna: The 1989 Eruption*. Giardini, Italy, 68.

CALVARI, S., COLTELLI, M., NERI, M., POMPILIO, M. & SCRIBANO, V. 1994. The 1991–1993 Etna eruption: chronology and lava flow-field evolution. *Acta Vulcanologica*, **4**, 1–14.

CARAPEZZA, M. L., GIAMMANCO, S., GURRIERI, S., HAUSER, S., NUCCIO, P. M., PARELLO, F. & VALENZA, M. 1990. Soil gas geochemsitry (a) CO_2. *In*: BARBERI, F., BERTAGNINI, A. & LANDI, P. (eds) *Mt. Etna: The 1989 Eruption*. Giardini, Italy, 62-64.

CASADEVALL, T. J. 1994. The 1989–1990 eruption of Redoubt Volcano, Alaska: impacts on aircraft operations. *Journal of Volcanology and Geothermal Research*, **62**, 301–316.

CHESTER, D. K., DUNCAN, A. M., GUEST, J. E. & KILBURN, C. R. J. 1985. *Mount Etna: The Anatomy of a Volcano*. Chapman & Hall, London.

CHOUET, B. A. 1996. New methods and future trends in seismological volcano monitoring. *In*: SCARPA, R. & TILLING, R. I. (eds) *Monitoring and Mitigation of Volcano Hazards*. Springer, Berlin, 23–98.

CHIODINI, G., CIONI, R., PESCIA, A., RACO, B. & TADDEUCCI, G. 1990. Soil gas geochemistry – H_2, CO and ^{222}Rn. *In*: BARBERI, F., BERTAGNINI, A. & LANDI, P. (eds) *Mt. Etna: The 1989 Eruption*. Giardini, Italy, 65–67.

COSENTINO, M., LOMBARDO, G., PATERNE, G., SCHICK, R. & SHARP, A. D. L. 1982. Seismological researches on Mount Etna: state-of-the-art and recent trends. *Memorie della Societa Geologica Italiana*, **23**, 159–202.

CRISTALDI, M., LANGER, H. & NUNNARI, G. 1997. Inverse and on-line modelling. *In*: Ferrucci, F. (ed.) *TEKVOLC – Technique and Method Innovation in Geophysical Research, Monitoring and Early Warning at Active Volcanoes*. Commission of European Communities Environment Program Interim Report Contract ENV4 CT95.

DAVIS, P. M. 1981. Gravity and tilt Earth tides measured on an active volcano, Mt. Etna, Sicily. *Journal of Volcanology and Geothermal Research*, **11**, 213–223.

DI MAIO, R. & PATELLA, D. 1994. Self-potential anomaly

generation in volcanic areas. The Mt. Etna case history. *Acta Vulcanologica*, **4**, 119–124.

DI PALMA, S., DRAGO, F., GALANTI, E. & PENNISI, V. 1994. Earthen barriers and explosion tests to delay the lava advance: the 1992 Mt. Etna experience. *Acta Vulcanologica*, **4**, 167–172.

D'OREYE, N., DUCARME, B., HENDRICKX, M., LAURENT, R., SOMERHAUSEN, A. & VAN RUYMBEKE, M. 1994. Tidal gravity observations at Mount Etna volcano. *In*: *Volcanic Deformation and Tidal Gravity Effects at Mt Etna, Sicily*. EU Science Project No. ERB40002PL900491 Final report, 60–80.

ÉLIE DE BEAUMONT, E. J. B. A. L. 1838. Recherches sur la structure et sur l'origine du Mont Etna. *In*: DUFRESNOY, M. & DE BEAUMONT, E. (eds) *Memoires pour Servir a une Description Geologique de la France*. Levrault, Paris, vol. 4, 1–226.

FERRUCCI, F. 1990. Seismicity. *In*: BARBERI, F., BERTAGNINI, A. & LANDI, P. (eds) *Mt. Etna: The 1989 Eruption*. Giardini, Italy, 36–43,

—— 1995. Seismic monitoring at active volcanoes. *In*: McGUIRE, W. J., KILBURN, C. & MURRAY, J. B. (eds) *Monitoring Active Volcanoes; Strategies, Procedures and Techniques*. UCL Press, London, 60–92.

FRANCIS, P. W. 1993. *Volcanoes, a Planetary Perspective*. Oxford University Press, Oxford.

——, Maciejewski, A. & Oppenheimer, C. 1996. Remote determination of SiF_4 in volcanic plumes: a new tool for volcano monitoring. *Geophysical Research Letters*, **23**, 249–252.

——, ——, ——, CHAFFIN, C. & CALTABIANO, T. 1995. SO_2:HCL ratios in the plumes from Mt Etna and Vulcano determined by Fourier transform spectroscopy. *Geophysical Research Letters*, **22**, 1717–1720.

——, WADGE, G. & MOUGINIS-MARK, P. J. 1996. Satellite monitoring of volcanoes. *In*: SCARPA, R. & TILLING, R. J. (eds) *Monitoring and Mmitigation of Volcano Hazards*. Springer, Berlin, 257–298.

HARRIS, A. J. L., BLAKE, S., ROTHERY, D. A. & STEVENS, N. F. 1997. A chronology of the 1991 to 1993 Mount Etna eruption using advanced very high resolution radiometer data: Implications for real time thermal volcano monitoring. *Journal of Geophysical Research*, **102**, 7985–8003.

KIEFFER, G. 1975. Sur l'existence d'une 'rift zone' a l'Etna (Sicile). *Compte Rendu de l'Academie des Sciences, Paris*, **280**, 263–266.

—— 1985. *Evolution Structurale et Dynamique d'un Grand Volcan Polygenique: Staeds d'Edification et Activité Actuelle de l'Etna (Sicile)*. PhD thesis. Universite de Clermont-Ferrand II, Clermont-Ferrand.

LESAGE, P., VANDEMEULEBROUCK, J. & HALBWACHS, M. 1995. Data aquisition and telemetry. *In*: McGuire, W. J., Kilburn, C. R. J. & Murray, J. B. (eds) *Monitoring Active Volcanoes*. UCL Press, London, 32–59.

LUONGO, G., DEL GAUDIO, C., OBRIZZO, F. & RICCO, C. 1990. Precision levelling. *In*: BARBERI, F., BERTAGNINI, A. & LANDI, P. (eds) *Mt. Etna: the 1989 Eruption*. Giardini, Italy, 52–55.

LYELL, C. 1858. On the structure of lavas which have

consolidated on steep slopes; with remarks on the mode of origin of Mount Etna and on the theory of 'craters of elevation'. *Philosophical Transactions of the Royal Society of London*, **148**, 703–786.

MALINCONICO, L. L. 1979. Fluctuations in SO_2 emission during recent eruptions of Etna. *Nature*, **278**, 43–45.

MASON, R .G., BILL, M. G. & MUNIRUZZAMANN, M. 1975. Microgravity and micro-earthquake studies. *In*: *U.K. Research on Mount Etna, 1974*. The Royal Society, London, 43.

MASSONNET, D., BRIOLE, P. & ARNAUD, A. 1995. Deflation of Mount Etna monitored by spaceborne interferometry. *Nature* , **375**, 567–570.

McGUIRE, W. J. & PULLEN, A. D. 1989. Location and orientation of eruptive fissures and feeder-dykes at Mt Etna; influences of gravitational and regional tectonic stress regimes. *Journal of Volcanology and Geothermal Research*, **38**, 325–344.

——, PULLEN, A. D. & SAUNDERS S. J. 1990. Recent dyke-induced large-scale block movement at Mt Etna and potential slope failure. *Nature* , **343**, 357-359.

McNUTT, S. R. 1996. Seismic monitoring and eruption forecasting of volcanoes: a review of the state-of-the-art. *In*: SCARPA, R. & TILLING, R. J. (eds) *Monitoring and Mitigation of Volcano Hazards*. Springer, Berlin, 99–146.

MELCHIOR, P. 1978, *The Tides of the Planet Earth* Pergamon Press, Oxford.

MORI, T., NOTSU, K., TOHJIMA, Y. & WAKITI, H. 1993. Remote detection of HCl and SO_2 in volcanic gas from Unzen volcano, Japan. *Geophysical Research Letters*, **20**, 1355–1358.

MURRAY, J. B. 1990. High-level magma transport at Mount Etna volcano, as deduced from ground deforation measurements. *In*: RYAN, M. P. (ed.) *Magma Transport and Storage*. Wiley, Chichester, 357–383.

—— 1994. Elastic model of the actively intruded dyke feeding the 1991–1993 eruption of Mt. Etna, derived from ground deformation measurements. *Acta Vulcanologica*, **4**, 97–100.

—— 1997. Deformation mega-network at Mt Etna Volcano. *Journal of Conference Abstracts*, **2**(1), 53.

—— & GUEST, J. E. 1982. Vertical ground deformation on Mount Etna, 1975–1980. *Geological Society of America Bulletin*, **93**, 1160–1175.

——, VOIGHT, B. & GLOT, J.-P. 1994. Slope movement crisis on the east flank of Mt. Etna volcano: Models for eruption triggering and forecasting. *Engineering Geology*, **38**, 245–259.

MURRAY, T. L., EWERT, J. W., LOCKHART, A. B. & LAHUSEN, R. G. 1996. The Integrated Mobile Volcano-Monitoring System used by the Volcano Disaster Assistance Program (VDAP). *In*: SCARPA, R. & TILLING, R. J. (eds) *Monitoring and Mitigation of Volcano Hazards*, Springer, Berlin, 315–364.

NUNNARI, G. & PUGLISI, G. 1995. GPS – monitoring volcanic deformation from space. *In*: McGUIRE, W. J., KILBURN, C. R. J. & MURRAY, J. B. (eds) *Monitoring Active Volcanoes*. UCL Press, London, 151–184.

PATANE, D., PRIVITERA, E., FERRUCCI, F. & GRESTA, S. 1994. Seismic activity leading to the 1991–1993 eruption of Mt. Etna and its tectonic implications. *Acta Vulcanologica*, **4**, 47–56.

PATELLA, D., TRAMACERE, A. & DI MAIO, R. 1990. Self potential anomalies. *In*: BARBERI, F., BERTAGNINI, A. & LANDI, P. (eds) *Mt. Etna: the 1989 Eruption*. Giardini, Italy, 58–61.

ROTHERY, D. A., OPPENHEIMER, C. & BONNEVILLE, A. 1995. Infrared thermal monitoring. *In*: McGUIRE, W. J., KILBURN, C. R. J. & MURRAY, J. (eds) *Monitoring Active Volcanoes*. UCL Press, London, 184–216.

RYMER, H., CASSIDY, J., LOCKE, C. A. & MURRAY, J. B. 1995. Magma movements in Etna volcano associated with the major 1991–1993 lava eruption: evidence from gravity and deformation. *Bulletin of Volcanology*, **57**, 451–461.

——, MURRAY, J. B., BROWN, G. C., FERRUCCI, F & McGUIRE, W. J. 1993. Mechanisms of magma eruption and emplacement at Mt. Etna between 1989 and 1992. *Nature,* **351**, 439–441.

SANDERSON, T. J. O. 1982. Direct gravimetric detection of magma movements at Mount Etna. *Nature*, **297**, 487–490.

——, BERRINO, G., CORRADO, G. & GRIMALDI, M. 1983. Ground deformation and gravity changes accompanying the March 1981 eruption of Mount Etna. *Journal of Volcanology and Geothermal Research*, **16**, 299–315.

SEIDEL, J. L. & MONNIN, M. 1984. Mesures de Radon-222 dans le sol de l'Etna (Sicile): 1980-1983. *Bulletin of Volcanology*, **22**, 1071–1077.

SCARPA, R. & TILLING, R. I. 1996. Preface. *In*: SCARPA, R. & TILLING, R. J. (eds) *Monitoring and mitigation of volcano hazards*. Springer, Berlin, v–xi.

SCHICK, R. & MUGIONO, R. 1991. Introduction. *In*: SCHICK, R. & MUGIONO, R. (eds) *Volcanic Tremor and Magma Flow*. Scientific Series of the International Bureau, Forschungszentrum Julich GmbH, Jülich, no. 4, 1–3.

SCROPE, G. J. P. 1825. *Considerations on Volcanoes, the Probable Causes of their Phenomena and their Connection with the Present State and Past History of the Globe; Leading to the Establishment of a New Theory of the Earth*. W. Phillips & G. Yard, London.

STEVENS, N. F., MURRAY, J. B. & WADGE, G. 1997. The volume and shape of the 1991–1993 lava flow field at Mt Etna, Sicily. *Bulletin of Volcanology*, **58**, 449–454.

VASSALE, R. 1994. The use of explosive for the diversion of the 1992 Mt. Etna lava flow. *Acta Vulcanologica*, **4**, 173–177.

VON BUCH, L. 1818–1819. *Uber die Zusammensetzung der basaltischen Inseln und über Erhebungskrater.* Abhandl. Preuss. Akad. Wiss., Berlin, 51–68.

WADGE, G. 1976. Deformation of Mount Etna 1971–1974. *Journal of Volcanology and Geothermal Research*, **1**, 237–263.

WILLIAMS, S. N. & DICK, R. 1997. The GASPEC remote sensor for quantification of CO_2 flux by volcanoes. *In*: *Pre-Congress Short Course on Volcanic Gases*. IAVCEI General Assembly, Mexico.

Earthquakes and Earth structure: a perspective since Hutton and Lyell

BRUCE A. BOLT

Department of Geology and Geophysics, University of California, Berkeley, CA 94720, USA

Abstract: Lyell's interest in earthquakes as part of the *Principles of Geology* continues to be justified many fold. A quarter century after Lyell's death, seismology began to open the window on the contemporary structure and tectonic deformation of the Earth. Detailed non-biased observations of the global distribution of earthquakes played a crucial role in the attack on pre-plate theories of Earth dynamics. There were three critical seismological assault tools: reliable hypocentre catalogues, uniform magnitude estimates, and fault source mechanisms. Previously used as evidence for plate tectonics, seismicity is now often taken as predicted by it. Nevertheless, earthquake occurrence remains unforecastable in definite temporal terms. Interplate and intraplate spatial patterns show complexity in macro-crustal and micro-crustal structures. In particular, the mechanism and dynamic implications of deep-focus earthquakes and subduction remain a challenge.

Local and global seismographic networks are increasingly enhanced by broadband digital seismometry. This modern instrumentation provides high resolution of strong ground shaking and crustal and deeper interior structure. Second-order structural variations are now being mapped in the upper mantle and more detailed boundary conditions for convection models are being resolved in the lithosphere and in the D″ mantle–core layer. Recently, seismological evidence for scattering anomalies throughout the mantle has become persuasive.

It is well known that Charles Lyell's *Principles of Geology* (1875) contains considerable descriptive material on earthquakes and links them with uplift and other deformation of the Earth's surface. Only eight years after its publication, Professor John Milne, then working in Japan, surmised (see Bolt 1993) that 'it was not unlikely that every large earthquake might with proper appliances be recorded at any point of the globe'.

This prediction was fulfilled in 1889 by the German physicist E. Von Rebeur Paschwitz, who 'was struck by the coincidence in time' between the arrival of singular waves which were registered by delicate horizontal pendulums at Potsdam and Wilhelmshaven in Germany and the time of a damaging earthquake that shook Tokyo at 2:07 am Greenwich Mean Time on 18 April. His conclusion was that 'the disturbances that were noticed in Germany were really due to the earthquake in Tokyo'. The significance of this identification – an early example of remote sensing – was that earthquakes in inhabited and uninhabited parts of the world alike could be monitored uniformly, and thus patterns of geological activity could be mapped without bias; an era in the quantitative study of earthquakes and geology not known to Lyell then began.

A principal aim of this paper is to provide historical illustrations and a short commentary on major ongoing problems in seismology. In tracing the historical evolution of knowledge in seismology and related tectonics from Lyell's day, I have been forced to select only four central topics on earthquakes: their tectonic causes; their wave motion; their prediction in time and location; and their use to image the three-dimensional structure of the deep interior. Even these subjects, each of interest to my own research, must be considered very briefly, with a narrow focus on recent debates. My textual reference to Lyell's writings is, for brevity only, the twelfth (and last) edition of *Principles of Geology*. Each successive edition of this seminal treatise incorporated 'important additions and corrections'. Nevertheless, Lyell comments that although between the first and twelfth editions numerous descriptions of recent earthquakes had been published, he doubted that they illustrated new principles.

Lyell's accounts of earthquakes

James Hutton wrote little on earthquakes (Bailey 1967). He did describe processes that had led to land surfaces above the sea surface. He concluded that 'the land in which we dwell' has been elevated 'by extreme heat and expanded with amazing force'. This belief led in turn to a consideration of volcanoes, active and extinct, with slight reference

BOLT, B. A. 1998. Earthquakes and Earth structure: a perspective since Hutton and Lyell. *In*: BLUNDELL, D. J. & SCOTT, A. C. (eds) *Lyell: the Past is the Key to the Present*. Geological Society, London, Special Publications, **143**, 349–361.

349

to earthquakes, neither of which were within his personal experience. He considered volcanic eruptions to be safety valves 'in order to prevent the unnecessary elevation of land and fatal effect of earthquakes'.

In contrast, Lyell emphasized the value of earthquake studies for geology. In the twelfth edition of the *Principles* he discusses volcanoes and earthquakes as constructive forces. These accounts reflect the prevailing view of a common underlying cause and intimate physical connections. Nevertheless, they still read well today, with many case histories and arguments based on the very limited geophysical measurements available. Lyell begins by regretting the deficiency of accounts of ancient earthquakes, almost all descriptions being restricted to damage and injury. His interest was in the *geological* aspect of earthquakes, particularly the coseismic changes in the Earth's crust that accompanied them.

By 1875, Lyell had available reports by Robert Mallet and the catalogues of Alexis Perry and others. There is little doubt that he developed a strong interest in seismology and he summarized published reports of major earthquakes in such widely distributed places as Jamaica (1692), Java (1699), Chile (1751), Lisbon (1755), Calabria (1783), Sicily (1790), Bengal (1792), Quito (1797), and New Madrid, Missouri (1811–1812). He took any opportunity to converse with engineers and others who had been eyewitnesses and these second-hand accounts are of continuing value.

There was no surface evidence of fault rupture genesis of many of the earthquakes discussed by Lyell. We see in his writings only the beginnings of the accumulated field evidence for the uniformitarianism of the seismic source of most tectonic earthquakes. It is of interest that the separate classification 'volcanic earthquakes' persisted well into this century in text books. Now they are regarded as also immediately produced by sudden elastic strain release in fractured rocks around the volcanic tubes and chambers.

A number of nineteenth century earthquakes described at length in the *Principles of Geology* have been the subject of much recent research. We might mention the 1835 elevations along the Chilean coast (nowadays described as being associated with subduction earthquakes) and the coseismic uplift along the coast during the prototype intraplate earthquake in the Rann of Kutch, India on 6 June 1819. In the latter, land rose by up to 10 feet over an area of radius 50 miles. The woodcuts in the *Principles of Geology* (p. 100) showing Sundree Fort before and after this earthquake are classics. Other intraplate earthquakes in an area specially visited by Lyell are the New Madrid earthquakes of 1811 and 1812; in

March 1846 he had an opportunity to visit the disturbed region of the Mississippi embayment and talk with eye-witnesses. The main geological conclusion reached by Lyell in his study of earthquakes was the contravention of the belief that significant changes of relative levels of land and sea had ceased: 'in the face of so many striking facts, it is vain to hope that this favourite dogma will be shaken'.

His celebrated description of an 1883 Italian earthquake series (pp. 113–144), which lasted for many months, continues to have a prominent place among seismological studies. These earthquakes in Calabria were powerful enough to destroy over 180 towns and villages and kill 30 000 people. They were accompanied by many striking geological phenomena, and furnished examples of many seismic effects common to earthquakes around the world. A special importance of the 1883 Calabrian earthquakes was, as Lyell states (p. 113), that they afforded 'the first example of a region visited during and after the convulsions, by men possessing sufficient leisure, zeal and scientific information to enable them to collect and describe with accuracy such physical facts as throw light on geological questions'. Lyell relied on the extensive field report of the Neapolitan Academy of Sciences to whom goes the credit for appointing the first scientific commission to investigate a great earthquake. He also quotes D. Vivenzio, who wrote the first monograph devoted to an earthquake disaster, and the report of the French geologist Déodat Gratet de Dolomieu. Some authors (Yeats *et al.* 1997) suggest that Gratet de Dolomieu who described a fissure several feet wide over 10 miles along the contour margin of the Aspromote massif, may have been the first to discover surface faulting which had led to an earthquake.

It should not be overlooked that Lyell includes some of the best descriptions of widespread liquefaction in his Calabria case analysis, including drawings and descriptions of sand 'blows' and 'boils'. Typically, he does not speculate on their physical basis in terms of the modern explanation involving shear strength and pore pressure of soils. Lyell remarks that the shocks caused no eruption of either of the nearby volcanoes Etna and Stromboli. He acutely concluded that therefore the 'sources of the Calabrian convulsions and the volcanic fires of Etna and Stromboli appear to be very independent of each other'.

Causes of tectonic earthquakes

As a result of direct geological and geodetic field measurements after the 1906 San Francisco earthquake, H. F. Reid propounded an elastic rebound theory of earthquake genesis: strains build up in the faulted rocks until a failure point is reached; rupture then takes place in the strained rock; each side of the fault rebounds under the elastic stress field until the strain is largely or wholly relieved. On this theory there is no direct connection between volcanic activity and the sudden emission of seismic waves; rather so-called volcanic earthquakes, often in swarms, may be associated with the movement of magma in subterranean ducts from one chamber to another. Conversely, large earthquakes in a volcanic region may produce seismic P and S waves energetic enough at regional sites to stimulate volcanic activity by means of shaking of the magma in underground chambers with consequent activation of superheated steam.

The theory of plate tectonics was the first to provide a global physical reason for the uneven geographical pattern of significant seismicity around the world. In brief, it explains why most earthquakes occur along the edges of the interacting tectonic plates (interplate earthquakes), and why the Wadati–Benioff zones along the ocean trenches coincide with the plate convergence that results in crustal rocks subducting into the mantle. In addition, the convergence rates match the seismic energy budget derived from the standardized earthquake observatory catalogues of the last 50 years (Bolt 1993). These show that earthquakes at convergent plate boundaries contribute more than 90 per cent of the Earth's release of seismic energy for shallow earthquakes, as well as most of the energy for intermediate and deep-focus earthquakes (down to 680 km depth). Most of the Earth's largest earthquakes (such as the 1960 and 1985 Chile earthquakes, the 1964 Alaskan earthquake, and the 1985 Mexican earthquake) originate in subduction zones. The high rate of seismicity occurring along undersea faults along the mid-oceanic ridges (unknown to Hutton and Lyell), is the consequence of the construction of tectonic plates by volcanic processes. Collision margins (such as the Himalayas and Caucasus) also generate energetic earthquakes with thrust mechanisms.

One example must suffice to illustrate how the present geological knowledge extends far beyond Lyell's scope. He spent considerable space (Chapter 28) discussing earthquakes in New Zealand: 'in no country, perhaps, have earthquakes, or to speak more correctly, the subterranean causes to which such movements are due, been so active in producing changes of geological interest as in New Zealand.' Yet tectonics and seismogenesis in New Zealand have occasioned considerable controversy over the years and research continues on a tectonic synthesis based on regional plate tectonic models (Berryman *et al.* 1992).

Two great earthquakes are addressed: 19 October 1848 and 23 January 1855. Secondhand accounts of the two are given in the *Principles of Geology*. Sir F. Weld informed Lyell that he had seen in 1848, in the northeast South Island, fissures extending for 60 miles, striking north-northeast in a line parallel to the mountain chain. Circumstantial evidence is that at least part of the 'great rent' described by Weld was fresh displacement on the Awatere Fault. Verification is complicated by the Marlborough earthquake of 1888 ($M = 7.3$) which is now ascribed to rupture of the Hope fault. Like the Awatere Fault, the latter branches from the major Alpine Fault of the South Island of New Zealand, but some hundred kilometres to the south. These large seismic sources are part of a rather unusual trench–trench boundary (Yeats *et al.* 1997), an active transform system of which the Alpine Fault is a part, although the main trace of the Alpine Fault has not generated a major earthquake since at least 1840 when European settlement began.

The most complete description by Lyell was of the West Wairarapa earthquake in 1855, which according to Lyell was felt by ships at sea 150 miles from the coast, with a strongly shaken area estimated at 360 000 square miles, 'an area three times as large as the British Isles.' In the vicinity of Wellington (Fig. 1) in the North Island, a tract of land comprising 4600 square miles was supposed to have been 'permanently' upraised by 1–9 feet. He repeats eye-witness descriptions of changes in geomorphology. These contain what may be the first instance of observed faulting generally known (Yeats *et al.* 1997). Later field work by M. Ongley and others maps the Wairarapa Fault as passing in the northeast direction about 15 miles east from Wellington, which suffered serious damage in the 1855 earthquake. The source rupture is now estimated to have uplifted the country to the west with an upthrow along the fault scarp of 3–10 feet (Ongley 1943).

The New Zealand geologist Alexander McKay, carrying a copy of the *Principles of Geology,* made an immediate field investigation of the subsequent and also damaging 1 September 1888 earthquake (see Richter 1958). Towards the northern end of the South Island, McKay observed new fault rupture giving freshly disturbed ground and shifting fences near the Hope River more than 2.6 m out of line, with the northern side displaced with left-lateral offsets. McKay is credited with being one of the first geologists to document strike-slip on a fault. The kinematics of the Alpine fault system accommodates the plate motion by right lateral offsets along the Alpine Fault and subsidiary faults, such as the Awatere and Hope Faults, together with shortening of geological structures to the east of the main fault system in response to the general convergence (Berryman *et al.* 1992).

Fig. 1. Aerial photo looking northwest into Wellington Harbour, New Zealand. The Wellington Fault strikes to the west of the harbour. (Courtesy DSIR Geology and Geophysics.)

Finally, it is of interest that while Lyell was not fortunate enough to visit the scenes of earthquakes where surface faulting was evident, he did correlate the 1855 earthquake in New Zealand with what he defined as a fault. He comments (p. 88), 'The geologist has rarely enjoyed so good an opportunity of observing one of the steps by which those great displacements of the rocks called 'faults' may be brought about.' His following remarks make clear, however, that he still held to the then usual belief that fault displacement is the result of the earthquake rather than the cause of it.

Predictions of ground shaking

We find in the *Principles of Geology* various accounts of the intensity of seismic ground shaking, but Lyell refrains from attempting quantitative dynamical explanations. Lyell was ever wary of what he called 'the spirit of exaggeration in which the vulgar are ever ready to indulge'. Very rarely is he physically gullible and his description of the passage of strong earthquake waves in the Calabrian earthquakes is sound. An exception is his recounting a secondhand story that rents and

chasms in the ground opened and closed alternatively 'so that houses, trees, cattle and men were first engulfed in the instant and then the sides of the fissures coming together again no vestige of them was to be seen on the surface'. In his account of the Calabrian earthquake he mentions observations of pavement stones 'bounding into the air' and coming down with their sides reversed and he ventures a dynamical argument for the occurrence. Typically, on matters of mechanics, he quotes Mallet who had much deeper engineering knowledge.

As with the other parts of earthquake studies, it was the operation of specially-designed seismographs to record on-scale the strongest shaking in earthquakes that led to the modern advances in strong motion seismology (see Fig. 2). In the 1970s and 1980s, particularly as wave interpretation of the recorded strong seismic waves became more reliable, mathematical work began on ways to solve the basic inverse problem of strong motion seismology: namely, the prediction of strong ground motions given the seismic source and site (Bolt 1996). An associated trend is a return to the prevailing concept in Lyell's time that engineers

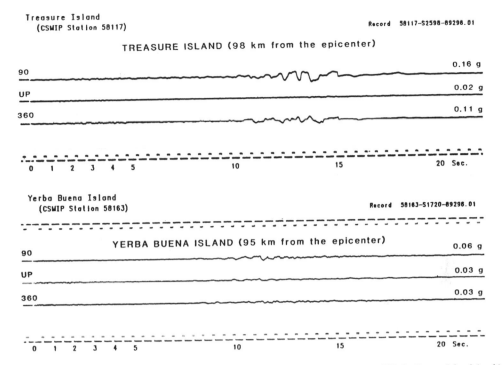

Fig. 2. Comparison of strong ground accelerations recorded at Treasure Island (soil) and Yerba Buena Island (rock) in the 1989 Loma Prieta earthquake, California.

and geologists share a common interest in earthquakes. It recognized that there is a significant overlap in expertise, even though it is not to be expected that seismologists and geologists will be specialists in engineering analysis or mechanics. The common physical understanding of strong ground motions and their effect on structures comes from a shared training in basic mechanics.

One geological aspect of recent studies in strong motion seismology recalls Lyell's interest in earthquake intensity distribution. It has now been demonstrated that deep crustal structure as well as surficial soil and sediments affects the distribution of the high seismic intensity around the earthquake source. The 1989 Loma Prieta earthquake ($M = 6.9$) in California provided perhaps the firmest yet quantitative evidence. The damage patterns and strong motion recordings showed significant spatial variations in the shaking (Lomax & Bolt 1992) around San Francisco Bay. Large coherent shear wave displacements observed at certain distances were in the period range of 2–5 s (see Fig. 2). These waves were amplified by factors of two within San Francisco and Oakland at distances of 50–60 km from the earthquake source when compared with similar wave amplitudes at much shorter distances.

Theoretical wave modelling with a realistic three-dimensional crustal structure demonstrated that the lateral refraction of the shear waves by the particular geological structures in the San Francisco Bay Area was responsible. The wave focusing was a consequence of the different rock types across the San Andreas Fault and deep sedimentary basins at the south end of San Francisco Bay. The study illustrated the present ability of realistic propagation path modelling to explain variations in shaking which are so clearly described in historical earthquakes by Lyell.

Global seismicity and its inferential value

The construction of the plate tectonic model of terrestrial deformation depended to a crucial extent on two seismological products: the uniform global mapping of earthquake foci and the estimation of source mechanisms. This information flowed from the punctilious earthquake surveillance work at seismographic stations around the world (see Bullen & Bolt 1985). By the 1950s, the enlightened proposals by Mallet and later by Milne had led to the Worldwide Standardized Seismographic Network of about 120 stations distributed in 60

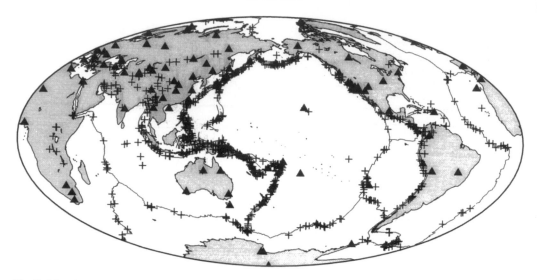

Fig. 3. Map showing the location (1996) of seismographic stations (triangles) with broadband digital instruments belonging to the Global Seismic Network. Crosses plot 979 shallow earthquakes with $M \geq 5.7$ occurring during 1988–1994 (Astiz *et al.* 1996).

countries (see Fig. 3). Global earthquake measurements, together with the concurrent introduction of high-speed digital computers, meant that seismologists were in the right place at the right time (see Oliver 1996). Precise hypocentre locations, magnitude estimates and source mechanisms became almost routine. By 1960 it was possible to have a broad global classification of seismogenesis caused by crustal convergence and divergence. These results were basic to the development of plate tectonics. Some peculiarities, however, remained and remain for additional detailed investigation (Yeats *et al.* 1997). Why on the plate tectonic description do many earthquakes, including major damaging ones, occur far from plate boundaries? Indeed, hinterland seismic activity occurs in all continents except Greenland (Bolt 1993). As mentioned already, Lyell described a number of these intraplate earthquakes.

There is one special class of earthquake which continues to attract research ingenuity. Deep-focus earthquakes have sources well below the crust (the deepest have foci at depths of about 680 km) and hypocentres within or along the boundary of the relatively cold subduction slab. The great pressures at such depths (up to 240 kbars) make it difficult to envisage their genesis by elastic rebound along a discontinuous surface in brittle rocks. Various alternative mechanisms have been suggested to settle this longtime dilemma, such as sudden dilatational change in volume of the rocks, perhaps from a sudden change in the phase state of the con-

stituent minerals. Another requires that the water of crystallization becomes mobilized at the high ambient temperatures and pressures and migrates throughout the rock pores; fractures are thus lubricated, allowing slip to become coherent. A third is that the phase transitions are localized to boundaries between rock lenses, where fluid conditions are particularly favourable to sudden transitions. Along these pre-existing grain boundaries, the crystal structures change rapidly, thus weakening the bonds across the discontinuities.

It is instructive to consider the extraordinary deep earthquake that occurred on 9 June 1994 at a depth of 637 km within the Nasca subduction slab under Bolivia. This earthquake ($M = 8.3$) is the largest earthquake at these extreme depths ever recorded. Strong ground motion was felt over much of Bolivia, but there were no deaths and relatively minor damage. One remarkable feature was that 10 to 20 minutes after the deep energy release, seismic waves were felt by many people in the Caribbean and in North American cities as distant as Chicago, Illinois and Toronto. The location of the focus placed it in 'the grand jog' of the subduction slabs between the Wadati–Benioff zone under Argentina near southern Bolivia and the zone under Peru and Brazil. These zones are nearly parallel, but offset from each other by 1000 km in the vicinity of this earthquake focus.

Cross-correlation between seismic wave forms recorded at a number of the newly positioned global digital seismographic stations (see Fig. 3) now allows theoretical wave forms computed from fault models to be compared directly with the observations. An analysis of this type (Antolik 1996) demonstrated that the source mechanism for this very deep earthquake source (with magnitude exceeding that of the 1906 San Francisco earthquake) is the same as that for shallower earthquakes in the crust. The solution in this study was a near-horizontal nodal plane (strike 302 degrees, dip 11 degrees), an average rupture velocity of 1.7 km/s over a roughly ellipsoidal area with major axis 12 km long. The average slip throughout the rupture area was about 5 m with a peak value of 16.5 m. Unfortunately, no uncertainties on these values are given and the solution is not unique.

The subduction slabs defined by deep-focus earthquakes bear in a critical way on one of the foremost geophysical questions, extending back to at least Lyell's time (Brush 1979), namely, the effect of viscous hydrodynamical conditions in the Earth's mantle. In recent times, the kinematics of plate tectonics led to the adoption of large-scale mantle convection as the driving force for subduction, mountain building, volcanic activity, and major earthquake genesis. The older question of rigidity versus fluidity (Brush 1979) shifted to a long-running debate on whether the mantle convects as a whole or whether it convects in a two shell system.

Recent special studies of earthquake waves have indicated a way to discriminate between the two models. Statistical reanalysis of the standard catalogues of earthquake locations and, hence,

reported travel times has provided a significantly more precise sample of seismic travel times for tomographic imaging. From such a database, inversions to the three-dimensional spherical structure of the mantle paths have indicated linear patterns of velocity anomalies which extend downwards from subduction slabs in the lithosphere (Van der Hilst *et al.* 1997). These linear patterns that map regions of relatively fast seismic waves can be interpreted as slabs extending at least 1000 km below the 400 km boundary in the upper mantle (see Table 1) at places around the Pacific. The consequence of the confirmation of such models would be that subduction slabs sink to the core boundary. There they may accumulate and perhaps remelt, mix and rise again as molten plumes to the surface.

Earthquake prediction

The hallmark of a scientific theory is its predictive power. Indeed, the attraction of the Newtonian dynamical theory, modified appropriately for relativity, and of the Darwinian theory of evolution are their ability to forecast beyond the available observations and descriptions. From early times, a belief in forerunners to large earthquakes and reported eye-witness accounts of them have been recorded. Lyell only refers to such forerunners in passing (p. 81). He mentions reported irregularities in the seasons preceding shocks, animals evincing extraordinary alarm, violent rains and other

Table 1. *Bullen's 1942 specification of average internal shells of the Earth*

Region	Level	Depth (km)	Features of region
–	Outer surface	–	–
A			Crustal layers
–	Base of crustal layers (distance R from the Earth's centre)	33	–
B			Steady positive P and S velocity gradients
–	$0.94R$	413	–
C			Transition region*
–	$0.85R$	984*	–
D			Steady positive P and S velocity gradients
–	$0.548R = R_1$	2898	–
E			Steady positive P velocity gradient
–	$0.40R_1$	4982	–
F			Negative P velocity gradient[†]
–	$0.36R_1$	5121	–
G			Small positive P velocity gradient
–	Earth's centre	6371	–

* Now estimated to be 660 km.
[†] In 1962 shown incorrect (Bolt 1982).

occurrences. Nevertheless, he confines his text 'almost entirely to the changes brought about by earthquakes and in the configurations of the Earth's crust'.

It is generally accepted that a serious earthquake prediction must also specify the location, origin time, and magnitude within specific known limits. Such certainty is impossible to justify because prediction is always based on a limited number of measurements, themselves imprecise. Consequently, scientific predictions must state the probability of the occurrence.

A great amount of research has been carried out on earthquake prediction in a number of countries, particularly Japan, the United States, the Soviet Union, China, and Italy in recent decades (Mogi 1985). Comparison of the earthquake predictions published in earthquake journals with the actual seismological record indicates no proclaimed method can be taken as proved or effective. In my view, the failure of the modern research on earthquake prediction raises doubt whether forecasts within strict bounds of time and place will ever be possible for most earthquakes, particularly the large damaging ones. Two contemporary cases give the flavour of the present hiatus in prediction programmes. The first was the inability of the Japanese seismological prediction programme, often regarded as a model, to give forewarnings of the devastating Kobe earthquake of 17 January 1995. The failure was exacerbated by the numerous public statements made over decades in Japan that 'precursors' had been recorded after earlier damaging earthquakes (Geller 1997). For example, within hours of the Kobe earthquake, there were announcements from some seismological quarters that instruments had recorded several small earthquakes with almost the same epicentre as the main shock, a few hours to weeks before it. There are, however, no scientific grounds that allowed any of these small earthquakes to be identified in advance as a forerunner of a major earthquake. Sharp criticisms in Japan of such hindsights based on coincidence were reflected in the public press.

The second illustration comes from ongoing debate of the validity of the predictions of earthquakes in Greece (Varotsos et al. 1996). These predictions used variations in the electrical field within a specified region to trigger alarms. The proponents argue strongly that their predictions are more successful than those provided by chance coincidences. For example, between 1987 and 1989 the number of events with magnitudes greater or equal to 4.7 was 39. During this interval, the method yielded 23 predictions with a claimed success of 38 per cent. The nominal duration of the alarm period was 23 days. There are many critics of these claims (e.g. Mulargia & Gasperini 1992).

Some doubt the instrumental ability to even detect, above the ambient electrical noise, the tiny electromagnetic variations due to static elastic straining in the seismogenic area; substantial electric signals of all kinds are prevalent in modern populated areas. Another evaluation approach is to accept the published observed correlations and subject them to strict statistical tests (Stark 1996, 1997). The most detailed tests along these lines to date indicate that the success of the claimed predictions is unsupported by the evidence.

My own inclination is to be skeptical of most prediction claims based on 'abnormal' seismicity, geodetic, geological or geophysical observations, or unexpected deviations from average regional parameters. One source of doubt is that the physical genesis of tectonic earthquakes denies the assumption that origin times, locations, and magnitudes of earthquakes are jointly independent. Most statistical analyses of seismicity catalogues assume the constant rate of a memoryless Poisson distribution; yet the elastic rebound theory states that there is an evolution of the stress field, interrupted by sudden strain decrease at the time of fault slip (Bullen & Bolt 1985). So, too, after the substantial rupture of a fault, like the San Andreas fault in California, the frictional distribution on the contiguous fault surfaces is likely to be irrevocably changed so that the traction memory is lost or weak. Although the mechanism of earthquake genesis is understood physically, the concomitant geological complexities and remoteness of the rupture surfaces have limited prediction abilities to a stage not much beyond the time of Lyell (see Geller 1997).

The three-dimensional geological structure of the Earth

Although many of his case histories describe large seismic events in regions where there are no volcanoes, Lyell continued to link them together. 'The regions convulsed by violent earthquakes include within them the site of all the active volcanoes. Earthquakes sometimes local, sometimes extending over vast areas, often precede volcanic eruptions.' He argues that the common cause is the passage of heat from the interior to the Earth's surface. His discussion of the physical condition of the Earth's interior uses arguments from mechanics, such as experiments with pendulums and the attraction of the Earth to the Moon. These investigations 'have shown that our planet is not an empty sphere, but on the contrary that its interior, whether solid or fluid, has a higher specific gravity than the exterior'. He also accepts that vibrations of the Moon indicate that the Earth has an increase of density from the surface towards the centre, the average value being about 5.5. It is actually 5.52 g/cm^3. He quotes Young on the effect of compression at the Earth's centre, which would compress steel into a quarter of its volume using the assumption that a terrestrial nucleus may be metallic with a specific gravity of iron of about 7. He is acute in judging that such extrapolations are uncertain because compressibilities of bodies differ from values obtained in the laboratory.

Finally, Lyell adopts the notion of a solid crust and refers to the effect of the attraction of the Moon on the Earth's precession. Lyell admitted that the inside of the Earth was hot, but in accordance with his uniformitarian view, denied its temperature had changed significantly in the course of geological history (pp. 210–240). He differed from the argument for an Earth model with a solid shell over a liquid nucleus, favoured by William Hopkins, the founder of British physical geology, who often opposed the uniformitarian school (see Fisher 1889). Lyell points out that theoretical precessional orbits do not agree with the astronomical observations Hopkin's relied upon unless the minimum thickness of the crust of the globe 'cannot be less than 1/4 or 1/5 of the Earth's radius'. In fact, the rigid mantle is about one half of the radius. Debates of this kind defined the status of global Earth structure at the end of Lyell's life (Brush 1979). Seismological tomography (Iyer & Hirahara 1993) was needed to estimate the detailed elastic structure and physical properties of the Earth's interior.

The worldwide network of seismographic stations allowed the production of standard travel times of the main seismic body waves through the Earth. The most notable estimates were obtained by Jeffreys and Bullen in 1940. These tables (Fig. 4) provided the solution of two basic geophysical inverse problems. Problem 1: if the location of an earthquake's epicentre is known, times of travel to observatories at any site can be calculated; the actual problem is to locate an epicentre anywhere in the world using readings of arrival times of earthquake waves at observatories. Problem 2: if the seismic wave velocity distribution at each depth in the Earth is known, one can calculate the travel time $T(D)$ to distance D. The inverse question is: given $T(D)$, calculate the variation in velocity v(r) in the Earth.

The use of the Jeffreys–Bullen travel-time tables in solving Problem 2 provided robust estimates of the unknown velocity structure but on the assumption of radial symmetry (Bullen & Bolt 1989). Their construction, accompanied by application of probability and inference theory, was so successful that the tables have been used widely up to today even though regional anomalies in the average times were known and were shown to be correlated with deviations from interior radial symmetry.

Estimation of the variation of seismic wave velocities, V_P and V_S, within the Earth provided the key to structure (Fig. 5). Bullen went on to define a nomenclature for radial shells based on the velocities (Bullen & Bolt 1985). The shells range from layer A, the crust, to layer G, the inner core of the Earth (see Table 1). The layers B and C incorporate what has become known as the upper mantle of the Earth where the largest deviations from spherical symmetries occur. The layer D is the solid lower mantle and E the liquid outer core.

As more earthquake observatories with better equipment were put into operation, particularly the broadband digital seismographs of the 1980s and 1990s, regional deviations from the radial dependence became more sharply focused. These deviations entailed significant heterogeneities in deep Earth structure, some of which correlate well with surface geology. (Three-dimensional maps of this heterogeneous structure really require colour projections of the globe (see for example Dziewonski & Woodhouse 1987; Bolt 1993).) Nevertheless, deviations from spherical symmetry become less marked, and are generally of the second order, with depth because of the strong influence of pressure on the mineralogy of the rocks. Below a low velocity lithosphere layer in which S and perhaps P velocities decrease somewhat, the velocities increase by about a percent or so down to depth of 220 km, often called the Lehmann discontinuity, where there is a jump of about 3 per cent. The depth of this discontinuity varies from region to region, and it is not detected in some studies in some places. The 410 and 660 km depth discontinuities are features of the upper mantle that have been observed worldwide. They have been confirmed by underside reflections which occur as precursors to the reflective core waves, although summarizing the structural trends of these discontinuities has proved difficult.

The next zone of interest is near the mantle–core boundary (MCB). High frequency (about 1 Hz) P and S waves are both recorded that are reflected from the upper side of the MCB; P waves are also reflected from its underside. These common observations are convincing evidence that the MCB is quite sharp. In the last seven years, much effort has been made to map the structural details of the layer called D″, at the mantle base. Bullen differentiated this sub-shell because the 1939 Jeffreys P and S velocity curves flattened for about 200 km above the core boundary. In fact, the constant slope was an artifact of the smoothing by Jeffreys of the inversion process, but the suggestion led to attempts to find more direct observational evidence for a transition shell. This partly comes from

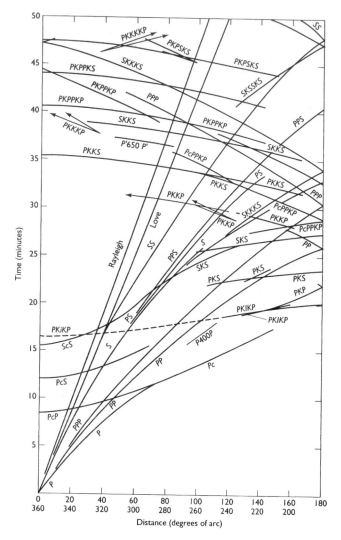

Fig. 4. Travel-time curves of various seismic wave types for a surface source to various angular distances on the Earth's surface (after Jeffreys & Bullen 1940).

diffracted earthquake waves (of both P and SH types), which run around the mantle core boundary (Bolt, 1982). It is now known that a radially spherical inversion for D″ structure does not yield a realistic description. There are relatively large-scale deviations in elastic properties within the layer, which affect waves according to their wave length. These azimuth and wave-length dependent effects are difficult to tie down and a number of competing solutions have been given. The most recent studies of high frequency core waves appear to require heterogeneous (non-radially symmetric) structures both in D″ and well above this layer in the mantle (Hedlin *et al.* 1997).

Future work

At present, the research situation in seismology is excellent. The worldwide network of broadband digital stations and the easy exchange of digital recordings of earthquakes through the Incorporated Research Institutions for Seismology (IRIS) programme has made accessible to researchers, in a straightforward way, very high quality seismic wave-form data. Even the debilitating uneven sampling for tomographic studies of the deep interior due to the predominantly continental locations of observatories is being addressed. The next decade will see more ocean-bottom seismographic stations at critical distances.

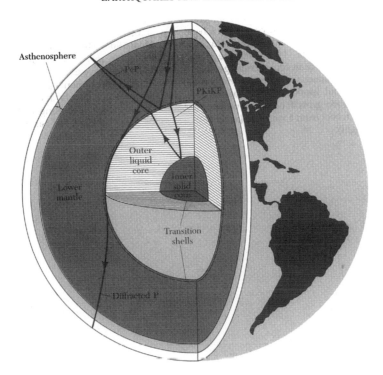

Fig. 5. A cross-section of a radially symmetric Earth model based on seismological evidence. The paths taken by three major kinds of earthquake waves are shown (Bolt 1982).

The digital earthquake data of the present seismographic networks are easily downloaded onto personal computers. On the interpretation side, there can be a drawback. The paper photographic records of yesteryear, that displayed 24 hour signals continuously, enabled students to examine the whole of an earthquake record, including the background noise in which it is embedded. Because the present methods tend to be restricted to short runs of the seismic evolutary time series, a narrow focus on the expected wave forms results rather than scanning for abnormal observations.

The picking by eye of the first onsets of seismic waves recorded on seismograms, such as P, S and their multiples, was the major tool at earthquake observatories to provide the basic data for the construction of travel-time curves for waves throughout the Earth's interior. A problem was always that at many observatories the selection of phases was highly correlated with the available charts of travel-time curves. Readers sought to find the appropriate identification by reference to available curves such as Jeffreys–Bullen, rather than picking values using the assumption of ignorance. More and more in the last decades, reported readings in the seismological catalogues were therefore biassed to the better known branches of the travel-time tables (see Fig. 4).

An alternative method now coming into vogue is to make use of the wave forms on the seismograms, rather than picking the first wave onsets. The idea is to superimpose ('stack') the digital seismograms provided by the global digital networks at common source–receiver distances and to average the result to obtain a composite record. Figure 6 gives an example from a recent paper that discusses the procedure (Astiz et al. 1996). The figure shows the stacking of transverse components of the recorded seismic ground motions after the signals were passed through a 30 s low-pass filter before the superposition was made. In this case, the procedure enhances the horizontally polarized shear waves (SH) passing through the Earth relative to the compressional P waves. It is instructive to compare the concentration of wave energy on this plot with the predicted theoretical times for the S and P phases in Fig. 4. Outstanding problems on Earth structure may achieve better resolution using such methods. An example of such a problem is the need to obtain a sharp resolution of the longitudinal velocity in the upper part of the outer core (Murtha 1984).

Transverse

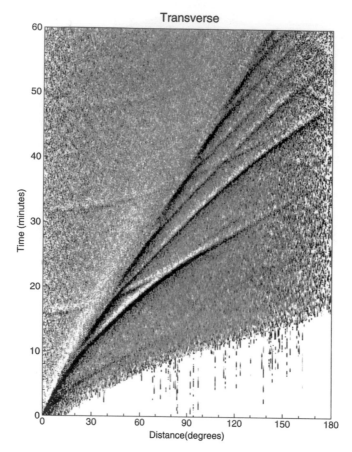

Fig. 6. A composite photograph produced by superimposing (stacking) over 33 000 digital seismograms. Only the transverse components of the wave motions have been used after 30 s low-pass filtering of each trace (Astiz *et al.* 1996).

Of course, much more than the velocity structure of the Earth's interior can be inferred from earthquake wave probes of the Earth. The attenuation of earthquake waves with distance gives a direct measure of the non-elastic (viscous) properties of the interior rocks. Wave polarization allows their anisotropic elastic parameters to be estimated. Strict bounds are also placed on the density distribution by considering the complete seismic response spectrum of the Earth. For example, from short period reflections (see Bolt 1991), it was demonstrated in 1970 that the average inner core density could not exceed about 14 g/cm^3.

There is much reason to be grateful to Hutton and, particularly, Lyell for their recognition of the crucial evidence on global tectonic processes provided by earthquakes. The geologically ephemeral duration of earthquakes did not preclude them informing the past. Lyell had no profound theories on the Earth's interior dynamics and his views on the immediate causal link between volcanoes and earthquakes were erroneous. Nevertheless, his emphasis on the value of field studies immediately after earthquakes and his judicious description of them remain important educationally even today.

I am grateful to D. Drager and P. Shearer for assistance with figures.

References

ASTIZ, L., EARLE, P. & SHEARER, P. 1996. Global stacking of broadband seismograms, *Seismological Research Letters*, **67**, 8–18.

ANTOLIK, M. S. 1996. *New Results from Studies of Three Outstanding Problems in Local, Regional, and Global Seismology.* PhD thesis. University of California, Berkeley.

BAILEY, E. B. 1967. *James Hutton – the Founder of Modern Geology*. Elsevier, Amsterdam.

BERRYMAN, K. R., BEANLAND, S., COOPER, A., CUTTEN, H., NORRIS, R. & WOOD, P. 1992. The Alpine fault, New Zealand: variation of quaternary structural style and geomorphic expression. *Annalis Tectonicae, (Suppl.)*, **6**, 126–163.

BOLT, B. A. 1982. *Inside the Earth*. W. H. Freeman, New York.

—— 1991. The precision of density estimation deep in the Earth. *Quarterly Journal of Royal Astronomical Society*, **32**, 367–388.

—— 1993. *Earthquakes and Geological Discovery*. Scientific American Library, New York.

—— 1996. From earthquake acceleration to seismic displacement, the Fifth Mallet–Milne Lecture. *Society for Earthquake and Civil Engineering Dynamics*. John Wiley & Sons, Chichester.

BRUSH, S. G. 1979. Nineteenth century debates about the inside of the Earth: solid, liquid or gas? *Annals of Science*, **35**, 225–254.

BULLEN, K. E. & BOLT, B. A. 1989. *Introduction to the Theory of Seismology*. 4th edn. Cambridge University Press, Cambridge, 1985.

DZIEWONSKI, A. M. & WOODHOUSE, J. H. 1987. Global images of the Earth's interior. *Science*, **236**, 37–48.

FISHER, O. 1889. *Physics of the Earth's Crust*. Macmillan, London.

GELLER, R. J. 1997. Predictable publicity, astronomy and geophysics. *Journal of the Royal Astronomical Society*, **38**, 16–18.

HEDLIN, M. A. H., SHEARER, P. M. & EARLE, P. S. 1997. Seismic evidence for small-scale heterogeneity throughout the Earth's mantle. *Nature*, **387**, 145–150.

IYER, H. M. & HIRAHARA, K. (eds) 1993. *Seismic Tomography: Theory and Practice*. Chapman & Hall, London.

JEFFREYS, H. & BULLEN, K. E. 1940. *Seismological Tables*. British Association Gray–Milne Trust, Edinburgh.

LOMAX, A. & BOLT, B. A. 1992. Broadband waveform modelling of anomalous strong ground motion in the 1989 Loma Prieta earthquake using three-dimensional geologic structures. *Geophysical Research Letters*, **19**, 1963–1966.

LYELL, C. 1875. *Principles of Geology*, 12th edn. 4 vols, London.

MOGI, K. 1985. *Earthquake Prediction*. Academic Press, Tokyo.

MULARGIA, F. & GASPERINI, P. 1992. Evaluating the statistical validity of 'VAN' earthquake precursors. *Geophysical Journal International*, **111**, 32–44.

MURTHA, P. E. 1984. *Seismic Velocities in the Upper Part of the Earth's Core*. PhD thesis. University of California, Berkeley.

OLIVER, J. 1996, *Shocks and Rocks: Seismology in the Plate Tectonics Revolution*. American Geophysical Union, Washington.

ONGLEY, M. 1943. Surface trace of the 1855 earthquake. *New Zealand Transactions*, **73**, 84–89.

RICHTER, C. 1958. *Elementary Seismology*. W. H. Freeman, New York.

STARK, P. B. 1996. A few considerations for ascribing statistical significance to earthquake predictions. *Geophysics Research Letters*, **23**, 1399–1402.

STARK, P. B. 1997. Earthquake prediction: the null hypothesis. *Geophysical Journal International,* (in press).

VAN DER HILST, R. D., WIDIYANTORO, S. & ENGDAHL, E. R. 1997. Evidence for deep mantle circulation from global tomography. *Nature*, **386**, 578–584.

VAROTSOS, P., EFTAXIAS, K., VALLIANATOS, F. & LAZARIDOU, M. 1996. Basic principles for evaluating an earthquake prediction method. *Geophysical Research Letters*, **23**, 1295–1298.

YEATS, R. S., SIEH, K. & ALLEN, C. R. 1997. *The Geology of Earthquakes*. Oxford University Press, Oxford.

Humanity and the modern environment

SIR JOHN KNILL

Highwood Farm, Long Lane, Shaw, Newbury, Berkshire RG14 2TB, UK

Abstract: The geological record is a chronicle of change in which long periods of apparent climatic stability were successively replaced by new environmental conditions. These broad cycles of change were punctuated by shorter-term events, some of which had catastrophic implications. Lyell recognized that this menu of environmental scenarios recorded in the past was well reflected by the range of environmental conditions which occur at the present day.

It is now recognized that human influence on climate and the environment is engendering a new event which has all the hallmarks of a phase which might, if preserved in the geological record, be interpreted as cataclysmic. However, the Earth system has proved remarkably effective in recovering from extremes of climate or devastating reductions in biodiversity.

Humankind has two alternatives. To accept the inevitable, adapting to such changes as can occur, whether of human or natural origin, and facing the possibility of ultimate extinction as a species, thereby leaving the Earth ultimately to recover. Or to attempt to control man-made change so effectively that eventually it will be possible to manage the Earth's natural processes. Our understanding of the past is therefore central to our ability to survive in the future.

We live in a world that is changing on all scales from the submicroscopic to the global. Hutton's great contribution lay in his identification of the cyclical nature of geological change so that, as geologists, change is something we accept as the norm. Lyell, in his turn, appreciated that the laws determining physical and chemical behaviour are constant, so that the processes which occur at the present day had their analogues in the past. Uniformitarianism, a misnomer much beloved by university examiners, applies rather to the invariable character of the natural laws rather than implying that the rates and nature of geological processes remained unchanged through time.

We now view a changing world in a different light, a world in which a full understanding of the consequences of humankind's interventions will be critical for the indefinite future. All decisions which affect our future have now to be viewed in the context of sustainable development and the precautionary principle, in order to ensure the maintenance of the quality of life, the conservation of the Earth's resources, and the preservation of biodiversity. Inter-generational equity is required to ensure that the resources which we deplete or endanger are matched with an equivalent legacy of physical assets and intellectual capital.

More than ever before, therefore, it is necessary to interrogate the processes which manipulate the natural environment and their consequential impacts. On all scales from the molecule to the solar system, scientists are exploring these processes in order to understand more about how the Earth system works. But this inquiry is not solely driven by the simple curiosity of scholarly research on the one hand, or by the emotive style of some environmental campaigning on the other. There is now seen to be a clear imperative driven by the need to deliver real answers, based on high-quality science and engineering, in a limited timescale, to questions which will have major social, economic and political consequences.

Traditionally, geologists have felt in tune with the environment, having spent much of their lives mapping and studying rocks in the open air. Many geologists who work in the real world see both the beneficial and detrimental consequences of their work through the development of geological resources such as water, energy and materials, or through construction. It has been argued that geologists have, through the duality of their responsibilities to the conservation of both physical and environmental resources, a special role in contributing to a world that has a changing environment. This offers both a challenge and an opportunity to our profession.

Global environmental change

Global environmental change is being driven by two factors: population growth causing a depletion in resources, and climate change resulting in modifications to the distribution and availability of resources, and an enhancement of environmental hazards.

Today's world population is 5.8 billion. The worst predictions have suggested that the population will double, possibly by about 2040. More

KNILL, SIR J. 1998. Humanity and the modern environment. *In*: BLUNDELL, D. J. & SCOTT, A. C. (eds) *Lyell: the Past is the Key to the Present*. Geological Society, London, Special Publications, **143**, 363–368.

likely projections suggest that population growth will peak at about 11 billion in the final decades of the next century. Effectively 90 per cent of the increase during this period, whatever the prediction, is expected to be located in the developing countries, and at least 60 per cent of this increased population will be living in megacities. There will, in the next century, be massive localised consumption of resources and accumulation of waste products in the urban populations of today's developing world. Whatever precautions are taken there will also be hundreds of millions of environmental refugees in the initial decades of the next century.

This increased population will require housing, food, clean water and effective waste disposal for a basic level of subsistence. It will require, at a minimum, an enhanced infrastructure of agriculture, transport, water and energy supply, waste disposal and pollution control. The population and its supporting infrastructures will need to be provided with stability through adequate protection against the consequences of environmental and man-made disasters and epidemics and institutional change. Beyond such primary needs, all humankind has a right to aspire to an enhanced quality of life within which there is opportunity for personal, family or community betterment.

Climate-related processes are primarily derived from man-made increases in emissions of greenhouse gases resulting in enhanced retention of solar energy in the oceans. The increase in the world's population is the fundamental driving force in causing, adding to or accelerating global change at the present time.

Scientific research is essential in order to provide for the quality of long-term prediction required to resolve the myriad of industrial, economic and social policy decisions with which the world is faced. Climate-related research is focused on achieving a fuller understanding of atmospheric and oceanic circulation, and of interactions which occur at the interfaces between the three main components of the Earth system: the atmosphere, the oceans and the solid Earth. The atmosphere transmits gaseous and energy fluxes around the Earth and has a remarkably short memory of a few tens of days. In contrast the ocean climate has a longer potential life of some tens of years whereas the solid Earth's cyclicity of change is measured in terms of millions or tens of millions of years.

However, it is at the interfaces between the atmosphere, ocean and solid earth that interchange occurs, life flourishes and physical processes are most active. The uppermost 3 m of the oceans contain as much heat as the hundred kilometres of height of the atmosphere. The oceans provide the heat store which moderates climate change. The

biological, chemical and physical exchanges and interactions which occur within a few metres of vertical height at the ocean–atmosphere and land–atmosphere interface are therefore critical factors in climate change. The elemental cycles of carbon, sulphur and nitrogen which occur within this zone are essential to the maintenance of life. It is within this narrow interface that we live, derive much of our sustenance, and dispose of most of our waste products – it is here that we must survive.

Central to this scientific research effort has been the fuller understanding of the circulation of the atmosphere and oceans which drive atmospheric change and which will thus lead to climatic evolution, and to potential extremes of wind, rain or drought. Meteorologists, oceanographers, glaciologists and ecologists have been remarkably successful in mounting major international community research programmes to study global environmental change through over-arching organisations such as the International Geosphere Biosphere Programme and the World Climate Research Programme. This work has had a major influence on the work of such bodies as the Intergovernmental Panel on Climate Change, and on national and international policies. What contribution, therefore, has been made by the geological sciences?

Role of the geological sciences

The geological record provides us with evidence for continual environmental change with extremes of climate, changes in atmospheric chemistry, catastrophic events and massive loss in biological species. Geological history, spread over hundreds of thousands and millions of years, offers analogues which have the potential to help resolve the scientific questions which we face today. Just under thirteen thousand years ago temperatures in the northern Atlantic area rose by over 15°C (Atkinson *et al.* 1987), to be followed by a rapid cooling of 7°C in 50 years between 11 000 and 10 000 BP (Dansgaard *et al.* 1989). This figure can be compared to the observed rise of 4°C in the Pacific associated with the initiation of the 1997 El Niño event. During the Cretaceous period the average global temperatures were the highest recorded over a considerable span of geological time, the atmospheric carbon dioxide content was several times the present level, and the oceans were covered with algal blooms. Global warming is not unique to the present day and it would appear almost self-evident that geologists have much to contribute to the scientific understanding of the subject. But geologists, in the main, have made a disappointingly small contribution to our understanding of these processes of recent change. Why is this so?

Whereas the rates of change within the atmosphere and oceans are observable and measurable, geological change is slow. Only in the case of relatively localized events such as landslides, volcanic eruptions or earthquakes does the timescale compare with that of atmospheric processes. Because most geological change is so slow it has been argued that, notwithstanding use of the geological record as a model for understanding present-day atmospheric processes, the geological sciences have little to contribute to the science of global change.

The Gaia concept (Lovelock 1995) provides a longer-term perspective of this issue in that the Earth is viewed as a self-regulating system where the biosphere, and the solid, liquid and gaseous geospheres interact together to behave with many of the characteristics of a living organism. Thus with time the Earth's physical and biological systems evolved together, maintaining, for the future, a healthy planet. Although the concept that the Earth is, in a sense, alive is as old as the human race, it was James Hutton in 1785 who first suggested that the Earth was analogous to a superorganism and was best studied scientifically in terms of physiology.

Rock weathering has a major influence on atmospheric carbon dioxide content. The biosphere acts as a pump drawing carbon dioxide into the soil, thereby enhancing the rate of decomposition of calcium silicate-containing minerals and resulting in the formation of calcium carbonate and silicic acid which are transported to, and deposited in, sinks on the sea floor. Such rocks are removed from the geological cycle until involved, tens or hundreds of millions of years later, in the processes of plate tectonics and mountain building. The $^{87}Sr/^{86}Sr$ ratio of continental rocks is significantly greater than that in normal sea water but it is recognized that an increased $^{87}Sr/^{86}Sr$ ratio occurs in marine sediments since 50 Ma. The logic follows that, since 50 Ma, when the development of the Alpine–Himalayan mountain chain started, there has been enhanced silicate weathering. During this period there has also been a reduction in global temperature which could be the result of a loss of atmospheric carbon dioxide due to its increased uptake as a result of intensified weathering associated with the creation of the mountain chain.

It is a tantalizing thought that accelerated rock weathering could be used as a means of artificial control on climate change, through the continuous, large-scale removal of the protective blanket of the existing regolith.

This example also illustrates the importance of geological processes in providing sinks for the components of greenhouse gases, notably as elemental carbon, limestones, and carbonaceous material. The relative fluxes of carbon, water and heat at the interfaces between the land, sea and atmosphere are critical. Spring plankton blooms in the oceans have a vital role in the uptake of carbon from sea water. The flux of carbon into the deep ocean is transferred to the sea bed, thereby preserving it for geological time. Probably somewhat less than half the man-made carbon dioxide is being absorbed within the oceans. If the seas could take up, or the forests absorb, the steady increase in man-made carbon dioxide, a significant component of the problem of global environmental change would be removed.

Impacts of global environmental change

Science can identify and model the processes that cause global environmental change, thereby providing for the prediction of the effects of continuing change. Science can also determine the impacts of that change on the environment and so offer potential solutions to such impacts. The masterly study by Working Group II of the Intergovernmental Panel on Climate Change (Watson *et al.* 1995) sets out the anticipated impacts and how it might be possible to adapt to, or mitigate against, these changes. Many of these impacts are more visual, and so far more compelling, than the invisible processes associated with global atmospheric chemistry:

(a) six million hectares of new desert being created each year;
(b) eleven million hectares of rain forest each year converted to crops and cattle ranching;
(c) evidence of acid rain in trees throughout much of the developed world far from sites of industrialisation and pollution;
(d) pollution of freshwater through algal blooms that are toxic or destructive to the local ecosystems;
(e) melting of polar ice sheets;
(f) consequences of intense fishing, whaling and sealing with the associated loss of the lower parts of the marine food chain;
(g) disasters in areas of population concentration which result in human suffering.

It is, of course, such events which today catch the public imagination, drive environmental pressure groups, and are persuasive in influencing governments to develop policies which recognise responsibilities nationally and globally.

Role of the geologist

What then, in this vast scenario, is the role of the geologist? We are not leading the science of global environmental change although there is potentially much to be offered. Equally there is still much to be gained from acknowledging the benefits of interdisciplinarity collaboration. The profession is not sufficiently anticipatory and proactive, with honourable exceptions, in identifying or leading projects or policies. How, therefore, can the geological profession undergo its own adaptation in response to global environmental change and to the challenges that lies ahead? The main opportunities lie in the following areas:

(a) reducing climate change;
(b) geological resources: quantum and impact;
(c) infrastructural development;
(d) waste management and pollution;
(e) extreme events;
(f) environmental vulnerability and conservation
(g) ethics.

Global warming can, in principle, be slowed down and reversed by two means: the adoption of methods of energy generation which do not generate greenhouse gases and are non-polluting, and/or by the enhancement of the role of geological sinks for carbon. Alternative energy sources, although locally important, have no long-term prospect of reversing the warming process, and clean coal-fired or gas power stations will only slow down the process. Nuclear power offers the only large-scale method of power production available at the present which could, at least theoretically, reverse the trend. But the lack of public confidence in nuclear energy would require resolution, reprocessing of spent fuel would have to stop, and the seemingly intractable issue of waste disposal settled.

Geological principles could be used to establish new sinks for atmospheric carbon. Deep geological reservoirs have the potential to provide storage for carbon dioxide extracted from the atmosphere. Geobiological methods could be used to develop aqueous environments which would provide geological sinks for carbon, as well as providing an alternative source of nutrients. As agriculture intensifies, so the geological understanding of soils, with their multiple roles as sinks and sources, will become increasingly important.

Many new opportunities will occur for the geological profession to respond to new demands placed on the resource industries such as water availability, pollution control and waste disposal, and the reduction of the impact of mineral extraction on land and at sea.

In the future most of the world's population will live in urban areas, concentrated in megacities located in regions most affected by natural hazards. Critical to the effective operation of such a city is the role of low cost transportation: for people, materials, food, water, energy and waste. Under-ground communication, and the use of underground space, will become commonplace. The resources of food, water and energy will need to be developed within the surrounding regions, or transported for very long distances. The disposal of waste and fixed sources of pollution will need to take place outside the anticipated ultimate limits of mega-cities. The infrastructural planning of such massive developments will require a fundamental under-pinning based upon knowledge of geological conditions, ground and surface water distribution, natural hazards and the regional environment. A study carried out by the Institution of Civil Engineers (1995) as part of the work of the International Decade of Natural Disaster Reduction established that for the three potential megacities of Jakarta, Karachi and Metro Manila there is, at present, no such basic database upon which infra-structural planning for a megacity can proceed.

Turning to pollution and waste management, it is recognized that the movement of pollutants in groundwater can be relatively rapid, and aquifers may also act as sinks for such pollutants. The philo-sophy of responsible waste disposal has moved on from that of 'dilute and disperse' to one of 'concentrate and contain' with the ability to 'disarm and destroy' some of the more intractable wastes. Protection zones defined on geological and hydrogeological criteria are required which ensure that the consequences of waste disposal on, or the dispersion of potential pollutants within, the environment can be minimized. Ultimately, how-ever, responsible waste disposal within the environ-ment relies on geological containment. The demon-stration of geological containment is central to effective long-term environmental waste manage-ment.

One of the most severe tests of the feasibility of the geological containment of waste relates to the disposal of long-lived radioactive waste. In the United Kingdom UK Nirex investigated from 1989 to 1997 a site for the disposal of transuranic wastes in Lower Palaeozoic volcanic rocks near Sellafield in Cumbria since 1989 . An application to construct an underground rock laboratory was rejected by the UK Government in March 1997 on planning grounds, but also because of 'the scientific uncertainties and deficiencies in the proposals presented by Nirex which would also justify refusal of the appeal' as well as concerns regarding the site selection process. The failure of Nirex to take the project forward, even to the next stage of investigation, raises important questions about the scientific rationale associated with the deep geological disposal of active wastes and thus the adequacy of the principle of geological contain-ment. A clear message has been delivered that public and political understanding and acceptance is unlikely to be achieved by offering elaborate models of inherently complex geologies and hydrogeologies.

There has been a growing recognition worldwide of the potential consequences of natural hazards associated with rapid onset events, and the purpose of the International Decade of Natural Disaster Reduction (IDNDR), established in 1990, was to ensure better preparedness for, and mitigation of, disasters. The menu of hazard processes defined by IDNDR included volcanoes, landslides, earthquakes and floods, all of which have a major geological dimension. IDNDR is an important demonstration of how geologists can effectively use their skills and experience, in a multidisciplinary environment, to reduce the adverse consequences of rapid onset environmental processes.

The environment is vulnerable both to natural processes and human impacts. The coastal zone, for example, is important not only because of the biological, physical and chemical processes which take place there, but as an area where many people live, draw their livelihood, and dispose of their waste. It is, for example, particularly exposed to catastrophic events, being also affected by sea level rise, providing thereby opportunities for non–intrusive, precautionary and low-cost engineering solutions.

The conservation of the natural environment also embraces the conservation and protection of the geological, geomorphological and landscape heritage which have underpinned the evolution of our science, contribute much to training and research, and provide a window by which the public can understand the geological sciences.

Most geologists in industry face a dichotomy in that the engineering processes with which they are professionally associated – drilling, underground mining, excavation, pumping, and disposal of solid or liquid waste or contaminated materials – have created some of the most extreme examples of environmental degradation. There are, of course, many exceptions to such a generalization, such as the construction of reservoirs which have created new habitats for wildfowl, the sympathetic restoration of gravel pits and open-cast mines, and the uniquely preserved wildlife corridors provided by some motorways. Many surface sites form important Earth science conservation sites, with the status of SSSI (Sites of Special Scientific Interest) and RIGS (Regionally Important Geological Sites), for training and research, while others can take visitor pressure away from the most vulnerable localities. Geological resources can be developed in a manner which is sympathetic to the environment. Directional drilling, developed so impressively at Wyche Farm, has enabled the extraction of hydro-carbons from small reservoirs located offshore from an onshore location through demanding applications of reservoir geology. Cuttings, embankments and waste disposal sites can be developed in ways which ensure that the completed structure simulates the natural landscape. Nevertheless, for many geologists such an end product may appear unattainable because of the nature of the project and the lack of influence which geology has, at board or management level, on overall project planning. The profession should see this as a trial of its own stature, and provide a lead in establishing guidelines and codes which define best practice.

The future identity of our profession is dependent on our accepting the challenges presented by environmental change, and by pursuing an approach which recognizes sustainable development, the precautionary principle, and a decision-making procedure based on risk and safety assessment. This will require a better understanding of our ethical responsibilities to the wider public, not least in relation to the development of natural resources and the use of land, incidentally an issue not addressed in the Geological Society's code of conduct (Reeves 1996).

Conclusion

How then would Hutton and Lyell interpret the present situation? Hutton might well feel pleased that scientific evidence for his perceptive recognition of the Gaia hypothesis is now being collected. Lyell might well be surprised at the extent to which human intervention is now influencing processes in the natural environment, and the degree to which rapid-onset extreme events are an almost daily occurrence. Both would be impressed by the holistic approach brought to the Earth sciences through our understanding of the role of plate tectonics, the techniques of Earth observation, and wider interdisciplinarity.

Humankind has two alternatives. To accept the inevitable, adapting to such changes as occur, whether of human or natural origin, through regulation or response to impacts. Political reality dictates that the attainment of the requisite degree of global regulation of population growth, power generation or pollution control necessary to achieve success is questionable. Thus, the species faces the possibility of decline and ultimate extinction, taking much biodiversity with it, and leaving the Earth to recover by natural means on a timescale too long to save humanity. This scenario has been repeated through geological time. The second, more radical, alternative is to attempt to control man-made change so effectively that it becomes possible to manage the Earth's natural processes through physical, chemical and biological engineering. Quite apart from the enormous technological advances required, this demands a degree of regional and global governance, political willpower and personal discipline unknown in the democratic world. Short-termism might well rule the day. Whichever pathway proves to be the case, our understanding of the past will be central to our ability to survive in the future.

References

ATKINSON, T. C., BRIFFA, K. R. & COOPE, D. R. 1987. Seasonal temperatures in Britain during the past 22,000 years reconstructed using beetle remains. *Nature*, **325**, 587–92.

DANSGAARD, W., WHITE, J. W. C. & JOHNSEN, S. J. 1989. The abrupt termination of the Younger Dryas climate event. *Nature*, **339**, 532–533.

INSTITUTION OF CIVIL ENGINEERS, 1995. *Megacities: reducing vulnerability to natural disasters*. Thomas Telford, 169 pp.

LOVELOCK, J. 1995. The Ages of Gaia, 2nd edn. Oxford University Press, 255 pp.

REEVES, G. M. 1996. The Geologist's Directory 1996, The Geological Society London, 568 pp.

WATSON, R. T., ZINYOWERA, M. C. & MOSS, R. H. 1996. *Climate Change 1995*. Impacts, Adaptations and Mitigation of Climate Change: Scientific-Technical Analyses, Contribution of Working Group II to the Second Assessment Report of the Intergovernmental Panel on Climate Change, Cambridge University Press, 879 pp.

Index

369